W9-CGX-334

CONVERSION FACTORS TO SI UNITS (*continued*)

English	SI	SI symbol	To convert from English to SI multiply by	To convert from SI to English multiply by
Power, Heat Rate				
horsepower	kilowatt	kW	0.7457	1.341
foot pound/sec	watt	W	1.356	0.7376
BTU/hour	watt	W	0.2929	3.414
Pressure				
pound/square inch	kilopascal	kPa	6.895	0.1450
pound/square foot	kilopascal	kPa	0.04788	20.89
inches of H_2O	kilopascal	kPa	0.2486	4.023
inches of Hg	kilopascal	kPa	3.374	0.2964
Temperature				
Fahrenheit	Celsius	°C	5/9(°F − 32)	9/5 × °C + 32
Fahrenheit	kelvin	K	5/9(°F + 460)	9/5 × K − 460
Velocity				
foot/second	meter/second	m/s	0.3048	3.281
mile/hour	meter/second	m/s	0.4470	2.237
mile/hour	kilometer/hour	km/h	1.609	0.6215
Acceleration				
foot/second squared	meter/second squared	m/s^2	0.3048	3.281
Torque				
pound-foot	newton-meter	N · m	1.356	0.7376
pound-inch	newton-meter	N · m	0.1130	8.85
Viscosity, Kinematic Viscosity				
pound-sec/square foot	newton-sec/square meter	$N · s/m^2$	47.88	0.02089
square foot/second	square meter/second	m^2/s	0.09290	10.76
Flow Rate				
cubic foot/second	cubic meter/second	m^3/s	0.02832	35.32
cubic foot/second	liter/second	L/s	28.32	0.03532

Mechanics
of
Fluids

Merle C. Potter
Michigan State University

David C. Wiggert
Michigan State University

Prentice Hall
Englewood Cliffs, NJ 07632

Library of Congress Cataloging-in-Publication Data

POTTER, MERLE C.
    Mechanics of fluids / by M.C. Potter & D. C. Wiggert.
        p.   cm.
    Includes bibliographical references and index.
    ISBN 0-13-572793-6
    1. Fluid mechanics.   I. Wiggert, D. C.   II. Title.
TA357.P725   1991
620.1'06—dc20   90-48055
                                                 CIP

Acquisitions editor: Doug Humphrey
Editorial/production supervision and
    interior design: Marianne Peters
Cover design: Ben Santora
Cover photo: M. Koochesfahani and P. Dimotakis,
''The interface pattern of pairing vortices in a two-stream shear layer.''
Prepress buyer: Linda Behrens
Manufacturing buyer: David Dickey
Photo editor: Lorinda Morris-Nantz
Photo research: Tobi Zausner
Chapter opening photo credits: **1:** United States Air Force and Canadian Pacific; **2:** Bureau of
Reclamation, Department of the Interior; **4:** Ron Church/Photo Researchers; **5:** Vincent E.
Camagna/Photo Researchers; **6:** Courtesy of Seattle, Seattle-King County Convention & Visitors
Bureau; **7:** Courtesy of Alaska Division of Tourism; **8:** Courtesy of Goodyear; **9:** Courtesy
Rockwell International; **10:** Bureau of Reclamation, photo by Joe Madrigal, Jr.; **11:** Spencer
Grant/Photo Researchers; **12:** Tennessee Valley Authority; **13:** Takeshi Takahara/Photo
Researchers.

 © 1991 by Prentice-Hall, Inc.
A Paramount Communications Company
Englewood Cliffs, New Jersey 07632

Printed in the United States of America
10   9   8   7   6

ISBN 0-13-572793-6

Prentice-Hall International (UK) Limited, *London*
Prentice-Hall of Australia Pty. Limited, *Sydney*
Prentice-Hall Canada Inc., *Toronto*
Prentice-Hall Hispanoamericana, S.A., *Mexico*
Prentice-Hall of India Private Limited, *New Delhi*
Prentice-Hall of Japan, Inc., *Tokyo*
Simon & Schuster Asia Pte. Ltd., *Singapore*
Editora Prentice-Hall do Brasil, Ltda., *Rio de Janeiro*

To
Michelle, Gloria,
Kara, and Kristie

Contents

Chapter 12   Turbomachinery   559

Chapter 13   Measurements in Fluid Mechanics   615

Appendix   652

Bibliography   672

Answers to Selected Problems   676

Index   683

Preface

The motivation to write a book is difficult to describe. Most often the authors suggest that the other texts on the subject have certain deficiencies that they will correct. They, of course, introduce other deficiencies that future authors hope to correct! And life goes on. This is another fluids book that has been written in hopes of presenting fluid mechanics so that the undergraduate can understand the physical concepts and follow the mathematics. This is not an easy task; fluid mechanics is a subject that contains many difficult-to-understand phenomena. For example, how would you explain the hole scooped out in the snow by the wind on the upwind side of a tree during a snow storm? Or the high concentration of smog contained in the Los Angeles area (it doesn't exist to the same level in New York)? Or the unexpected strong wind around the corner of a tall building in Chicago? Or the vibration and subsequent collapse of a large concrete-steel bridge due to the wind? We have attempted to present fluid mechanics so that the student can understand and analyze the many important phenomena encountered by the engineer.

The mathematical level of this text is based on the previous mathematics courses required in all engineering curricula. We will use solutions to differential equations and vector algebra. Some use will be made of vector calculus with the use of the gradient operator, but this will be kept to a minimum since it tends to obscure the physics involved.

Most popular texts in fluid mechanics have not presented fluid flows as fields. That is, they have presented primarily those flows that can be approximated as one-dimensional flows and have treated other flows using experimental data. We must recognize that when a fluid flows around an object, such as a building or an abutment, its velocity possesses all three components which depend on all three space variables and, possibly, time. If we present the equations that describe such a general flow, the equations are referred to as field equations and velocity and pressure fields become of interest. This is quite analogous to electrical and magnetic fields in electrical engineering. In order for the difficult problems of the future, such as large-scale environmental pollution, to be analyzed by engineers, it is imperative that we understand fluid fields.

Thus, in Chapter 5 we introduce the equations and discuss several solutions for some relatively simple geometries. The more conventional manner of treating the flows as one-dimensional is provided as an alternate route for those who wish this approach.

The introductory material included in Chapters 1 through 8 of this text has been carefully selected to introduce the students to all of the fundamental areas of fluid mechanics. Not all of the material in each chapter need be covered in an introductory course. The instructor can fit the material to a selected course outline. Some sections at the end of each chapter may be omitted without loss of continuity in later chapters. In fact, Chapter 5 can be omitted in its entirety if it is decided to exclude field equations in the introductory course, a relatively common decision. That chapter can then be included in an intermediate fluid mechanics course. After the introductory material has been presented, there is sufficient material to present an additional course under the semester system, or two additional courses under the quarter system. An additional course or courses could include material that had been omitted in the introductory course and combinations of material from the more specialized chapters, 9 through 13. Much of the material is of interest to all engineers while several of the chapters are of interest to only particular disciplines.

We have included example problems worked out in detail to illustrate each important concept presented in the text material. Numerous home problems then provide the student with ample opportunity to gain experience solving problems of various levels of difficulty. After studying the material, reviewing the examples, and working several of the home problems, the student should gain the needed capability to work many of the problems encountered in actual engineering situations. Of course, there are numerous classes of problems that are extremely difficult to solve, even for an experienced engineer. To solve these more difficult problems, the engineer must gain considerably more information than is included in this introductory text. There are, however, many problems that can be successfully solved using the material and concepts presented herein.

The text is written emphasizing SI units; however, all properties and dimensional constants are given in English units also. Approximately one third of the examples and problems are duplicated using English units.

There are many interesting films and videocassettes that have been developed in the area of fluid mechanics; it is often quite helpful for the student to observe the phenomena being described. A listing of some of the available films and videos are included in the Appendix.

The authors are very much indebted to both their former professors and to their present colleagues. Professors C.S. Yih and V.L. Streeter of the University of Michigan demanded that each of us learn this subject well! Professor R. Bouwmeester of Michigan State University has been of special help in preparing this manuscript. Chapter 10 was written with inspiration from Professor F.M. Henderson's book titled "Open Channel Flow" (1966), and Professor D. Wood of the University of Kentucky encouraged us to incorporate comprehensive material on pipe network analysis in Chapter 11. Professor A.R.D. Thorley provided some of the problems at the end of Chapter 12. Several colleagues, among them M. Koochesfahani, A. Mueller, S. Stuckenbruck, and R. Wallace, either taught from the class notes that led to the text, or reviewed portions of the manuscript. They provided many valuable suggestions, as did the numerous reviewers selected by the publisher. The manuscript was typed by C. Sovis.

ONE

Basic Considerations

1.1 INTRODUCTION

A proper understanding of the mechanics of fluids is extremely important in many areas of engineering. In biomechanics the flow of blood and cerebral fluid are of particular interest; in meteorology and ocean engineering an understanding of the motions of air movements and ocean currents requires a knowledge of the mechanics of fluids; chemical engineers must understand fluid mechanics to design the many different kinds of chemical-processing equipment; aeronautical engineers use their knowledge of fluids to maximize lift and minimize drag on aircraft and to design fan-jet engines; mechanical engineers design pumps, turbines, internal combustion engines, air compressors, air-conditioning equipment, pollution-control equipment, and power plants using a proper understanding of fluid mechanics; and civil engineers must also utilize the results obtained from a study of the mechanics of fluids to understand the transport of river sediment and erosion, the pollution of the air and water, and to design piping systems, sewage treatment plants, irrigation channels, flood control systems, dams, and domed athletic stadiums.

It is not possible to present the mechanics of fluids in such a way that all of the foregoing subjects can be treated specifically; it is possible, however, to present the fundamentals of the mechanics of fluids so that engineers are able to understand the role that the fluid plays in a particular application. This role may involve the proper sizing of a pump (the horsepower and flow rate) or the calculation of a force acting on a structure.

In this book we present the general equations, both integral and differential, that result from the conservation of mass principle, Newton's second law, and the first law of thermodynamics. From these a number of particular situations will be considered that are of special interest. After studying this book the engineer should be able to apply the basic principles of the mechanics of fluids to new and different situations.

In this chapter topics are presented that are directly or indirectly relevant to all subsequent chapters. We include a macroscopic description of fluids, fluid properties, physical laws dominating fluid mechanics, and a summary of units and dimensions of important physical quantities. Before we can discuss quantities of interest, we must present the units and dimensions that will be used in our study of fluid mechanics.

1.2 DIMENSIONS, UNITS, AND PHYSICAL QUANTITIES

Before we begin the more detailed studies of the mechanics of fluids, let us discuss the dimensions and units that will be used throughout the book. Physical quantities require quantitative descriptions when solving an engineering problem. Density is one such physical quantity. It is a measure of the mass contained in a unit volume. Density does not, however, represent a fundamental dimension. There are nine quantities that are considered to be fundamental dimensions: length, mass, time, temperature, amount of a substance, electric current, luminous intensity, plane angle, and solid angle. The dimensions of all other quantities can be expressed in terms of the fundamental dimensions. For example, the quantity "force" can be related to the fundamental dimensions of mass, length, and time. To do this, we use Newton's second law, expressed in simplified form in one direction as

$$F = ma \tag{1.2.1}$$

Using brackets to denote "the dimension of," this is written dimensionally as

$$[F] = [m][a]$$
$$= \frac{ML}{T^2} \tag{1.2.2}$$

where M, L, and T are the dimensions of mass, length, and time, respectively. If force had been selected as a fundamental dimension rather than mass, a common alternative, mass would have dimensions of

$$[m] = \frac{[F]}{[a]}$$
$$= \frac{FT^2}{L} \tag{1.2.3}$$

where F is the dimension[1] of force.

There are systems of dimensions in which both mass and force are selected as fundamental dimensions. In such systems conversion factors, such as a gravitational constant, are required; we do not consider these types of systems in this book, so they will not be discussed.

To give the dimensions of a quantity a numerical value, a set of units must be selected. In the United States presently, two primary sets of units are being used, English units and SI (Système International) units. SI are the preferred units and are used internationally; the United States is the only major country not using the metric SI units, but there is now a program of conversion in most industries to the predominant use of SI units. Following this trend, we have used primarily SI units. However, as English units are still in use, some examples and problems are presented in these units as well.

The fundamental dimensions and their units are presented in Table 1.1; some derived units appropriate to fluid mechanics are given in Table

[1]Unfortunately, the quantity force F and the dimension of force $[F]$ use the same symbol.

TABLE 1.1 FUNDAMENTAL DIMENSIONS AND THEIR UNITS

Quantity	Dimension	SI Units		English Units	
Length l	L	meter	m	foot	ft
Mass m	M	kilogram	kg	slug	slug
Time t	T	second	s	second	sec
Electric current i		ampere	A	ampere	A
Temperature T	Θ	kelvin	K	Rankine	°R
Amount of substance	M	kg-mole	kg-mol	lb-mole	lb-mol
Luminous intensity		candela	cd	candela	cd
Plane angle		radian	rad	radian	rad
Solid angle		steradian	sr	steradian	sr

1.2. Other units that are acceptable are the hectare (ha), which is 10 000 m², used for large areas; the metric ton (t), which is 1000 kg, used for large masses; and the liter (L), which is 0.001 m³. Also, density is occasionally expressed as grams per liter (g/L).

TABLE 1.2 DERIVED UNITS

Quantity	Dimensions	SI units	English units
Area A	L^2	m²	ft²
Volume $V\!\!\!/$	L^3	m³	ft³
		L (liter)	
Velocity V	L/T	m/s	ft/sec
Acceleration a	L/T^2	m/s²	ft/sec²
Angular velocity ω	T^{-1}	s⁻¹	sec⁻¹
Force F	ML/T^2	kg · m/s²	slug-ft/sec²
		N (newton)	lb (pound)
Density ρ	M/L^3	kg/m³	slug/ft³
Specific weight γ	M/L^2T^2	N/m³	lb/ft³
Frequency f	T^{-1}	s⁻¹	sec⁻¹
Pressure p	M/LT^2	N/m²	lb/ft²
		Pa (pascal)	
Stress τ	M/LT^2	N/m²	lb/ft²
		Pa (pascal)	
Surface tension σ	M/T^2	N/m	lb/ft
Work W	ML^2/T^2	N · m	ft-lb
		J (joule)	
Energy E	ML^2T^2	N · m	ft-lb
		J (joule)	
Heat rate $\dot{Q}$	ML^2/T^3	J/s	Btu/sec
Torque T	ML^2/T^2	N · m	ft-lb
Power P	ML^2/T^3	J/s	ft-lb/sec
		W (watt)	
Viscosity μ	M/LT	N · s/m²	lb-sec/ft²
Mass flux $\dot{m}$	M/T	kg/s	slug/sec
Flow rate Q	L^3/T	m³/s	ft³/sec
Specific heat c	$L^2/T^2\Theta$	J/kg · K	Btu/slug-°R
Conductivity K	$ML/T^2\Theta$	W/m · K	lb/sec-°R

TABLE 1.3 SI PREFIXES

Multiplication factor	Prefix	Symbol
10^{12}	tera	T
10^{9}	giga	G
10^{6}	mega	M
10^{3}	kilo	k
10^{-2}	centi[a]	c
10^{-3}	milli	m
10^{-6}	micro	μ
10^{-9}	nano	n
10^{-12}	pico	p

[a]Permissible if used alone as cm, cm², or cm³.

In chemical calculations the mole is often a more convenient unit than the kilogram. In some cases it is also useful in fluid mechanics. For gases the kilogram-mole (kg-mol) is the quantity that fills the same volume as 32 kilograms of oxygen at the same temperature and pressure. The mass (in kilograms) of a gas filling that volume is equal to the molecular weight of the gas; for example, the mass of 1 kg-mol of nitrogen is 28 kilograms.

When expressing a quantity with a numerical value and a unit, prefixes have been defined so that the numerical value may be between 0.1 and 1000. These prefixes are presented in Table 1.3. Using scientific notation, however, we use powers of 10 rather than prefixes (e.g., 2×10^6 N rather than 2 MN). If larger numbers are written the comma is not used; twenty thousand would be written as 20 000 with a space and no comma.

Newton's second law relates a net force acting on a rigid body to its mass and acceleration. This is expressed as

$$\Sigma \mathbf{F} = m\mathbf{a} \tag{1.2.4}$$

Consequently, the force needed to accelerate a mass of 1 kilogram at 1 meter per second squared in the direction of the net force is 1 newton; using English units, the force needed to accelerate a mass of 1 slug at 1 foot per second squared in the direction of the net force is 1 pound. This allows us to relate the units by

$$N = kg \cdot m/s^2 \qquad lb = slug\text{-}ft/sec^2 \tag{1.2.5}$$

which are included in Table 1.2. These relationships between units are often used in the conversion of units.

In the SI system, weight is always expressed in newtons, never in kilograms. In the English system, mass is always expressed in slugs, never in pounds. To relate weight to mass, we use

$$W = mg \qquad (1.2.6)$$

where g is the local gravity. The standard value for gravity is 9.80665 m/s² (32.174 ft/sec²) and it varies from a minimum of 9.77 m/s² to a maximum of 9.83 m/s². A nominal value of 9.81 m/s² (32.2 ft/sec²) will be used unless otherwise stated.

Finally, a note on significant figures. In engineering calculations we often do not have confidence in a calculation beyond three significant numbers since the information given in the problem statement is often not known to more than three significant numbers; in fact, viscosity and other fluid properties may not be known to even three significant numbers. The diameter of a pipe may be stated as 2 cm; this would, in general, not be as precise as 2.000 cm would imply. If information used in the solution of a problem is known to only two significant numbers, it is incorrect to express a result to more than two significant numbers. In the examples and problems we will assume that all information given is known to three significant numbers, and the results will be expressed accordingly.

EXAMPLE 1.1

A mass of 100 kg is acted on by a 400-N force acting vertically upward and a 600-N force acting upward at a 45° angle. Calculate the vertical component of the acceleration. The local vertical component of the acceleration of gravity is 9.81 m/s².

Solution

The first step in solving a problem involving forces is to draw a free-body diagram with all forces acting on it. It appears as follows:

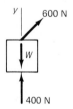

Newton's second law relates the net force acting on a mass to the acceleration, expressed as

$$\Sigma F_y = ma_y$$

Using the appropriate components, we have

$$400 + 600 \sin 45° - 100 \times 9.81 = 100a_y$$

$$\therefore \quad a_y = -1.57 \text{ m/s}^2$$

Note: We have used only three significant numbers in the answer since the information given in the problem is assumed known to three significant numbers.

EXAMPLE 1.1 (English)

A mass of 20 slug is acted on by a 200-lb force acting vertically upward and a 300-lb force acting upward at a 45° angle. Calculate the vertical component of the acceleration. The local acceleration of gravity is 32.2 ft/sec^2.

Solution

The first step in solving a problem involving forces is to draw a free-body diagram with all forces acting on it. It appears as follows:

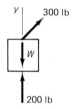

Newton's second law relates the net force acting on a mass to the acceleration, expressed as

$$\Sigma\, F_y = ma_y$$

Using the appropriate components, we have

$$200 + 300 \sin 45° - 20 \times 32.2 = 20a_y$$

$$\therefore \quad a_y = -11.6 \text{ ft/sec}^2$$

Note: We have used only three significant numbers in the answer since the information given in the problem is assumed known to three significant numbers.

1.3 CONTINUUM VIEW OF GASES AND LIQUIDS

Substances referred to as fluids may be liquids or gases. The definition[2] of a **liquid** is:

> A state of matter in which the molecules are relatively free to change their positions with respect to each other but restricted by cohesive forces so as to maintain a relatively fixed volume.

---

[2]*Handbook of Chemistry and Physics,* 40th ed., CRC Press, Boca Raton, Fla.

A **gas** is defined to be:

> A state of matter in which the molecules are practically unrestricted by cohesive forces. A gas has neither definite shape nor volume.

In a first course in the mechanics of fluids we restrict the liquids that are studied. Before we state the restriction, we must define a shearing stress. A force ΔF that acts on an area ΔA can be decomposed into a normal component ΔF_n and a tangential component ΔF_t, as shown in Fig. 1.1. The force divided by the area upon which it acts is called a **stress.** The normal component of force divided by the area is a **normal stress,** and the tangential force divided by the area is a **shear stress.** In this discussion we are interested in the shear stress τ. Mathematically, it is defined as

$$\tau = \lim_{\Delta A \to 0} \frac{\Delta F_t}{\Delta A} \tag{1.3.1}$$

Our restricted family of fluids may now be identified; the fluids considered in this book are *those liquids and gases that move under the action of a shear stress, no matter how small that shear stress may be.* This means that even a very small shear stress results in motion in the fluid. Gases obviously fall within this category of fluids. Some substances, such as plastics, may resist small shear stresses without moving; a study of these substances are not included in this book.

It is worthwhile to consider the microscopic behavior of fluids in more detail. Consider the molecules of a gas in a container. These molecules are not stationary but move about in space with very high velocities. They collide with each other and strike the walls of the container in which they are confined, giving rise to the pressure exerted by the gas. If the volume of the container is increased while the temperature is maintained constant, the number of molecules impacting on a given area is decreased and as a result the pressure decreases. If the temperature of a gas in a given volume increases (i.e., the velocities of the molecules increase), the pressure increases.

Molecular forces in liquids are relatively high, as can be inferred from the following example. The pressure necessary to compress 18 grams (g) of water vapor at 20°C into 18 cm³, assuming that no molecular forces exist, is equal to 1340 times the atmospheric pressure. Of course, this pressure is not required because 18 g of water occupies 18 cm³. It follows that the cohesive forces in the liquid phase must be very large.

Figure 1.1 The normal and tangential components of a force.

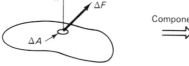
Components

Despite the high molecular attractive forces in a liquid, some of the molecules at the surface escape into the space above. If the liquid is contained, an equilibrium is established between outgoing and incoming molecules. The presence of molecules above the liquid surface leads to a so-called vapor pressure. This pressure increases with temperature. For water at 20°C this pressure is 0.02 times the atmospheric pressure.

In our study of the mechanics of fluids it is convenient to assume that both gases and liquids are continuously distributed throughout a region of interest, that is, the fluid is treated as a **continuum.** The primary property used to determine if the continuum assumption is appropriate is the **density** ρ, defined by

$$\rho = \lim_{\Delta V \to 0} \frac{\Delta m}{\Delta V} \qquad (1.3.2)$$

where Δm is the incremental mass contained in the incremental volume ΔV. The density for air at **standard atmospheric conditions,** that is, at a pressure of 101.3 kPa (14.7 psi) and a temperature of 15°C (59°F), is 1.23 kg/m³ (0.00238 slug/ft³). For water, we use the nominal value of 1000 kg/m³ (1.94 slug/ft³).

Physically, we cannot let $\Delta V \to 0$ since, as ΔV gets extremely small, the mass contained in ΔV would vary discontinuously depending on the number of molecules in ΔV; this is shown graphically in Fig. 1.2. Actually, the zero in the definition of density should be replaced by some small volume ε, below which the continuum assumption fails. For most engineering applications, the small volume ε shown in Fig. 1.2 is extremely small. For example, there are 2.7×10^{16} molecules contained in a cubic millimeter of air at standard conditions; hence, ε is much smaller than a cubic millimeter. An appropriate way to determine if the continuum model is acceptable is to compare a characteristic length l of the device or object of interest with the **mean free path** λ, the average distance a molecule travels before it collides with another molecule; if $l \gg \lambda$, the continuum model is acceptable. The mean free path is derived in molecular

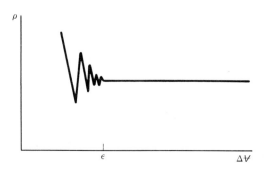

Figure 1.2   Density at a point in a continuum.

theory. It is

$$\lambda = 0.225 \frac{m}{\rho d^2} \tag{1.3.3}$$

where m is the mass (kg) of a molecule, ρ the density (kg/m³) and d the diameter (m) of a molecule. For air $m = 4.8 \times 10^{-26}$ kg and $d = 3.7 \times 10^{-10}$ m. At standard atmospheric conditions the mean free path is approximately 6×10^{-6} cm, at an elevation of 100 km it is 10 cm, and at 160 km it is 5000 cm. Obviously, at higher elevations the continuum assumption is not acceptable and the theory of rarefied gas dynamics (or free molecular flow) must be utilized.

With the continuum assumption, fluid properties can be assumed to apply uniformly at all points in the region at any particular instant in time. For example, the density ρ can be defined at all points in the fluid; it may vary from point to point and from instant to instant; that is, in Cartesian coordinates ρ is a function of x, y, z, and t, written as $\rho(x, y, z, t)$.

1.4 PRESSURE AND TEMPERATURE SCALES

In fluid mechanics pressure results from a normal compressive force acting on an area. The **pressure** p is defined as (see Fig. 1.3)

$$p = \lim_{\Delta A \to 0} \frac{\Delta F_n}{\Delta A} \tag{1.4.1}$$

where ΔF_n is the incremental normal compressive force acting on the incremental area ΔA. The metric units to be used on pressure are newtons per square meter (N/m²) or pascal (Pa). Since the pascal is a very small unit of pressure, it is more conventional to express pressure in units of kilopascal (kPa). For example, standard atmospheric pressure at sea level is 101.3 kPa. The English units for pressure are pounds per square inch (psi) or pounds per square foot (lb/ft²).

Both pressure and temperature are physical quantities that can be measured using different scales. There exist absolute scales for pressure and temperature and there are scales that measure these quantities relative to selected reference points. In many thermodynamic relationships

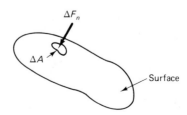

Figure 1.3 Definition of pressure.

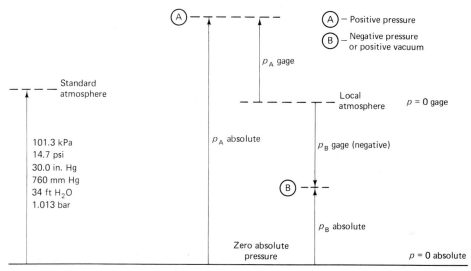

Figure 1.4   Gage pressure and absolute pressure.

(see Section 1.6) absolute scales must be used for pressure and temperature. Figures 1.4 and 1.5 summarize the commonly used scales.

   The **absolute pressure** reaches zero when an ideal vacuum is achieved, that is, when no molecules are left in a space; consequently, a negative absolute pressure is an impossibility. A second scale is defined by measuring pressures relative to the local atmospheric pressure. This pressure is a so-called **gage pressure.** A conversion from gage pressure to absolute pressure can be carried out using the following equation:

$$p_{\text{absolute}} = p_{\text{atmospheric}} + p_{\text{gage}} \qquad (1.4.2)$$

Note that the atmospheric pressure in Eq. 1.4.2 is the local atmospheric pressure, which may change with time, particularly when a weather "front" moves through. However, if the local atmospheric pressure is not given, we use the value given for a particular elevation, as given in Table B.3 of Appendix B. The gage pressure is negative whenever the absolute pressure is less than atmospheric pressure; it may then be called a **vacuum.** In this book the word "absolute" will generally follow the pressure value if the pressure is given as an absolute pressure (e.g., $p = 50$ kPa absolute). If it were stated as $p = 50$ kPa, the pressure would be taken as a gage pressure, except that atmospheric pressure is always an absolute pressure. Most often in fluid mechanics gage pressure is used.

   Two temperature scales are commonly used, the Celsius (C) and Fahrenheit (F) scales. Both scales are based on the ice point and steam point of water at an atmospheric pressure of 101.3 kPa (14.7 psi). Figure 1.5 shows that the ice and steam point are 0 and 100°C on the Celsius scale and 32 and 212°F on the Fahrenheit scale. There are two corresponding

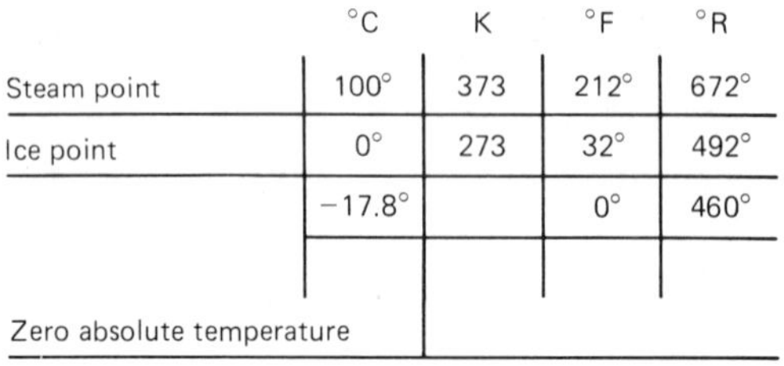

	°C	K	°F	°R
Steam point	100°	373	212°	672°
Ice point	0°	273	32°	492°
	−17.8°		0°	460°

Figure 1.5   Temperature scales.   Zero absolute temperature

absolute temperature scales. The absolute scale corresponding to the Celsius scale is the kelvin (K) scale. The relation between these scales is

$$K = °C + 273.15 \qquad (1.4.3)$$

The absolute scale corresponding to the Fahrenheit scale is the Rankine scale (°R). The relation between these scales is

$$°R = °F + 459.67 \qquad (1.4.4)$$

Note that in the SI system we do not write 100°K but simply 100 K, which is read "100 kelvins," similar to other units.

Reference will often be made to "standard atmospheric conditions" or "standard temperature and pressure." This refers to sea-level conditions at 40° latitude, which are taken to be 101.3 kPa (14.7 psi) for pressure and 15°C (59°F) for temperature.

EXAMPLE 1.2

A pressure gage attached to a rigid tank measures a vacuum of 42 kPa inside the tank, which is situated at a site in Colorado where the elevation is 2000 m. Determine the absolute pressure inside the tank.

Solution

To determine the absolute pressure, the atmospheric pressure must be known. If the elevation were not given, we would assume the standard atmospheric pressure of 101 kPa. However, with the elevation given, the atmospheric pressure is found from Table B.3 in Appendix B to be 78.5 kPa. Thus

$$p = -42 + 78.5 = 36.5 \text{ kPa absolute}$$

Note: A vacuum is always a negative gage pressure.

1.5 FLUID PROPERTIES

In this section we present several of the more common fluid properties. If density variation or heat transfer is significant, several additional properties, not presented here, become important.

1.5.1 Density and Specific Weight

Fluid density was defined in Eq. 1.3.1 as mass per unit volume. A fluid property directly related to density is the **specific weight** γ or weight per unit volume. It is defined by

$$\gamma = \rho g \qquad (1.5.1)$$

where g is the local gravity. The units of specific weight are N/m^3 (lb/ft^3).

The **specific gravity** S is often used to determine the specific weight or density of a fluid (usually a liquid). It is defined as the ratio of the density of a substance to that of water at a reference temperature of 4°C:

$$S = \frac{\rho}{\rho_{water}} = \frac{\gamma}{\gamma_{water}} \qquad (1.5.2)$$

For example, the specific gravity of mercury is 13.6, a dimensionless number; that is, the mass of mercury is 13.6 times that of water for the same volume. The density, specific weight, and specific gravity of air and water at standard conditions are given in Table 1.4.

1.5.2 Viscosity

Viscosity can be thought of as the internal stickiness of a fluid. It is one of the properties that controls the amount of fluid that can be transported in a pipeline during a specific period of time. It accounts for the energy losses associated with the transport of fluids in ducts, channels, and pipes. Further, viscosity plays a primary role in the generation of turbulence. Needless to say, viscosity is an extremely important fluid property in our study of fluid flows.

The rate of deformation of a fluid is directly linked to the viscosity of the fluid. For a given stress, a highly viscous fluid deforms at a slower rate than a fluid with a low viscosity. Consider a flow in which the fluid particles move in the x-direction at different speeds, so that particle velocities u vary with the y-coordinate. Figure 1.6 shows two particle positions at different times; observe how the particles move relative to one another. For such a simple flow field, in which $u = u(y)$, we can define the

TABLE 1.4 DENSITY, SPECIFIC WEIGHT, AND SPECIFIC GRAVITY OF AIR AND WATER AT STANDARD CONDITIONS

	Density ρ		Specific weight γ		
	kg/m³	slug/ft³	N/m³	lb/ft³	Specific gravity S
Air	1.23	0.0024	12.1	0.077	0.00123
Water	1000	1.94	9810	62.4	1

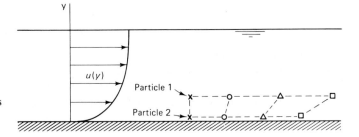

Figure 1.6 Relative movement of two fluid particles in the presence of shear stresses.

viscosity μ of the fluid by the relationship

$$\tau = \mu \frac{du}{dy} \tag{1.5.3}$$

where τ is the shear stress and u is the velocity in the x-direction. The units of τ are N/m² or Pa (lb/ft²), and of μ are N · s/m² (lb-sec/ft²). The quantity du/dy is a velocity gradient and can be interpreted as a **strain rate.** Stress velocity–gradient relationships for more complicated flow situations are presented in Chapter 7.

The concept of viscosity and velocity gradients can also be illustrated by considering a fluid within the small gap between two concentric cylinders, as shown in Fig. 1.7. A torque is necessary to rotate the inner cylinder at constant speed while the outer cylinder remains stationary. This resistance to the rotation of the cylinder is due to viscosity. The only stress that exists to resist the applied torque for this simple flow is a shear stress, which is observed to depend directly on the velocity gradient; that is,

$$\tau = \mu \left| \frac{du}{dr} \right| \tag{1.5.4}$$

where du/dr is the velocity gradient and u is the tangential velocity component, which depends only on r. For a small gap ($h \ll R$), this gradient can be approximated by assuming a linear velocity distribution[3] in the gap. Thus

$$\left| \frac{du}{dr} \right| = \frac{\omega R}{h} \tag{1.5.5}$$

---

[3]If the gap is not small relative to R, the velocity distribution will not be linear (see Section 7.5).

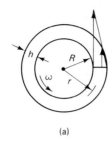

(a)

(c)

(b)

Figure 1.7 Fluid being sheared between cylinders with a small gap: (a) rotating inner cylinder; (b) velocity distribution; (c) the inner cylinder. The outer cylinder is fixed and the inner cylinder is rotating.

where h is the gap width. We can thus relate the applied torque T to the viscosity and other parameters by the equation

$$T = \text{stress} \times \text{area} \times \text{moment arm}$$

$$= \tau \times 2\pi RL \times R \qquad\qquad (1.5.6)$$

$$= \mu \frac{\omega R}{h} \times 2\pi RL \times R = \frac{2\pi R^3 \omega L \mu}{h}$$

where we have neglected the shearing stress acting on the ends of the cylinder; L represents the length of the rotating cylinder. Note that the torque depends directly on the viscosity, thus the cylinders could be used as a **viscometer,** a device that measures the viscosity of a fluid.

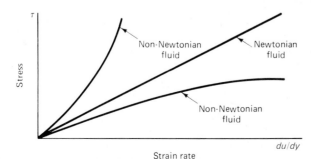

Figure 1.8 Newtonian and non-Newtonian fluids.

If the shear stress of a fluid is directly proportional to the velocity gradient, as was assumed in Eqs. 1.5.3 and 1.5.4, the fluid is said to be a **Newtonian fluid.** Fortunately, many common fluids, such as air, water, and oil, are Newtonian. **Non-Newtonian fluids,** with shear stress versus strain rate relationships as shown in Fig. 1.8, often have a complex molecular composition. Examples of non-Newtonion fluids are liquid plastics, blood, slurries, paints, and toothpaste.

An important effect of viscosity is to cause the fluid to adhere to the surface; this is known as the **no-slip condition.** This was assumed in the

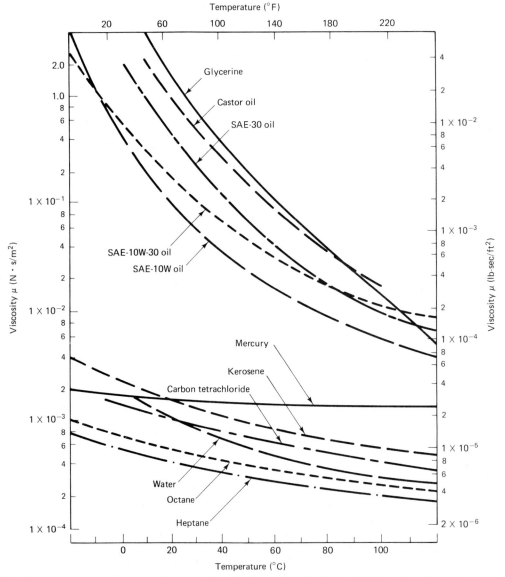

Figure 1.9   Viscosity versus temperature for several liquids. From Fox, R. W. and McDonald, A. T., *Introduction to Fluid Mechanics,* 2nd ed., John Wiley & Sons, 1978.

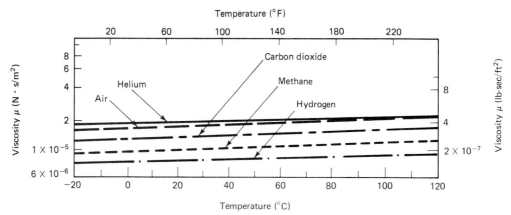

Figure 1.10   Viscosity versus temperature for several gases. From Fox, R. W. and McDonald, A. T., *Introduction to Fluid Mechanics*, 2nd ed., John Wiley & Sons, 1978.

example above. The velocity of the fluid at the rotating cylinder was taken to be ωR and the velocity of the fluid at the stationary cylinder was set equal to zero, as shown in Fig. 1.7b.

The viscosity is very dependent on temperature in liquids in which cohesive forces play a dominant role; note that the viscosity of liquids decreases with increased temperature, as shown in Fig. 1.9. For a gas it is molecular collisions that provide the internal stresses, so that as the temperature increases, resulting in increased molecular activity, the viscosity increases. This can be observed in Fig. 1.10. Note, however, that the percentage change of viscosity in a liquid is much greater than in a gas for the same temperature difference. Also, one can show that cohesive forces and molecular activity are quite insensitive to pressure, so that $\mu = \mu(T)$ only.

Since the viscosity is often divided by the density in the derivation of equations, it has become useful and customary to define **kinematic viscosity** to be

$$\nu = \frac{\mu}{\rho} \tag{1.5.7}$$

where the units of ν are m^2/s (ft^2/sec). Note that for a gas, the kinematic viscosity will also depend on the pressure since the density is pressure sensitive.

EXAMPLE 1.3

A viscometer is constructed with two 30-cm-long concentric cylinders, one 20.0 cm in diameter and the other 20.2 cm in diameter. A torque of 0.13 N·m is required to rotate the inner cylinder at 400 rpm (revolutions per minute). Calculate the viscosity.

Solution

The applied torque is just balanced by a resisting torque due to the shear stresses (see Fig. 1.7c). This is expressed by Eq. 1.5.6. The viscosity is

$$\mu = \frac{Th}{2\pi R^3 \omega L}$$

$$= \frac{0.13(0.001)}{2\pi(0.1)^3[400(2\pi)/60](0.3)} = 0.00165 \text{ N} \cdot \text{s/m}^2$$

Note: All lengths are in meters and the angular velocity must have units of rad/s.

EXAMPLE 1.3 (English)

A viscometer is constructed with two 12-in.-long concentric cylinders, one 8.00 in. in diameter and the other 8.08 in. in diameter. A torque of 0.1 ft-lb is required to rotate the inner cylinder at 400 rpm. Calculate the viscosity.

Solution

The applied torque is just balanced by a resisting torque due to the shear stresses (see Fig. 1.7c). This is expressed by Eq. 1.5.6. The viscosity is

$$\mu = \frac{Th}{2\pi R^3 \omega L}$$

$$= \frac{0.1 \times 0.04/12}{2\pi(4/12)^3[400(2\pi)/60](12/12)} = 3.42 \times 10^{-5} \text{ lb-sec/ft}^2$$

Note: All lengths are in feet and the angular velocity must have units of rad/sec.

1.5.3 Compressibility

In the preceding section we discussed the deformation of fluids that results from shear stresses. In this section we discuss the deformation that results from pressure changes. All fluids compress if the pressure increases, resulting in an increase in density. A common way to describe the compressibility of a fluid is by the following definition of the **bulk modulus of elasticity** B:

$$B = \lim_{\Delta V \to 0} -\frac{\Delta p}{\Delta V / V}\bigg|_T = \lim_{\Delta p \to 0} \frac{\Delta p}{\Delta \rho / \rho}\bigg|_T$$

$$= -V\frac{\partial p}{\partial V}\bigg|_T = \rho\frac{\partial p}{\partial \rho}\bigg|_T \tag{1.5.8}$$

In words, the bulk modulus is defined as the ratio of the change in pressure (Δp) to relative change in density ($\Delta \rho / \rho$) while the temperature remains constant. The bulk modulus obviously has the same units as pressure.

The bulk modulus for water at standard conditions is approximately 2100 MPa (310,000 psi), or 21 000 times the atmospheric pressure. For air at standard conditions, B is equal to 1 atm. In general, B for a gas is equal to the pressure of the gas. To cause a 1% change in the density of water a pressure of 21 MPa (210 atm) is required. This is an extremely large pressure needed to cause such a small change; thus liquids are often assumed to be incompressible. For gases, if significant changes in density occur, say 4%, they should be considered as compressible; for small density changes they may also be treated as incompressible.

Small density changes in liquids can be very significant when large pressure changes are present. For example, they account for "water hammer," which can be heard shortly after the sudden closing of a valve in a pipeline; when the valve is closed an internal pressure wave propagates down the pipe, producing a hammering sound due to pipe motion when the wave reflects from the closed valve.

The bulk modulus can also be used to calculate the speed of sound in a liquid; it is given by

$$c = \sqrt{\frac{B}{\rho}} \qquad (1.5.9)$$

This yields approximately 1450 m/s (4800 ft/sec) for the speed of sound in water at standard conditions.

1.5.4 Surface Tension

Surface tension is a property that results from the attractive forces between molecules. As such, it manifests itself only in liquids. The forces between molecules in the bulk of a liquid are equal in all directions, and as a result, no net force is exerted on the molecules. However, at the surface the molecules exert a force that has a resultant in the surface layer. This force holds a drop of water suspended on a rod and limits the size of the drop that may be held. It also causes the small drops from a sprayer or atomizer to assume spherical shapes.

Surface tension has units of force per unit length, N/m (lb/ft). The force due to surface tension results from a length multiplied by the surface tension; the length to use is the length of fluid in contact with a solid, or the circumference in the case of a bubble. A surface tension effect can be illustrated by considering the freebody diagrams of half a droplet and half a bubble as shown in Fig. 1.11. The droplet has one surface and the bubble is composed of a thin film of liquid with an inside surface and an outside surface. The pressure inside the droplet and bubble can now be calculated.

The pressure force $p\pi R^2$ in the droplet balances the surface tension

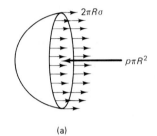
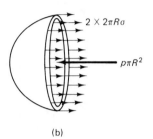

Figure 1.11 Internal forces in
(a) a droplet and
(b) a bubble.   (a)   (b)

force around the circumference. Hence

$$p\pi R^2 = 2\pi R\sigma$$

$$\therefore \quad p = \frac{2\sigma}{R} \tag{1.5.10}$$

Similarly, the pressure force in the bubble is balanced by the surface tension forces on the two circumferences. Therefore,

$$p\pi R^2 = 2(2\pi R\sigma)$$

$$\therefore \quad p = \frac{4\sigma}{R} \tag{1.5.11}$$

From Eqs. 1.5.10 and 1.5.11 we can conclude that the internal pressure in a bubble is twice as large as that in a droplet of the same size.

Figure 1.12 shows the rise of a liquid in a clean glass capillary tube due to surface tension. The liquid makes a contact angle β with the glass tube. Experiments have shown that this angle for water and most liquids is zero. There are also cases for which this angle is greater than 90° (e.g., mercury); such liquids have a capillary drop. If h is the capillary rise, D the diameter, and ρ the density, σ can be determined from equating the surface tension force to the weight of the liquid column. This will be illustrated in Example 1.4.

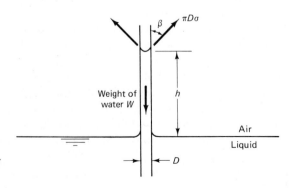

Figure 1.12 Rise in a capillary
tube.

   Surface tension may influence engineering problems when, for example, laboratory modeling of waves is conducted at a scale that surface tension forces are of the same order of magnitude as gravitational forces.

EXAMPLE 1.4

A 2-mm-diameter clean glass tube is inserted, as shown, in water at 15°C. Determine the height that the water will climb up the tube. The water makes a contact angle of 0° with the clean glass.

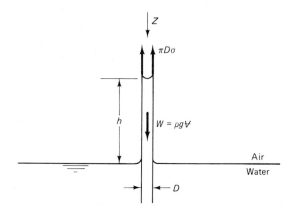

Solution

A free-body diagram of the water shows that the upward surface-tension force is equal and opposite to the weight. Writing the surface-tension force as surface tension times distance, we have

$$\Sigma F_z = 0$$

$$\sigma \pi D - \rho g \frac{\pi D^2}{4} h = 0$$

or

$$h = \frac{4\sigma}{\rho g D} = \frac{4(0.0741)}{1000(9.81)(0.002)} = 0.0151 \text{ m} \quad \text{or} \quad 15.1 \text{ mm}$$

The numerical values for σ and ρ were obtained from Table B.1 in Appendix B. Note that the nominal value used for the density of water is 1000 kg/m³.

1.5.5 Vapor Pressure

When a small quantity of liquid is placed in a closed container, a certain fraction of the liquid will vaporize. Vaporization will terminate when equilibrium is reached between the liquid and gaseous states of the substance in the container—in other words, when the number of molecules escaping from the water surface is equal to the number of incoming molecules. The pressure resulting from molecules in the gaseous state is the **vapor pressure.**

The vapor pressure is different from one liquid to another. For example, the vapor pressure of water at standard conditions (15°C, 101.3 kPa) is 1.70 kPa absolute and for ammonia it is 33.8 kPa absolute. At 59°F and 14.7 psi, it is 0.025 psi absolute for water and for ammonia it is 4.92 psi absolute.

The vapor pressure is highly dependent on pressure and temperature; it increases significantly when the temperature increases. For example, the vapor pressure of water increases to 101.3 kPa (14.7 psi) if the temperature reaches 100°C (212°F). Water vapor pressures for other temperatures at atmospheric pressure are given in Appendix B.

It is, of course, no coincidence that the water vapor pressure at 100°C is equal to the standard atmospheric pressure. At that temperature the water is **boiling;** that is, the liquid state of the water can no longer be sustained because the attractive forces are not sufficient to contain the molecules in a liquid phase. In general, a transition from the liquid state to the gaseous state occurs if the local absolute pressure is less than the vapor pressure of the liquid. In liquid flows, conditions can be created that lead to a pressure below the vapor pressure of the liquid. When this happens, bubbles are formed locally. This phenomenon, called **cavitation,** can be very damaging when these bubbles are transported by the flow to higher-pressure regions. What happens is that the bubbles collapse upon entering the higher-pressure region, and this collapse produces local pressure spikes which have the potential of damaging a pipe wall or a ship's propeller. Additional information on cavitation is included in Section 7.3.4.

EXAMPLE 1.5

Calculate the vacuum necessary to cause cavitation in a water flow at a temperature of 80°C in Colorado where the elevation is 3000 m.

Solution

The vapor pressure of water at 80°C is given in Table B.1. It is 46.4 kPa. The atmospheric pressure is found by interpolation using Table B.3 to be 70.6 kPa. The required pressure is then

$$p = 46.4 - (70.6) = -24.2 \text{ kPa} \quad \text{or} \quad 24.2 \text{ kPa vacuum}$$

1.6 CONSERVATION LAWS

From experience it has been found that fundamental laws exist that appear exact; that is, if experiments are conducted with the utmost precision and care, deviations from these laws are very small and in fact, the deviations would be even smaller if improved experimental techniques were employed. Three such laws form the basis for our study of fluid mechanics. The first is the **conservation of mass,** which states that matter is indestructible. Even though Einstein's theory of relativity postulates that under certain conditions, matter is convertible into energy and leads to the statement that the extraordinary quantities of radiation from the sun are associated with a conversion of 3.3×10^{14} kg of matter per day into energy, the destructibility of matter under typical engineering conditions is not measurable and does not violate the conservation of mass principle.

For the second and third laws it is convenient to introduce the concept of a system. A **system** is defined as a fixed quantity of matter upon which attention is focused. Everything external to the system is separated by the system boundaries. These boundaries may be fixed or movable. With this definition we can now present our second law, the **conservation of momentum:** The momentum of a system remains constant if no external forces are acting on the system. A more specific law based on this principle is **Newton's second law:** The sum of all external forces acting on a system is equal to the time rate of change of linear momentum of the system. A parallel law exists for the moment of momentum: The rate of change of angular momentum is equal to the sum of all torques acting on the system.

The third law is the **conservation of energy,** which is also known as the **first law of thermodynamics:** The total energy of an isolated system remains constant. If a system is in contact with the surroundings, its energy increases only if the energy of the surroundings experiences a corresponding decrease. It is noted that the total energy consists of potential, kinetic, and internal energy, the latter being the energy content due to the temperature of the system. Other forms of energy are not considered in fluid mechanics. The first law of thermodynamics and other thermodynamic relationships are presented in the following section.

1.7 THERMODYNAMIC PROPERTIES AND RELATIONSHIPS

For incompressible fluids, the three laws mentioned in the preceding section suffice. This is usually true for liquids but also for gases if insignificant pressure, density, and temperature changes occur. However, for a compressible fluid, it may be necessary to introduce other relationships, so that density, temperature, and pressure changes are properly taken into account. An example is the prediction of changes in density, pressure, and temperature when compressed gas is released from a container.

Thermodynamic properties, quantities that define the state of a system, either depend on the system's mass or are independent of the mass. The former is called an **extensive property** and the latter is called an **intensive property.** An intensive property can be obtained by dividing the extensive property by the mass of the system. Temperature and pressure are intensive properties; momentum and energy are extensive properties.

1.7.1 Properties of an Ideal Gas

The behavior of gases in most engineering applications can be described by the ideal-gas law, also called the perfect-gas law. When the temperature is relatively low and/or the pressure relatively high, caution should be exercised and real-gas laws should be applied. For air with temperatures higher than $-50°C$ $(-58°F)$ the ideal-gas law approximates the behavior of air to an acceptable degree provided that the pressure is not extremely high.

The **ideal-gas law** is given by

$$p = \rho RT \tag{1.7.1}$$

where p is the absolute pressure, ρ the density, T the absolute temperature, and R the gas constant. The gas constant is related to the universal gas constant R_u by the relationship

$$R = \frac{R_u}{M} \tag{1.7.2}$$

where M is the molar mass, values of which are tabulated in Table B.4 in Appendix B. The value of R_u is

$$R_u = 8.314 \text{ kJ/kg-mol} \cdot \text{K}$$
$$= 49{,}710 \text{ ft-lb/slug-mol} \cdot °\text{R} \tag{1.7.3}$$

For air $M = 28.97$ kg/kg-mol (28.97 slug/slug-mol), so that for air $R = 0.287$ kJ/kg·K (1716 ft-lb/slug-°R), a value used extensively in calculations involving air.

Other forms that the ideal-gas law takes are

$$p\Psi = mRT \tag{1.7.4}$$

and

$$p\Psi = nR_uT \tag{1.7.5}$$

where n is the number of moles.

EXAMPLE 1.6

A tank with a volume of 0.2 m³ contains 0.5 kg of nitrogen (molar mass is 28 kg/kg-mol). The temperature is 20°C. What is the pressure?

Solution

Assume an ideal gas and use Eq. 1.7.1; we obtain

$$p = \rho R T$$

$$= \frac{0.5 \text{ kg}}{0.2 \text{ m}^3} \times \frac{8.314}{28} \frac{\text{kJ}}{\text{kg} \cdot \text{K}} (273 + 20)\text{K} = 218 \text{ kPa absolute}$$

Note: The resulting units are $\text{kJ/m}^3 = \text{kN} \cdot \text{m/m}^3 = \text{kN/m}^2 = \text{kPa}$. The ideal-gas law requires the pressure and temperature to be in absolute units.

1.7.2 First Law of Thermodynamics

In the study of incompressible fluids, the first law of thermodynamics is particularly important. The **first law of thermodynamics** states that when a system, which is a fixed quantity of fluid, changes from state 1 to state 2, its energy content changes from E_1 to E_2 by energy exchange with its surroundings. The energy exchange is in the form of heat transfer or work. If we define heat transfer to the system as positive and work done by the system as positive,[4] the first law of thermodynamics can be expressed as

$$Q_{1-2} - W_{1-2} = E_2 - E_1 \tag{1.7.6}$$

where Q_{1-2} is the amount of heat transfer and W_{1-2} is the amount of work done by the system. The energy E represents the total energy, which consists of kinetic energy ($mV^2/2$), potential energy (mgz), and internal energy ($m\bar{u}$), where $\bar{u}$ is the internal energy per unit mass; hence

$$E = m\left(\frac{V^2}{2} + gz + \bar{u}\right) \tag{1.7.7}$$

Note that $V^2/2$, gz, and $\bar{u}$ are all intensive properties and E is an extensive property.

For an isolated system, one that is thermodynamically disconnected from the surroundings (i.e., $Q_{1-2} = W_{1-2} = 0$), Eq. 1.7.6 becomes

$$E_1 = E_2 \tag{1.7.8}$$

This equation represents the **conservation of energy.**

The work term in Eq. 1.7.6 results from a force F moving through a distance as it acts on the system's boundary; if the force is due to pres-

---

[4]In some presentations the work done on the system is positive, so that Eq. 1.7.6 would appear as $Q + W = \Delta E$. Either choice is acceptable.

sure, it is given by

$$W_{1-2} = \int_{l_1}^{l_2} F \, dl$$

$$= \int_{l_1}^{l_2} pA \, dl = \int_{\Psi_1}^{\Psi_2} p \, d\Psi$$

(1.7.9)

where $A \, dl = d\Psi$. An example that demonstrates an application of the first law of thermodynamics follows.

EXAMPLE 1.7

A cart with a mass of 20 kg is pushed up a ramp with an initial force of 400 N. The force decreases according to

$$F = 40(10 - l)N$$

If the cart starts from rest at $l = 0$, determine its velocity after it has traveled 6 m along the ramp. Neglect friction.

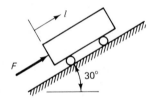

Solution

The energy equation (1.7.6) allows us to relate the quantities of interest. For zero heat transfer we have

$$-W_{1-2} = E_2 - E_1$$

Recognizing that the force is doing work on the system, the work is negative. Hence the energy equation becomes

$$\int_0^6 40(10 - l) \, dl = m\left(\frac{V_2^2}{2} + gz_2\right) - m\left(\cancel{\frac{V_1^2}{2}}^{0} + \cancel{gz_1}^{0}\right)$$

Taking the datum as $z_1 = 0$, we have $z_2 = 6 \sin 30° = 3$ m. Thus

$$400 \times 6 - 40 \times \frac{6^2}{2} = 20\left(\frac{V_2^2}{2} + 9.81 \times 3\right)$$

$$\therefore \quad V_2 = 10.4 \text{ m/s}$$

Note: We have assumed no internal energy change and no heat transfer.

EXAMPLE 1.7 (English)

A cart with a mass of 2 slug is pushed up a ramp with an initial force of 100 lb. The force decreases according to

$$F = 5(20 - l)\text{lb}$$

If the cart starts from rest at $l = 0$, determine its velocity after it has traveled 20 ft along the ramp. Neglect friction.

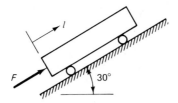

Solution

The energy equation (1.7.6) allows us to relate the quantities of interest. For zero heat transfer we have

$$-W_{1-2} = E_2 - E_1$$

Recognizing that the force is doing work on the system, the work is negative. Hence the energy equation becomes

$$\int_0^{20} 5(20 - l)\, dl = m\left(\frac{V_2^2}{2} + gz_2\right) - m\left(\cancel{\frac{V_1^2}{2}}^{0} + \cancel{gz_1}^{0}\right)$$

Taking the datum as $z_1 = 0$, we have $z_2 = 20 \sin 30° = 10$ ft. Thus

$$100 \times 20 - 5 \times \frac{20^2}{2} = 2\left(\frac{V_2^2}{2} + 32.2 \times 10\right)$$

$$\therefore \quad V_2 = 18.9 \text{ ft/sec}$$

Note: We have assumed no internal energy change and no heat transfer.

1.7.3 Other Thermodynamic Quantities

In compressible fluids it is useful to define thermodynamic quantities that are combinations of other thermodynamic quantities. One such situation is demonstrated below. Consider a fixed quantity of matter that is compressed such that the pressure remains constant. Figure 1.13 depicts such a process. The heat transfer during the process can be predicted from Eq. 1.7.6. If we assume that the potential and kinetic energy remain the same,

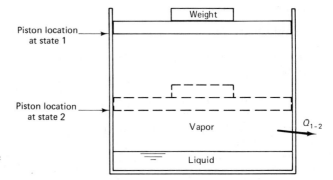

Figure 1.13 Constant-pressure process.

then

$$Q_{1-2} = m(\bar{u}_2 - \bar{u}_1) + W_{1-2} \qquad (1.7.10)$$

The work necessary to change the volume and/or pressure of the system can be calculated using Eq. 1.7.9. Since the pressure remains constant, $p = p_1 = p_2$ and

$$\begin{aligned}
W_{1-2} &= \int_{V_1}^{V} p \, dV = p(V_2 - V_1) \\
&= p_2 V_2 - p_1 V_1
\end{aligned} \qquad (1.7.11)$$

Therefore,

$$Q_{1-2} = (m\bar{u}_2 + p_2 V_2) - (m\bar{u}_1 + p_1 V_1) \qquad (1.7.12)$$

The sum $(m\bar{u} + pV)$ can be considered a system property and is also encountered in other thermodynamic processes. This property is defined as **enthalpy** H:

$$H = m\bar{u} + pV \qquad (1.7.13)$$

The corresponding intensive property (H/m) is

$$h = \bar{u} + \frac{p}{\rho} \qquad (1.7.14)$$

Other useful thermodynamic quantities are the **constant-pressure specific heat** c_p and the **constant-volume specific heat** c_v; they are used to calculate the enthalpy and the internal energy changes in an ideal gas as follows:

$$\Delta h = \int c_p \, dT \qquad (1.7.15)$$

and

$$\Delta \tilde{u} = \int c_v \, dT \qquad (1.7.16)$$

For many situations we can assume constant specific heats in the forego-ing relationships. Specific heats for common gases are listed in Table B.4. For an ideal gas c_p is related to c_v by using Eq. 1.7.14 in differential form:

$$c_p = c_v + R \qquad (1.7.17)$$

The **ratio of specific heats** k is often of use for an ideal gas; it is expressed as

$$k = \frac{c_p}{c_v} \qquad (1.7.18)$$

A process in which pressure, temperature, and other properties are essentially constant at any instant throughout the system is called a **quasi-equilibrium** or **quasi-static** process. Examples of such processes are com-pression and expansion in an internal combustion engine as well as the process of Fig. 1.13. If, in addition, no heat is transferred ($Q_{1-2} = 0$), the process is called an **adiabatic,** quasi-equilibrium process or an **isentropic** process. For such an isentropic[5] process the following relationships may be used:

$$\frac{p_1}{p_2} = \left(\frac{\rho_1}{\rho_2}\right)^k \qquad \frac{T_1}{T_2} = \left(\frac{p_1}{p_2}\right)^{(k-1)/k} \qquad \frac{T_1}{T_2} = \left(\frac{\rho_1}{\rho_2}\right)^{k-1} \qquad (1.7.19)$$

For a small pressure wave traveling in a gas at relatively low frequency, the wave speed is given by

$$c = \sqrt{\frac{dp}{d\rho}}\bigg|_s = \sqrt{kRT} \qquad (1.7.20)$$

If the frequency is relatively high, we use

$$c = \sqrt{\frac{dp}{d\rho}}\bigg|_T = \sqrt{RT} \qquad (1.7.21)$$

EXAMPLE 1.8

A cylinder fitted with a piston has an initial volume of 0.5 m³. It contains 2.0 kg of air at 400 kPa absolute. Heat is transferred to the air while the pressure remains

[5]An isentropic process occurs when the entropy is constant. We will not define or calculate entropy here; it is discussed in Section 9.1.

constant until the temperature is 300°C. Calculate the heat transfer and the work done. Assume constant specific heats.

Solution

Using Eq. 1.7.12 and the definition of enthalpy, we see that

$$Q_{1-2} = m\bar{u}_2 + p_2\Psi_2 - (m\bar{u}_1 + p_1\Psi_1)$$
$$= H_2 - H_1 = m(h_2 - h_1) = mc_p(T_2 - T_1)$$

The initial temperature is

$$T_1 = \frac{p_1\Psi_1}{mR}$$

$$= \frac{400 \times 0.5}{2.0 \times 0.287} = 348.4 \text{ K}$$

Thus the heat transfer is (c_p is found in Table B.4)

$$Q_{1-2} = 2.0 \times 1.0[(300 + 273) - 348.4] = 449 \text{ kJ}$$

The final volume is found using the ideal-gas law:

$$\Psi_2 = \frac{mRT_2}{p_2} = \frac{2 \times 0.287 \times 573}{400} = 0.822 \text{ m}^3$$

The work done for the constant-pressure process is, using Eq. 1.7.9,

$$W_{1-2} = p(\Psi_2 - \Psi_1)$$
$$= 400(0.822 - 0.5) = 129 \text{ kJ}$$

EXAMPLE 1.9

The temperature on a cold winter day in the mountains of Wyoming is −30°C at an elevation of 3000 m. Calculate the density of the air assuming the same pressure as in the normal atmosphere, and also find the speed of sound.

Solution

From Table B.3 we find the atmospheric pressure at an elevation of 3000 m to be 70.1 kPa absolute. The absolute temperature is found to be

$$T = 273 - 30 = 243 \text{ K}$$

Using the ideal-gas law, the density is calculated as

$$\rho = \frac{p}{RT}$$

$$= \frac{70.1}{0.287(243)} = 1.005 \text{ kg/m}^3$$

The speed of sound, using Eq. 1.7.20, is determined to be

$$c = \sqrt{kRT}$$

$$= \sqrt{1.4(287)(243)} = 312 \text{ m/s}$$

Note: The gas constant in this equation for the speed of sound must have units of $J/kg \cdot K$ so that the appropriate units result. Express $kg = N \cdot s^2/m$ to observe that this is true.

EXAMPLE 1.9 (English)

The temperature on a cold winter day in the mountains of Wyoming is $-22°F$ at an elevation of 10,000 ft. Calculate the density of the air assuming the same pressure as in the normal atmosphere; also find the speed of sound.

Solution

From Table B.3 we find the atmospheric pressure at an elevation of 10,000 ft to be 10.1 psi. The absolute temperature is found to be

$$T = 460 - 22 = 438°R$$

Using the ideal-gas law, the density is calculated as

$$\rho = \frac{p}{RT}$$

$$= \frac{10.1 \times 144}{1716 \times 438} = 0.00194 \text{ slug/ft}^3$$

The speed of sound, using Eq. 1.7.20, is determined to be

$$c = \sqrt{kRT}$$

$$= \sqrt{1.4 \times 1716 \times 438} = 1026 \text{ ft/sec}$$

Note: The gas constant in the foregoing equations has units of ft-lb/slug-°R so that the appropriate units result. Express $slug = lb\text{-}sec^2/ft$ to observe that this is true.

PROBLEMS

1.1. State the three basic laws that are used in the study of the mechanics of fluids. State at least one global (integral) quantity that occurs in each. State at least one quantity that is defined at a point that occurs in each.

Dimensions, Units, and Physical Quantities

1.2. Verify the dimensions given in Table 1.2 for the following quantities.
(a) Density
(b) Pressure
(c) Power
(d) Energy
(e) Mass flux
(f) Flow rate

1.3. Express the dimensions of the following quantities using the *F-L-T* system.
(a) Density
(b) Pressure
(c) Power
(d) Energy
(e) Mass flux
(f) Flow rate

1.4. State the SI units of Table 1.1 on each of the following.
(a) Pressure   **(d)** Viscosity
(b) Energy   **(e)** Heat flux
(c) Power   **(f)** Specific heat

1.5. Write the following with the use of prefixes.
(a) 2.5×10^5 N   **(d)** 1.76×10^{-5} m^3
(b) 5.72×10^{11} Pa   **(e)** 1.2×10^{-4} m^2
(c) 4.2×10^{-8} Pa   **(f)** 7.6×10^{-8} m^3

1.6. Write the following with the use of powers. Do not use a prefix.
(a) 125 MN   **(d)** 0.0056 mm^3
(b) 32.1 μs   **(e)** 520 cm^2
(c) 0.67 GPa   **(f)** 7.8 km^3

1.7. What net force is needed to accelerate a 10-kg mass at the rate of 40 m/s^2:
(a) Horizontally?
(b) Vertically upward?
(c) On an upward slope of 30°?

1.8. A particular body weighs 250 N on earth. Calculate its weight on the moon where $g \approx 1.6$ m/s^2.

1.8E. A particular body weighs 60 lb on earth. Calculate its weight on the moon, where $g \approx 5.4$ ft/sec^2.

Pressure

1.9. A gage pressure of 52.3 kPa is read on a gage. Find the absolute pressure if the elevation is:
(a) At sea level.
(b) 1000 m.
(c) 5000 m.
(d) 10 000 m.
(e) 30 000 m.

1.10. A vacuum of 31 kPa is measured in an airflow. Find the absolute pressure in:
(a) kPa.
(b) mm Hg.
(c) psi.
(d) ft H$_2$O.
(e) in. Hg.

1.11. A force of 26.5 MN is distributed uniformly over a 152-cm^2 area; however, it acts at an angle of 42° with respect to a normal vector. If it produces a compressive stress, calculate the resulting pressure.

Density and Specific Weight

1.12. Calculate the density and specific weight of water if 0.1 kg occupies 100 cm^3.

1.12E. Calculate the density and specific weight of water if 0.2 slug occupies 180 in^3.

1.13. Determine the density and the specific gravity of air at standard conditions (that is, 15°C and 101.3 kPa absolute).

1.14. Calculate the density of air inside a house and outside a house using 20°C inside and −25°C outside. Use an atmospheric pressure of 85 kPa. Do you think there would be a movement of air from the inside to the outside (infiltration), even without a wind? Explain.

Viscosity

1.15. A velocity distribution in a 2-cm-diameter pipe is measured to be $u(r) = 1000(0.01 - r^2)$ m/s, with r measured in meters. Calculate the shear stress at the wall if water at 25°C is flowing.

1.15E. A velocity distribution in a 2-in.-diameter pipe is measured to be $u(r) = 3600(1/12 - r^2)$ ft/sec, with r measured in feet. Calculate the shear stress at the wall if water at 75°F is flowing.

1.16. For two 0.2-m-long rotating cylinders, the velocity distribution is given by $u(r) = 0.004r - 1000 \, r$ m/s, with r measured in meters. If the diameters of the cylinders are 2 cm and 4 cm, respectively, calculate the fluid viscosity if the torque on the inner cylinder is measured to be 0.0026 N · m.

1.17. A 1.2-m-long, 2-cm-diameter shaft rotates inside an equally long cylinder that is 2.06 cm in diameter. Calculate the torque required to rotate the inner shaft at 2000 rpm if SAE-30 oil at 20°C fills the gap. Also, calculate the horsepower required. Assume symmetric motion.

1.17E. A 4-ft-long, 1-in.-diameter shaft rotates inside an equally long cylinder that is 1.02 in. in diameter. Calculate the torque required to rotate the inner shaft at 2000 rpm if SAE-30 oil at 70°F fills the gap. Also, calculate the horsepower required. Assume symmetric motion.

1.18. A 60-cm-wide belt moves as shown. Calculate the horsepower requirement assuming a linear velocity profile in the 10°C water.

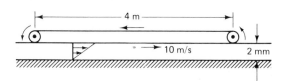

1.19. A 25-cm-diameter horizontal disk rotates a distance of 2 mm above a solid surface. Water at 10°C fills the gap. Estimate the torque required to rotate the disk at 400 rpm.

1.19E. A 6-in.-diameter horizontal disk rotates a distance of 0.08 in. above a solid surface. Water at 60°F fills the gap. Estimate the torque required to rotate the disk at 400 rpm.

1.20. The velocity distribution in a 1.0-cm-diameter pipe is given by $u(r) = 4(1 - 4r^2)$ m/s, where r is measured in centimeters. Calculate the shearing stress at the centerline, at $r = 0.25$ cm, and at the wall if water at 20°C is flowing.

Compressibility

1.21. Show that $d\rho/\rho = -d\Psi/\Psi$, as was assumed in Eq. 1.5.8.

1.22. What is the volume change of 2 m³ of water at 20°C due to an applied pressure of 10 MPa?

1.23 A pressure is applied to 20 L of water. The volume is observed to decrease to 18.7 L. Calculate the applied pressure.

1.24. Calculate the speed of propagation of a small-amplitude wave through water at:
(a) 10°C.
(b) 50°C.
(c) 95°C.

1.24E. Calculate the speed of propagation of a small-amplitude wave through water at:
(a) 40°F.
(b) 100°F.
(c) 200°F.

1.25. The change in volume of a liquid is given by $\Delta\Psi = \alpha_T \Psi \Delta T$, where α_T is the **coefficient of thermal expansion**. For water $\alpha_T = 1.53 \times 100^{-4}\ K^{-1}$. What is the volume change of 1 m³ of 40°C water if $\Delta T = -20$°C? What pressure change would be needed to cause that same volume change?

Surface Tension

1.26. Calculate the pressure in the small 10-μm-diameter droplets that are formed by spray machines. Assume the properties to be the same as water at 15°C. Calculate the pressure for bubbles of the same size.

1.27. Determine the height that 20°C water would climb in a vertical 0.02-cm-diameter tube if it attaches to the wall with an angle β of 30° to the vertical.

1.28. Mercury makes an angle of 130° (β in Fig. 1.12) when in contact with clean glass. What distance will mercury depress in a vertical, 2-mm-diameter glass tube? Use $\sigma = 0.5$ N/m.

1.28E. Mercury makes an angle of 130° (β in Fig. 1.12) when in contact with clean glass. What distance will mercury depress in a vertical, 0.08-in.-diameter glass tube. Use $\sigma = 0.032$ lb/ft.

1.29. Calculate the rise of liquid between two parallel plates a distance t apart. Use a contact angle β and surface tension σ.

1.30. Find an expression for the vertical force F needed to lift a wire ring of diameter D slowly from a liquid with surface tension σ.

1.31. Two flat plates are positioned as shown with a small angle α in an open container with a small amount of liquid. The plates are vertical and the liquid rises between the plates. Find an expression for the location $h(x)$ of the surface of the liquid assuming that $\beta = 0$.

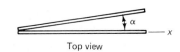

Top view

Vapor Pressure

1.32. Water is transported through a pipe such that a vacuum of 80 kPa exists at a particular location. What is the maximum possible temperature of the water? Use p_{atm} = 92 kPa.

1.33. Oil is transported through a pipeline by a series of pumps that can produce a pressure of 10 MPa in the oil leaving each pump. The losses in the pipeline cause a pressure drop of 600 kPa each kilometer. What is the maximum possible spacing of the pumps?

Ideal Gas

1.34. A 5-m³ air tank is pressurized to 5 MPa absolute. When the temperature reaches 8°C, calculate the density and the air mass.

1.34E. A 15-ft³ air tank is pressurized to 750 psia.

When the temperature reaches 10°F, calculate the density and the air mass.

1.35. A 5-m³ air tank is evacuated until the pressure is 20 Pa absolute. Find the density and the air mass if the temperature is 18°C.

First Law

1.36. A body falls from rest. Determine its velocity after 3 m and 6 m, using the energy equation.

1.36E. A body falls from rest. Determine its velocity after 10 ft and 20 ft, using the energy equation.

1.37. Determine the final velocity of a 15-kg mass moving horizontally if it starts at 10 m/s and moves a distance of 10 m while the following net force acts in the direction of motion:
(a) 200 N.
(b) 20s N.   where s is the distance in the
(c) 200 cos $s\pi/20$ N.   direction of motion.

1.38. A 10-kg mass is traveling at 40 m/s and strikes a plunger that is attached to a piston. The piston compresses 0.2 kg of air contained in a cylinder. If the

mass is brought to rest, calculate the maximum rise of temperature in the air. What effects could lead to a lower temperature rise?

1.39. A 1500-kg automobile, traveling at 100 km/h, is suddenly grabbed by a hook and all of its kinetic energy is dissipated in a hydraulic absorber containing 2000 cm³ of water. Calculate the maximum temperature rise in the water.

1.40. A fuel mass of 0.2 kg contains 40 MJ/kg of energy. Calculate the temperature rise of 100 kg of water if complete combustion occurs and the water, which surrounds the fuel, is completely insulated from the surroundings.

Isentropic Flow

1.41. Air flows from a tank maintained at 5 MPa absolute and 20°C. It exits a hole and reaches a pressure of 500 kPa absolute. Assuming an adiabatic, quasi-equilibrium process, calculate the exiting temperature.

1.42. An airstream flows with no heat transfer such that the temperature changes from 20°C to 150°C. If the initial pressure is measured to be 150 kPa, estimate the final pressure.

Speed of Sound

1.43. Compare the speed of sound at an elevation of 10 000 m with that at sea level by calculating a percentage decrease.

1.44. A lumberman, off in the distance, is chopping

with an axe. An observer, using her digital stopwatch, measures a time of 8.32 s from the instant the axe strikes the tree until the sound is heard. How far is the observer from the lumberman?

TWO

Fluid Statics

2.1 INTRODUCTION

Fluid statics is the study of fluids in which there is no relative motion between fluid particles. If there is no relative motion, no shearing stresses exist, since velocity gradients, such as du/dy, are required for such shearing stresses to be present. The only stress that exists is a normal stress, the pressure, so it is the pressure that is of primary interest in fluid statics.

Three situations, depicted in Fig. 2.1, involving fluid statics will be investigated. These include fluids at rest, such as water in a reservoir, fluids contained in devices that undergo linear acceleration, and fluids contained in rotating cylinders. In each of these three situations the fluid is in static equilibrium with respect to a reference frame attached to the boundary surrounding the fluid. In addition to the examples shown for fluids at rest, we consider instruments called manometers and investigate the forces of buoyancy. Finally, the stability of floating bodies such as ships will also be presented.

2.2 PRESSURE AT A POINT

We have defined pressure as being the infinitesimal normal compressive force divided by the infinitesimal area over which it acts. This defines the pressure at a point. One might question whether the pressure, at a given point, varies as the normal to the area changes direction. To show that this is not the case, even for fluids in motion, consider the wedge-shaped element of unit depth (in the z-direction) shown in Fig. 2.2. Assume that a pressure p acts on the hypotenuse and that a different pressure acts on each of the other areas, as shown. Since the forces on the two end faces are in the z-direction, we have not included them on the element. Now, let us apply Newton's second law to the element, for both the x- and y-directions:

$\Sigma F_x = ma_x$:

$$p_x \, \Delta y - p \, \Delta s \sin \theta = \rho \, \frac{\Delta x \, \Delta y}{2} \, a_x$$

$\Sigma F_y = ma_y$:

$$p_y \, \Delta x - \rho g \, \frac{\Delta x \, \Delta y}{2} - p \, \Delta s \cos \theta = \rho \, \frac{\Delta x \, \Delta y}{2} \, a_y$$

(2.2.1)

Figure 2.1 Examples included in fluid statics: (a) liquids at rest; (b) linear acceleration; (c) angular rotation.

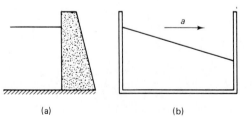

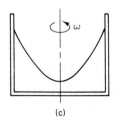

(a)   (b)   (c)

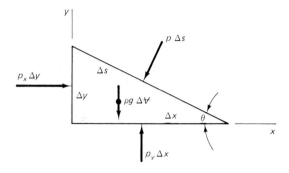

Figure 2.2 Pressure at a point in a fluid.

where we have used $\Delta V = \Delta x\, \Delta y/2$. The pressures shown are due to the surrounding fluid and are the average pressure on the areas. Substituting

$$\Delta s \sin \theta = \Delta y \qquad \Delta s \cos \theta = \Delta x \qquad (2.2.2)$$

we see that Eqs. 2.2.1 take the form

$$p_x - p = \frac{\rho a_x \, \Delta x}{2}$$

$$p_y - p = \frac{\rho(a_y + g)\, \Delta y}{2} \qquad (2.2.3)$$

Note that in the limit as the element shrinks to a point, $\Delta x \to 0$ and $\Delta y \to 0$. Hence the right-hand sides in the equations above go to zero, even for fluids in motion, providing us with the result that, at a point,

$$p_x = p_y = p \qquad (2.2.4)$$

Since θ is arbitrary, this relationship holds for all angles at a point. We could have analyzed an element in the xz-plane and concluded that $p_x = p_z = p$. Thus we conclude that the pressure in a fluid is constant at a point; that is, pressure is a scalar function. It acts equally in all directions at a given point for both a static fluid and a fluid that is in motion.

Alternatively, we could recall **Mohr's circle,** if we have studied solid mechanics. Recall that normal stress[1] was plotted against shear stress, as in Fig. 2.3. If p_x is different from p_y (normal stress on perpendicular faces is plotted at points 180° apart), shear stress must be present. If shear stress is absent, as in fluid statics, Mohr's circle degenerates to a point and $p_x = p_y$. The same argument holds for $p_x = p_z$.

2.3 PRESSURE VARIATION

A general equation is derived to predict the pressure variation of fluids at rest or fluids undergoing an acceleration while the relative position of fluid

[1]Pressure is a negative normal stress.

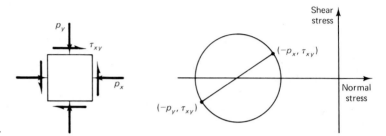

Figure 2.3 Mohr's circle.

elements to one another remains the same. To determine the pressure variation in such fluids, consider the infinitesimal element displayed in Fig. 2.4, where the z-axis is in the vertical direction. The pressure variation from one point to another will be determined by applying Newton's second law; that is, the sum of the forces acting on the fluid element is equal to the mass times the acceleration of the element.

If we assume that a pressure p exists at the center of this element, the pressures at each of the sides can be expressed by using a first-order Taylor series[2] expansion with $p(x, y, z)$:

$$dp = \frac{\partial p}{\partial x}\, dx + \frac{\partial p}{\partial y}\, dy + \frac{\partial p}{\partial z}\, dz \qquad (2.3.1)$$

If we move from the center to a face a distance $(dx/2)$ away, we see that the pressure is

$$p\left(x + \frac{dx}{2}, y, z\right) = p(x, y, z) + \frac{\partial p}{\partial x}\frac{dx}{2} \qquad (2.3.2)$$

The pressures at all faces are expressed in this manner, as shown in Fig. 2.4. Newton's second law is written in vector form for a constant-mass system as

$$\Sigma \mathbf{F} = m\mathbf{a} \qquad (2.3.3)$$

[2]The change in a function $p(x, y, z)$ from a point (x, y, z) to a neighboring point $(x + dx, y + dy, z + dz)$ is given by

$$dp = \frac{\partial p}{\partial x}\, dx + \frac{\partial p}{\partial y}\, dy + \frac{\partial p}{\partial z}\, dz + \text{higher-order terms}$$

The higher-order terms include terms like

$$\frac{1}{2}\frac{\partial^2 p}{\partial x^2}(dx)^2, \quad \frac{1}{2}\frac{\partial^2 p}{\partial x\,\partial y}\, dx\, dy, \quad \frac{1}{2}\frac{\partial^2 p}{\partial z^2}(dz)^2, \quad \frac{1}{6}\frac{\partial^3 p}{\partial x^3}(dx)^3, \quad \text{etc.}$$

Obviously, these higher-order terms are much smaller than the first-order terms and are neglected.

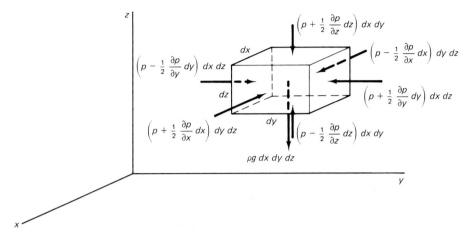

Figure 2.4   Forces acting on an infinitesimal element that is at rest in the xyz-reference frame.

This results in the three component equations, using the mass as $\rho\ dx\ dy\ dz$,

$$-\frac{\partial p}{\partial x}\ dx\ dy\ dz = \rho a_x\ dx\ dy\ dz$$

$$-\frac{\partial p}{\partial y}\ dx\ dy\ dz = \rho a_y\ dx\ dy\ dz \qquad (2.3.4)$$

$$-\frac{\partial p}{\partial z}\ dx\ dy\ dz = \rho(a_z + g)\ dx\ dy\ dz$$

where a_x, a_y, and a_z are the components of the acceleration of the element. Division by the element's volume $dx\ dy\ dz$ yields

$$\frac{\partial p}{\partial x} = -\rho a_x$$

$$\frac{\partial p}{\partial y} = -\rho a_y \qquad (2.3.5)$$

$$\frac{\partial p}{\partial z} = -\rho(a_z + g)$$

The pressure differential in any direction can now be determined from Eq. 2.3.1 as

$$dp = -\rho a_x\ dx - \rho a_y\ dy - \rho(a_z + g)\ dz \qquad (2.3.6)$$

Pressure differences between specified points can be found by integrating Eq. 2.3.6. This equation is useful in a variety of problems, as will be demonstrated in the remaining sections of this chapter.

2.4 FLUIDS AT REST

A fluid at rest does not undergo any acceleration. Therefore, Eq 2.3.6 reduces to

$$dp = -\rho g \, dz \qquad (2.4.1)$$

or

$$\frac{dp}{dz} = -\gamma \qquad (2.4.2)$$

This equation implies that there is no pressure variation in the x- and y-directions, that is, in the horizontal plane. The pressure varies in the z-direction only. Also note that dp is negative if dz is positive; that is, the pressure decreases as we move up and increases as we move down, a rather obvious result.

2.4.1 Pressures in Liquids at Rest

If the density can be assumed constant, Eq. 2.4.2 is integrated to yield

$$p + \gamma z = \text{constant} \qquad \text{or} \qquad \frac{p}{\gamma} + z = \text{constant} \qquad (2.4.3)$$

so that pressure increases with depth. Note that z is positive in the upward direction. The quantity $(p/\gamma + z)$ is often referred to as the **piezometric head.** If the point of interest were a distance h below a **free surface** (a surface separating a gas from a liquid), as shown in Fig. 2.5, Eq. 2.4.3 would result in

$$p = \gamma h \qquad (2.4.4)$$

where $p = 0$ at $h = 0$. This equation will be quite useful in converting pressure to an equivalent height of liquid. For example, atmospheric pressure is often expressed as millimeters of mercury; that is, the atmospheric pressure is equal to the pressure at a certain depth in a mercury column, and by knowing the specific weight of mercury, we can then determine that depth using Eq. 2.4.4.

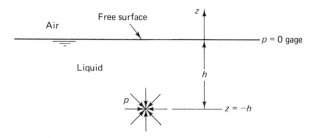

Figure 2.5 Pressure below a free surface.

2.4.2 Pressures in the Atmosphere

For the atmosphere where the density depends on height [i.e., $\rho = \rho(z)$], we must integrate Eq. 2.4.1 along a vertical path. The atmosphere is divided into four layers: the **troposphere** (nearest Earth), the **stratosphere**, the **ionosphere**, and the **exosphere**. Because conditions change with time and latitude in the atmosphere with the layers being thicker at the equator and thinner at the poles, we base calculations on the **standard atmosphere,** which is at 40° latitude. In the standard atmosphere the temperature in the troposphere varies linearly with elevation, $T(z) = T_0 - \alpha z$, where the **lapse rate** $\alpha = 0.0065$ K/m and T_0 is 288 K. In the lower part of the stratosphere, between 11 and 20 km, the temperature is constant at $-56.5°$C. The temperature then increases again and reaches a maximum near 50 km; it then decreases to the edge of the ionosphere. The standard atmosphere is sketched in Fig. 2.6. Because the density of the air in the ionosphere and the exosphere is so low, it is possible for satellites to orbit the earth in either of these layers.

   To determine the pressure variation of the troposphere, we can use the ideal-gas law $p = \rho RT$ and Eq. 2.4.1; there results

$$\frac{dp}{dz} = -\rho g$$

$$= -\frac{pg}{RT}$$

(2.4.5)

This can be integrated, between sea level and an elevation z, to yield the following:

$$\int_{P_{atm}}^{p} \frac{dp}{p} = -\frac{g}{R} \int_{0}^{z} \frac{dz}{T_0 - \alpha z}$$

(2.4.6)

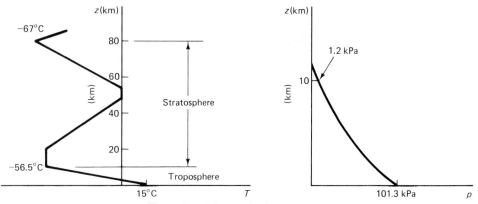

Figure 2.6 The standard atmosphere.

Upon integration this gives

$$\ln \frac{p}{p_{\text{atm}}} = \frac{g}{\alpha R} \ln \frac{T_0 - \alpha z}{T_0} \tag{2.4.7}$$

which can be put in the form

$$p = p_{\text{atm}} \left(\frac{T_0 - \alpha z}{T_0} \right)^{g/\alpha R} \tag{2.4.8}$$

In the stratosphere, where the temperature is constant, Eq. 2.4.5 is integrated again as follows:

$$\int_{p_s}^{p} \frac{dp}{p} = -\frac{g}{RT_s} \int_{z_s}^{z} dz \tag{2.4.9}$$

$$\ln \frac{p}{p_s} = -\frac{g}{RT_s} (z - z_s) \tag{2.4.10}$$

or

$$p = p_s \exp \left[\frac{g}{RT_s} (z_s - z) \right] \tag{2.4.11}$$

The subscript s denotes conditions at the troposphere–stratosphere interface.

EXAMPLE 2.1

The atmospheric pressure is given as 680 mm Hg at a mountain location. Convert this to kilopascals and meters of water.

Solution

Use Eq. 2.4.4 and find, using $S_{\text{Hg}} = 13.6$,

$$p = \gamma_{\text{Hg}} h$$
$$= (9810 \times 13.6) \times 0.680 = 90\ 700 \text{ Pa} \quad \text{or} \quad 90.7 \text{ kPa}$$

To convert this to meters of water, we have

$$h = \frac{p}{\gamma_{\text{H}_2\text{O}}}$$
$$= \frac{90\ 700}{9810} = 9.25 \text{ m of water}$$

EXAMPLE 2.2

Assume an isothermal atmosphere of $-20°C$ and approximate the pressure at 10 000 m. Calculate the percentage error when compared with the actual value from Appendix B.3.

Solution

The ideal-gas equation gives the density as

$$\rho = \frac{p}{RT}$$

Equation 2.4.1 is manipulated as follows:

$$dp = -\frac{p}{RT} g \, dz$$

$$\therefore \quad \frac{dp}{p} = -\frac{g}{RT} dz$$

This is then integrated, assuming that T is constant, as follows:

$$\int_{101}^{p} \frac{dp}{p} = -\frac{g}{RT} \int_{0}^{z} dz$$

$$\ln \frac{p}{101} = -\frac{gz}{RT} \quad \text{or} \quad p = 101e^{-gz/RT}$$

Substituting $z = 10\,000$ m and $T = 253$ K, there results

$$p = 101e^{-9.81 \times 10\,000/(287 \times 253)}$$

$$= 26.16 \text{ kPa}$$

The actual pressure at 10 000 m is found from Appendix B.3 to be 26.50 kPa. Hence the percent error is

$$\% \text{ error} = \left(\frac{26.16 - 26.50}{26.50}\right) \times 100 = -1.3\%$$

2.4.3 Manometers

Manometers are instruments that use columns of liquids to measure pressures. Three such instruments, shown in Fig. 2.7, are discussed to illustrate their use. Part (a) displays a U-tube manometer, used to measure relatively small pressures. In this case the pressure in the pipe can be determined by defining a point 1 at the center of the pipe and a point 2 at

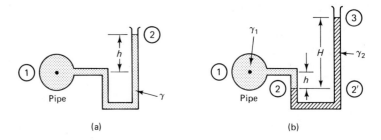

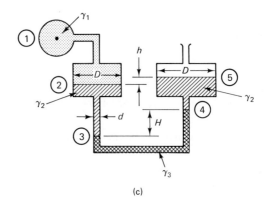

Figure 2.7 Manometers: (a) U-tube manometer (small pressures); (b) U-tube manometer (large pressures); (c) micromanometer (very small pressure changes).

the surface of the right column. Then, using Eq. 2.4.3,

$$p_1 + \gamma z_1 = p_2 + \gamma z_2$$

where the datum from which z_1 and z_2 are measured is located at any desired position, such as through point 1. Since $p_2 = 0$ (gage pressure is selected; if absolute pressure is desired, we would select $p_2 = p_{atm}$) and $z_2 - z_1 = h$,

$$p_1 = \gamma h \tag{2.4.12}$$

Figure 2.7b shows a manometer used to measure relatively large pressures since we can select γ_2 to be quite large; for example, we could select γ_2 to be that of mercury so that $\gamma_2 = 13.6\gamma_{water}$. The pressure can be determined by introducing three points as indicated. This is necessary because Eq. 2.4.3 applies throughout one fluid; γ must be constant. The value of γ changes abruptly at point 2; hence we write

$$p_1 - p_2 = \gamma_1(z_2 - z_1)$$
$$p_2 - p_3 = \gamma_2(z_3 - z_2) \tag{2.4.13}$$

Addition of these equations and setting $p_3 = 0$ (gage pressure is used) results in

$$p_1 = \gamma_1(z_2 - z_1) + \gamma_2(z_3 - z_2)$$
$$= -\gamma_1 h + \gamma_2 H \tag{2.4.14}$$

Here we recognize that the pressure at point 2′ is equal to the pressure at point 2, since points 2 and 2′ are at the same elevation above a datum passing through the horizontal bottom of the manometer.

Figure 2.7c shows a micromanometer that is used to measure very small pressure changes. Introducing five points as indicated, we can write

$$p_1 - p_2 = \gamma_1(z_2 - z_1)$$
$$p_2 - p_3 = \gamma_2(z_3 - z_2)$$
$$p_3 - p_4 = \gamma_3(z_4 - z_3) \tag{2.4.15}$$
$$p_4 - p_5 = \gamma_2(z_5 - z_4)$$

Addition of these equations and setting $p_5 = 0$ lead to

$$p_1 = \gamma_1(z_2 - z_1) + \gamma_2(z_3 - z_2) + \gamma_3(z_4 - z_3) + \gamma_2(z_5 - z_4)$$
$$= \gamma_1(z_2 - z_1) + \gamma_2(z_5 - z_2) + (\gamma_3 - \gamma_2)(z_4 - z_3) \tag{2.4.16}$$
$$= \gamma_1(z_2 - z_1) + \gamma_2 h + (\gamma_3 - \gamma_2)H$$

Note that in all of the equations above for all three manometers, we have identified all interfaces with a point. This is always necessary when analyzing a manometer. The micromanometer is capable of measuring small pressure changes because a small pressure change in p_1 results in a relatively large deflection H. The change in H due to a change in p_1 can be determined using Eq. 2.4.16. Suppose that p_1 increases by Δp_1 and, as a result, z_2 decreases by Δz; then h and H also change. Using the fact that a decrease in z_2 is accompanied by an increase in z_5 leads to an increase in h of $2\Delta z$ and, similarly, assuming that the volumes are conserved, it can be shown that H increases by $2\Delta z D^2/d^2$. Hence a pressure change Δp_1 can be evaluated from changes in deflections as follows:

$$\Delta p_1 = \gamma_1(-\Delta z) + \gamma_2(2\,\Delta z) + \frac{(\gamma_3 - \gamma_2)2\,\Delta z\, D^2}{d^2} \tag{2.4.17}$$

The rate of change in H with p_1 is

$$\frac{\Delta H}{\Delta p_1} = \frac{2\,\Delta z\, D^2/d^2}{\Delta p_1} \tag{2.4.18}$$

Using Eq. 2.4.17 we have

$$\frac{\Delta H}{\Delta p_1} = \frac{2D^2/d^2}{-\gamma_1 + 2\gamma_2 + 2(\gamma_3 - \gamma_2)D^2/d^2} \tag{2.4.19}$$

An example of this type of manometer is given below.

EXAMPLE 2.3

For a given condition the liquid levels in Fig. 2.7c are $z_1 = 0.95$ m, $z_2 = 0.70$ m, $z_3 = 0.52$ m, $z_4 = 0.65$ m, and $z_5 = 0.72$ m. Further, $\gamma_1 = 9810$ N/m^3, $\gamma_2 = 11\ 500$ N/m^3, and $\gamma_3 = 14\ 000$ N/m^3. The diameters are $D = 0.2$ m and $d = 0.01$ m. (a) Calculate the pressure p_1 in the pipe, (b) calculate the change in H if p_1 increases by 100 Pa, and (c) compare the change in part (b) with the variation in the liquid level of the manometer of Fig. 2.7a if $h = 0.5$ m of water.

Solution

(a) Referring to Fig. 2.7c, we have

$$h = 0.72 - 0.70 = 0.02 \text{ m}$$

$$H = 0.65 - 0.52 = 0.13 \text{ m}$$

Substituting the given values into Eq. (2.4.16) leads to

$$p_1 = \gamma_1(z_2 - z_1) + \gamma_2(h) + (\gamma_3 - \gamma_2)H$$

$$= 9810(0.70 - 0.95) + 11\ 500(0.02) + (14\ 000 - 11\ 500)(0.13)$$

$$= -1898 \text{ Pa}$$

(b) If the pressure p_1 is increased by 100 Pa to $p_1 = -1798$ Pa, the change in H is, using Eq. 2.4.19,

$$\Delta H = \Delta p_1 \frac{2D^2/d^2}{-\gamma_1 + 2\gamma_2 + 2(\gamma_3 - \gamma_2)D^2/d^2}$$

$$\Delta H = 100 \frac{2(20^2)}{-9810 + 2(11\ 500) + 2(14\ 000 - 11\ 500) \times 20^2} = 0.0397 \text{ m}$$

Thus H increases by 3.97 cm as a result of increasing the pressure by 100 Pa.
(c) For the manometer in Fig. 2.7a, the pressure p_1 is given by $p = \gamma h$. Assume that initially $h = 0.50$ m. Thus the pressure initially is

$$p_1 = 9810 \times 0.50 = 4905 \text{ Pa}$$

Now if p_1 is increased by 100 Pa, h can be found:

$$p_1 = \gamma h$$

$$h = \frac{p_1}{\gamma} = \frac{5005}{9810} = 0.510 \text{ m} \qquad \Delta h = 0.510 - 0.5 = 0.01 \text{ m}$$

Thus an increase of 100 Pa increases h by 1 cm in the manometer shown in part (a), only 25% of the change in the micromanometer.

EXAMPLE 2.4

Water and oil flow in horizontal pipelines. A double U-tube manometer is connected between the pipelines, as shown. Calculate the pressure difference between the water pipe and the oil pipe.

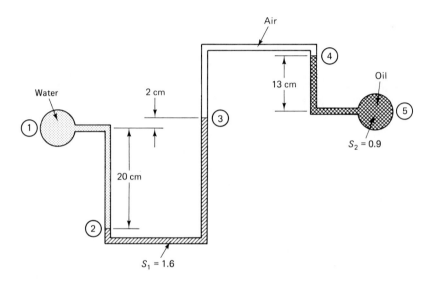

Solution

We first identify the relevant points 1, 2, 3, 4, and 5 as shown in the figure. We can then find the pressure difference from

$$p_1 - p_5 = (p_1 - p_2) + (p_2 - p_3) + (p_3 - p_4) + (p_4 - p_5)$$
$$= \gamma(z_2 - z_1) + \gamma S_1(z_3 - z_2) + \gamma S_{air}(z_4 - z_3) + \gamma S_2(z_5 - z_4)$$

where $\gamma = 9810 \text{ N/m}^3$, $S_1 = 1.6$, $S_2 = 0.9$, and $S_{air} \approx 0$. Thus

$$p_1 - p_5 = 9810(-0.20 + 1.6 \times 0.22 + 0 \times 0.13 + 0.9 \times -0.13)$$
$$= 343 \text{ Pa}$$

Note that by neglecting the weight of the air, the pressure at point 3 is equal to the pressure at point 4.

An alternative method of interpreting the manometer is to start at the left water pipe, add pressure when the elevation decreases, and subtract pressure when the elevation increases until the pipe at the right is encountered. This would result in

$$p_1 + 9810 \times 0.2 - (1.6 \times 9810) \times 0.22 + (0.9 \times 9810) \times 0.13 = p_5$$

This yields the same answer.

EXAMPLE 2.4 (English)

Water and oil flow in horizontal pipelines. A double U-tube manometer is connected between the pipelines, as shown. Calculate the pressure difference between the water pipe and the oil pipe.

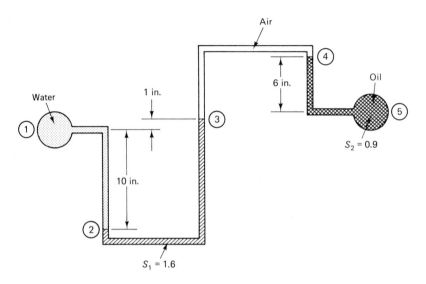

Solution

We can first identify the relevant points 1, 2, 3, 4, and 5 as shown in the figure. We can then find the pressure difference from

$$p_1 - p_5 = (p_1 - p_2) + (p_2 - p_3) + (p_3 - p_4) + (p_4 - p_5)$$

$$= \gamma(z_2 - z_1) + \gamma S_1(z_3 - z_2) + \gamma S_{air}(z_4 - z_3) + \gamma S_2(z_5 - z_4)$$

where $\gamma = 62.4$ lb/ft^3, $S_1 = 1.6$, $S_2 = 0.9$, and $S_{air} \approx 0$. Thus

$$p_1 - p_5 = 62.4(-\tfrac{10}{12} + 1.6 \times \tfrac{11}{12} + 0 \times \tfrac{6}{12} + 0.9 \times -\tfrac{6}{12})$$

$$= 11.44 \text{ lb/ft}^2 \quad \text{or} \quad 0.0794 \text{ psi}$$

Note that by neglecting the weight of the air, the pressure at point 3 is equal to the pressure at point 4.

   An alternative method of interpreting the manometer is to start at the left water pipe, add pressure when the elevation decreases, and subtract pressure when the elevation increases until the pipe at the right is encountered. This would result in

$$p_1 + 62.4 \times \tfrac{10}{12} - (1.6 \times 62.4) \times \tfrac{11}{12} + (0.9 \times 62.4) \times \tfrac{6}{12} = p_5$$

This yields the same answer.

2.4.4 Forces on Plane Areas

In the design of devices and objects that are submerged, such as dams, flow obstructions, surfaces on ships, and holding tanks, it is necessary to calculate the magnitudes and locations of forces that act on both plane and curved surfaces. In this section we consider only plane surfaces, such as the plane surface of general shape shown in Fig. 2.8. Note that a side view is given as well as a view showing the shape of the plane. The total force of the liquid on the plane surface is found by integrating the pressure over the area, that is,

$$F = \int_A p \; dA \tag{2.4.20}$$

The x and y coordinates are in the plane of the plane surface, as shown. Assuming that $p = 0$ at $h = 0$, we know that

$$p = \gamma h$$
$$= \gamma y \sin \alpha \tag{2.4.21}$$

where h is measured vertically down from the free surface to the elemental area dA and y is measured from point 0 on the free surface. The force may then be expressed as

$$F = \int_A \gamma h \; dA$$
$$= \gamma \sin \alpha \int_A y \; dA \tag{2.4.22}$$

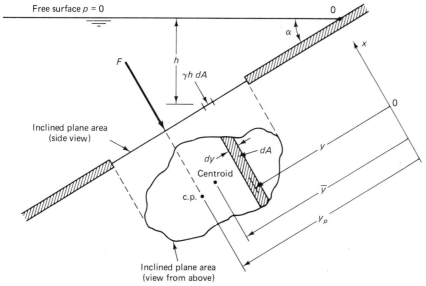

Figure 2.8 Force on an inclined plane area.

The distance to a centroid is defined as

$$\bar{y} = \frac{1}{A} \int_A y \, dA \tag{2.4.23}$$

The expression for the force then becomes

$$F = \gamma \bar{y} A \sin \alpha$$
$$= \gamma \bar{h} A = p_c A \tag{2.4.24}$$

where $\bar{h}$ is the vertical distance from the free surface to the centroid of the area and p_c is the pressure at the centroid. Thus we see that the magnitude of the force on a plane surface is the pressure at the centroid multiplied by the area. The force does not, in general, act at the centroid.

To find the location of the resultant force F, we note that the sum of the moments of all the infinitesimal pressure forces acting on the area A must equal the moment of the resultant force. Let the force F act at the point (x_p, y_p), the **center of pressure** (c.p.). The value of y_p can be obtained by equating moments about the x-axis:

$$y_p F = \int_A yp \, dA$$
$$= \gamma \sin \alpha \int_A y^2 \, dA = \gamma I_x \sin \alpha \tag{2.4.25}$$

where the second moment of the area about the x-axis is

$$I_x = \int_A y^2 \, dA \tag{2.4.26}$$

The second moment of an area can be determined from the second moment of an area $\bar{I}$ about the centroidal axis by the parallel-axis-transfer theorem,

$$I_x = \bar{I} + A\bar{y}^2 \tag{2.4.27}$$

Substitute Eqs. 2.4.24 and 2.4.27 into Eq. 2.4.25, and obtain

$$y_p = \frac{\gamma(\bar{I} + A\bar{y}^2)\sin \alpha}{\gamma \bar{y} A \sin \alpha}$$
$$= \bar{y} + \frac{\bar{I}}{A\bar{y}} \tag{2.4.28}$$

Centroids and moments for several areas are presented in Appendix C. Using the expression above, we can show that the force on a rectangu-

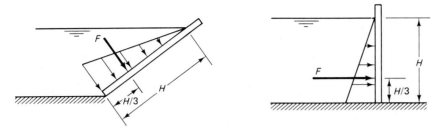

Figure 2.9   Force on a plane area with top edge in a free surface.

lar gate, with the top edge even with the liquid surface, acts two-thirds of the way down. This is also obvious considering the triangular pressure distribution acting on the gate. Note that Eq. 2.4.28 shows us that y_p is always greater than $\bar{y}$; that is, the resultant force of the liquid on a plane surface always acts below the centroid of the area, except on a horizontal area for which $\bar{y} = \infty$; then the center of pressure and the centroid coincide.

Similarly, to locate the x-coordinate x_p of the c.p., we write

$$x_p F = \int_A xp \, dA$$
$$= \gamma \sin \alpha \int_A xy \, dA = \gamma I_{xy} \sin \alpha \tag{2.4.29}$$

where the product of inertia of the area A is

$$I_{xy} = \int_A xy \, dA \tag{2.4.30}$$

Using the transfer theorem for the product of inertia,

$$I_{xy} = \bar{I}_{xy} + A\bar{x}\bar{y} \tag{2.4.31}$$

Equation 2.4.29 becomes

$$x_p = \bar{x} + \frac{\bar{I}_{xy}}{A\bar{y}} \tag{2.4.32}$$

We now have expressions for the coordinates locating the center of pressure.

EXAMPLE 2.5

Find the location of the resultant force F of the water on the triangular gate and the force P necessary to hold the gate in the position shown.

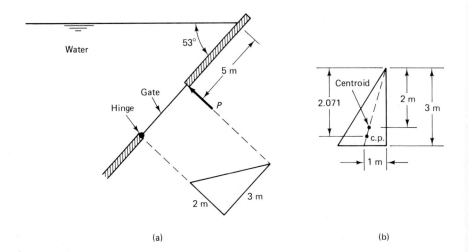

(a)

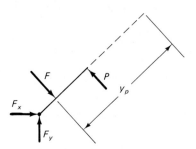

(b)

Solution

First we draw a free-body diagram of the gate, including all the forces acting on the gate. This is shown below. We have neglected the weight of the gate.

The y-coordinate of the location of the resultant F can be found using Eq. 2.4.28 as follows:

$$\bar{y} = 2 + 5 = 7$$

$$y_p = \bar{y} + \frac{\bar{I}}{A\bar{y}}$$

$$= 7 + \frac{2 \times 3^3/36}{3 \times 7} = 7.071 \text{ m}$$

To find x_p we could use Eq. 2.4.32. Rather than that, we recognize that the resultant force must act on a line connecting the vertex and the midpoint of the opposite side since each infinitesimal force acts on this line. Thus using similar triangles we have

$$\frac{x_p}{1} = \frac{2.071}{3}$$

$$\therefore \quad x_p = 0.690 \text{ m}$$

The coordinates x_p and y_p locate where the force due to the water acts on the gate. If we take moments about the hinge, we can determine the force P necessary to hold the gate in the position shown:

$$\Sigma M_{\text{hinge}} = 0$$
$$\therefore \quad 3 \times P = (3 - 2.071)F$$
$$= 0.929 \times \gamma \bar{h} A$$
$$= 0.929 \times 9810 \times (7 \sin 53°) \times 3$$

Hence

$$P = 50\,900 \text{ N} \quad \text{or} \quad 50.9 \text{ kN}$$

2.4.5 Forces on Curved Surfaces

We do not use a direct method of integration to find the force due to the hydrostatic pressure on a curved surface. Rather, a free-body diagram that contains the curved surface and the liquids directly above or below the curved surface is identified. Such a free-body diagram contains only plane surfaces upon which unknown fluid forces act; these unknown forces can be found as in the preceding section.

As an example, let us determine the force of the curved gate on the stop, shown in Fig. 2.10a. The free-body diagram, which includes the gate and some of the water contained directly above the gate, is shown in Fig. 2.10b; the forces F_x and F_y are the horizontal and vertical components, respectively, of the force acting on the hinge; F_1 and F_2 are due to the

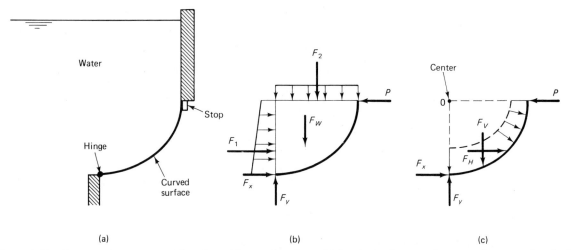

                (a)   (b)   (c)

Figure 2.10   Forces acting on a curved surface: (a) curved surface; (b) free-body diagram of water and gate; (c) free-body diagram of gate only.

surrounding water and are the resultant forces of the pressure distributions shown; the body force F_W is due to the weight of the water shown.
By summing moments about an axis passing through the hinge, we can
determine the force P acting on the stop.

If the curved surface is a quarter circle, the problem can be greatly
simplified. This is observed by considering a free-body diagram of the gate
only (see Fig. 2.10c). The horizontal force F_H acting on the gate is
equal to F_1 of Figure 2.10b, and the component F_V is equal to the combined force $F_2 + F_W$ of Figure 2.10b. Now, F_H and F_V are due to the
differential pressure forces acting on the circular arc; each differential
pressure force acts through the center of the circular arc. Hence the
resultant force $\mathbf{F}_H + \mathbf{F}_V$ (this is a vector addition) must act through the
center. Consequently, we can locate the components F_H and F_V at
the center of the quarter circle, resulting in a much simpler problem.
Example 2.6 will illustrate.

If the pressure on the free surface is p_0, we can simply add a depth of
liquid necessary to provide p_0 at the location of the free surface, and then
work the resulting problem, with a fictitious free surface located the appropriate distance above the original free surface.

EXAMPLE 2.6

Calculate the force P necessary to hold the 4-m-wide gate in the position shown.
Neglect the weight of the gate.

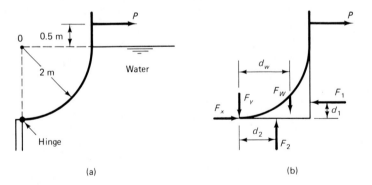

(a)   (b)

Solution

The first step is to draw a free-body diagram of the gate and the water directly
below the gate as shown. To calculate P, we must determine F_1, F_2, F_W, d_1, d_2,
and d_w; then moments about the hinge will allow us to find P. The force components are given by

$$F_1 = \gamma \bar{h} A$$

$$= 9810 \times 1 \times 8 = 78\,480 \text{ N}$$

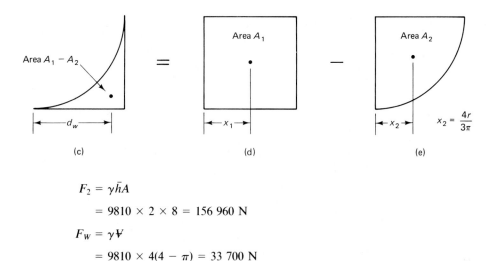

(c)   (d)   (e)

$$F_2 = \gamma \bar{h} A$$

$$= 9810 \times 2 \times 8 = 156\ 960 \text{ N}$$

$$F_W = \gamma \Psi$$

$$= 9810 \times 4(4 - \pi) = 33\ 700 \text{ N}$$

The distance d_w is the distance to the centroid of the volume. It can be determined by considering the area as the difference of a square and a quarter circle as shown. Moments of areas yield

$$d_w(A_1 - A_2) = x_1 A_1 - x_2 A_2$$

$$d_w = \frac{x_1 A_1 - x_2 A_2}{A_1 - A_2}$$

$$= \frac{1 \times 4 - (4 \times 2/3\pi) \times \pi}{4 - \pi} = 1.553 \text{ m}$$

The distance $d_2 = 1$ m and d_1 is given by

$$d_1 = 2 - y_p$$

$$= 2 - \left(\bar{y} + \frac{\bar{I}}{A\bar{y}}\right) = 2 - \left(1 + \frac{4 \times 2^3/12}{8 \times 1}\right) = 0.667 \text{ m}$$

Summing moments about the frictionless hinge gives

$$2.5P = d_1 F_1 + d_2 F_2 - d_w F_w$$

$$P = \frac{0.667 \times 78.5 + 1 \times 157.0 - 1.553 \times 33.7}{2.5} = 62.8 \text{ kN}$$

Rather than the somewhat tedious procedure above, we could observe that all the infinitesimal forces that make up the resultant force ($\mathbf{F}_H + \mathbf{F}_W$) acting on the circular arc pass through the center O, as noted in Fig. 2.10c. Since each infinitesimal force passes through the center, the resultant force must also pass through the center. Hence we could have located the resultant force ($\mathbf{F}_H + \mathbf{F}_V$) at point O. If F_V and F_H were located at O, F_V would pass through the hinge, producing no moment about the hinge. Then, realizing that $F_H = F_1$ and summing moments about the hinge gives

$$2.5P = 2F_H$$

Therefore,

$$P = 2 \times \frac{78.5}{2.5} = 62.8 \text{ kN}$$

This was obviously much simpler. All we needed to do was calculate F_H and then sum moments!

2.4.6 Buoyancy

The law of buoyancy, known as Archimedes' principle, dates back some 2200 years to the Greek philosopher Archimedes. Legend has it that Hiero, king of Syracuse, suspected that his new gold crown may have been constructed of materials other than pure gold, so he asked Archimedes to test it. Archimedes probably made a lump of pure gold that weighed the same as the crown. The lump was discovered to weigh more in water than the crown weighed in water, thereby proving to Archimedes that the crown was not pure gold. The fake material possessed a larger volume to have the same weight as gold, hence it displaced more water. **Archimedes' principle** is: There is a buoyancy force on an object equal to the weight of displaced liquid.

To prove the law of buoyancy, consider the submerged body shown in Fig. 2.11a. In part (b) a cylindrical free-body diagram is shown that includes the submerged body with weight W and liquid having a weight F_w; the cross-sectional area A is the maximum cross-sectional area of the body. From the diagram we see that the resultant vertical force acting on

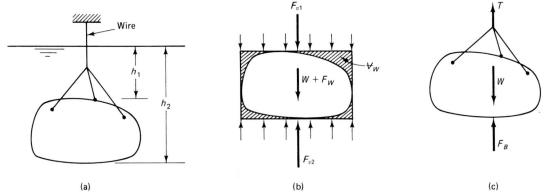

Figure 2.11 Forces on a submerged body: (a) submerged body; (b) free-body diagram; (c) free body showing F_B.

the free-body diagram (do not include W) is equal to

$$\Sigma F = F_{v2} - F_{v1} - F_w \qquad (2.4.33)$$

This force is by definition the **buoyant force** F_B. It can be expressed as

$$F_B = \gamma(h_2 A - h_1 A - \Psi_w) \qquad (2.4.34)$$

where Ψ_w is the liquid volume included in the free-body diagram. Recognizing that the volume of the submerged body is

$$\Psi_B = (h_2 - h_1)A - \Psi_w \qquad (2.4.35)$$

we see that

$$F_B = \gamma \Psi_{\text{displaced liquid}} \qquad (2.4.36)$$

thereby proving the law of buoyancy.

The force necessary to hold the submerged body in place (see Fig. 2.11c) is equal to

$$T = W - F_B \qquad (2.4.37)$$

where W is the weight of the submerged body.

For a floating object, as in Fig. 2.12, the buoyant force is

$$F_B = \gamma \Psi_{\text{displaced liquid}} \qquad (2.4.38)$$

Obviously, $T = 0$, so that Eq. 2.4.36 gives

$$F_B = W \qquad (2.4.39)$$

where W is the weight of the floating object.

From the foregoing analysis it is apparent that the buoyant force F_B acts through the centroid of the displaced liquid volume. For the floating object, the weight of the object acts through its center of gravity, so the center of gravity of the object must lie on the same vertical line as the centroid of the liquid volume.

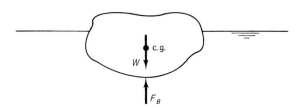

Figure 2.12 Forces on a floating object.

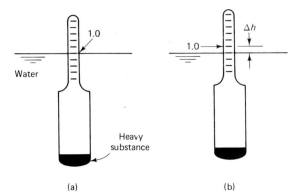

Figure 2.13 A hydrometer: (a) in water; (b) in an unknown liquid.

(a)   (b)

A **hydrometer,** an instrument used to measure the specific gravity of liquids, operates on the principle of buoyancy. A sketch is shown in Fig. 2.13. When placed in pure water the specific gravity is marked to read 1.0. The force balance is

$$W = \gamma_{\text{water}} V \qquad (2.4.40)$$

where W is the weight of the hydrometer and V is the submerged volume. In an unknown liquid of specific weight γ_x, a force balance would be

$$W = \gamma_x(V - A\,\Delta h) \qquad (2.4.41)$$

where A is the cross-sectional area of the stem. Equating these two expressions gives

$$\Delta h = \frac{V}{A}\left(1 - \frac{1}{S_x}\right) \qquad (2.4.42)$$

where $S_x = \gamma_x/\gamma_{\text{water}}$. For a given hydrometer, V and A are fixed so that the quantity Δh is dependent only on the specific gravity S_x. Thus the stem can be calibrated to read S_x directly.

EXAMPLE 2.7

The specific weight and the specific gravity of an unknown object are desired. Its weight in air is found to be 400 N and in water it weighs 300 N.

Solution

The volume is found from a force balance when submerged as follows (see Fig. 2.11c):

$$T = W - F_B$$

$$300 = 400 - 9810 \times V \qquad \therefore \quad V = 0.0102 \text{ m}^3$$

The specific weight is then

$$\gamma = \frac{W}{V}$$

$$= \frac{400}{0.0102} = 39\ 200 \text{ N/m}^3$$

The specific gravity is found to be

$$S = \frac{\gamma}{\gamma_{water}}$$

$$= \frac{39\ 200}{9810} = 4.00$$

EXAMPLE 2.7 (English)

The specific weight and the specific gravity of an unknown object are desired. Its weight in air is found to be 200 lb and in water it weighs 150 lb.

Solution

The volume is found from a force balance when submerged as follows (see Fig. 2.11c):

$$T = W - F_B$$

$$150 = 200 - 62.4V \qquad \therefore \quad V = 0.801 \text{ ft}^3$$

The specific weight is then

$$\gamma = \frac{W}{V}$$

$$= \frac{200}{0.801} = 250 \text{ lb/ft}^3$$

The specific gravity is found to be

$$S = \frac{\gamma}{\gamma_{water}}$$

$$= \frac{250}{62.40} = 4.00$$

2.4.7 Stability

The notion of stability can be demonstrated by considering the vertical stability of a floating object. If the object is raised a small distance, the

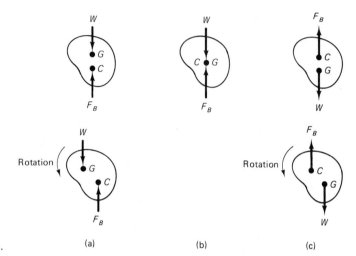

Figure 2.14 Stability of a submerged body: (a) unstable; (b) neutral; (c) stable.

buoyant force decreases and the object's weight returns the object to its original position. Conversely, if a floating object is lowered slightly, the buoyant force increases and the larger buoyant force returns the object to its original position. Thus a floating object has vertical stability since a small departure from equilibrium results in a restoring force.

Consider now the rotational stability of a submerged body, shown in Fig. 2.14. In part (a) the center of gravity G of the body is above the centroid C (also referred to as the **center of buoyancy**) of the displaced volume and a small angular rotation results in a moment that will continue to increase the rotation; hence the body is unstable and overturning would result. If the center of gravity is below the centroid, as in part (c), a small angular rotation provides a restoring moment and the body is stable. Part (b) shows neutral stability for a body in which the center of gravity and the centroid coincide, a situation that is encountered whenever the density is constant throughout the submerged body.

Next, consider the stability of a floating body, sketched in Fig. 2.15a. If the center of gravity is below the centroid, the body is always

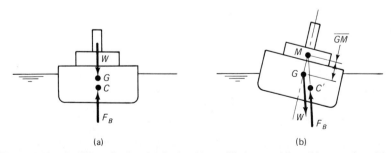

Figure 2.15 Stability of a floating body: (a) equilibrium position; (b) rotated position.

stable, as with submerged bodies. The body may be stable, though, even if the center of gravity is above the centroid, as sketched. When the body rotates the centroid of the volume of displaced liquid moves to the new location C', shown in part (b). If the centroid C' moves sufficiently far, a restoring moment develops and the body is stable. This is determined by the **metacentric height** $\overline{GM}$, defined as the point of intersection of the buoyant force before rotation with the buoyant force after rotation. If $\overline{GM}$ is positive, as shown, the body is stable; if $\overline{GM}$ is negative (M lies below G), the body is unstable.

To determine a quantitative relationship for the distance $\overline{GM}$ refer to the sketch of Fig. 2.16, which shows the uniform cross section. Let us find an expression for $\bar{x}$, the x-coordinate of the centroid of the displaced volume. It can be found by considering the volume to be the original volume plus the added wedge with cross-sectional area DOE minus the subtracted wedge with cross-sectional area AOB; to locate the centroid of a composite volume, we take moments as follows:

$$\bar{x}\mathcal{V} = \bar{x}_0\mathcal{V}_0 + \bar{x}_1\mathcal{V}_1 - \bar{x}_2\mathcal{V}_2 \qquad (2.4.43)$$

where $\mathcal{V}_0$ is the original volume below the water line, $\mathcal{V}_1$ is the area DOE times the length, $\mathcal{V}_2$ is the area AOB times the length; the cross section is assumed to be uniform so that the length l is constant for the body. The quantity $\bar{x}_0$, the x-coordinate of point C, is zero. The remaining two terms can best be represented by integrals so that

$$\bar{x}\mathcal{V} = \int_{\mathcal{V}_1} x\, d\mathcal{V} - \int_{\mathcal{V}_2} x\, d\mathcal{V} \qquad (2.4.44)$$

Then $d\mathcal{V} = x \tan\alpha\, dA$ in volume 1 and $d\mathcal{V} = -x \tan\alpha\, dA$ in volume 2, where $dA = l\, dx$, l being the constant length of the body. The equation

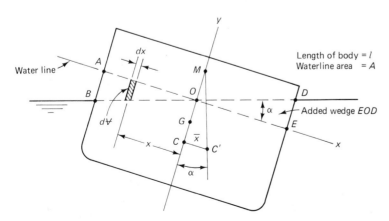

Figure 2.16 Uniform cross section of a float- ing body.

above becomes

$$\bar{x}\mathcal{V} = \tan \alpha \int_{A_1} x^2 \, dA + \tan \alpha \int_{A_2} x^2 \, dA$$

$$= \tan \alpha \int_{A} x^2 \, dA \qquad\qquad (2.4.45)$$

$$= \tan \alpha \, I_o$$

where I_o is the second moment (moment of inertia) of the waterline area about an axis passing through the origin O. The waterline area would be the length $\overline{AE}$ times the length l of the body if l were of constant length. Using $\bar{x} = \overline{CM} \tan \alpha$, we can write

$$\overline{CM} \, \mathcal{V} = I_o \qquad\qquad (2.4.46)$$

or, with $\overline{CG} + \overline{GM} = \overline{CM}$, we have

$$\overline{GM} = \frac{I_o}{\mathcal{V}} - \overline{CG} \qquad\qquad (2.4.47)$$

For a given body orientation, if $\overline{GM}$ is positive, the body is stable. Even though this relationship (2.4.47) is derived for a floating body with uniform cross section, it is applicable for floating bodies in general. We will apply it to a floating cylinder in the following example.

EXAMPLE 2.8

A 0.25-m-diameter cylinder is 0.25 m long and composed of material with specific weight 8000 N/m³. Will it float in water with the ends horizontal?

Solution

With the ends horizontal, I_o will be the second moment of the circular cross section,

$$I_o = \frac{\pi d^4}{64} = \frac{\pi \times 0.25^4}{64} = 0.000192 \text{ m}^4$$

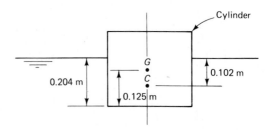

The displaced volume will be

$$V = \frac{W}{\gamma_{water}}$$

$$= \frac{8000 \times \pi/4 \times 0.25^2 \times 0.25}{9810} = 0.0100 \text{ m}^3$$

The depth the cylinder sinks in the water is

$$\text{depth} = \frac{V}{A}$$

$$= \frac{0.01}{\pi \times 0.25^2/4} = 0.204 \text{ m}$$

Hence, the distance $\overline{CG}$, as shown in the sketch, is

$$\overline{CG} = 0.125 - \frac{0.204}{2} = 0.023 \text{ m}$$

Finally,

$$\overline{GM} = \frac{0.000192}{0.01} - 0.023 = -0.004 \text{ m}$$

This is a negative value showing that the cylinder will not float with ends horizontal. It would undoubtedly float on its side.

2.5 LINEARLY ACCELERATING CONTAINERS

In this section the fluid will be at rest relative to a reference frame that is linearly accelerating with a horizontal component a_x and a vertical component a_z. Then Eq. 2.3.6 simplifies to

$$dp = -\rho a_x \, dx - \rho(g + a_z) \, dz \qquad (2.5.1)$$

Integrating between two arbitrary points 1 and 2 results in

$$p_2 - p_1 = -\rho a_x(x_2 - x_1) - \rho(g + a_z)(z_2 - z_1) \qquad (2.5.2)$$

If points 1 and 2 lie on a constant-pressure line, such as the free surface in Fig. 2.17, $p_2 - p_1 = 0$ and we have

$$\frac{z_1 - z_2}{x_2 - x_1} = \tan \alpha = \frac{a_x}{g + a_z} \qquad (2.5.3)$$

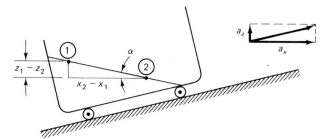

Figure 2.17 Linearly accelerating tank.

where α is the angle that the constant-pressure line makes with the horizontal.

In the solution of problems involving liquids, we must often utilize the conservation of mass and equate the volumes before and after the acceleration is applied. After the acceleration is initially applied, sloshing may occur. Our analysis will assume that sloshing is not present; either sufficient time passes to dampen out time-dependent motions, or the acceleration is applied in such a way that such motions are minimal.

EXAMPLE 2.9

The tank shown is accelerated to the right. Calculate the acceleration a_x needed to cause the free surface, shown by the dashed line, to touch point A. Also, find p_B and the total force acting on the bottom of the tank if the width is 1 m.

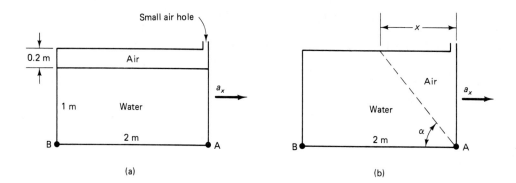

Solution

The angle the free surface takes is found by equating the air volume before and after since no water spills out; assuming constant width,

$$0.2 \times 2 = \tfrac{1}{2}(1.2x)$$

$$x = 0.667 \text{ m}$$

The quantity $\tan \alpha$ is now known. It is

$$\tan \alpha = \frac{1.2}{0.667} = 1.8$$

Using Eq. 2.5.3, we find a_x to be, letting $a_z = 0$,

$$a_x = g \tan \alpha$$
$$= 9.81 \times 1.8 = 17.66 \text{ m/s}^2$$

We can find the pressure at B by noting the pressure dependence on x. At A, the pressure is zero. Hence, Eq. 2.5.2 yields

$$p_B - p_A = -\rho a_x (x_B - x_A)$$
$$p_B = -1000 \times 17.66(0 - 2)$$
$$= 35\,300 \text{ Pa} \quad \text{or} \quad 35.3 \text{ kPa}$$

To find the total force acting on the bottom of the tank, we realize that the pressure distribution is decreasing linearly from $p = 35.3$ kPa at B to $p = 0$ kPa at A. Hence

$$F = \frac{p_B + p_A}{2} \times \text{area}$$

$$= \frac{35\,300}{2} \times 2 \times 1 = 35\,300 \text{ N}$$

2.6 ROTATING CONTAINERS

In this section we consider the situation of a liquid contained in a rotating container, such as that shown in Fig. 2.18. After a relatively short time the

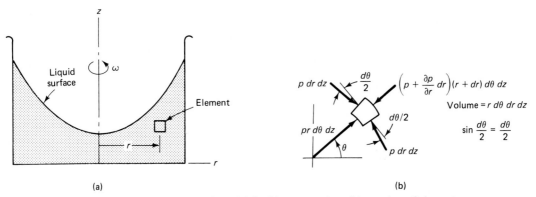

Figure 2.18   Rotating container: (a) liquid cross section; (b) top view of element.

liquid reaches static equilibrium with respect to the container and the rotating rz-reference frame. The horizontal rotation will not alter the pressure distribution in the vertical direction. There will be no variation of pressure with respect to the θ-coordinate. Applying Newton's second law ($\Sigma F_r = ma_r$) in the r-direction to the element shown, using $\sin d\theta/2 = d\theta/2$, yields

$$- \frac{\partial p}{\partial r} \, dr \, rd\theta \, dz - prd\theta \, dz - p \, dr \, d\theta \, dz - \frac{\partial p}{\partial r} \, (dr)^2 d\theta \, dz$$
$$+ 2\frac{d\theta}{2} p \, dr \, dz + prd\theta \, dz = -\rho \, dr \, rd\theta \, dz \, r\omega^2 \tag{2.6.1}$$

where the acceleration is $r\omega^2$ toward the center of rotation. Simplify and divide by the volume $rd\theta \, dr \, dz$; then

$$\frac{\partial p}{\partial r} = \rho r\omega^2 \tag{2.6.2}$$

where we have neglected the higher-order term that contains the differential dr. The pressure differential then becomes

$$dp = \frac{\partial p}{\partial r} \, dr + \frac{\partial p}{\partial z} \, dz$$
$$= \rho r\omega^2 \, dr - \rho g \, dz \tag{2.6.3}$$

where we have used the static pressure variation given by Eq. 2.4.1. We can now integrate between any two points (r_1, z_1) and (r_2, z_2) to obtain

$$p_2 - p_1 = \frac{\rho\omega^2}{2} (r_2^2 - r_1^2) - \rho g(z_2 - z_1) \tag{2.6.4}$$

If the two points are on a constant-pressure surface, such as the free surface, putting point 1 on the z-axis so that $r_1 = 0$, there results

$$\frac{\omega^2 r_2^2}{2} = g(z_2 - z_1) \tag{2.6.5}$$

which is the equation of a parabola. Hence the free surface is a parabaloid of revolution. The equations above can now, with the conservation of mass, be used to solve problems of interest.

EXAMPLE 2.10

The cylinder shown is rotated about its centerline. What rotational speed is necessary for the water to just touch the origin O. Also, calculate the pressure at A and B.

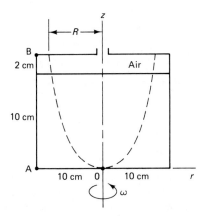

Solution

Since no water spills from the container, the air volume remains constant, that is,

$$\pi \times 10^2 \times 2 = \tfrac{1}{2}\pi R^2 \times 12$$

where we have used the fact that the volume of a paraboloid of revolution is one-half that of a circular cylinder with the same height and radius. This gives the value

$$R = 5.77 \text{ cm}$$

Using Eq. 2.6.5 with $r_2 = R$, we have

$$\frac{\omega^2 \times 0.0577^2}{2} = 9.81 \times 0.12$$

$$\omega = 26.6 \text{ rad/s}$$

To find the pressure at point A, we simply calculate the pressure difference between A and O. Using Eq. 2.6.4 with $r_2 = r_A = 0.1$ m and $r_1 = r_0 = 0$, there results

$$p_A = \frac{\rho\omega^2}{2}(r_A^2 - r_0^2)$$

$$= \frac{1000 \times 26.6^2}{2} \times 0.1^2$$

$$= 3540 \text{ Pa} \quad \text{or} \quad 3.54 \text{ kPa}$$

The pressure at B can be found by applying Eq. 2.6.4 to points A and B. This equation simplifies to

$$p_B - p_A = -\rho g(z_B - z_A)$$

Hence

$$p_B = 3540 - 1000 \times 9.81 \times 0.12$$

$$= 2360 \text{ Pa} \quad \text{or} \quad 2.36 \text{ kPa}$$

PROBLEMS

Pressure

2.1. Calculate the pressure at a depth of 10 m in a liquid with specific gravity of:
(a) 1.0.
(b) 0.8.
(c) 13.6.
(d) 1.59.
(e) 0.68.

2.2. What depth is necessary in a liquid to produce a pressure of 250 kPa if the specific gravity is:
(a) 1.0?
(b) 0.8?
(c) 13.6?
(d) 1.59?
(e) 0.68?

2.3. A pressure of 136 kPa is measured at a depth of 6 m. Calculate the specific gravity and the density of the liquid if $p = 0$ on the surface.

2.3E. A pressure of 20 psi is measured at a depth of 20 ft. Calculate the specific gravity and the density of the liquid if $p = 0$ on the surface.

2.4. How many meters of water are equivalent to:
(a) 760 mm Hg?
(b) 75 cm Hg?
(c) 10 mm Hg?

2.5. Determine the pressure at the bottom of a tank if it contains layers of
(a) 20 cm of water and 2 cm of mercury.
(b) 52 mm of water and 26 mm of carbon tetra-chloride.
(c) 3 m of oil, 2 m of water, and 10 cm of mercury.

2.6. Assuming the density of air to be constant at 1.23 kg/m^3, calculate the pressure change from the top of a mountain to its base if the elevation change is 3000 m.

2.6E. Assuming the density of air to be constant at 0.0024 slug/ft^3, calculate the pressure change from the top of a mountain to its base if the elevation change is 10,000 ft.

2.7. Estimate the pressure at 10 000 m assuming an isothermal atmosphere with temperature:
(a) 0°C.
(b) 15°C.
(c) −15°C.

2.8. The temperature in the atmosphere is approximated by $T(z) = 15 - 0.0065z$ °C for elevations less than 11 000 m. Calculate the pressure at elevations of:
(a) 3000 m.
(b) 6000 m.
(c) 9000 m.
(d) 11 000 m.

2.9. Determine the elevation where $p = 1.0$ Pa absolute assuming that the temperature at sea level is 15°C for an isothermal atmosphere with $T = -20$°C.

2.9E. Determine the elevation where $p = 0.0001$ psia assuming that the temperature at sea level is 60°F for an isothermal atmosphere with $T = -5$°F.

Manometers

2.10. Calculate the pressure in a pipe transporting air if a U-tube manometer measures 25 cm Hg. Note that the weight of air in the manometer is negligible.

2.11. If the pressure of air in a pipe is 450 kPa, what will a U-tube manometer with mercury read? Use $h = 1.5$ cm in Fig. 2.7b.
(a) Neglect the weight of the air column.
(b) Include the weight of the air column, assuming that $T_{air} = 20$°C, and calculate the percent error of part (a).

2.12. Oil with $S = 0.86$ is being transported in a pipe. Calculate the pressure if a U-tube manometer reads 19.7 cm Hg. The oil in the manometer is depressed 10 cm below the pipe centerline.

2.12E. Oil with $S = 0.86$ is being transported in a pipe. Calculate the pressure if a U-tube manometer reads 9.5 in. Hg. The oil in the manometer is depressed 5 in. below the pipe centerline.

2.13. Determine the pressure difference between the water pipe and the oil pipe.

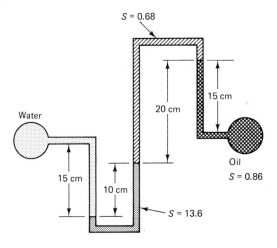

2.14. What is the pressure in the oil pipe if the pressure in the water pipe is 15 kPa?

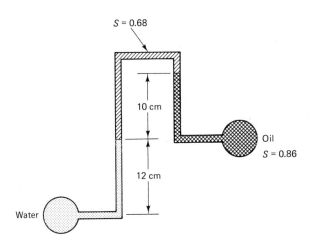

2.15. Determine the reading of the pressure gage if:
(a) $H = 2$ m, $h = 10$ cm.
(b) $H = 0.8$ m, $h = 20$ cm.
(c) $H = 6$ ft, $h = 4$ in.
(d) $H = 2$ ft, $h = 8$ in.

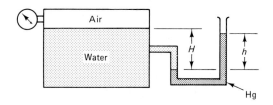

2.16. If $H = 16$ cm, what will the pressure gage read?

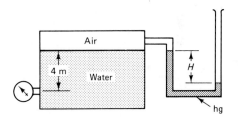

2.17. The pressure in the water pipe of Fig. 2.7b is 8.2 kPa, with $h = 25$ cm and $S_2 = 1.59$. Find the pressure in the water pipe if the H reading increases by 27.3 cm.

2.18. Find the pressure in the water pipe.

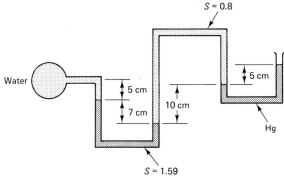

2.19. With the manometer top open the mercury level is 20 cm below the air pipe; there is no pressure in the air pipe. The manometer top is then sealed. Calculate the manometer reading H for a pressure of 200 kPa in the air pipe. Assume an isothermal process for the air in the sealed tube.

2.19E. With the manometer top open the mercury level is 8 in. below the air pipe, there is no pressure in the air pipe. The manometer top is then sealed. Calculate the manometer reading H for a pressure of 30 psi in the air pipe. Assume an isothermal process for the air in the sealed tube.

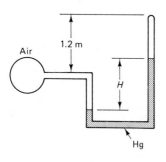

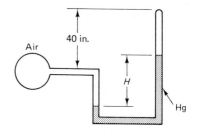

Forces on Plane Areas

2.20. Calculate the force acting on a 30-cm-diameter porthole of a ship if the center of the porthole is 10 m below water level.

2.21. A 2-m-wide, 3-m-high vertical rectangular gate has its top edge 2 m below the water level. It is hinged along its lower edge. What force, acting on the top edge, is necessary to hold the gate shut?

2.21E. A 6-ft-wide, 10-ft-high vertical rectangular gate has its top edge 6 ft below the water level. It is hinged along its lower edge. What force, acting on the top edge, is necessary to hold the gate shut?

2.22. Determine the force P needed to hold the 4-m-wide gate in the position shown.

2.23. Calculate the force P necessary to hold the 4-m-wide gate in the position shown.

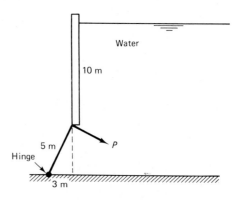

2.24. The triangular gate shown has its 3-m side parallel to and 10 m below the water surface. Calculate the magnitude and location of the force acting on the gate if it is:

(a) Vertical.

(b) Horizontal.

(c) On a 45° angle sloped upward.

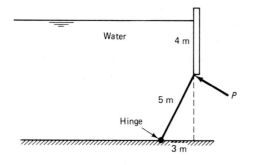

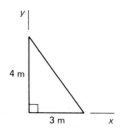

2.24E. The triangular gate shown has its 6-ft side parallel and 30 ft below the water surface. Calculate the magnitude and location of the force acting on the gate if it is:

(a) Vertical.

(b) Horizontal.

(c) On a 45° angle sloped upward.

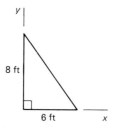

2.25. Find the force P needed to hold the 3-m-wide rectangular gate as shown.

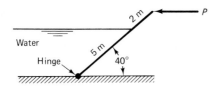

2.26. At what height H will the rigid gate, hinged at a central point as shown, open up if h is:

(a) 0.6 m?

(b) 0.8 m?

(c) 1.0 m?

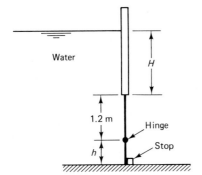

2.27. Calculate the height H that will result in the gate opening automatically if (neglect the weight of the gate):

(a) $l = 2$ m.

(b) $l = 1$ m.

(c) $l = 6$ ft.

(d) $l = 3$ ft.

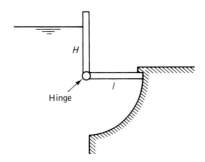

2.28. The top of each gate shown lies 4 m below the water surface. Find the location and magnitude of the force acting on one side assuming a vertical orientation.

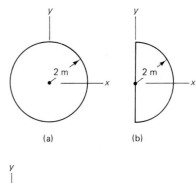

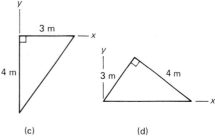

2.29. The pressure distribution over the base of the concrete ($S = 2.4$) dam varies linearly, as shown, producing an **uplift.** Will the dam topple over? Use:
(a) $H = 45$ m.
(b) $H = 60$ m.
(c) $H = 75$ m.

2.30. Assume a linear pressure distribution over the base of the concrete ($S = 2.4$) dam shown. Will the dam topple over? Use:
(a) $H = 45$ m.
(b) $H = 60$ m.
(c) $H = 75$ m.

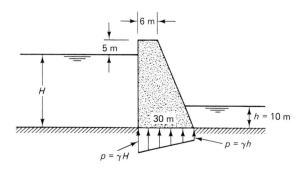

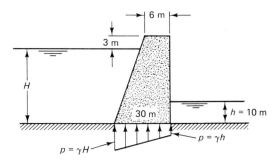

Forces on Curved Surfaces

2.31. Find the force P needed to hold the 10-m-long cylindrical object in position.

2.33. What P is needed to hold the 4-m-wide gate closed?

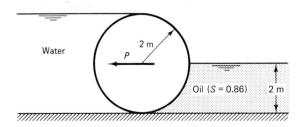

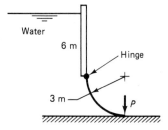

2.32. Find the force P needed to just open the gate if:
(a) $H = 6$ m, $R = 2$ m, and the gate is 4 m wide.
(b) $H = 20$ ft, $R = 6$ ft, and the gate is 12 ft wide.

2.34. Find the force P required to hold the gate in the position shown. The gate is 5 m wide.

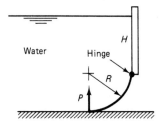

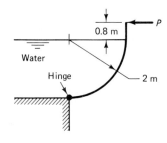

2.35. The quarter-circle, cylindrical gate ($S = 0.2$) is in equilibrium as shown, calculate the value of γ_x using:
(a) SI units.
(b) English units.

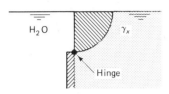

2.36. The log is in equilibrium, as shown. Calculate the force pushing it against the dam and the specific gravity of the log if:
(a) Its length is 6 m and $R = 0.6$ m.
(b) Its length is 20 ft and $R = 2$ ft.

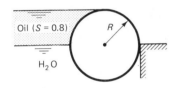

2.37. Find the force on the weld if:
(a) Air fills the hemisphere.
(b) Oil fills the hemisphere.

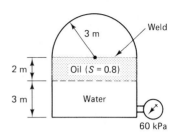

2.38. Find the force P if the parabolic gate is:
(a) 2 m wide and $H = 2$ m.
(b) 4 ft wide and $H = 8$ ft.

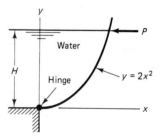

Buoyancy

2.39. The 3-m-wide barge weighs 20 kN empty. It is proposed that it carry a 250-kN load. Predict the draft in:
(a) Fresh water.
(b) Salt water ($S = 1.03$).

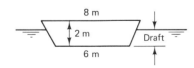

2.40. An object weighs 100 N in the air and 25 N when submerged in water. Calculate its volume and specific weight.

2.41. A car ferry is essentially rectangular with dimensions 8 m wide and 100 m long. If 60 cars, with an average mass per car of 1400 kg, are loaded on the ferry, how much farther will it sink into the water?

2.41E. A car ferry is essentially rectangular with dimensions 25 ft wide and 300 ft long. If 60 cars, with an average weight per car of 3000 lb, are loaded on the ferry, how much farther will it sink into the water?

2.42. A 30-m-long vessel, with cross section shown, is to carry a load of 6000 kN. How far will the water level be from the top of the vessel if its mass is 100 000 kg?

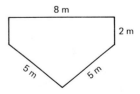

2.43. A body, with a volume of 2 m³, weighs 40 kN. Determine its weight when submerged in a liquid with $S = 1.59$.

2.44. The plug and empty cylinder weigh 6000 N. Calculate the height h needed to lift the plug if the radius R of the 4-m-long cylinder is:
(a) 30 cm.
(b) 40 cm.
(c) 50 cm.

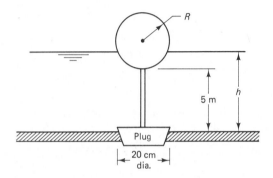

2.44E. The plug and empty cylinder weigh 1500 lb. Calculate the height h needed to lift the plug if the radius R of the 10-ft-long cylinder is:
(a) 12 in.
(b) 16 in.
(c) 20 in.

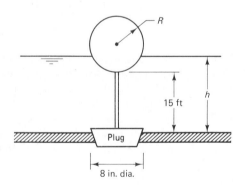

2.45. The hydrometer shown with no mercury has a mass of 0.01 kg. It is designed to float at the midpoint of the 12-cm-long stem in pure water.
(a) Calculate the mass of mercury needed.
(b) What is the specific gravity of the liquid if the hydrometer is just submerged?
(c) What is the specific gravity of the liquid if the stem of the hydrometer is completely exposed?

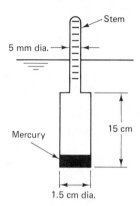

2.46. The hydrometer of Problem 2.45 is weighted so that in fresh water the stem is just submerged.
(a) What is the maximum specific gravity that can be read?
(b) What mass of mercury is required?

Stability

2.47. A 25-cm-diameter cylinder is composed of material with specific gravity 0.8. Will it float in water with the ends horizontal if its length is:
(a) 30 cm?
(b) 25 cm?
(c) 20 cm?

2.47E. A 10-in-diameter cylinder is composed of material with specific gravity 0.8. Will it float in water with ends horizontal if its length is:
(a) 12 in.?
(b) 10 in.?
(c) 8 in.?

2.48. Over what range of specific weights will a homogeneous cube float with sides horizontal and vertical?

2.49. For the object shown, calculate S_A for neutral stability when submerged.

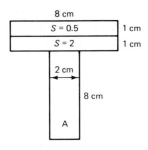

2.50. Orient the object for rotational stability when submerged if:
(a) $t = 2$ cm.
(b) $t = 1.0$ in.

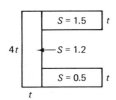

2.51. The barge shown is loaded such that the center of gravity of the barge and the load is at the waterline. Is the barge stable?

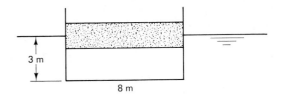

2.52. Is the barge shown stable? The center of gravity of the barge and load is located as shown.

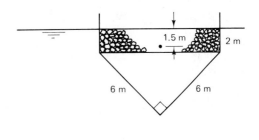

Linearly Accelerating Containers

2.53. The tank shown is completely filled with water and accelerated. Calculate the maximum pressure in the tank if:
(a) $a_x = 20$ m/s², $a_z = 0$, $L = 2$ m.
(b) $a_x = 0$, $a_z = 20$ m/s², $L = 2$ m.
(c) $a_x = 60$ ft/sec², $a_z = 60$ ft/sec², $L = 6$ ft.
(d) $a_x = 0$, $a_z = 60$ ft/sec², $L = 6$ ft.

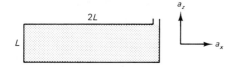

2.54. The tank shown is accelerated to the right at 10 m/s². Find:
(a) p_A.
(b) p_B.
(c) p_C.

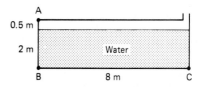

2.55. The tank of Problem 2.54 is accelerated so that $p_B = 60$ kPa. Find a_x assuming that:
(a) $a_z = 0$.
(b) $a_z = 10$ m/s².
(c) $a_z = 5$ m/s².

2.56. The tank shown is filled with water and accelerated. Find the pressure at A if:
(a) $a = 20$ m/s^2, $L = 1$ m.
(b) $a = 10$ m/s^2, $L = 1.5$ m.
(c) $a = 60$ ft/sec^2, $L = 3$ ft.
(d) $a = 30$ ft/sec^2, $L = 4$ ft.

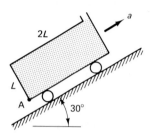

2.57. The tank of Problem 2.54 is 4 m wide. Find the force acting on:
(a) The end AB.
(b) The bottom.
(c) The top.

2.58. The tank of Problem 2.56(a) is 1.5 m wide. Calculate the force on:
(a) The bottom.
(b) The top.
(c) The left end.

2.59. Determine the pressure at points A, B, and C if:
(a) $a_x = 0$, $a_z = 10$ m/s^2, $L = 60$ cm.
(b) $a_x = 10$ m/s^2, $a_z = 0$, $L = 60$ cm.
(c) $a_x = 20$ m/s^2, $a_z = 10$ m/s^2, $L = 60$ cm.
(d) $a_x = 0$, $a_z = -60$ ft/sec^2, $L = 25$ in.
(e) $a_x = 60$ ft/sec^2, $a_z = 0$, $L = 25$ in.
(f) $a_x = -30$ ft/sec^2, $a_z = 30$ ft/sec^2, $L = 25$ in.

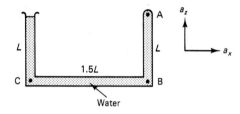

Rotating Containers

2.60. The U-tube of Problem 2.59 is rotated about the left leg at 50 rpm. Find p_A, p_B, and p_C if:
(a) $L = 60$ cm.
(b) $L = 40$ cm.
(c) $L = 25$ in.
(d) $L = 15$ in.

2.61. The U-tube of Problem 2.59 is rotated about the right leg at 10 rad/s. Find the pressures at points A, B, and C if:
(a) $L = 60$ cm.
(b) $L = 40$ cm.
(c) $L = 25$ in.
(d) $L = 15$ in.

2.62. The U-tube of Problem 2.59 is rotated about the center of the horizontal leg so that the pressure at the center is zero. Calculate ω if:
(a) $L = 60$ cm.
(b) $L = 40$ cm.
(c) $L = 25$ in.
(d) $L = 15$ in.

2.63. Determine the pressure at point A for a rotational speed of:
(a) 5 rad/s.
(b) 7 rad/s.
(c) 10 rad/s.
(d) 20 rad/s.

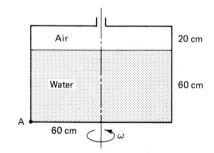

2.64. The hole in the cylinder of Problem 2.63 is closed and the air pressurized to 25 kPa. Find the pressure at point A if the rotational speed is:
(a) 5 rad/s.
(b) 7 rad/s.
(c) 10 rad/s.
(d) 20 rad/s.

2.65. Find the force on the bottom of the cylinder of:
(a) Problem 2.63(a).
(b) Problem 2.63(b).
(c) Problem 2.63(c).
(d) Problem 2.63(d).

THREE

Introduction to Fluids in Motion

3.1 INTRODUCTION

This chapter serves as an introduction to all the following chapters that deal with fluid motions. Fluid motions manifest themselves in many different ways. Some can be described very easily, while others require a thorough understanding of physical laws. In engineering applications, it is important to describe the fluid motions as simply as can be justified. This usually depends on the required accuracy. Typically, accuracies of $\pm 10\%$ are acceptable, although in some applications higher accuracies have to be achieved. It is the engineer's responsibility to know which simplifying assumptions can be made. This, of course, requires experience and, more important, an understanding of the physics involved.

Some common assumptions used to simplify a flow situation are related to fluid properties. For example, under certain conditions, the viscosity can affect the flow significantly; in others, viscous effects can be neglected without significantly altering the predictions. It is well known that the compressibility of a gas in motion should be taken into account if the velocities are very high. But compressibility effects do not have to be taken into account to predict wind forces on buildings or to predict any other physical quantity that is a direct effect of wind. Wind speeds are simply not high enough. Numerous examples could be cited. After our study of fluid motions, the appropriate assumptions used should become more obvious.

This chapter has three sections. In the first section we introduce the reader to some important general approaches used to analyze fluid mechanics problems. In the second section we give a brief overview of different types of flow, such as compressible and incompressible flows, and viscous and inviscid flows. Detailed discussions of each of these flow types follow in later chapters. The third section introduces the reader to the commonly used Bernoulli equation, an equation that establishes how pressures vary in a flow field. The use of this equation requires many simplifying assumptions, and its application is, therefore, limited.

3.2 DESCRIPTION OF FLUID MOTION

The analysis of complex fluid flow problems is often aided by the visualization of flow patterns, which permit the development of a better intuitive understanding and help in formulating the mathematical problem. In Section 3.2.1 we discuss different flow lines that are useful in this regard. The second topic in this section is the description of physical quantities as a function of space and time coordinates.

3.2.1 Pathlines, Streaklines, and Streamlines

Three different lines help us in describing a flow field. A **pathline** is the locus of points traversed by a given particle as it travels in a field of flow; the pathline provides us with a ''history'' of the particle's locations. A

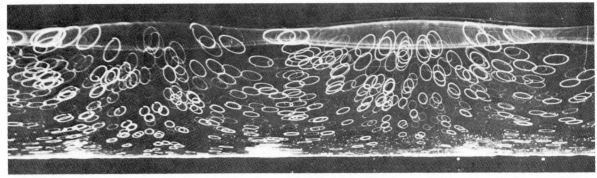

Figure 3.1   Pathlines underneath a wave in a tank of water. (Photograph by Wallet and Ruellan. Courtesy of M. C. Vasseur.)

photograph of a pathline would require a time exposure of an illuminated particle. A photograph showing pathlines of particles below a water surface with waves is given in Fig. 3.1.

A **streakline** is defined as an instantaneous line whose points are occupied by all particles originating from some specified point in the flow field. Streaklines tell us where the particles are "right now." A photograph of a streakline would be a snapshot of the set of illuminated particles that passed a certain point. Figure 3.2 shows streaklines produced by the continuous release of smoke in flow around a cylinder.

A **streamline** is a line in the flow possessing the following property: the velocity vector of each particle occupying a point on the streamline is tangent to the streamline. This is shown graphically in Fig. 3.3. An equation which expresses that the velocity vector is tangent to a streamline is

$$\mathbf{V} \times d\mathbf{r} = 0 \tag{3.2.1}$$

Figure 3.2   Streaklines in the unsteady flow around a cylinder. (Photograph by Sadatashi Taneda.)

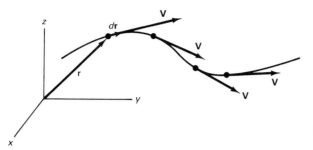

Figure 3.3 Streamline in a flow field.

since **V** and *d***r** are in the same direction, as shown; recall that the cross product of two vectors in the same direction is zero. This equation will be used in future chapters as the mathematical expression of a streamline. A photograph of a streamline cannot be made directly. For a general unsteady flow the streamlines can be inferred from photographs of short pathlines of a large number of particles.

In a steady flow, pathlines, streaklines, and streamlines are all coincident. All particles passing a given point will continue to trace out the same path since nothing changes with time; hence the pathlines and streaklines coincide. In addition, the velocity vector of a particle at a given point will be tangent to the line that the particle is moving along; thus the line is also a streamline. Since the flows that we observe in laboratories are invariably steady flows, we call the lines that we observe streamlines even though they may actually be streaklines, or for the case of time exposures, pathlines.

3.2.2 Lagrangian and Eulerian Descriptions of Motion

In the description of a flow field, it is convenient to think of individual particles each of which is considered to be a small mass of fluid, consisting of a large number of molecules, that occupies a small volume $\Delta \Psi$ that moves with the flow. If the fluid is incompressible, the volume does not change in magnitude but may deform. If the fluid is compressible, as the volume deforms, it also changes its magnitude. In both cases the particles are considered to move through a flow field as an entity.

In the study of particle mechanics, where attention is focused on an individual particle, the particle motion is observed as a function of time. Its position, velocity, and acceleration are listed as $\mathbf{s}(t)$, $\mathbf{V}(t)$, and $\mathbf{a}(t)$ and quantities of interest can be calculated. This is a **Lagrangian** description of motion. In the Lagrangian description many particles can be followed and their influence on one another noted. This becomes, however, a tremendous task as the number of particles becomes extremely large, as in a fluid flow.

An alternative to following each fluid particle separately is to identify a point in space and then observe the velocity of particles passing the point; we can observe the rate of change of velocity as the particles pass the point, that is, $\partial \mathbf{V}/\partial x$, $\partial \mathbf{V}/\partial y$, and $\partial \mathbf{V}/\partial z$, and we can observe if the velocity is changing with time at that particular point, that is, $\partial \mathbf{V}/\partial t$. In this **Eulerian** description of motion, the flow properties, such as velocity, are functions of both space and time. In rectangular, Cartesian coordinates the velocity is expressed as $\mathbf{V} = \mathbf{V}(x, y, z, t)$. The region of flow that is being considered is called a **flow field.**

An example may clarify these two ways of describing motion. An engineering firm is hired to make recommendations that would improve the traffic flow in a large city. The engineering firm has two alternatives: Hire college students to travel in automobiles throughout the city recording the appropriate observations (the Lagrangian approach), or hire college students to stand at the intersections and record the required information (the Eulerian approach). A correct interpretation of each set of data would lead to the same set of recommendations, that is, the same solution. In this example it may not be obvious which approach would be preferred; in fluids, however, the Eulerian description is used exclusively since the physical laws using the Eulerian description are easier to apply to actual situations.

If the quantities of interest do not depend on time, that is, $\mathbf{V} = \mathbf{V}(x, y, z)$, the flow is said to be a **steady flow.** Most of the flows of interest in this introductory textbook are steady flows. For a steady flow all flow quantities are independent of time, that is,

$$\frac{\partial \mathbf{V}}{\partial t} = 0 \qquad \frac{\partial p}{\partial t} = 0 \qquad \frac{\partial \rho}{\partial t} = 0 \qquad (3.2.2)$$

to list a few.

3.2.3 Acceleration

The acceleration of a fluid particle is found by considering a particular particle shown in Fig. 3.4. Its velocity changes from $\mathbf{V}(t)$ at time t to

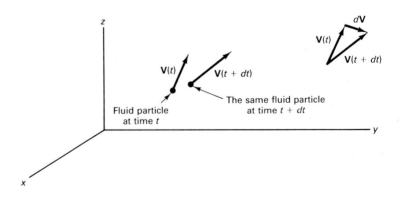

Figure 3.4 Velocity of a fluid particle.

$\mathbf{V}(t + dt)$ at time $t + dt$. The acceleration is, by definition,

$$\mathbf{a} = \frac{d\mathbf{V}}{dt} \qquad (3.2.3)$$

where $d\mathbf{V}$ is shown in Fig. 3.4. The velocity vector $\mathbf{V}$ is given in component form as

$$\mathbf{V} = u\hat{i} + v\hat{j} + w\hat{k} \qquad (3.2.4)$$

The quantity $d\mathbf{V}$ is, using a first-order Taylor series expansion (see Eq. 2.3.1) with $\mathbf{V} = \mathbf{V}(x, y, z, t)$,

$$d\mathbf{V} = \frac{\partial \mathbf{V}}{\partial x} dx + \frac{\partial \mathbf{V}}{\partial y} dy + \frac{\partial \mathbf{V}}{\partial z} dz + \frac{\partial \mathbf{V}}{\partial t} dt \qquad (3.2.5)$$

This gives the acceleration as

$$\mathbf{a} = \left(\frac{\partial \mathbf{V}}{\partial x} \frac{dx}{dt} + \frac{\partial \mathbf{V}}{\partial y} \frac{dy}{dt} + \frac{\partial \mathbf{V}}{\partial z} \frac{dz}{dt} + \frac{\partial \mathbf{V}}{\partial t} \right) \qquad (3.2.6)$$

Since we have followed a particular particle, we recognize that

$$\frac{dx}{dt} = u \qquad \frac{dy}{dt} = v \qquad \frac{dz}{dt} = w \qquad (3.2.7)$$

where (u, v, w) are the velocity components in the x-, y-, and z-direction, respectively. The acceleration is then expressed as

$$\mathbf{a} = u \frac{\partial \mathbf{V}}{\partial x} + v \frac{\partial \mathbf{V}}{\partial y} + w \frac{\partial \mathbf{V}}{\partial z} + \frac{\partial \mathbf{V}}{\partial t} \qquad (3.2.8)$$

The scalar component equations for rectangular coordinates are written as

$$a_x = \frac{\partial u}{\partial t} + u \frac{\partial u}{\partial x} + v \frac{\partial u}{\partial y} + w \frac{\partial u}{\partial z}$$

$$a_y = \frac{\partial v}{\partial t} + u \frac{\partial v}{\partial x} + v \frac{\partial v}{\partial y} + w \frac{\partial v}{\partial z} \qquad (3.2.9)$$

$$a_z = \frac{\partial w}{\partial t} + u \frac{\partial w}{\partial x} + v \frac{\partial w}{\partial y} + w \frac{\partial w}{\partial z}$$

We often write Eq. 3.2.8 in a simplified form as

$$\mathbf{a} = \frac{D\mathbf{V}}{Dt} \tag{3.2.10}$$

where, in rectangular coordinates,

$$\frac{D}{Dt} = u \frac{\partial}{\partial x} + v \frac{\partial}{\partial y} + w \frac{\partial}{\partial z} + \frac{\partial}{\partial t} \tag{3.2.11}$$

This derivative is called the **substantial derivative,** or **material derivative.** It is given a special name and special symbol (D/Dt instead of d/dt) because we followed a particular fluid particle, that is, we followed the substance (or material). It can be used with other dependent variables; for example, DT/Dt would represent the rate of change of the temperature of a fluid particle as we followed the particle along.

In cylindrical and spherical coordinate systems, the substantial derivative has the following respective forms:

$$\frac{D}{Dt} = v_r \frac{\partial}{\partial r} + \frac{v_\theta}{r} \frac{\partial}{\partial \theta} + v_z \frac{\partial}{\partial z} + \frac{\partial}{\partial t}$$

$$\frac{D}{Dt} = v_r \frac{\partial}{\partial r} + \frac{v_\theta}{r} \frac{\partial}{\partial \theta} + \frac{v_\phi}{r \sin \theta} \frac{\partial}{\partial \phi} + \frac{\partial}{\partial t} \tag{3.2.12}$$

where (v_r, v_θ, v_z) are the cylindrical components of $\mathbf{V}$, and (v_r, v_θ, v_ϕ) are the spherical components of $\mathbf{V}$. In cylindrical coordinates the acceleration components are

$$a_r = \frac{\partial v_r}{\partial t} + v_r \frac{\partial v_r}{\partial r} + \frac{v_\theta}{r} \frac{\partial v_r}{\partial \theta} + v_z \frac{\partial v_r}{\partial z} - \frac{v_\theta^2}{r}$$

$$a_\theta = \frac{\partial v_\theta}{\partial t} + v_r \frac{\partial v_\theta}{\partial r} + \frac{v_\theta}{r} \frac{\partial v_\theta}{\partial \theta} + v_z \frac{\partial v_\theta}{\partial z} + \frac{v_r v_\theta}{r} \tag{3.2.13}$$

$$a_z = \frac{\partial v_z}{\partial t} + v_r \frac{\partial v_z}{\partial r} + \frac{v_\theta}{r} \frac{\partial v_z}{\partial \theta} + v_z \frac{\partial v_z}{\partial z}$$

In spherical coordinates the acceleration components are

$$a_r = \frac{\partial v_r}{\partial t} + v_r \frac{\partial v_r}{\partial r} + \frac{v_\theta}{r} \frac{\partial v_r}{\partial \theta} + \frac{v_\phi}{r \sin \theta} \frac{\partial v_r}{\partial \phi} - \frac{v_\theta^2 + v_\phi^2}{r}$$

$$a_\theta = \frac{\partial v_\theta}{\partial t} + v_r \frac{\partial v_\theta}{\partial r} + \frac{v_\theta}{r} \frac{\partial v_\theta}{\partial \theta} + \frac{v_\phi}{r \sin \theta} \frac{\partial v_\theta}{\partial \phi} + \frac{v_r v_\theta - v_\phi^2 \cot \theta}{r} \tag{3.2.14}$$

$$a_\phi = \frac{\partial v_\phi}{\partial t} + v_r \frac{\partial v_\phi}{\partial r} + \frac{v_\theta}{r} \frac{\partial v_\phi}{\partial \theta} + \frac{v_\phi}{r \sin \theta} \frac{\partial v_\phi}{\partial \phi} + \frac{v_r v_\phi + v_\theta v_\phi \cot \theta}{r}$$

The time-derivative term on the right side of each of the preceding equations for the acceleration is called the **local acceleration** and the remaining terms on the right side in each equation form the **convective acceleration.** Hence the acceleration of a fluid particle is the sum of the local acceleration and convective acceleration. In a pipe, local acceleration results if, for example, a valve is being opened or closed; and convective acceleration occurs in the vicinity of a change in the pipe geometry, such as a pipe contraction or an elbow. In both cases fluid particles change speed, but for very different reasons.

We must note that the foregoing expressions for acceleration give the acceleration relative to an observer in the observer's reference frame only. In certain situations the observer's reference frame may be accelerating; then the acceleration of a particle relative to a fixed reference frame may be needed. It is given by

$$\mathbf{A} = \mathbf{a} + \underset{\substack{\text{acceleration of}\\\text{reference frame}}}{\frac{d^2\mathbf{S}}{dt^2}} + \underset{\substack{\text{Coriolis}\\\text{acceleration}}}{2\boldsymbol{\omega} \times \mathbf{V}} + \underset{\substack{\text{normal}\\\text{acceleration}}}{\boldsymbol{\omega} \times (\boldsymbol{\omega} \times \mathbf{r})} + \underset{\substack{\text{angular}\\\text{acceleration}}}{\frac{d\boldsymbol{\omega}}{dt} \times \mathbf{r}}$$

$$(3.2.15)$$

where $\mathbf{a}$ is given by Eq. 3.2.9, $d^2\mathbf{S}/dt^2$ is the acceleration of the observer's reference frame, $\mathbf{V}$ is the velocity in the observer's reference frame, $\mathbf{r}$ positions the particle, and $\boldsymbol{\omega}$ is the angular velocity of the observer's reference frame (see Fig. 3.5). Note that all vectors are written using the unit vectors of the XYZ-reference frame. For most engineering applications, reference frames attached to the earth yield $\mathbf{A} = \mathbf{a}$, since the other terms in Eq. 3.2.15 are often negligible with respect to $\mathbf{a}$.

3.2.4 Angular Velocity and Vorticity

A fluid flow may be thought of as the motion of a collection of fluid particles. As a particle travels along it may rotate or deform. The rotation and deformation of the fluid particles are of particular interest in our study

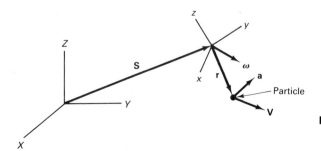

Figure 3.5 Motion relative to a noninertial reference frame.

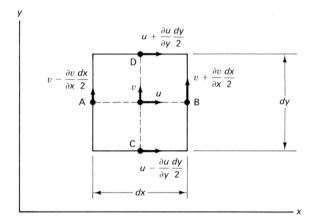

Figure 3.6 Fluid particle occupying an infinitesimal parallelepiped at a particular instant.

of fluid mechanics. There are certain flows, or regions of a flow, in which the fluid particles do not rotate; such flows are of special importance, particularly in flows around objects, and are referred to as **irrotational flows.** In general, fluid particles deform as they travel along; we do not concern ourselves with flows with zero deformation. Let us consider a small fluid particle that occupies an infinitesimal volume that has the xy-face as shown in Fig. 3.6. The **angular velocity** about the z-axis, Ω_z, is the average of the angular velocity of line segment AB and line segment CD. The two angular velocities, counterclockwise being positive, are

$$\Omega_{AB} = \frac{v_B - v_A}{dx}$$
$$= \left[v + \frac{\partial v}{\partial x}\frac{dx}{2} - \left(v - \frac{\partial v}{\partial x}\frac{dx}{2} \right) \right] \Big/ dx = \frac{\partial v}{\partial x} \tag{3.2.16}$$

$$\Omega_{CD} = -\frac{u_D - u_C}{dy}$$
$$= -\left[u + \frac{\partial u}{\partial y}\frac{dy}{2} - \left(u - \frac{\partial u}{\partial y}\frac{dy}{2} \right) \right] \Big/ dy = -\frac{\partial u}{\partial y} \tag{3.2.17}$$

Consequently, the angular velocity Ω_z of the fluid particle is

$$\Omega_z = \tfrac{1}{2}(\Omega_{AB} + \Omega_{CD})$$
$$= \frac{1}{2}\left(\frac{\partial v}{\partial x} - \frac{\partial u}{\partial y} \right) \tag{3.2.18}$$

If we had considered the xz-face, we would have found the angular velocity about the y-axis to be

$$\Omega_y = \frac{1}{2}\left(\frac{\partial u}{\partial z} - \frac{\partial w}{\partial x} \right) \tag{3.2.19}$$

and the *yz*-face would provide us with the angular velocity about the *x*-axis:

$$\Omega_x = \frac{1}{2}\left(\frac{\partial w}{\partial y} - \frac{\partial v}{\partial z}\right) \tag{3.2.20}$$

These are the three components of the angular velocity vector. A cork placed in a water flow in a wide channel (the *xy*-plane) would rotate with the angular velocity given by Eq. 3.2.18.

It is common to define the **vorticity** ω to be twice the angular velocity; its three components are then

$$\omega_x = \frac{\partial w}{\partial y} - \frac{\partial v}{\partial z} \qquad \omega_y = \frac{\partial u}{\partial z} - \frac{\partial w}{\partial x} \qquad \omega_z = \frac{\partial v}{\partial x} - \frac{\partial u}{\partial y} \tag{3.2.21}$$

An irrotational flow possesses no vorticity. We consider this special flow in Section 8.5.

In cylindrical and spherical coordinates, the vorticity components are, respectively:

$$\omega_r = \frac{1}{r}\left(\frac{\partial v_z}{\partial \theta} - \frac{\partial v_\theta}{\partial z}\right) \qquad \omega_\theta = \frac{\partial v_r}{\partial z} - \frac{\partial v_z}{\partial r} \qquad \omega_z = \frac{1}{r}\frac{\partial(rv_\theta)}{\partial r} - \frac{1}{r}\frac{\partial v_r}{\partial \theta} \tag{3.2.22}$$

$$\omega_r = \frac{1}{r \sin \theta}\left[\frac{\partial}{\partial \theta}(v_\phi \sin \theta) - \frac{\partial v_\theta}{\partial \phi}\right] \qquad \omega_\phi = \frac{1}{r}\left[\frac{\partial}{\partial r}(rv_\theta) - \frac{\partial v_r}{\partial \theta}\right]$$

$$\tag{3.2.23}$$

$$\omega_\theta = \frac{1}{r}\left[\frac{1}{\sin \theta}\frac{\partial v_r}{\partial \phi} - \frac{\partial}{\partial r}(rv_\phi)\right]$$

The deformation of the particle of Fig. 3.6 is the rate of change of the angle that line segment AB makes with line segment CD. If AB is rotating with an angular velocity different from that of CD, the particle is deforming. The deformation is represented by the **rate-of-strain** tensor; its component ε_{xy} in the *xy*-plane is given by

$$\varepsilon_{xy} = \tfrac{1}{2}(\Omega_{AB} - \Omega_{CD})$$

$$= \frac{1}{2}\left(\frac{\partial v}{\partial x} + \frac{\partial u}{\partial y}\right) \tag{3.2.24}$$

For the *xz*-plane and the *yz*-plane we have

$$\varepsilon_{xz} = \frac{1}{2}\left(\frac{\partial w}{\partial x} + \frac{\partial u}{\partial z}\right) \qquad \varepsilon_{yz} = \frac{1}{2}\left(\frac{\partial w}{\partial y} + \frac{\partial v}{\partial z}\right) \tag{3.2.25}$$

The fluid particle could also deform by being stretched or compressed in a particular direction. For example, if point B of Fig. 3.6 is moving faster than point A, the particle would be stretching in the *x*-direction. This normal rate of strain is measured by

$$\varepsilon_{xx} = \frac{u_B - u_A}{dx}$$

$$= \left[u + \frac{\partial u}{\partial x}\frac{dx}{2} - \left(u - \frac{\partial u}{\partial x}\frac{dx}{2} \right) \right] \Big/ dx = \frac{\partial u}{\partial x}$$

(3.2.26)

Similarly, in the y- and z-directions we would find that

$$\varepsilon_{yy} = \frac{\partial v}{\partial y} \qquad \varepsilon_{zz} = \frac{\partial w}{\partial z}$$

(3.2.27)

We will see in a later chapter that the normal and shear stress components in a flow are related to the foregoing rate-of-strain components. In fact, in the one-dimensional flow of Fig. 1.6, the shear stress was related to $\partial u/\partial y$; note that $\partial u/\partial y$ is twice the rate-of-strain component given by Eq. 3.2.24 with $v = 0$.

EXAMPLE 3.1

A velocity field in a particular flow is given by $\mathbf{V} = 20y^2\hat{i} - 20xy\hat{j}$ m/s, where $\hat{i}$ and $\hat{j}$ are unit vectors in the x- and y-direction, respectively. Calculate the acceleration, the angular velocity, and the vorticity vector at the point $(1, -1, 2)$.

Solution

We could use Eq. 3.2.9 and find each component of the acceleration, or we could use Eq. 3.2.8 and find a vector expression. Using Eq. 3.2.8, we have

$$\mathbf{a} = u\frac{\partial \mathbf{V}}{\partial x} + v\frac{\partial \mathbf{V}}{\partial y} + w\overset{0}{\cancel{\frac{\partial \mathbf{V}}{\partial z}}} + \overset{0}{\cancel{\frac{\partial \mathbf{V}}{\partial t}}}$$

$$= 20y^2(-20y\hat{j}) - 20xy(40y\hat{i} - 20x\hat{j})$$

$$= -800xy^2\hat{i} - 400(y^3 - x^2y)\hat{j}$$

All particles passing the point $(1, -1, 2)$ have the acceleration

$$\mathbf{a} = -800\hat{i} \text{ m/s}^2$$

The angular velocity has a z-component only; it is, at the point $(1, -1, 2)$,

$$\Omega_z = \frac{1}{2}\left(\frac{\partial v}{\partial x} - \frac{\partial u}{\partial y} \right)$$

$$= \tfrac{1}{2}(-20y - 40y) = 30 \text{ rad/s}$$

The vorticity vector is twice the angular velocity vector:

$$\boldsymbol{\omega} = 60\hat{k} \text{ rad/s}$$

3.3 CLASSIFICATION OF FLUID FLOWS

In this section we provide an overview of some of the aspects of fluid mechanics that are considered in more depth in subsequent chapters and sections. Although most of the notions presented here are redefined and discussed in detail later, it will be helpful at this point to introduce the general classification of fluid flows.

3.3.1 One-, Two-, and Three-Dimensional Flows

In the Eulerian description of motion the velocity vector, in general, depends on three space variables and time, that is, $\mathbf{V} = \mathbf{V}(x, y, z, t)$. Such a flow is a **three-dimensional flow,** because the velocity vector depends on three space coordinates. The solutions to problems in such a flow are very difficult and are beyond the scope of an introductory course. Even if the flow could be assumed to be steady [i.e., $\mathbf{V} = \mathbf{V}(x, y, z)$], it would remain a three-dimensional flow.

Often a three-dimensional flow can be approximated as a two-dimensional flow. For example, the flow over a wide dam is three-dimensional because of the end conditions, but the flow in the central portion away from the ends can be treated as two-dimensional. In general, a **two-dimensional flow** is a flow in which the velocity vector depends on only two space variables. An example is a **plane flow,** in which the velocity vector depends on two spatial coordinates, x and y, but not z [i.e., $\mathbf{V} = \mathbf{V}(x, y)$]. A particular flow is shown in Fig. 3.7. This flow is normal to a plane surface; the fluid decelerates and comes to rest at the **stagnation point.** The two velocity components, u and v, depend only on x and y; that is, $u = u(x, y)$ and $v = v(x, y)$ in a plane flow.

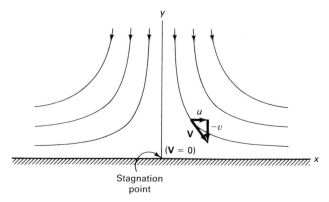

Stagnation
point

Figure 3.7 Plane, stagnation point flow.

Figure 3.8   One-dimensional flow: (a) flow in a pipe; (b) flow between parallel plates.

A **one-dimensional flow** is a flow in which the velocity vector depends on only one space variable. Such flows occur in long, straight pipes or between parallel plates, as shown in Fig. 3.8. The velocity in the pipe varies only with r: $u = u(r)$. The velocity between parallel plates varies only with the coordinate y: $u = u(y)$. Even if the flow is unsteady so that $u = u(y, t)$, as would be the situation during startup, the flow is one-dimensional. The flows shown may also be referred to as **developed flows**; that is, the velocity profiles do not vary with respect to the space coordinate in the direction of flow. This demands that the region of interest be a substantial distance from an entrance or a sudden change in geometry.

There are many engineering problems in fluid mechanics in which a flow field is simplified to a **uniform flow:** the velocity, and other fluid properties, are constant over the area. This simplification is made when the velocity is essentially constant over the area. Examples of such flows are high speed flow in a pipe section, and flow in a stream. The average velocity may change from one section to another; the flow conditions depend only on the space variable in the flow direction. The schematic representation of the velocity is shown in Figure 3.9. For large conduits, however, it may be necessary to consider hydrostatic variation in the pressure normal to the streamlines.

3.3.2 Viscous and Inviscid Flows

A fluid flow may be broadly classified as either a viscous flow or an inviscid flow. An **inviscid flow** is one in which viscous effects do not significantly influence the flow and are thus neglected. In a **viscous flow** the effects of viscosity are important and cannot be ignored.

To model an inviscid flow analytically, we can simply let the viscosity be zero; this will obviously make all viscous effects zero. It is more difficult to create an inviscid flow experimentally, because all fluids of interest (such as water and air) have viscosity. The question then becomes: Are there flows of interest in which the viscous effects are negligibly small? The answer is "yes, if the shear stresses in the flow are small

Figure 3.9   Uniform velocity profiles.

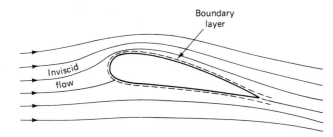

Figure 3.10   Flow around an airfoil.

and act over such small areas that they do not significantly affect the flow field.'' This statement is very general, of course, and it will take consider-able analysis to justify the inviscid flow assumption.

Based on experience, it has been found that the primary class of flows, which can be modeled as inviscid flows, are **external flows,** that is, flows which exist exterior to a body, such as flow around an airfoil or a hydrofoil. Any viscous effects that may exist are confined to a thin layer, called a **boundary layer,** that is attached to the boundary, such as that shown in Fig. 3.10. Boundary layers are so thin, for many flow situations, that they can simply be ignored when studying the gross features of a flow. For example, the inviscid flow solution provides an excellent predic-tion to the flow around the airfoil, except possibly near the trailing edge.

Viscous flows include the broad class of **internal flows**, such as flows in pipes and conduits and in open channels. In such flows viscous effects cause substantial ''losses'' and account for the huge amounts of energy that must be used to transport oil and gas in pipelines.

3.3.3 Laminar and Turbulent Flows

A viscous flow can be classified as either a laminar flow or a turbulent flow. In a **laminar flow** the fluid flows with no significant mixing of neigh-boring fluid particles. If dye were injected into the flow, it would not mix with the neighboring fluid except by molecular activity; it would retain its identity for a relatively long period of time. Viscous shear stresses always influence a laminar flow. The flow may be highly time dependent, as shown by the output of a velocity probe in Fig. 3.11a, or it may be steady, as shown in Fig. 3.11b.

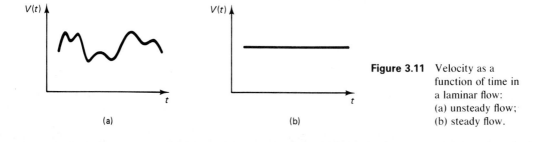

Figure 3.11   Velocity as a function of time in a laminar flow: (a) unsteady flow; (b) steady flow.

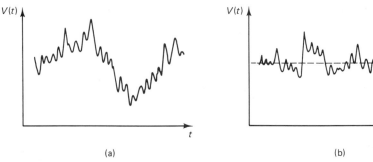

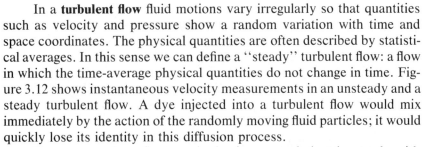

(a)   (b)

Figure 3.12   Velocity as a function of time in a turbulent flow: (a) unsteady flow; (b) "steady" flow.

In a **turbulent flow** fluid motions vary irregularly so that quantities such as velocity and pressure show a random variation with time and space coordinates. The physical quantities are often described by statistical averages. In this sense we can define a "steady" turbulent flow: a flow in which the time-average physical quantities do not change in time. Figure 3.12 shows instantaneous velocity measurements in an unsteady and a steady turbulent flow. A dye injected into a turbulent flow would mix immediately by the action of the randomly moving fluid particles; it would quickly lose its identity in this diffusion process.

The reason why a flow can be laminar or turbulent has to do with what happens to a small flow disturbance, a perturbation to the velocity components. A flow disturbance can either increase or decrease in size. If a flow disturbance in a laminar flow increases (i.e., the flow is unstable), the flow may become turbulent; if the disturbance decreases, the flow remains laminar. In certain situations the flow may develop into a different laminar flow, as is the case between rotating cylinders shown in Fig. 3.13.

The flow regime depends on three physical parameters describing the flow conditions. The first parameter is a length scale of the flow field, such as the thickness of a boundary layer or the diameter of a pipe. If this length scale is sufficiently large, a flow disturbance may increase and the flow may be turbulent. The second parameter is a velocity scale such as a spatial average of the velocity; for a large enough velocity the flow may be turbulent. The third parameter is the kinematic viscosity; for a small enough viscosity the flow may be turbulent.

The three parameters can be combined into a single parameter that can serve as a tool to predict the flow regime. This quantity is the

Figure 3.13   Laminar flow between rotating cylinders. A secondary flow occurs as regularly spaced toroidal vortices. (Photograph by Burkhalter and Koschmieder.)

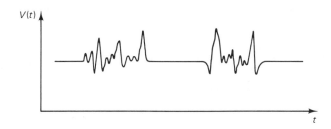

Figure 3.14 Velocity versus time signal from a velocity probe in an intermittent flow.

Reynolds number, a dimensionless parameter, defined as

$$\text{Re} = \frac{VL}{\nu} \tag{3.3.1}$$

where L and V are a characteristic length and velocity, respectively, and ν is the kinematic viscosity; for example, in a pipe flow L could be the pipe diameter and V could be the average velocity. If the Reynolds number is relatively small, the flow regime is laminar; if it is large, the flow regime is turbulent. This is more precisely stated by defining a **critical Reynolds number,** Re_{crit}, so that the flow is laminar if $\text{Re} < \text{Re}_{\text{crit}}$. For example, in a flow inside a rough-walled pipe it is found that $\text{Re}_{\text{crit}} \approx 2000$. This is the minimum critical Reynolds number and is used for most engineering applications. If the pipe wall is extremely smooth, the critical Reynolds number can be increased as the fluctuation level in the flow is decreased; values in excess of 40 000 have been measured. The flow can also be intermittently turbulent and laminar; this is called an **intermittent flow.** This phenomenon can occur when the Reynolds number is close to Re_{crit}. Figure 3.14 shows the output of a velocity probe for such a flow.

In a boundary layer that exists on a flat plate, due to a constant-velocity fluid stream, as shown in Fig. 3.15, the length scale changes with distance from the upstream edge. A Reynolds number is calculated using the length x as the characteristic length. For a certain x_T, Re becomes Re_{crit} and the flow undergoes transition from laminar to turbulent. For a smooth plate in a uniform flow with a low free-stream fluctuation level, values as high as $\text{Re}_{\text{crit}} = 10^6$ have been observed. In most engineering applications we assume a rough wall, or high free-stream fluctuation level, with an associated critical Reynolds number of approximtely 3×10^5.

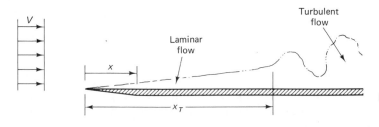

Figure 3.15 Boundary layer flow on a flat plate.

EXAMPLE 3.2

A 2-cm-diameter pipe is used to transport water at 20°C. What is the maximum average velocity that may exist in the pipe for a laminar flow?

Solution

The kinematic viscosity is found in Appendix B to be $\nu = 10^{-6}$ N · s/m². Using a maximum Reynolds number of 2000 for the laminar flow, we find that

$$V = \frac{2000\nu}{D}$$

$$= \frac{2000 \times 10^{-6}}{0.02} = 0.1 \text{ m/s}$$

This average velocity is quite small. Velocities this small are not usually encountered in actual situations; hence laminar flow is seldom of engineering interest. Most internal flows are turbulent flows and thus the study of turbulence gains much attention.

3.3.4 Incompressible and Compressible Flows

The last major classification of fluid flows to be considered in this chapter separates flows into incompressible and compressible flows. An **incompressible flow** exists if the density of each fluid particle remains relatively constant as it moves through the flow field, that is,

$$\frac{D\rho}{Dt} = 0 \tag{3.3.2}$$

This does not demand that the density is everywhere constant. If the density is constant, then obviously, the flow is incompressible, but that would be a more restrictive condition. Atmospheric flow, in which $\rho = \rho(z)$, where z is vertical, and flows that involve adjacent layers of fresh and salt water are examples of incompressible flows in which the density varies.

   In addition to liquid flows, low-speed gas flows, such as the atmospheric flow referred to above, are also considered to be incompressible flows. The **Mach number** is defined as

$$\mathrm{M} = \frac{V}{c} \tag{3.3.3}$$

where V is the gas speed and the wave speed $c = \sqrt{kRT}$; Eq. 3.3.3 is useful in deciding whether a particular gas flow can be studied as an incompressible flow. If M < 0.3, density variations are at most 3% and the flow is assumed to be incompressible for standard air; this corresponds to a velocity below about 100 m/s or 300 ft/sec. If M > 0.3, the density variations influence the flow and compressibility effects should be accounted for; such flows are **compressible flows** and are considered in Chapter 9.

Incompressible gas flows include atmospheric flows, the aerodynamics of landing and takeoff of commercial aircraft, heating and air-conditioning airflows, flow around automobiles and through radiators, and the flow of air around buildings, to name a few. Compressible flows include the aerodynamics of high-speed aircraft, airflow through jet engines, steam flow through the turbine in a power plant, airflow in a compressor, and the flow of the air–gas mixture in an internal combustion engine.

3.4 THE BERNOULLI EQUATION

In this section we present an equation that is probably used more often in fluid flow applications than any other equation. It is also often misused; it is thus important to understand its limitations. Its limitations are a result of several assumptions made in the derivation. One of the assumptions is that viscous effects are neglected. In other words, in view of Eq. 1.5.3, shear stresses introduced by velocity gradients are not taken into consideration. These stresses are often very small compared with pressure differences in the flow field. Locally, these stresses have little effect on the flow field and the assumption is justified. However, over long distances or in regions of high-velocity gradients, these stresses may affect the flow conditions so that viscous effects must be included.

The derivation of the Bernoulli equation starts with the application of Newton's second law to a fluid particle. Let us use a particle positioned as shown in Fig. 3.16, with length ds and cross-sectional area dA. The forces acting on the particle are the pressure forces and the weight, as shown. Summing forces in the direction of motion, the s-direction, there

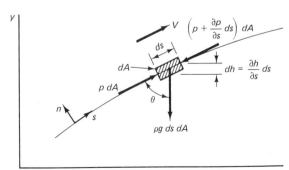

Figure 3.16   Particle moving along a streamline.

results

$$p \, dA - \left(p + \frac{\partial p}{\partial s} \, ds\right) dA - \rho g \, ds \, dA \cos \theta = \rho \, ds \, dA \, a_s \quad (3.4.1)$$

where a_s is the acceleration of the particle in the s-direction. It is given by[1]

$$a_s = V \frac{\partial V}{\partial s} + \cancelto{0}{\frac{\partial V}{\partial t}} \quad (3.4.2)$$

where $\partial V/\partial t = 0$ since we will assume steady flow. Also, we see that

$$dh = ds \cos \theta = \frac{\partial h}{\partial s} \, ds \quad (3.4.3)$$

so that

$$\cos \theta = \frac{\partial h}{\partial s} \quad (3.4.4)$$

Then, after dividing by $ds \, dA$, Eq. 3.4.1 takes the form

$$-\frac{\partial p}{\partial s} - \rho g \frac{\partial h}{\partial s} = \rho V \frac{\partial V}{\partial s} \quad (3.4.5)$$

Now, we assume constant density and note that $V \, \partial V/\partial s = \partial(V^2/2)/\partial s$; then we can write Eq. 3.4.5 as

$$\frac{\partial}{\partial s} \left(\frac{V^2}{2} + \frac{p}{\rho} + gh\right) = 0 \quad (3.4.6)$$

This is satisfied if, along the streamline,

$$\frac{V^2}{2} + \frac{p}{\rho} + gh = \text{const} \quad (3.4.7)$$

or, between two points on the same streamline,

$$\frac{V_1^2}{2} + \frac{p_1}{\rho} + gh_1 = \frac{V_2^2}{2} + \frac{p_2}{\rho} + gh_2 \quad (3.4.8)$$

---

[1]This can be verified by considering Eq. 3.2.9a, assuming that $v = w = 0$. Think of the x-direction being tangent to the streamline at the instant shown, so that $u = V$.

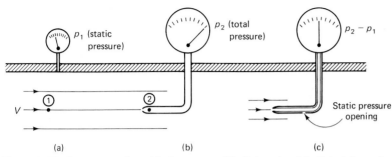

Figure 3.17   Pressure probes: (a) piezometer; (b) pitot probe; (c) pitot-static probe.

This is the well-known **Bernoulli equation.** Note the assumptions:

Inviscid flow (no shear stresses).
Steady flow ($\partial V/\partial t = 0$).
Along a streamline ($a_s = V\, \partial V/\partial s$).
Constant density ($\partial \rho/\partial s = 0$).
Inertial reference frame ($\mathbf{A} = \mathbf{a}$).

If Eq. 3.4.8 is divided by g, this equation becomes

$$\frac{V_1^2}{2g} + \frac{p_1}{\gamma} + h_1 = \frac{V_2^2}{2g} + \frac{p_2}{\gamma} + h_2 \qquad (3.4.9)$$

The sum of the two terms ($p/\gamma + h$) is called the **piezometric head** and the sum of the three terms the **total head.** The pressure p is often referred to as **static pressure** and the sum of the two terms

$$p + \rho\,\frac{V^2}{2} = p_T \qquad (3.4.10)$$

is called the **total pressure** p_T or **stagnation pressure.**

The static pressure in a pipe can be measured simply by installing a so-called **piezometer,** shown[2] in Fig. 3.17a. A device, known as a **pitot probe,** sketched in Fig. 3.17b, is used to measure the total pressure in a fluid flow. Point 2 just inside the pitot tube is a stagnation point; the velocity there is zero. The difference between the readouts can be used to determine the velocity at point 1. A **pitot-static probe** is also used to measure the difference between total and static pressure with one probe

---

[2]When drilling the hole in the wall necessary for the piezometer, burrs are often formed on the inner surface. It is important that such burrs be removed since they may cause errors as high as 30% in the pressure reading.

(Fig. 3.17c). The velocity at point 1 (using the readings of the piezometer and pitot probes, or the reading from the pitot-static probe) can be determined by applying the Bernoulli equation between points 1 and 2:

$$\frac{V_1^2}{2g} + \frac{p_1}{\gamma} = \frac{p_2}{\gamma} \tag{3.4.11}$$

where we have assumed point 2 to be a stagnation point so that $V_2 = 0$. This gives

$$V_1 = \sqrt{\frac{2}{\rho}(p_2 - p_1)} \tag{3.4.12}$$

We will find many uses for Bernoulli's equation in our study of fluids. We must be careful, however, never to use it in an unsteady flow or if viscous effects are significant (the primary reasons for making Bernoulli's equation inapplicable).

The Bernoulli equation can be used to determine how high the water from a fireman's hose will reach, to find the pressure on the surface of a low-speed airfoil,[3] and to find the wind force on the window in a house. These examples are all external flows, flows around objects submerged in the fluid.

Another class of problems where inviscid flow can be assumed and where the Bernoulli equation finds frequent application involves internal flows over relatively short distances, for example, flow through a contraction, as shown in Fig. 3.18a, or flow from a plenum, as shown in Fig. 3.18b. For a given velocity profile entering the short contraction, the pressure drop $(p_1 - p_2)$ and the velocity profile at section 2 can be approximated assuming an inviscid flow. Viscous effects are typically very small and require substantial distances and areas over which to operate in order to become significant; so in situations such as those shown in Fig. 3.18, viscous effects can often be neglected.

Inviscid flow does not always give a good approximation to the actual flow that exists around a body. Consider the inviscid flow around the sphere, shown in Fig. 3.19. A stagnation point where $V = 0$ exists at both the front and the back of the sphere. Bernoulli's equation predicts a maximum pressure at the stagnation points A and C because the velocity is zero at such points. A maximum velocity, and thus a minimum pressure, would exist at point B. In the inviscid flow of part (a) the fluid flowing from B to C must flow from the low-pressure region near B to the high-pressure region near C. In the actual flow there exists a thin boundary layer in which the velocity goes to zero at the surface of the sphere.

---

[3]To consider the flow around an aircraft as a steady flow, we simply make the aircraft stationary and move the air, as is done in model studies using a wind tunnel. The pressures and forces remain unchanged.

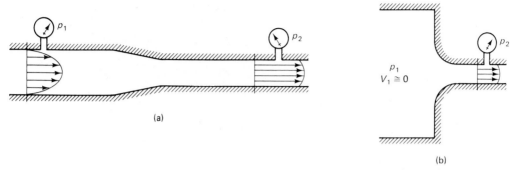

Figure 3.18   Internal inviscid flows: (a) flow through a contraction; (b) flow from a plenum.

This slow-moving fluid near the boundary does not have sufficient momentum to move into the higher-pressure region near C; the result is that the fluid **separates** from the boundary—the boundary streamline leaves the boundary—creating a separated region as shown in the actual flow of part (b). The pressure does not increase but remains relatively low over the rear part of the sphere. The high pressure that exists near the front stagnation point is never recovered on the rear of the sphere, resulting in a relatively large drag force in the direction of flow. A similar situation occurs in the flow around an automobile.

The flow on the front of the sphere is well approximated by an inviscid flow; but it is obvious that the flow over the rear of the sphere deviates radically from an inviscid flow. Viscous effects in the boundary layer have led to a separated flow, a phenomenon that is often undesirable. For example, separated flow on an airfoil is called **stall** and must never occur, except on the wings of special stunt planes. On the blades of a turbine separated flows lead to substantially reduced efficiency. The air deflector on the roof of the cab of a semitruck reduces the separated region, thereby reducing drag and fuel consumption.

If viscous effects are negligible in a steady liquid flow, we can use Bernoulli's equation to locate points of possible **cavitation.** This condition

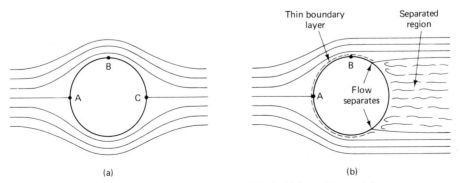

Figure 3.19   Flow around a sphere: (a) inviscid flow; (b) actual flow.

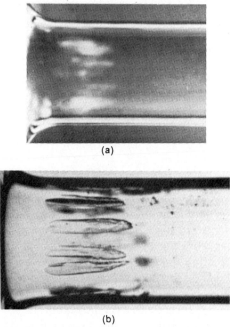

(a)

(b)

Figure 3.20   Cavitation in a venturi nozzle, with water flowing at a velocity of 15 m/s: (a) incandescent lamp, exposure time $\frac{1}{30}$ s; and (b) strobe light exposure time 5 μs. (Photograph courtesy of the Japan Society of Mechanical Engineers and Pergamon Press).

occurs when the local pressure becomes equal to the vapor pressure of the liquid. It is to be avoided, if at all possible, because of damage to solid surfaces. Figure 3.20 shows cavitating flow just downstream of a contraction in a pipe. At the point where cavitation occurs, small vapor bubbles are generated and these bubbles collapse when they enter a higher-pressure region. The collapse is accompanied by very large local pressures that last for only a small fraction of a second. These pressure spikes may reach a wall, where they can, after repeated applications, result in significant damage.

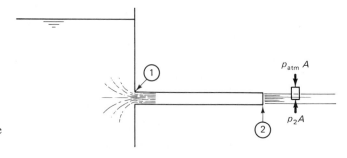

Figure 3.21   Exit flow into the atmosphere.

One last comment regarding pressure changes in a fluid. We often encounter entrances and exits to a pipe or conduit. Consider a flow from a reservoir through a pipe, as sketched in Fig. 3.21. At the entrance the streamlines are curved and the pressure is not constant across section 1. At the exit, however, the streamlines are straight, so that no acceleration exists normal to the streamlines; hence the pressure forces acting on the ends of the small cylindrical control volume must be equal. We write this as $p_2 = p_{atm}$ or $p_2 = 0$ gage pressure.

EXAMPLE 3.3

The wind reaches a speed of 100 km/h in a storm. Calculate the force acting on a 1 m × 2 m window facing the storm. The window is in a high-rise building, so the wind speed is not reduced due to ground effects. Use $\rho = 1.2$ kg/m^3.

Solution

The window facing the storm will be in a stagnation region where the wind speed is brought to zero. Working with gage pressures, the pressure p upstream in the wind is zero. The velocity V must have units of m/s. It is

$$V = 100 \, \frac{km}{h} \times \frac{1 \, h}{3600 \, s} \times \frac{1000 \, m}{1 \, km} = 27.8 \text{ m/s}$$

Bernoulli's equation then allows us to calculate the pressure on the window as follows:

$$\cancel{\frac{V_2^2}{2g}}_0 + \frac{p_2}{\gamma} + h_2 = \frac{V_1^2}{2g} + \cancel{\frac{p_1}{\gamma}}_0 + h_1$$

$$\therefore \quad p_2 = \frac{\rho V_1^2}{2}$$

$$= \frac{1.2 \times (27.8)^2}{2} = 464 \text{ N/m}^2$$

where we have used $h_2 = h_1$, $p_1 = 0$, $V_2 = 0$, and $\gamma = \rho g$. Multiply by the area and find the force to be

$$F = pA$$

$$= 464 \times 1 \times 2 = 928 \text{ N}$$

EXAMPLE 3.3 (English)

The wind reaches a speed of 65 mph in a storm. Calculate the force acting on a 3 ft × 6 ft window facing the storm. The window is in a high-rise building, so the wind speed is not reduced due to ground effects. Use $\rho = 0.0024$ slug/ft^3.

Solution

The window facing the storm will be in a stagnation region where the wind speed is brought to zero. Working with gage pressures, the pressure p upstream in the wind is zero. The velocity V must have units of ft/sec. It is

$$V = 65 \, \frac{\text{mi}}{\text{hr}} \times \frac{1 \text{ hr}}{3600 \text{ sec}} \times \frac{5280 \text{ ft}}{1 \text{ mi}} = 95.3 \text{ ft/sec}$$

Bernoulli's equation then allows us to calculate the pressure on the window as follows:

$$\cancel{\frac{V_2^2}{2g}}_0 + \frac{p_2}{\gamma} + h_2 = \frac{V_1^2}{2g} + \cancel{\frac{p_1}{\gamma}}_0 + h_1$$

$$\therefore \quad p_2 = \frac{\rho V_1^2}{2}$$

$$= \frac{0.0024 \times 95.3^2}{2} = 10.9 \text{ lb/ft}^2$$

where we have used $h_2 = h_1$, $p_1 = 0$, $V_2 = 0$, and $\gamma = \rho g$. Multiply by the area and find the force to be

$$F = pA$$

$$= 10.9 \times 3 \times 6 = 196 \text{ lb}$$

EXAMPLE 3.4

The static pressure head in an air pipe is measured with a piezometer as 16 mm of water. A pitot probe at the same location indicates 24 mm of water. Calculate the velocity of the 20°C air. Also, calculate the Mach number and comment as to the compressibility of the flow.

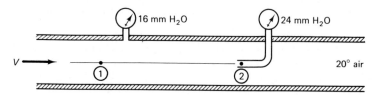

Solution

Bernoulli's equation is applied between two points on the streamline that terminates at the stagnation point of the pitot probe. Point 1 is upstream and p_2 is the total pressure; then

$$\frac{V_1^2}{2g} + \frac{p_1}{\gamma} = \frac{p_T}{\gamma}$$

We use the ideal gas law to calculate the density:

$$\rho = \frac{p}{RT}$$

$$= \frac{9810 \times 0.016 + 101\ 000}{287 \times (273 + 20)} = 1.20 \text{ kg/m}^3$$

where the pressure has been found using $p = \gamma h$ and atmospheric pressure, which is 101 000 Pa, is added since absolute pressure is needed in the preceding equation. The velocity is then

$$V_1 = \sqrt{\frac{2}{\rho}(p_T - p_1)}$$

$$= \sqrt{\frac{2(0.024 - 0.016) \times 9810}{1.20}} = 11.44 \text{ m/s}$$

To find the Mach number, we must calculate the speed of sound. It is

$$c = \sqrt{kRT}$$

$$= \sqrt{1.4 \times 287 \times 293} = 343 \text{ m/s}$$

The Mach number is then

$$M = \frac{V}{c} = \frac{11.44}{343} = 0.0334$$

Obviously, the flow can be assumed to be incompressible since $M < 0.3$. The velocity would have to be much higher before compressibility would be important.

EXAMPLE 3.5

Bernoulli's equation, in the form of Eq. 3.4.8, looks very much like the energy equation developed in thermodynamics for a control volume. Discuss the differences between the two equations.

Solution

From thermodynamics we recall that the steady-flow energy equation for a control volume with one inlet and one outlet takes the form (see Section 4.4)

$$\dot{Q} - \dot{W}_s = \dot{m}\left(\frac{V_2^2}{2} + \frac{p_2}{\rho_2} + \tilde{u}_2 + gz_2\right) - \dot{m}\left(\frac{V_1^2}{2} + \frac{p_1}{\rho_1} + \tilde{u}_1 + gz_1\right)$$

This becomes, after dividing through by g,

$$\frac{V_2^2}{2g} + \frac{p_2}{\gamma} + z_2 = \frac{V_1^2}{2g} + \frac{p_1}{\gamma} + z_1$$

where we have made the following assumptions:

No heat transfer ($\dot{Q} = 0$)

No shaft work ($\dot{W}_s = 0$)

No temperature change ($\bar{u}_2 = \bar{u}_1$, i.e., no losses due to shear stresses)

Uniform velocity profiles at the two sections

Steady flow

Constant density

Even though several of these assumptions are the same as those made in the derivation of the Bernoulli equation (steady flow, constant density, and no shear stress), we must not confuse the two equations; the Bernoulli equation is derived from Newton's second law and is valid along a streamline, whereas the energy equation is derived from the first law of thermodynamics and is valid between two sections in a fluid flow. The energy equation can be used across a pump to determine the horsepower required to provide a particular pressure rise; the Bernoulli equation can be used along a stagnation streamline to determine the pressure at a stagnation point, a point where the velocity is zero. The equations are quite different, and just because the energy equation degenerates to the Bernoulli equation for particular situations, the two should not be used out of context.

PROBLEMS

Flow Fields

3.1. A fire is started and the smoke from the chimney goes straight up; no wind is present. After a few minutes a wind arises but the smoke continues to rise slowly. Sketch the streakline of the smoke, the pathline of the first particles leaving the chimney, and a few streamlines, assuming that the wind has been blowing for a while.

3.2. A person has a large number of small flotation devices each equipped with a battery and light bulb. Explain how she would determine the pathlines and streaklines near the surface of a stream with some unknown currents that vary with time.

3.3. A little boy chases his dad around the yard with a water hose. Sketch a pathline and a streakline if the boy is running perpendicular to the water jet.

3.4. Using rectangular coordinates, express the z-component of Eq. 3.2.1.

3.5. The traffic situation on Mackinac Island, Michigan, where no automobiles are allowed (bikes are permitted), is to be studied. Comment on how such a study could be performed using a Lagrangian approach and a Eulerian approach.

3.6. Find the acceleration vector field for a fluid flow that possesses the following velocity field. Evaluate the acceleration at $(2, -1, 3)$ at $t = 2$ s.
   (a) $\mathbf{V} = 20(1 - y^2)\hat{i}$.
   (b) $\mathbf{V} = 2x\hat{i} + 2y\hat{j}$.
   (c) $\mathbf{V} = x^2 t\hat{i} + 2xyt\hat{j} + 2yzt\hat{k}$.
   (d) $\mathbf{V} = x\hat{i} - 2xyz\hat{j} + tz\hat{k}$.

3.7. Find the angular velocity vector for the flow fields of Problem 3.6. Evaluate the angular velocity at $(2, -1, 3)$ at $t = 2$ s.

3.8. Find the vorticity vector for the flow fields of Problem 3.6. Evalute the vorticity at $(2, -1, 3)$ at $t = 2$ s.

3.9. The velocity components in cylindrical coordinates are given by

$$v_r = \left(10 - \frac{40}{r^2}\right)\cos\theta \qquad v_\theta = -\left(10 + \frac{40}{r^2}\right)\sin\theta$$

   (a) Calculate the acceleration of a fluid particle occupying the point $(4, 180°)$.
   (b) Calculate the vorticity component at $(4, 180°)$.

3.10. The velocity components in spherical coordinates are given by

$$v_r = \left(10 - \frac{80}{r^3}\right) \cos\theta \qquad v_\theta = -\left(10 + \frac{80}{r^3}\right) \sin\theta$$

(a) Calculate the acceleration of a fluid particle occupying the point (4, 180°).
(b) Calculate the vorticity component at (4, 180°).

3.11. The temperature changes periodically in a flow according to $T(y, t) = 20(1 - y^2) \cos \pi t/100$ °C. If the velocity is given by $u = 2(1 - y^2)$ m/s, determine the rate of change of the temperature of a fluid particle located at $y = 0$ if $t = 20$ s.

3.12. Density of air in the atmosphere varies according to $\rho(z) = 1.23e^{-10^{-4}z}$ kg/m³. Air flowing over a mountain has the velocity vector $\mathbf{V} = 20\hat{i} + 10\hat{k}$ m/s at a location of interest where $z = 3000$ m. Find the rate at which a particle's density is changing at that location.

3.13. The density variation with elevation is given by $\rho(z) = 1000(1 - z/4)$ kg/m³. At a location where $\mathbf{V} = 10\hat{i} + 10\hat{k}$ m/s, find $D\rho/Dt$.

3.14. Salt is slowly added to water in a pipe so that $\partial\rho/\partial x = -0.01$ kg/m⁴. Determine $D\rho/Dt$ if the velocity is uniform at 4 m/s.

3.15. Relative to a fixed reference frame, find the acceleration of a fluid particle at:
(a) point A.
(b) point B.
The water at B makes an angle of 30° with respect to the ground and the sprinkler arm is horizontal.

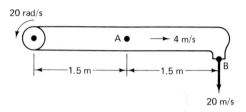

3.16. A river is flowing due south at 5 m/s at a latitude of 45°. Calculate the acceleration of a particle floating with the river relative to a fixed reference frame. The radius of earth of 6000 km.

Classification of Fluid Flows

3.17. Consider each of the following flows and state whether it could be approximated as a one-, two-, or three-dimensional flow or as a uniform flow.
(a) Flow from a horizontal nozzle striking a vertical wall.
(b) Flow near the entrance of a pipe.
(c) Flow around a rocket.
(d) Flow around an automobile.
(e) Flow in an irrigation channel.
(f) Flow through an artery.

3.18. State whether each of the flows in Problem 3.17 should be considered primarily as an inviscid flow or as a viscous flow.

3.19. From which of the following objects would you expect the flow to separate and form a substantial separated region?
(a) A golf ball.
(b) A telephone wire.
(c) A wind machine blade.
(d) A 20-mm-dia. wire in a low-speed wind tunnel.
(e) An automobile.
(f) An aircraft.
Note: separation occurs whenever the Reynolds number exceeds a value around 20.

3.20. The 32°C water exiting a 1.5-cm-dia. faucet has an average velocity of 2 m/s. Would you expect the flow to be laminar or turbulent?

3.20E. The 90°F water exiting a 0.5-in.-dia. faucet has an average velocity of 6 ft/sec. Would you expect the flow to be laminar or turbulent?

3.21. The Red Cedar River flows placidly through Michigan State University's campus. In a certain section the depth is 0.8 m and the average velocity is 0.2 m/s. Is the flow laminar or turbulent?

3.22. Air at 40°C flows in a rectangular 30 cm × 6 cm heating duct at an average velocity of 4 m/s. Is the flow laminar or turbulent?

3.23. The airfoil on a commercial airliner is approximated as a flat plate. How long would you expect the laminar portion of the boundary layer to be if it is flying a) at an altitude of 10 000 m and a speed of 900 kph. b) at an altitude of 30,000 ft and a speed of 600 mph.

3.24. A leaf keeps cool by transpiration, a process in which water flows from the leaf to the atmosphere. An experimenter wonders if the boundary layer on a leaf influences the transpiration so an "experimental" leaf is set up in the laboratory and air is blown over it at 6 m/s. Comment as to whether the boundary layer is expected to be laminar or turbulent.

3.25. For the following situations state whether a compressible flow is required or if the flow can be approximated with an incompressible flow.
(a) An aircraft flying at 100 m/s at an elevation of 8000 m.
(b) A golf ball traveling at 80 m/s.
(c) Flow around an object being studied in a high-temperature wind tunnel if the temperature is 100°C and the air velocity is 100 m/s.

Bernoulli's Equation

3.26. A pitot tube is used to measure the velocity of a small aircraft flying at 1000 m. Calculate its velocity if the pitot tube measures
(a) 2000 Pa,   **(b)** 6 kPa,   **(c)** 600 Pa.
3.26E. A pitot tube is used to measure the velocity of a small aircraft flying at 3000 ft. Calculate its velocity if the pitot tube measures
(a) 0.3 psi,   **(b)** 0.9 psi,   **(c)** 0.09 psi.
3.27. Approximate the force acting on a 15-cm-dia. headlight of an automobile traveling at 120 kph.

3.28. A vacuum cleaner is capable of creating a vacuum of 2 kPa just inside the hose. What maximum velocity would be expected in the hose?

3.29. To what maximum velocity can water be accelerated before it reaches the turbine blades of a hydro-turbine if it enters with relatively low velocity at
(a) 600 kPa,   **(c)** 80 psi,
(b) 300 kPa,   **(d)** 40 psi.

3.30. Water exists in a city's water system at a pressure of 500 kPa at a particular location. How high could a hill be, above that location, for the system to supply water to the other side of the hill?(A vacuum is not allowed.)

3.31. Air flows between the radial disks shown. Estimate the pressure in the 2-cm-dia. pipe if the air exits to the atmosphere. Neglect viscous effects and assume $\rho = 1.23$ kg/m^3.

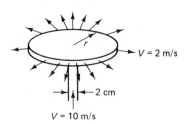

3.32. Estimate the pressure at r = 10 cm if the velocity there is 8 m/s in Prob. 3.31.

FOUR

The Integral Forms of the Fundamental Laws

4.1 INTRODUCTION

Quantities of interest to engineers can often be expressed in terms of integrals. For example, volume flow rate is the integral of the velocity over an area; heat transfer is the integral of the heat flux over an area; force is the integral of a stress over an area; mass is the integral of the density over a volume; and kinetic energy is the integral of $V^2/2$ over each mass element in a volume. There are, of course, many other integral quantities. To determine an integral quantity the integrand must be known, or information must be available so that a good approximation to the integrand can be made. If the integrand is not known or cannot be approximated with any degree of certainty, appropriate differential equations (see Chapter 5) must be solved yielding the needed integrand; the integration is then performed giving the engineer the desired integral quantity.

In this chapter we present the integral quantities of interest, develop equations that relate the integral quantities, and work a number of problems for which the integrands are given or can be approximated. This includes a surprisingly large variety of problems. There are, however, many integral quantities that cannot be determined since the integrands are unknown. These would include the lift and drag on an airfoil, the torque on the blades of a wind machine, and the kinetic energy in the wake of a submarine. To determine such integrands, it would be necessary to solve the appropriate differential equations, a task that is often quite difficult; some relatively simple situations are considered in subsequent chapters.

Also, there are many quantites of interest that are not integral in nature. Included would be the point of separation of the flow around a body, the concentration of a pollutant in a stream at a certain location, the pressure distribution on the side of a building, and the wave–shore interaction along a lake. To study subjects such as these, it is necessary to consider the differential equations that describe the flow situation. Most of the topics mentioned are relegated to specialized graduate courses; however, some topics that require the solution of the more easily solvable differential equations are included in this book.

The integral quantities of primary interest in fluid mechanics are contained in three basic laws: conservation of mass, first law of thermodynamics, and Newton's second law. These basic laws are expressed in terms of a **system,** a fixed collection of material particles. For example, if we consider flow through a pipe, we could identify a fixed quantity of fluid at time t as the system (Fig. 4.1); this system would then move due to velocity to a downstream location at time $t + \Delta t$. Any of the three basic laws could be applied to this system. This is, however, not an easy task. First let us state the basic laws in their general form.

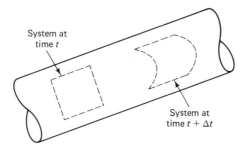

System at
time t

System at
time $t + \Delta t$

Figure 4.1 Example of a system in fluid mechanics.

Conservation of Mass.   The law stating that mass must be conserved is:

> The mass of a system remains constant.

The mass of a fluid particle is $\rho \, d\Psi$, where $d\Psi$ is the volume occupied by the particle and ρ is its density. Knowing that the density can change from point to point in the system, the conservation of mass can be expressed in integral form as

$$\frac{D}{Dt} \int_{\text{sys}} \rho \, d\Psi = 0 \qquad (4.1.1)$$

where D/Dt is used since we are following a specified group of material particles, a system.

First Law of Thermodynamics.   The law that relates heat transfer, work, and energy change is the first law of thermodynamics or, more simply, the energy equation; it states:

> The rate of heat transfer to a system minus the rate at which the system does work equals the rate at which the energy of the system is changing.

Recognizing that both density and specific energy may change from point to point in the system, it may be expressed as

$$\dot{Q} - \dot{W} = \frac{D}{Dt} \int_{\text{sys}} e\rho \, d\Psi \qquad (4.1.2)$$

where the **specific energy** e accounts for kinetic energy, potential energy, and internal energy. Other forms of energy (chemical, electrical, nuclear) are not included in an elementary course in fluid mechanics. In its basic form stated here, the first law of thermodynamics applies only to a system, a collection of fluid particles; therefore, D/Dt is used.

Newton's Second Law.   Newton's second law, also called the momentum equation, states:

> The resultant force acting on a system equals the rate at which the momentum of the system is changing.

The momentum of a fluid particle of mass $\rho\, d\Psi$ is a vector quantity given by $\mathbf{V}\rho\, d\Psi$; consequently, Newton's second law may be expressed as

$$\Sigma\, \mathbf{F} = \frac{D}{Dt} \int_{\text{sys}} \mathbf{V}\rho\, d\Psi \qquad (4.1.3)$$

recognizing that both density and velocity may change from point to point in the system. This equation reduces to $\Sigma\, \mathbf{F} = m\mathbf{a}$ if $\mathbf{V}$ and ρ are constant throughout the entire system; ρ is often a constant, but in fluid mechanics the velocity vector invariably changes from point to point. Again, D/Dt is used to provide the rate-of-change, since Newton's second law is applied to a system.

Moment-of-Momentum   Equation.   The   moment-of-momentum equation results from Newton's second law; it states:

> The resultant moment acting on a system equals the rate of change of the angular momentum of the system.

In equation form this becomes

$$\Sigma\, \mathbf{M} = \frac{D}{Dt} \int_{\text{sys}} \mathbf{r} \times \mathbf{V}\rho\, d\Psi \qquad (4.1.4)$$

where $\mathbf{r} \times \mathbf{V}\rho\, d\Psi$ represents the angular momentum of a fluid particle with mass $\rho\, d\Psi$. The vector $\mathbf{r}$ locates the volume element $d\Psi$ and is measured from the origin of the coordinate axes, the point relative to which the resultant moment is measured.

Note that in each of the basic laws the integral quantity is an extensive property of the system (see Section 1.7). We will use the symbol N_{sys} to denote this extensive property; for example, N_{sys} could be the mass, the momentum, or the energy of the system. The left-hand side of Eq. 4.1.1 and the right-hand sides of Eqs. 4.1.2, 4.1.3, and 4.1.4 may all be expressed as

$$\frac{DN_{\text{sys}}}{Dt} \qquad (4.1.5)$$

where N_{sys} represents an integral quantity.

It is also useful to introduce the variable η for the intensive property, the property of the system per unit mass. The relation between N_{sys} and η is given by

$$N_{\text{sys}} = \int_{\text{sys}} \eta\rho \, d\Psi \qquad (4.1.6)$$

As an example, the extensive property of Newton's second law is the momentum

$$\textbf{momentum}_{\text{system}} = \int_{\text{sys}} \textbf{V}\rho \, d\Psi \qquad (4.1.7)$$

which is a vector quantity. The corresponding intensive property would be the velocity vector **V**. Note that the density and velocity, which may vary from point to point within the system, may also be a function of time, as in unsteady flow.

Our interest is most often focused on a device, or a region of space, into which fluid enters and/or from which fluid leaves; we identify this region as a **control volume.** An example of a fixed control volume is shown in Fig. 4.2a. A control volume need not be fixed; it could deform. We will, however, consider only fixed control volumes in this book.

The difference between a control volume and a system is illustrated in Fig. 4.2b. The figure indicates that the system occupies the control volume at time t and has partially moved out of it at time $t + \Delta t$. Since it is often more convenient to focus on a control volume (e.g., a pump) rather than on a system, the first order of business is to find a transformation that will allow us to express the substantial derivative of a system in terms of quantities associated with a control volume so that the basic laws can be applied directly to a control volume. This will be done in general and then applied to the specific laws.

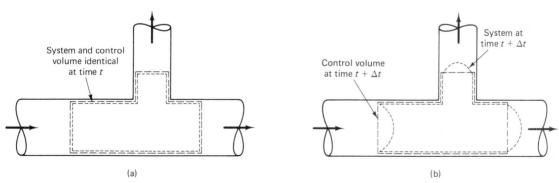

Figure 4.2   Example of a fixed control volume and a system: (a) time t; (b) time $t + \Delta t$.

4.2 SYSTEM-TO-CONTROL-VOLUME TRANSFORMATION

We are interested in the time rate of change of the extensive property N_{sys} as we follow the system along, that is, DN_{sys}/Dt, and we would like to express this in terms of quantities that pertain to the control volume. In this section we present the derivation of the transformation.

The derivation involves fluxes of the extensive property in and out of the control volume. A flux is a measure of the rate at which an extensive property crosses an area; for example, a mass flux is the rate at which mass crosses an area. It is useful to introduce vector notation to describe these fluxes. Consider an area element dA of the **control surface,** the surface area that completely encloses the control volume. The property flux across an elemental area dA (see Fig. 4.3) may be expressed by

$$\text{flux across } dA = \eta\rho\hat{n} \cdot \mathbf{V}\, dA \qquad (4.2.1)$$

where $\hat{n}$, a unit vector normal to the area element dA, always points out of the control volume, and η represents the intensive property associated with N_{sys}. Note that this expression yields a negative value if it concerns a property influx. Only the normal component $\hat{n} \cdot \mathbf{V}$ of the velocity vector contributes to this flux term. If there is no normal component of velocity on a particular area, such as the wall of a pipe, no flux occurs across that area. If $\hat{n} \cdot \mathbf{V}$ is positive then there is a flux out of the volume; if $\hat{n} \cdot \mathbf{V}$ is negative, that is, $\mathbf{V}$ has a component in the opposite direction of $\hat{n}$, a flux occurs into the volume. We must always use $\hat{n}$ pointing out of the volume. The velocity vector $\mathbf{V}$ may be at some angle to the unit vector $\hat{n}$; the dot product $\hat{n} \cdot \mathbf{V}$ accounts for the appropriate component of $\mathbf{V}$ that produces a flux through the area.

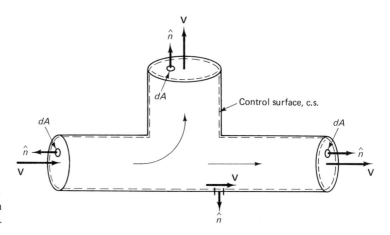

Figure 4.3 Illustration used to show the flux of an extensive property.

The net property flux out of the entire control surface is then obtained by integrating over the entire control surface:

$$\text{net flux of property} = \int_{c.s.} \eta\rho\hat{n} \cdot \mathbf{V}\, dA \qquad (4.2.2)$$

If the net flux is positive, the flux out is larger than the flux in.

Let us return now to the derivative DN_{sys}/Dt. The definition of a derivative allows us to write

$$\frac{DN_{sys}}{Dt} = \lim_{\Delta t \to 0} \frac{N_{sys}(t + \Delta t) - N_{sys}(t)}{\Delta t} \qquad (4.2.3)$$

The system is shown in Fig. 4.4 at times t and $t + \Delta t$. Assume that the system occupies the full control volume at time t; if we were considering a device, such as a pump, the particles of the system would just fill the device at time t. Since the device, the control volume shown in Fig. 4.4, is assumed to be fixed in space, the system will move through the device. Equation 4.2.3 can then be written

$$\frac{DN_{sys}}{Dt} = \lim_{\Delta t \to 0} \frac{N_3(t + \Delta t) + N_2(t + \Delta t) - N_2(t) - N_1(t)}{\Delta t}$$

$$= \lim_{\Delta t \to 0} \frac{N_2(t + \Delta t) + N_1(t + \Delta t) - N_2(t) - N_1(t)}{\Delta t} \qquad (4.2.4)$$

$$+ \lim_{\Delta t \to 0} \frac{N_3(t + \Delta t) - N_1(t + \Delta t)}{\Delta t}$$

where, in this second expression, we have simply added and subtracted $N_1(t + \Delta t)$ in the numerator. In the equations above, the numerical subscript denotes the region; for example, $N_2(t)$ signifies the extensive property in region 2 at time t. Now, we observe that the first limit on the right

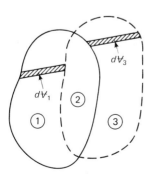

Fixed control volume occupies ① and ②.

System at time t occupies volumes ① and ②.

System at time $t + \Delta t$ occupies volumes ② and ③.

Figure 4.4 The system and fixed control volume.

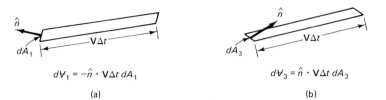

Figure 4.5 Differential volume elements.

$$dV_1 = -\hat{n} \cdot \mathbf{V}\Delta t\, dA_1$$

(a)

$$dV_3 = \hat{n} \cdot \mathbf{V}\Delta t\, dA_3$$

(b)

hand side refers to the control volume, so we can write

$$\frac{DN_{sys}}{Dt} = \lim_{\Delta t \to 0} \frac{N_{c.v.}(t + \Delta t) - N_{c.v.}(t)}{\Delta t}$$
$$+ \lim_{\Delta t \to 0} \frac{N_3(t + \Delta t) - N_1(t + \Delta t)}{\Delta t} \qquad (4.2.5)$$

The first ratio on the right-hand side is $dN_{c.v.}/dt$, where we use an ordinary derivative since we are not following specific fluid particles. Thus there results

$$\frac{DN_{sys}}{Dt} = \frac{dN_{c.v.}}{dt} + \lim_{\Delta t \to 0} \frac{N_3(t + \Delta t) - N_1(t + \Delta t)}{\Delta t} \qquad (4.2.6)$$

Now, we must find expressions for the extensive quantities $N_3(t + \Delta t)$ and $N_1(t + \Delta t)$. They, of course, depend on the mass contained in the volume elements shown in Fig. 4.4 and enlarged in Fig. 4.5. Note that the unit vector $\hat{n}$ always points out of the volume, and hence to obtain a positive differential volume a negative sign is required for region 1. Also, note that the cosine of the angle between the velocity vector and the normal vector is required,[1] thus the presence of the dot product. Referring to Fig. 4.5, we have

$$N_3(t + \Delta t) = \int_{A_3} \eta\rho\hat{n} \cdot \mathbf{V}\,\Delta t\, dA_3$$
$$N_1(t + \Delta t) = -\int_{A_1} \eta\rho\hat{n} \cdot \mathbf{V}\,\Delta t\, dA_1 \qquad (4.2.7)$$

Recognizing that A_3 plus A_1 completely surrounds the control volume, we combine the two integrals into one integral. That is,

$$N_3(t + \Delta t) - N_1(t + \Delta t) = \int_{c.s.} \eta\rho\hat{n} \cdot \mathbf{V}\,\Delta t\, dA \qquad (4.2.8)$$

[1]To obtain the volume of a box, we multiply the height by the area of the base, provided that the box is upright. If it is completely collapsed, its volume is zero. Hence for some intermediate position the volume is the height times the area of the base times the cosine of the appropriate angle.

where the control surface, denoted by c.s., is an area that completely surrounds the control volume. Substituting Eq. 4.2.8 back into Eq. 4.2.6 yields the desired result, the system-to-control-volume transformation, or equivalently, the **Reynolds transport theorem:**

$$\frac{DN_{sys}}{Dt} = \frac{d}{dt}\int_{c.v.} \eta\rho \; d\Psi + \int_{c.s.} \eta\rho\hat{n}\cdot\mathbf{V} \; dA \qquad (4.2.9)$$

The first integral represents the rate of change of the extensive property in the control volume. The second integral represents the flux of the extensive property across the control surface; it may be nonzero only where fluid crosses the control surface. We study this flux term in considerable detail in the following sections. Thus we can now express the basic laws in terms of a fixed volume in space. We will do this in subsequent sections for each of the basic laws.

We can move the time derivative of the control volume term inside the integral since the limits on the volume integral are independent of time (it is a fixed control volume) and write

$$\frac{DN_{sys}}{Dt} = \int_{c.v.} \frac{\partial}{\partial t}(\rho\eta) \; d\Psi + \int_{c.s.} \eta\rho\hat{n}\cdot\mathbf{V} \; dA \qquad (4.2.10)$$

In this form we have used $\partial/\partial t$ since ρ and η are, in general, dependent on the position variables.

4.2.1 Simplifications of the System-to-Control-Volume Transformation

Many flows of interest are steady flows, so that $\partial(\eta\rho)/\partial t = 0$. Our system-to-control-volume transformation then takes the form

$$\frac{DN_{sys}}{Dt} = \int_{c.s.} \eta\rho\hat{n}\cdot\mathbf{V} \; dA \qquad (4.2.11)$$

Furthermore, there is often only one area A_1 across which fluid enters the control volume and one area A_2 across which fluid leaves the control volume; assuming that the velocity vector is normal to the area (see Fig. 4.6), we can write $\hat{n}\cdot\mathbf{V}_1 = -V_1$ over area A_1 and $\hat{n}\cdot\mathbf{V}_2 = V_2$ over area A_2. Then Eq. 4.2.11 becomes

$$\frac{DN_{sys}}{Dt} = \int_{A_2} \eta_2\rho_2 V_2 \; dA - \int_{A_1} \eta_1\rho_1 V_1 \; dA \qquad (4.2.12)$$

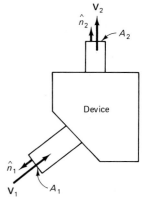

Figure 4.6 Flow into and from a device.

Finally, there are many situations that are modeled acceptably by assuming uniform properties over each plane area (see Fig. 3.9); then the equation simplifies to

$$\frac{DN_{\text{sys}}}{Dt} = \eta_2 \rho_2 V_2 A_2 - \eta_1 \rho_1 V_1 A_1 \qquad (4.2.13)$$

We will find that the system-to-control-volume transformation in this simplified form is most often used in the application of the basic laws to problems of interest in the introductory course in fluid mechanics. Some applications will, however, be included that will illustrate nonuniform distributions and unsteady flows.

If we generalize Eq. 4.2.13 to include several areas across which the fluid flows, we could write

$$\frac{DN_{\text{sys}}}{Dt} = \sum_{i=1}^{N} \eta_i \rho_i \mathbf{V}_i \cdot \hat{n}_i A_i \qquad (4.2.14)$$

where N is the number of areas. The dot product $\hat{n} \cdot \mathbf{V}$ would provide us with the appropriate sign at each area; for an inlet area, $\hat{n} \cdot \mathbf{V}$ introduces a negative sign, and for an exit area, $\hat{n} \cdot \mathbf{V}$ introduces a positive sign.

For an unsteady flow in which flow properties are assumed to be uniform throughout the control volume, the system-to-control-volume equation takes the form

$$\frac{DN_{\text{sys}}}{Dt} = V_{\text{c.v.}} \frac{d(\eta \rho)}{dt} + \eta_2 \rho_2 V_2 A_2 - \eta_1 \rho_1 V_1 A_1 \qquad (4.2.15)$$

for one inlet and one outlet with uniform properties.

4.3 CONSERVATION OF MASS

A system is a given collection of fluid particles; hence its mass remains fixed:

$$\frac{Dm_{\text{sys}}}{Dt} = \frac{D}{Dt} \int_{\text{sys}} \rho \, dV = 0 \qquad (4.3.1)$$

In Eq. 4.1.6, N_{sys} represents the mass of the system, so we simply let $\eta = 1$. Thus the conservation of mass, referring to Eq. 4.2.9, becomes

$$0 = \frac{d}{dt} \int_{\text{c.v.}} \rho \, dV + \int_{\text{c.s.}} \rho \hat{n} \cdot \mathbf{V} \, dA \qquad (4.3.2)$$

or, if we prefer,

$$0 = \int_{\text{c.v.}} \frac{\partial \rho}{\partial t} \, dV + \int_{\text{c.s.}} \rho \hat{n} \cdot \mathbf{V} \, dA \qquad (4.3.3)$$

Either of the equations above is called the **continuity equation.**

If the flow is steady, there results

$$\int_{c.s.} \rho \hat{n} \cdot \mathbf{V} \, dA = 0 \tag{4.3.4}$$

which, for a uniform flow with one entrance and one exit, takes the form

$$\rho_2 A_2 V_2 = \rho_1 A_1 V_1 \tag{4.3.5}$$

where we have used $\hat{n}_1 \cdot \mathbf{V}_1 = -V_1$ and $\hat{n} \cdot \mathbf{V}_2 = V_2$.

If the density is constant in the control volume, the derivative $\partial \rho / \partial t = 0$ even if the flow is unsteady. The continuity equation (4.3.3) then reduces to

$$A_1 V_1 = A_2 V_2 \tag{4.3.6}$$

This form of the continuity equation is used quite often, particularly with liquids and low-speed gas flows.

At this point we wish to discuss again the use of uniform velocity profiles (see also Section 3.3.1). Suppose that the velocity profiles at the entrance and the exit are not uniform, such as sketched in Fig. 4.7. Furthermore, suppose that the density is uniform over each area. Then the continuity equation takes the form

$$\rho_1 \int_{A_1} V_1 \, dA = \rho_2 \int_{A_2} V_2 \, dA \tag{4.3.7}$$

or, letting an overbar denote an average, we can write

$$\rho_1 \bar{V}_1 A_1 = \rho_2 \bar{V}_2 A_2 \tag{4.3.8}$$

where $\bar{V}_1$ and $\bar{V}_2$ are the **average velocities** at sections 1 and 2, respectively. In examples and problems the overbar is often omitted. It should be kept in mind, however, that actual velocity profiles are usually not uniform.

Before presenting some examples applying the continuity equation, two fluxes are defined that will be useful in specifying the quantity of flow.

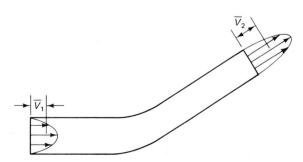

Figure 4.7 Nonuniform velocity profiles.

The **mass flux** $\dot{m}$, the mass rate of flow, is

$$\dot{m} = \int_A \rho V_n \, dA \qquad (4.3.9)$$

and has units of kg/s (slug/sec); V_n is the normal component of velocity. The **flow rate** Q, the volume rate of flow[2], is

$$Q = \int_A V_n \, dA \qquad (4.3.10)$$

and has units of m³/s (ft³/sec). The mass flux is usually used in specifying the quantity of flow for a compressible flow and the flow rate for an incompressible flow.

In terms of average velocity, we have

$$Q = A\bar{V} \qquad (4.3.11)$$

$$\dot{m} = \rho A\bar{V} \qquad (4.3.12)$$

where for the mass flux we assume a uniform density profile; we also assume that the velocity is normal to the area.

The following examples are solved by first selecting a control volume. If you study the examples carefully, you will notice that often there is only one proper choice for the control volume. We must position the inlet and exit areas at locations where the integrands are either known or where they can be approximated; also, the quantity being sought is often included at an inlet or exit area. In a few cases there may be more freedom in the selection of the control volume (Example 4.6).

This first example represents the primary use of the continuity equation. It allows us to calculate the velocity at one section if it is known at another section.

---

[2]The heat transfer rate is denoted by $\dot{Q}$, whereas the volume rate of flow is denoted by Q. These are not related quantities and should not be confused.

EXAMPLE 4.1

Water flows at a uniform velocity of 3 m/s into a nozzle that reduces the diameter from 10 cm to 2 cm. Calculate the water's velocity leaving the nozzle and the flow rate.

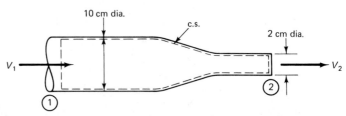

Solution

The control volume is selected to be the inside of the nozzle as shown. Flow enters the control volumn at section 1 and leaves at section 2. The simplified continuity equation (4.3.6) is used:

$$A_1 V_1 = A_2 V_2$$

$$\therefore \quad V_2 = V_1 \frac{A_1}{A_2}$$

$$= 3 \frac{\pi \times 0.1^2/4}{\pi \times 0.02^2/4} = 75 \text{ m/s}$$

The flow rate is found to be

$$Q = V_1 A_1$$

$$= 3 \times \pi \times 0.1^2/4 = 0.0236 \text{ m}^3/\text{s}$$

EXAMPLE 4.2

Water flows in and out of a device as shown. Calculate the rate of change of the mass of water (dm/dt) in the device.

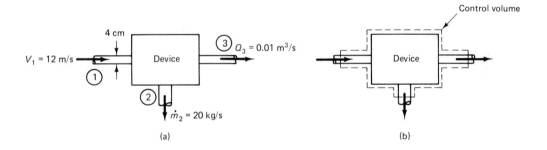

(a)   (b)

Solution

The control volume selected is shown. For the control surface surrounding the device, the continuity equation, with three surfaces across which water flows, can be written as (Eq. 4.3.2)

$$0 = \frac{d}{dt} \int_{c.v.} \rho \, d\mathcal{V} + \int_{c.s.} \rho \hat{n} \cdot \mathbf{V} \, dA$$

or

$$0 = \frac{dm}{dt} - \rho_1 A_1 V_1 + \rho_2 A_2 V_2 + \rho_3 A_3 V_3$$

where we have used $\mathbf{V}_1 \cdot \hat{n}_1 = -V_1$ since $\hat{n}_1$ points out of the volume, opposite to the direction of $\mathbf{V}_1$. In terms of the quantities given, the above can be expressed as

$$0 = \frac{dm}{dt} - \rho_1 A_1 V_1 + \dot{m}_2 + \rho_3 Q_3$$

$$= \frac{dm}{dt} - 1000 \times \pi \times 0.02^2 \times 12 + 20 + 1000 \times 0.01$$

This is solved to yield

$$\frac{dm}{dt} = -14.9 \text{ kg/s}$$

Hence, the mass is decreasing at a rate of 14.9 kg/s.

EXAMPLE 4.2 (English)

Water flows in and out of a device as shown. Calculate the rate of change of the mass of water (dm/dt) in the device.

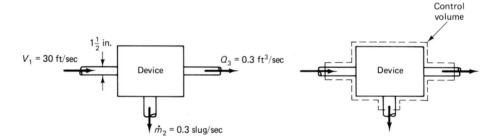

Solution

The control volume selected is shown. For the control surface surrounding the device, the continuity equation, with three surfaces across which water flows, can be written (Eq. 4.3.2)

$$0 = \frac{d}{dt} \int_{c.v.} \rho \, dV + \int_{c.s.} \rho \hat{n} \cdot \mathbf{V} \, dA$$

$$0 = \frac{dm}{dt} - \rho_1 A_1 V_1 + \rho_2 A_2 V_2 + \rho_3 A_3 V_3$$

where we have used $\mathbf{V}_1 \cdot \hat{n} = -V_1$ since $\hat{n}_1$ points out of the volume, opposite to the direction of $\mathbf{V}_1$. In terms of the quantities given, the above can be expressed as

$$0 = \frac{dm}{dt} - \rho_1 A_1 V_1 + \dot{m}_2 + \rho_3 Q_3$$

$$= \frac{dm}{dt} - 1.94 \times \pi \times \frac{0.75^2}{144} \times 30 + 0.3 + 1.94 \times 0.3$$

This is solved to yield

$$\frac{dm}{dt} = -0.1678 \text{ slug/sec}$$

Hence the mass is decreasing at the rate of 0.1678 slug/sec.

EXAMPLE 4.3

Uniform flow approaches a cylinder as shown. The velocity distribution at the location shown downstream in the wake of the cylinder is approximated by

$$u(y) = 1.25 + \frac{y^2}{4} \qquad -1 < y < 1$$

where $u(y)$ is in m/s and y is in meters. Determine the mass flux across the surface AB per meter of depth. Use $\rho = 1.23$ kg/m³.

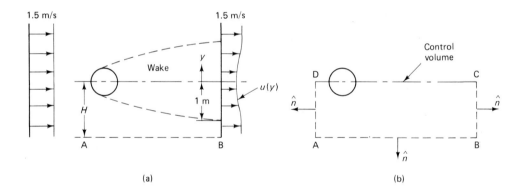

(a)   (b)

Solution

Select ABCD as the control volume. Outside the wake (a region of retarded flow) the velocity is constant at 1.5 m/s. Hence the velocity normal to plane AD is 1.5 m/s. Obviously, no mass flux crosses the surface CD. Assuming a steady flow, the continuity equation becomes

$$0 = \int_{c.s.} \rho \mathbf{V} \cdot \hat{n} \, dA$$

Mass flux occurs across three surfaces: AB, BC, and AD. Thus the equation above takes the form

$$0 = \int_{A_{AB}} \rho \mathbf{V} \cdot \hat{n} \, dA + \int_{A_{BC}} \rho \mathbf{V} \cdot \hat{n} \, dA + \int_{A_{AD}} \rho \mathbf{V} \cdot \hat{n} \, dA$$

$$= \dot{m}_{AB} + \int_0^H \rho u(y) \, dy - \rho H \times 1.5$$

where the negative sign for surface AD results from the fact that the unit vector points out of the volume to the left while the velocity vector points to the right. It is noted that a negative sign in the steady-flow continuity equation is always associated with an influx and a positive sign with an outflux. Now, we integrate out to 1 m instead of H, since the mass that enters on the left beyond 1 m simply leaves on the right with no net gain or loss. So, letting $H = 1$ m, we have

$$0 = \dot{m}_{AB} + \int_0^1 1.23\left(1.25 + \frac{y^2}{4}\right) dy - 1.23 \times 1 \times 1.5$$

There results

$$\dot{m}_{AB} = 0.205 \text{ kg/s per meter}$$

EXAMPLE 4.4

A balloon is being inflated with a water supply of 0.6 m³/s. Find the rate of growth of the radius at the instant when $R = 0.5$ m.

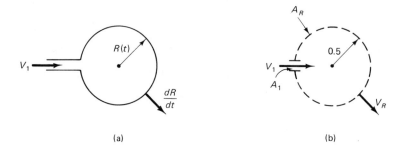

(a)   (b)

Solution

The question is to find dR/dt when the radius $R = 0.5$ m. This growth rate is the same as the water velocity normal to the wall of the balloon. Therefore, we select as our control volume a sphere with a radius of 0.5 m so that we can calculate the velocity of the water next to the surface moving radially out at $R = 0.5$ m. The continuity equation is written as

$$0 = \int_{c.v.} \frac{\partial \rho}{\partial t} dV + \int_{c.s.} \rho \mathbf{V} \cdot \hat{n} \, dA$$

The first term is zero because the density of water inside the control volume does not change in time. Further, the water crosses two areas: the inlet area A_1 with a velocity V_1 and the remainder of the sphere surface with a velocity V_R. We will assume that $A_1 \ll A_R$. The continuity equation then takes the form

$$0 = -\rho A_1 V_1 + \rho A_R V_R$$

Since the flow rate into the volume is $A_1 V_1 = 0.6$ m³/s and $A_R = 4\pi R^2$, we can solve for V_R. At $R = 0.5$ m

$$V_R = \frac{0.6}{4\pi \times 0.5^2} = 0.191 \text{ m/s}$$

$$\therefore \quad \frac{dR}{dt} = 0.191 \text{ m/s}$$

We have used a fixed control volume and allowed the moving surface of the balloon to pass through it at the instant considered. With this approach it is possible to model situations in which surfaces, such as a piston, are allowed to move.

EXAMPLE 4.5

A 1.4-m³ container is filled with pressurized air at 20°C. At time $t = 0$ air escapes out of a small 0.1-cm² tube in the side of the tank. The velocity out of the tube may be approximated by $V = \sqrt{2(p - p_{atm})/\rho}$, where p is the absolute tank pressure. Determine the time at which the pressure (gage) in the tank is 20 kPa if the initial pressure is 200 kPa (gage). Assume that the air temperature remains at 20°C.

Solution

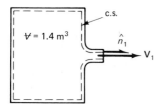

The control volume is shown with the control surface coinciding with the interior wall. The air density is uniform; therefore, the continuity equation (4.3.2) applied to this control volume is

$$\frac{d(\rho \mathcal{V})}{dt} + \rho V_1 A_1 = 0$$

or, since $\mathcal{V}$ is constant,

$$\mathcal{V} \frac{d\rho}{dt} + \rho V_1 A_1 = 0$$

The velocity is given as a function of pressure, while the density in the continuity equation may be expressed in terms of pressure using the ideal gas law $p = \rho RT$.

Thus, substituting for p,

$$\frac{\mbox{\Vcenter}\ dp}{RT\ dt} + \frac{pA_1}{RT}\sqrt{\frac{2(p - p_{atm})RT}{p}} = 0$$

This is a first-order differential equation with p as the unknown dependent variable. It can be written

$$\frac{dp}{\sqrt{p(p - p_{atm})}} = -\frac{A_1\sqrt{2RT}}{\mbox{\Vcenter}}\ dt$$

Using the initial condition $p = 200\ 000$ Pa (gage) or $p = 301\ 000$ Pa absolute, we integrate until $p = 20\ 000$ Pa (gage) or $121\ 000$ Pa absolute. Using integral tables, we find that

$$2\ \ln[\sqrt{p} + \sqrt{(p - p_{atm})}]^{121\ 000}_{301\ 000} = -\frac{A_1\sqrt{2RT}}{\mbox{\Vcenter}}\ t$$

Using $\mbox{\Vcenter} = 1.4$ m^3, $A_1 = 10^{-5}$ m^2, $T = 273 + 20 = 293$ K, $R = 287$ J/kg·K, this takes the form

$$2\ \ln(\sqrt{121\ 000} + \sqrt{20\ 000}) - 2\ \ln(\sqrt{301\ 000} + \sqrt{200\ 000}) = -0.00293t$$

Solving for t, we find that

$$t = 485\ \text{s}$$

EXAMPLE 4.6

This example shows that there may be more than one good choice for a control volume. We want to determine the rate at which the water level rises in an open container if the water coming in through a 0.10-m^2 pipe has a velocity of 0.5 m/s and the flow rate going out is 0.2 m^3/s. The container has a circular cross section with a diameter of 0.5 m.

Solution

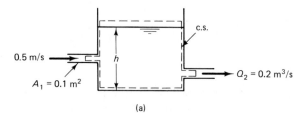

(a)

First we select a control volume that extends above the water surface as shown. Apply the continuity equation (Eq. 4.3.2)

$$\frac{d}{dt}\int_{c.v.} \rho \, d\mathbb{V} + \rho(-V_1)A_1 + \rho V_2 A_2 = 0$$

in which the first term describes the rate of change of water mass in the control volume. Hence

$$\frac{d(\rho\pi D^2 h/4)}{dt} - \rho V_1 A_1 + \rho Q_2 = 0$$

or, dividing by ρ,

$$\frac{\pi D^2}{4}\frac{dh}{dt} - V_1 A_1 + Q_2 = 0$$

The rate at which the water level is rising is then

$$\frac{dh}{dt} = \frac{V_1 A_1 - Q_2}{\pi D^2/4}$$

Thus

$$\frac{dh}{dt} = \frac{0.5 \times 0.1 - 0.2}{\pi \times 0.5^2/4} = -0.764 \text{ m/s}$$

The negative sign indicates that the water level is actually going down.
        Another choice for the control volume is one with its top surface below the water level. The velocity at the top surface is then equal to the rate at which the surface rises.

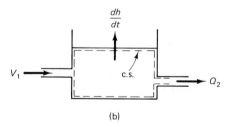

(b)

The flow conditions inside the control volume are steady. Hence we can apply Eq. 4.3.4. There are three areas across which fluid flows. On the third area, the velocity is dh/dt; hence the continuity equation takes the form

$$\rho(-V_1)A_1 + \rho Q_2 + \rho \frac{dh}{dt}\frac{\pi}{4}D^2 = 0$$

so that

$$\frac{dh}{dt} = \frac{V_1 A_1 - Q_2}{\pi D^2/4}$$

This is the same result as given above.

4.4 ENERGY EQUATION

Many problems involving fluid motion demand that the first law of thermodynamics, often referred to as the **energy equation,** be used to relate quantities of interest. If the heat transferred to a device, or the work done by a device, is desired, the energy equation is obviously needed. It is also used to relate pressures and velocities when Bernoulli's equation is not applicable; this is the case whenever viscous effects cannot be neglected. Let us express the energy equation in control volume form. For a system it is

$$\dot{Q} - \dot{W} = \frac{D}{Dt} \int_{sys} e\rho \, d\Psi \tag{4.4.1}$$

where the specific energy e includes specific kinetic energy $V^2/2$, specific potential energy gz, and specific internal energy $\tilde{u}$; that is,

$$e = \frac{V^2}{2} + gz + \tilde{u} \tag{4.4.2}$$

We will not include other forms of energy, such as energy due to magnetic or electric field-flow field interactions or those due to chemical reactions. In terms of a control volume, Eq. 4.4.1 becomes

$$\dot{Q} - \dot{W} = \frac{d}{dt} \int_{c.v.} e\rho \, d\Psi + \int_{c.s.} \rho e \mathbf{V} \cdot \hat{n} \, dA \tag{4.4.3}$$

This can be put in simplified forms for certain restricted flows, but first let us discuss the rate-of-heat transfer term $\dot{Q}$ and the work-rate term $\dot{W}$.

The term $\dot{Q}$ represents the rate-of-energy transfer across the control surface due to a temperature difference. The rate-of-heat transfer term is either given or results from using Eq. 4.4.3. The calculation of $\dot{Q}$ is the objective of a course in heat transfer. The work-rate term is discussed in detail in the following section.

4.4.1 Work-Rate Term

The work-rate term results from work being done by the system. Or, since we consider the instant that the system occupies the control volume, we can also state that the work-rate term results from work being done by the control volume. Work is due to a force moving through a distance while it acts on the control volume. The rate of doing work $\dot{W}$, or power, is given by the dot product of a force $\mathbf{F}$ with its velocity:

$$\dot{W} = -\mathbf{F} \cdot \mathbf{V}_I \qquad (4.4.4)$$

where $\mathbf{V}_I$ is the velocity measured with respect to a fixed reference frame. The negative sign results because we suppose that the force is acting on the control volume. A positive force would act on the surroundings.

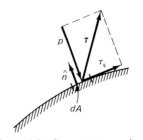

Figure 4.8   Stress vector acting on the control surface.

If the force results from a variable stress acting over the control surface, we must integrate,

$$\dot{W} = -\int_{\text{c.s.}} \boldsymbol{\tau} \cdot \mathbf{V}_I \, dA \qquad (4.4.5)$$

where $\boldsymbol{\tau}$ is the stress vector acting on the elemental area dA, the differential force being represented by $d\mathbf{F} = \boldsymbol{\tau} \, dA$, as shown in Fig. 4.8.

For a moving control volume, such as a car, we have to evaluate the velocity with respect to a fixed reference frame. For example, let us consider an automobile traveling at constant speed (see Example 4.10). If we want to apply the energy equation, we could make the car the control volume. In that case, the velocity in Eq. 4.4.4 would be measured relative to a fixed reference and not relative to the car. If the velocity relative to the car were used, the drag force would have a zero velocity which would result in no work done; but we know that at high speed, energy from the gasoline goes to overcome the drag. Thus the stationary reference frame is needed.

In general, for moving control volumes the velocity vector $\mathbf{V}_I$ is related to a relative velocity $\mathbf{V}$, observed in a reference frame attached to the control volume by

$$\mathbf{V}_I = \mathbf{V} + \dot{\mathbf{S}} + \boldsymbol{\omega} \times \mathbf{r} \qquad (4.4.6)$$

where $\dot{\mathbf{S}}$ is the velocity of the control volume (see Fig. 3.5). We can now write the work rate as

$$\dot{W} = -\int \boldsymbol{\tau} \cdot \mathbf{V} \, dA + \dot{W}_I \qquad (4.4.7)$$

where the "inertial work-rate" term is given by

$$\dot{W}_I = -\int_{\text{c.s}} \boldsymbol{\tau} \cdot (\dot{\mathbf{S}} + \boldsymbol{\omega} \times \mathbf{r}) \, dA \qquad (4.4.8)$$

Next, express the stress vector as the sum of a normal component and a shear component, that is,

$$\boldsymbol{\tau} = -p\hat{n} + \boldsymbol{\tau}_s \qquad (4.4.9)$$

where the pressure p is assumed to be positive in a compressive state, as shown in Fig. 4.8. Then

$$\dot{W} = \int_{c.s.} p\hat{n} \cdot \mathbf{V} \, dA - \int_{c.s.} \boldsymbol{\tau}_s \cdot \mathbf{V} \, dA + \dot{W}_I \qquad (4.4.10)$$

We will allow the shear stress term to consist of two parts. One part accounts for work transmitted by rotating shafts, called **shaft work** $\dot{W}_S$; this term is important when we deal with flows in pumps and turbines; the other will be denoted **shear work** $\dot{W}_{shear}$ and results from moving boundaries; this term is required if the control surface itself moves relative to the control volume as occurs with a moving belt.

Hence the work-rate term becomes

$$\dot{W} = \int_{c.s.} p\hat{n} \cdot \mathbf{V} \, dA + \dot{W}_S + \dot{W}_{shear} + \dot{W}_I \qquad (4.4.11)$$

The terms are summarized as follows:

$\int p\hat{n} \cdot \mathbf{V} \, dA$ work rate resulting from the force due to pressure moving at the control surface. It is often referred to as **flow work.**

$\dot{W}_S$ work rate resulting from rotating shafts such as that of a pump or turbine, or the equivalent electric power.

$\dot{W}_{shear}$ work rate due to a moving boundary such as a moving belt.

$\dot{W}_I$ work rate that occurs when the control volume moves relative to a fixed reference frame.

We should note that the work-rate terms $\dot{W}_{shear}$ and $\dot{W}_I$ are seldom encountered in problems in an introductory course and are often omitted from textbooks. They are included here for completeness.

4.4.2 General Energy Equation

When the work-rate term of Eq. 4.4.11 is substituted into Eq. 4.4.3, we obtain the energy equation in the form

$$\dot{Q} - \dot{W}_S - \dot{W}_{shear} - \dot{W}_I = \frac{d}{dt} \int_{c.v.} e\rho \, d\mathbf{V} + \int_{c.s.} \left(e + \frac{p}{\rho} \right) p\hat{n} \cdot \mathbf{V} \, dA$$

$$(4.4.12)$$

Note that the flow work term has been moved to the right-hand side and is treated like an energy flux term.

Substitution of Eq. 4.4.2 and rearrangement of the terms results in

$$\dot{Q} - \dot{W}_S - \dot{W}_{shear} - \dot{W}_I = \frac{d}{dt} \int_{c.v.} \left(\frac{V_I^2}{2} + gz + \tilde{u}\right)\rho\, d\Psi$$
$$+ \int_{c.s.} \left(\frac{V_I^2}{2} + gz + \tilde{u} + \frac{p}{\rho}\right)\rho\mathbf{V}\cdot\hat{n}\, dA \qquad (4.4.13)$$

This general form of the energy equation is useful in analyzing fluid flow problems that may include time-dependent effects and nonuniform profiles. Before we simplify the equation for steady flow and uniform profiles, let us introduce the notion of "losses."

In many fluid flows, useful forms of energy (kinetic energy and potential energy) and flow work are converted into unusable energy forms (internal energy or heat transfer). If we assume that the temperature of the control volume remains unchanged, the internal energy does not change and the losses are balanced by heat transfer across the control surface. This heat transfer can be a result of convection, radiation, or conduction at the control surfaces. The theory of heat transfer is aimed at the detailed description of these effects. However, in an introductory fluid mechanics course the sum of these effects are lumped together and denoted $\dot{Q}$. Thus we define **losses** as the sum of all the terms representing unusable forms of energy:

$$\text{losses} = -\dot{Q} + \frac{d}{dt}\int_{c.v.} \tilde{u}\rho\, dV + \int_{c.s.} \tilde{u}\rho\, \mathbf{V}\cdot\hat{n}\, dA \qquad (4.4.14)$$

We can now rewrite the energy equation as

$$-\dot{W}_S - \dot{W}_{shear} - \dot{W}_I = \frac{d}{dt} \int_{c.v.} \left(\frac{V_I^2}{2} + gz\right)\rho\, d\Psi$$
$$+ \int_{c.s.} \left(\frac{V_I^2}{2} + gz + \frac{p}{\rho}\right)\rho\mathbf{V}\cdot\hat{n}\, dA + \text{losses} \qquad (4.4.15)$$

Losses are due to two primary effects:

1. Viscosity causes internal friction that results in increased internal energy (temperature increase) or heat transfer.
2. Changes in geometry result in separated flows that require useful energy to maintain the resulting secondary motions that are generated.

In a conduit, the losses due to viscous effects are distributed over the entire length, whereas the loss due to a geometry change (a valve, an elbow, an enlargement) is concentrated in the vicinity of the geometry change.

It turns out that the analytical calculation of losses is rather difficult, particularly when the flow is turbulent. In general, the prediction of losses is based on empirical formulas. Such formulas will be given in subsequent chapters. In this chapter we discuss losses qualitatively and, in examples and problems, losses will be given. For a pump or a turbine the losses are expressed in terms of the efficiency. For example, if the efficiency of a pump is 80%, the losses would be 20% of the energy input to the pump.

It may be that the objective in a particular fluid flow is to change the internal energy of the fluid, such as in the steam generator (boiler) of a power plant, by the transfer of heat; then the definition of losses above must be altered so that the internal energy terms are retained and the loss term includes only the dissipative effects of the viscosity of the fluid. Generally, for problems of interest in fluid mechanics, Eq. 4.4.15 is acceptable.

4.4.3 Steady, Uniform Flow

Consider a steady-flow situation in which there is one entrance and one exit across which uniform profiles can be assumed. Also, assume that $\dot{W}_{shear} = \dot{W}_I = 0$. For such a flow the term $(V^2/2 + gz + p/\rho)$ in Eq. 4.4.15 is constant across the cross section because V is constant (we assume a uniform velocity profile) and the sum of $p/\rho + gz$ is constant if the streamlines at each section are parallel. The energy equation (Eq. 4.4.15) then simplifies to

$$-\dot{W}_S = \rho_2 V_2 A_2 \left(\frac{V_2^2}{2} + \frac{p_2}{\rho_2} + gz_2 \right)$$
$$- \rho_1 V_1 A_1 \left(\frac{V_1^2}{2} + \frac{p_1}{\rho_1} + gz_1 \right) + \text{losses} \tag{4.4.16}$$

where the subscripts 1 and 2 refer to the entrance and exit, respectively. The mass flux is given by $\dot{m} = \rho_1 A_1 V_1 = \rho_2 A_2 V_2$. After dividing by $\dot{m}g$, we have

$$-\frac{\dot{W}_S}{\dot{m}g} = \frac{V_2^2 - V_1^2}{2g} + \frac{p_2}{\gamma_2} - \frac{p_1}{\gamma_1} + z_2 - z_1 + h_L \tag{4.4.17}$$

where we have introduced the **head loss** h_L, defined to be

$$h_L = -\frac{\dot{Q}}{\dot{m}g} + \frac{\bar{u}_2 - \bar{u}_1}{g} \tag{4.4.18}$$

It is often written in terms of a **loss coefficient** K as

$$h_L = K \frac{V^2}{2g} \qquad (4.4.19)$$

where V may be either V_1 or V_2; if it is not obvious, it will be specified. Loss coefficients are discussed in some detail in Chapter 7 and are tabulated in Table 7.2.

The head loss is referred to as a "head" since it has dimensions of length. We could also refer to $V^2/2g$ as the **velocity head** and p/γ as the **pressure head** since those terms also have dimensions of length. Also recall from Chapter 3 that $p/\gamma + z$ is called the **piezometric head**. Further, the sum of the piezometric head and the velocity head is called the **total head**.

The energy equation, in the form of Eq. 4.4.17, is useful in many applications and is, perhaps, the most often used form of the energy equation. If the losses are negligible, if there is no shaft work and if the flow is incompressible, we note that the energy equation takes the form

$$\frac{V_2^2}{2g} + \frac{p_2}{\gamma} + z_2 = \frac{V_1^2}{2g} + \frac{p_1}{\gamma} + z_1 \qquad (4.4.20)$$

Observe that the energy equation has been reduced to a form identical with Bernoulli's equation. We must remember, however, that Bernoulli's equation is a momentum equation applicable along a streamline and the equation above is an energy equation applied between two sections of a flow. It is not surprising that both should predict identical results from the conditions stated because the velocity head is constant over a cross section and the sum of pressure head and elevation remains constant over a cross section.

The energy equation (4.4.17) may be applied to any steady, uniform flow with one entrance and one exit. The control volume is usually selected such that the entrance and exit sections have a uniform total head. For example, it may be applied to water flow through a long pipeline; the total head at the entrance and exit may then be evaluated conveniently at the center of the pipe entrance and exit. The energy equation may be applied to the flow passing a gate (Fig. 4.9). An appropriate control volume is shown. The total head at the entrance and exit can be evaluated at any point at the entrance and exit, respectively. However, a convenient choice would be the points at the water surface. Thus the energy equation becomes

$$\frac{V_1^2}{2g} + \cancel{\frac{p_1}{\gamma}}^{0} + h_1 = \frac{V_2^2}{2g} + \cancel{\frac{p_2}{\gamma}}^{0} + h_2 + h_L \qquad (4.4.21)$$

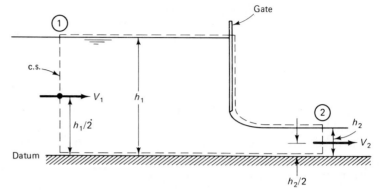

Figure 4.9 Application of the energy equation to a gate in an open channel.

where the shaft work has been set to zero. If we had picked the centroids of the entrance and exit, as shown in Fig. 4.9, we would have obtained

$$\frac{V_1^2}{2g} + \frac{p_1}{\gamma} + \frac{h_1}{2} = \frac{V_2^2}{2g} + \frac{p_2}{\gamma} + \frac{h_2}{2} + h_L \qquad (4.4.22)$$

We see that this result is the same as in Eq. 4.4.21 if we substitute $p_1 = h_1\gamma/2$ and $p_2 = h_2\gamma/2$. For completeness, it is noted that the losses between 1 and 2 in Fig. 4.9 could be neglected because internal viscous effects act only over a relatively short distance and no significant secondary flows are generated.

The energy equation (4.4.15) can be applied to any control volume. For example, the energy equation for steady, uniform incompressible flow through a T-section in a pipe (Fig. 4.10) in which there is one entrance and two exits can be written for the mass flux that exits section 2 and for the mass flux that exits section 3:

$$\frac{V_1^2}{2g} + \frac{p_1}{\gamma} + z_1 = \frac{V_2^2}{2g} + \frac{p_2}{\gamma} + z_2 + h_{L_{1-2}}$$

$$\qquad\qquad (4.4.23)$$

$$\frac{V_1^2}{2g} + \frac{p_1}{\gamma} + z_1 = \frac{V_3^2}{2g} + \frac{p_3}{\gamma} + z_3 + h_{L_{1-3}}$$

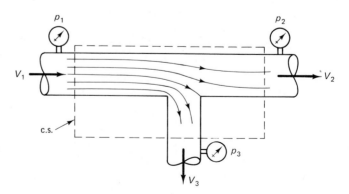

Figure 4.10 Application of the energy equation to a T-section.

where the loss terms include the losses between the inlet and the respective exits. If the losses are negligible, the energy equation reduces to a form similar to the Bernoulli equation being applied along a streamline going from 1 to 2 or a streamline going from 1 to 3.

A final note to this section regards nomenclature for pumps and turbines in a flow system. It is often conventional to call the energy term $(\dot{W}_S/\dot{m}g)$ associated with a pump the **pump head** H_P, and the term $(\dot{W}_S/\dot{m}g)$ associated with a turbine the **turbine head** H_T. Then the energy equation takes the form

$$H_P + \frac{V_1^2}{2g} + \frac{p_1}{\gamma} + z_1 = H_T + \frac{V_2^2}{2g} + \frac{p_2}{\gamma} + z_2 + h_L \qquad (4.4.24)$$

In this form we have equated the energy at the inlet plus added energy to the energy at the exit plus extracted energy (energy per unit weight, of course). If any of the quantities is zero (e.g., there is no pump), the appropriate term is simply omitted.

In some situations the turbine head or the pump head is known. If this is the case, the power generated by the turbine with an efficiency of η_T is simply

$$\dot{W}_T = \dot{m}gH_T\eta_T = Q\gamma H_T\eta_T \qquad (4.4.25)$$

The power requirement by a pump with an efficiency of η_P would be

$$\dot{W}_P = \frac{\dot{m}gH_P}{\eta_P} = \frac{Q\gamma H_P}{\eta_P} \qquad (4.4.26)$$

We will calculate power in watts, ft-lb/sec, or horsepower. Recall that one horsepower is equivalent to 746 W or 550 ft-lb/sec.

4.4.4 Steady, Nonuniform Flow

If the assumption of uniform velocity profiles is not acceptable for a problem of interest, as is sometimes the situation, we have to consider the control-surface integral in Eq. 4.4.15 with the proper expression for the velocity distribution. In practice, a velocity distribution can be accounted for by introducing the **kinetic-energy correction factor** α, defined by

$$\alpha = \frac{\int V^3 \, dA}{\bar{V}^3 A} \qquad (4.4.27)$$

where $\bar{V}$ is the average velocity over the area A. Then the term that accounts for the flux of kinetic energy is

$$\rho \int_A V^3 \, dA = \alpha \rho \bar{V}^3 A \qquad (4.4.28)$$

Using this factor, we can account for nonuniform velocity distributions by modifying Eq. 4.4.24 to read

$$H_P + \alpha_1 \frac{\bar{V}_1^2}{2g} + \frac{p_1}{\gamma} + z_1 = H_T + \alpha_2 \frac{\bar{V}_2^2}{2g} + \frac{p_2}{\gamma} + z_2 + h_L \quad (4.4.29)$$

where $\bar{V}_1$ and $\bar{V}_2$ are the average velocities at sections 1 and 2, respectively. For a flow with a parabolic profile in a pipe we can calculate $\alpha =$ 2.0 (see Example 4.9). For most internal turbulent flows, however, the profile is nearly uniform with $\alpha \simeq 1.05$. Hence we simply let $\alpha = 1$ since it is so close to unity; this will always be done unless otherwise stated.

EXAMPLE 4.7

Water flows from a reservoir through a 0.8-m-diameter pipeline to a turbine-generator unit and exits to a river that is 30 m below the reservoir surface. If the flow rate is 3 m³/s and the turbine-generator efficiency is 80%, calculate the power output. Assume the loss coefficient in the pipeline (including the exit) to be $K = 2$.

Solution

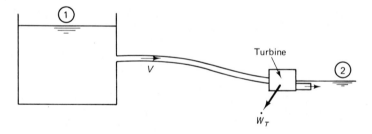

The control volume to be used extends from section 1 to section 2; we consider the water surface of the left reservoir to be the entrance and the water surface of the river to be the exit. Because we assume the water surfaces to be large, the velocities at the surfaces are negligible. The velocity in the pipe is

$$V = \frac{Q}{A}$$

$$= \frac{3}{\pi \times 0.8^2/4} = 5.968 \text{ m/s}$$

Now, consider the energy equation. We will use gage pressures so that $p_1 = p_2 = 0$; the datum is placed through the lower section 2 so that $z_2 = 0$; the velocities V_1 and V_2 are negligibly small; K is assumed to be based on the 0.8-m-diameter pipe velocity. The energy equation (4.4.24) then becomes

$$\cancel{H_P}^{0} + \cancel{\frac{V_1^2}{2g}}^{0} + \cancel{\frac{p_1}{\gamma}}^{0} + z_1 = H_T + \cancel{\frac{V_2^2}{2g}}^{0} + \cancel{\frac{p_2}{\gamma}}^{0} + \cancel{z_2}^{0} + K\frac{V^2}{2g}$$

$$30 = H_T + 2\frac{5.968^2}{2 \times 9.81}$$

$$\therefore H_T = 26.4 \text{ m}$$

From this the power output is found, using Eq. 4.4.25, to be

$$\dot{W}_T = Q\gamma H_T \eta_T$$

$$= 3 \times 9810 \times 26.4 \times 0.8 = 622\,000 \text{ W} \quad \text{or} \quad 622 \text{ kW}$$

In this example we have used gage pressure; the potential-energy datum was assumed to be placed through section 2, V_1 and V_2 were assumed to be insignificantly small, and K was assumed to be based on the 0.8 m-diameter pipe velocity.

EXAMPLE 4.7 (English)

Water flows from a reservoir through a 2.5 ft-diameter pipeline to a turbine-generator unit and exits to a river that is 100 ft below the reservoir surface. If the flow rate is 90 ft³/sec, and the turbine-generator efficiency is 80%, calculate the power output. Assume the loss coefficient in the pipeline (including the exit) to be $K = 2$.

Solution

Referring to the figure in Example 4.7, we select the control volume to extend from section 1 to section 2 on the reservoir and river surfaces, where we know the velocities, pressures, and elevations; we consider the water surface of the left reservoir to be the entrance and the water surface of the river to be the exit. The velocity in the pipe is

$$V = \frac{Q}{A}$$

$$= \frac{90}{\pi \times 2.5^2/4} = 18.3 \text{ ft/sec}$$

Now, consider the energy equation. We will use gage pressures so that $p_1 = p_2 = 0$; the datum is placed through the lower section 2 so that $z_2 = 0$; the velocities V_1 and V_2 are negligibly small; K is assumed to be based on the 2.5-ft-diameter pipe velocity. The energy equation (4.4.24) then becomes

$$\cancel{H_P}^{0} + \cancel{\frac{V_1^2}{2g}}^{0} + \cancel{\frac{p_1}{\gamma}}^{0} + z_1 = H_T + \cancel{\frac{V_2^2}{2g}}^{0} + \cancel{\frac{p_2}{\gamma}}^{0} + \cancel{z_2}^{0} + K\frac{V^2}{2g}$$

$$100 = H_T + 2\,\frac{18.3^2}{2 \times 32.2}$$

$$\therefore H_T = 89.6 \text{ ft}$$

From this the power output is found to be

$$\dot{W}_T = Q\gamma H_T \eta_T$$

$$= 90 \times 62.4 \times 89.6 \times 0.8 = 403\,000 \text{ ft-lb/sec} \quad \text{or} \quad 733 \text{ hp}$$

In this example we have used gage pressure; the potential-energy datum was assumed to be placed through section 2, V_1 and V_2 were assumed to be insignificantly small, and K was assumed to be based on the 2.5 ft-diameter pipe velocity.

EXAMPLE 4.8

A venturi meter reduces the pipe diameter from 10 cm to 5 cm. Calculate the flow rate and the mass flux assuming ideal conditions.

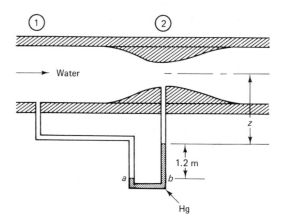

Solution

The control volume is selected such that the entrance and exit correspond to the sections where the pressure information of the manometer can be applied. The manometer's reading is interpreted as follows:

$$p_1 + \gamma(z + 1.2) = p_2 + \gamma z + 13.6\gamma \times 1.2$$

where z is the distance from the pipe centerline to the top of the mercury column. The manometer then gives

$$\frac{p_1 - p_2}{\gamma} = 15.12 \text{ m}$$

Continuity allows us to relate V_2 to V_1 as

$$V_1 A_1 = V_2 A_2$$

$$\therefore \quad V_2 = 4V_1$$

The energy equation with no losses takes the form

$$0 = \frac{V_2^2 - V_1^2}{2g} + \frac{p_2 - p_1}{\gamma} + (z_2 \!\!\!\!\diagup\!\!\!\! z_1)^0$$

$$= \frac{16V_1^2 - V_1^2}{2g} - 15.12$$

$$\therefore \quad V_1 = 4.45 \text{ m/s}$$

The flow rate is

$$Q = A_1 V_1$$

$$= \pi \times 0.05^2 \times 4.45 = 0.0350 \text{ m}^3/\text{s}$$

The mass flux is

$$\dot{m} = \rho Q$$

$$= 1000 \times 0.035 = 35.0 \text{ kg/s}$$

EXAMPLE 4.9

The velocity distribution for a certain flow in a pipe is $V(r) = V_{\text{max}} (1 - r^2/r_0^2)$, where r_0 is the pipe radius, as shown. Determine the kinetic-energy correction factor.

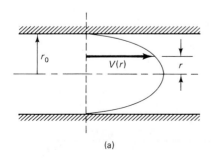

(a)

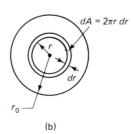

(b)

Solution

To find the kinetic-energy correction factor α, we must know the average velocity. It is

$$\bar{V} = \frac{\int V \, dA}{A}$$

$$= \frac{1}{\pi r_0^2} \int_0^{r_0} V_{\text{max}} \left(1 - \frac{r^2}{r_0^2}\right) 2\pi r \, dr = \frac{2\pi V_{\text{max}}}{\pi r_0^2} \int_0^{r_0} \left(r - \frac{r^3}{r_0^2}\right) dr$$

$$= \frac{2 V_{\text{max}}}{r_0^2} \left(\frac{r_0^2}{2} - \frac{r_0^4}{4r_0^2}\right) = \frac{1}{2} V_{\text{max}}$$

Using Eq. 4.4.27, there results

$$\alpha = \frac{\int V^3 \, dA}{\bar{V}^3 A}$$

$$= \frac{\int_0^{r_0} V_{\text{max}}^3 (1 - r^2/r_0^2)^3 \cdot 2\pi r \, dr}{(\tfrac{1}{2} V_{\text{max}})^3 \pi r_0^2} = \frac{16}{r_0^2} \int_0^{r_0} \left(1 - \frac{3r^2}{r_0^2} + \frac{3r^4}{r_0^4} - \frac{r^6}{r_0^6}\right) r \, dr$$

$$= \frac{16}{r_0^2} \left(\frac{r_0^2}{2} - \frac{3r_0^2}{4} + \frac{3r_0^2}{6} - \frac{r_0^2}{8}\right) = 2$$

Consequently, the kinetic energy flux associated with a parabolic velocity distribution across a circular area is given by

$$\int \rho \mathbf{V} \cdot \hat{n} \, \frac{V^2}{2} \, dA = 2 \frac{\dot{m} \bar{V}^2}{2}$$

EXAMPLE 4.10

The drag force on an automobile is approximated by the expression $0.15 \rho V_\infty^2 A$, where A is the projected cross-sectional area and V_∞ is the automobile's speed. If $A = 1.2 \text{ m}^2$, calculate the efficiency η of the engine if the rate of fuel consumption $\dot{f}$ (the gas mileage) is 15 km/L and the automobile travels at 90 km/h. Assume that the fuel releases 44 000 kJ/kg during combustion. Neglect the energy contained in the exhaust gases and assume that the only resistance to motion is the drag force. Use $\rho_{\text{air}} = 1.12 \text{ kg/m}^3$ and $\rho_f = 0.93 \text{ kg/L}$.

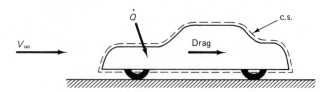

Solution

If the car is taken as the moving control volume (note that the control volume is fixed), as shown, we can simplify the energy equation (Eq. 4.4.3 in combination with 4.4.11) to

$$\dot{Q} - \dot{W}_I = 0$$

since all other terms are negligible; there is no velocity crossing the control volume, so $\mathbf{V} \cdot \hat{n} = 0$ (neglect the energy of the exhaust gases); there is no shear or shaft work; the energy of the c.v. remains constant. The energy input $\dot{Q}$ which accomplishes useful work is η times the energy released during combustion, that is,

$$\dot{Q} = \dot{m}_f \times 44\ 000\eta \text{ kJ/s}$$

where $\dot{m}_f$ is the mass flux of the fuel. The mass flux of fuel is determined knowing the rate of fuel consumption $\dot{f}$ and the density of fuel as 0.93 kg/L, as follows:

$$\dot{f} = \frac{\text{distance}}{\text{volume}} = \frac{V_\infty \times \text{time}}{Q \times \text{time}} = \frac{V_\infty}{\dot{m}_f/\rho_f} = \frac{\rho_f V_\infty}{\dot{m}_f}$$

with $V_\infty = 90\ 000/3600 = 25$ m/s, we have, using $\dot{f} = 15 \times 1000$ m/L,

$$15 \times 1000 = \frac{0.93 \times 25}{\dot{m}_f}$$

$$\therefore \dot{m}_f = 0.00155 \text{ kg/s}$$

The inertial work-rate term is

$$\dot{W}_I = V_\infty \times \text{drag}$$

$$= 0.15\rho V_\infty^3 A = 0.15 \times 1.12 \times 25^3 \times 1.2 = 3150 \text{ J/s}$$

Equating $\dot{Q} = \dot{W}_I$, we have

$$44\ 000\eta \times 0.00155 = 3.15$$

$$\therefore \eta = 0.0462 \quad \text{or} \quad 4.62\%$$

This is obviously a very low percentage, perhaps surprisingly low to the reader. Very little power (3.15 kJ/s = 4.22 hp) is actually needed to propel the automobile at 90 km/h. The relatively large engine, needed primarily for acceleration, is quite inefficient when simply propelling the automobile. Note the importance of using a stationary reference frame. The reference frame attached to the automobile is an inertial reference frame since it is moving at constant velocity. Yet the energy equation demands a stationary reference frame so that the energy required by the drag force can properly be included.

4.5 MOMENTUM EQUATION

4.5.1 General Equation

Newton's second law, often called the **momentum equation,** states that the resultant force acting on a system equals the rate of change of momentum of the system when measured in an inertial reference frame; that is,

$$\Sigma \, \mathbf{F} = \frac{D}{Dt} \int_{\text{sys}} \rho \mathbf{V} \, d\Psi \qquad (4.5.1)$$

Using Eq. 4.2.9, with η replaced by $\mathbf{V}$, this is written for a control volume as

$$\Sigma \, \mathbf{F} = \frac{d}{dt} \int_{\text{c.v.}} \rho \mathbf{V} \, d\Psi + \int_{\text{c.s.}} \rho \mathbf{V}(\mathbf{V} \cdot \hat{n}) \, dA \qquad (4.5.2)$$

where the quantity in parentheses is simply a scalar for each differential area dA.

When applying Newton's second law the quantity $\Sigma \, \mathbf{F}$ represents all forces acting on the control volume. The forces include the surface forces resulting from the surroundings acting on the control surface and body forces that result from gravity and magnetic fields. The momentum equation is often used to determine the forces induced by the flow. For example, the equation allows us to calculate the force on the support of an elbow in a pipeline or the force on a submerged body in a free-surface flow.

When we apply the momentum equation the surrounding fluid and sometimes the entire conduit or container is separated from the control volume. For example, in the horizontal nozzle of Fig. 4.11a, the nozzle

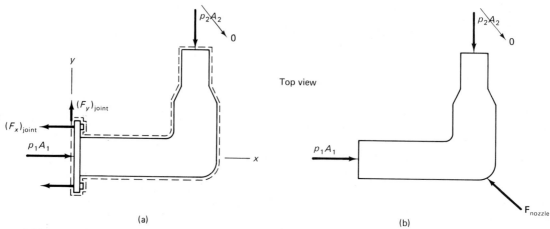

(a)   (b)

Figure 4.11   Forces acting on the control volume of a horizontal nozzle: (a) control volume includes nozzle and fluid in nozzle; (b) control volume includes fluid in nozzle only.

and the fluid in the nozzle are isolated. Thus care must be taken to include the pressure forces shown and the force $\mathbf{F}_{\text{joint}}$. It is convenient to use gage pressures so that the pressure acting on the exterior of the pipe is then zero, and its effect can be ignored. Alternatively, we could have selected a control volume that includes only the fluid in the nozzle (Fig. 4.11b). In that case we have to consider the two pressure forces at the entrance and exit and the resultant pressure force $\mathbf{F}_{\text{nozzle}}$ of the interior wall of the nozzle on the fluid. Of course, the force $\mathbf{F}_{\text{joint}}$ and $\mathbf{F}_{\text{nozzle}}$ are equal, as is obvious from a free body of the nozzle excluding the fluid. If the problem is to determine the force exerted by the flow on the nozzle (Fig. 4.11a), we have to reverse the direction of the calculated force $\mathbf{F}_{\text{nozzle}}$. Examples at the end of this section illustrate this.

4.5.2 Steady Uniform Flow

Equation 4.5.2 can be simplified considerably if a device has entrances and exits across which the velocity may be assumed to be uniform and if the flow is steady. Then there results

$$\Sigma\,\mathbf{F} = \sum_{i=1}^{N} \rho_i A_i \mathbf{V}_i (\mathbf{V}_i \cdot \hat{n}) \qquad (4.5.3)$$

where N is the number of flow exit/entrance areas.

At an entrance $\mathbf{V} \cdot \hat{n} = -V$ since the unit vector points out of the volume and at the exit $\mathbf{V} \cdot \hat{n} = V$. If there is only one entrance and one exit, as in Fig. 4.11, the momentum equation becomes

$$\Sigma\,\mathbf{F} = \rho_2 A_2 V_2 \mathbf{V}_2 - \rho_1 A_1 V_1 \mathbf{V}_1 \qquad (4.5.4)$$

Using continuity,

$$\dot{m} = \rho_1 A_1 V_1 = \rho_2 A_2 V_2 \qquad (4.5.5)$$

the momentum equation takes the simplified form

$$\Sigma\,\mathbf{F} = \dot{m}(\mathbf{V}_2 - \mathbf{V}_1) \qquad (4.5.6)$$

Note that the momentum equation is a vector equation. If we consider the nozzle of Fig. 4.11a and we want to determine the x-component of the force of the joint on the nozzle, $(\mathbf{V}_1)_x = V_1$ and $(\mathbf{V}_2)_x = 0$ the momentum equation for x-direction becomes

$$\Sigma\,F_x = -(F_x)_{\text{joint}} + p_1 A_1 = -\dot{m} V_1 \qquad (4.5.7)$$

Similarly, we could write the y-component equation and find an expression for $(F_y)_{\text{joint}}$.

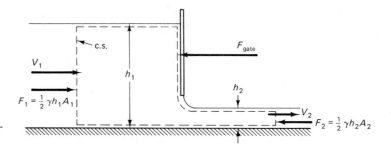

Figure 4.12 Force of the flow on a gate in a free-surface flow.

An example of a free-surface flow in a rectangular channel is shown in Fig. 4.12. If we want to determine the force of the gate on the flow, the following expression can be derived from the momentum equation:

$$\Sigma F_x = -F_{\text{gate}} + F_1 - F_2 = \dot{m}(V_2 - V_1) \tag{4.5.8}$$

where F_1 and F_2 are pressure forces (see Fig. 4.12).

4.5.3 Steady Nonuniform Flow

If we cannot assume uniform velocity profiles, we can let

$$\int_A V^2 \, dA = \beta \bar{V}^2 A \tag{4.5.9}$$

where we have introduced the **momentum-correction factor** β, expressed explicitly as

$$\beta = \frac{\int V^2 \, dA}{\bar{V}^2 A} \tag{4.5.10}$$

The momentum equation, for a steady flow, can then be written as

$$\Sigma \mathbf{F} = \sum_{i=1}^{N} \rho_i \beta_i A_i \mathbf{V}_i (\mathbf{V}_i \cdot \hat{n}) \tag{4.5.11}$$

For a laminar flow with a parabolic profile in a circular pipe, $\beta = \frac{4}{3}$. If a profile is given, however, it is usually simply integrated using Eq. 4.5.2.

4.5.4 Noninertial Reference Frames

In certain situations it may be necessary to choose a noninertial reference frame in which the velocity is measured. This would be the case if we were to study the flow through a dishwasher arm, around a turbine blade, or from a rocket. Relative to a noninertial reference frame, Newton's second law takes the form (refer to Eq. 3.2.15)

$$\Sigma \mathbf{F} = \frac{D}{Dt} \int_{\text{sys}} \rho \mathbf{V} \, d\mathbb{V}$$

$$+ \int_{\text{sys}} \left[\frac{d^2 \mathbf{S}}{dt^2} + 2\boldsymbol{\omega} \times \mathbf{V} + \boldsymbol{\omega} \times (\boldsymbol{\omega} \times \mathbf{r}) + \frac{d\boldsymbol{\omega}}{dt} \times \mathbf{r} \right] \rho \, d\mathbb{V}$$

$$(4.5.12)$$

where $\mathbf{V}$ is the velocity relative to the noninertial frame and where the acceleration $\mathbf{a}$ of each particle in the system is already accounted for in the first integral. Equation 4.5.12 is often written as

$$\Sigma \mathbf{F} - \mathbf{F}_I = \frac{D}{Dt} \int_{\text{sys}} \rho \mathbf{V} \, d\mathbb{V}$$

$$= \frac{d}{dt} \int_{\text{c.v.}} \rho \mathbf{V} \, d\mathbb{V} + \int_{\text{c.s.}} \rho \mathbf{V}(\mathbf{V} \cdot \hat{n}) \, dA$$

$$(4.5.13)$$

where $\mathbf{F}_I$ is called the "inertial body force," given by

$$\mathbf{F}_I = \int_{\text{sys}} \left[\frac{d^2 \mathbf{S}}{dt^2} + 2\boldsymbol{\omega} \times \mathbf{V} + \boldsymbol{\omega} \times (\boldsymbol{\omega} \times \mathbf{r}) + \frac{d\boldsymbol{\omega}}{dt} \times \mathbf{r} \right] \rho \, d\mathbb{V} \quad (4.5.14)$$

Since the system and control volume are identical at time t the system integration can be replaced with a control volume integration in the integral of Eq. 4.5.14. Example 4.19 will illustrate the use of a nonintertial reference frame.

4.5.5 Momentum Equation Applied to Deflectors

The application of the momentum equation to deflectors forms an integral part of the analysis of many turbomachines, such as turbines, pumps, and compressors. In this section we illustrate the steps in such an analysis. It will be separated into two parts: jets deflected by stationary deflectors and jets deflected by moving deflectors. For both problems we will assume the following:

— The pressure external to the jets is everywhere constant so that the pressure in the fluid entering the deflector is the same as that in the fluid exiting the deflector.
— The frictional resistance due to the fluid–deflector interaction is negligible so that the relative speed between the deflector surface and the jet stream remains unchanged, a result of Bernoulli's equation.
— Lateral spreading of a plane jet is neglected.
— Body forces are small and will be neglected.

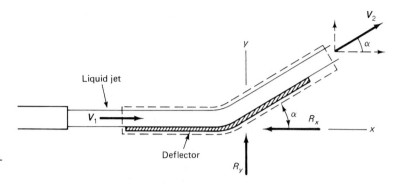

Figure 4.13 Stationary deflector.

Stationary Deflector

Let us first consider the stationary deflector, illustrated in Fig. 4.13. Bernoulli's equation allows us to conclude that $V_2 = V_1$ since the pressure is assumed to be constant external to the fluid jet and elevation changes are negligible (see Eq. 3.4.9). Assuming steady, uniform flow the momentum equation takes the form of Eq. 4.5.6, which for the x- and y-directions becomes

$$-R_x = \dot{m}(V_2 \cos \alpha - V_1) = \dot{m}V_1(\cos \alpha - 1)$$
$$R_y = \dot{m}V_2 \sin \alpha = \dot{m}V_1 \sin \alpha$$

$$(4.5.15)$$

For given jet conditions the reactive force components can be calculated.

Moving Deflectors

The situation involving a moving deflector depends on whether a single deflector is moving (a water scoop used to slow a high-speed train) or whether a series of deflectors is moving (the vanes on a turbine). Let us first consider that the single deflector shown in Fig. 4.14 is moving in the positive x-direction with the speed V_B. In a reference frame attached to the stationary nozzle, from which the fluid jet issues, the flow is unsteady; that is, at a particular point in space, the flow situation varies with time.[3] A steady flow is observed, however, from a reference frame attached to the deflector. From this inertial reference frame, moving with the constant velocity V_B, we observe the relative speed V_r entering the control volume to be $V_1 - V_B$, as shown in Fig. 4.14. It is this speed that remains constant as the fluid flows relative to the deflector. Hence the momentum

---

[3]This is a rather subtle point. To determine whether a flow is steady, we observe the flow at a given point in space. If a flow property changes with time at that point, the flow is unsteady. In this situation, if we focus our attention on a particular point, first there is no flow, then the blade and the jet pass through the point; then there is again no flow. This is an unsteady flow.

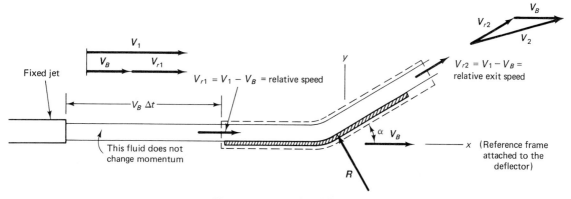

Figure 4.14 Moving deflector.

equation takes the forms

$$-R_x = \dot{m}_r(V_1 - V_B)(\cos \alpha - 1)$$

$$R_y = \dot{m}_r(V_1 - V_B) \sin \alpha$$

(4.5.16)

where $\dot{m}_r$ represents only that part of the mass flux exiting the fixed jet that has its momentum changed; since the deflector moves away from the fixed jet some of the fluid that exits the fixed jet never experiences a momentum change; this fluid is represented by the distance $V_B \, \Delta t$, shown in Fig. 4.14. Hence

$$\dot{m}_r = \rho A(V_1 - V_B)$$

(4.5.17)

where the relative speed $(V_1 - V_B)$ is used in the calculation; the mass flux $\rho A V_B$ is subtracted from the exiting mass flux $\rho A V_1$ to provide the mass flux $\dot{m}_r$.

For a series of vanes (a cascade) the jets may be oriented to the side, as shown in Fig. 4.15. The actual force on a particular vane would be zero until the jet strikes the vane; then the force would increase to a maximum and decrease to zero as the vane leaves the jet. We will idealize the situation as follows: Assume that, on the average, the jet is deflected by the vanes as shown in Figs. 4.15 and 4.16a as viewed from a stationary reference frame; the fluid jet enters the vanes with an angle β_1 and exits with an angle β_2. What is desired, however, is that the relative velocity enter the vanes tangent to the leading edge of the vanes, that is, V_{r1} in Fig. 4.16b is at the angle α_1. The relative speed then remains constant as it travels over the vane with the exiting relative velocity V_{r2} leaving with the vane angle α_2. The relative and absolute velocities are related with the velocity polygons of Fig. 4.16b and c. Assuming that all of the mass exiting the fixed jet has its momentum changed, we can write the momen-

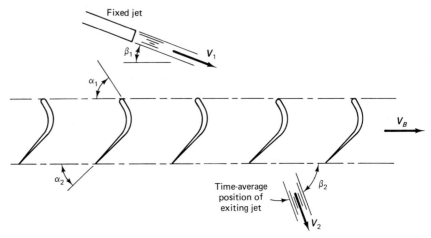

Figure 4.15 Fluid striking a series of vanes.

tum equation as

$$-R_x = \dot{m}(V_{2x} - V_{1x})$$ (4.5.18)

Example 4.14 will illustrate the details.

Interest is usually focused on the x-component of force since it is this component that is related to the power output (or requirement). The power would be found by multiplying the x-component force by the blade speed for each jet; this takes the form

$$\dot{W} = NR_x V_B$$ (4.5.19)

where N represents the number of jets.

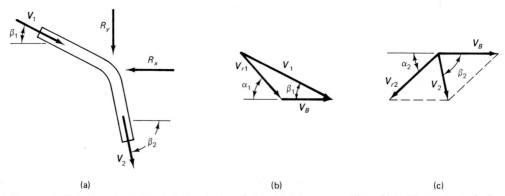

Figure 4.16 Detail of the flow situation involving a series of vanes: (a) average position of jet; (b) entrance velocity polygon; (c) exit velocity polygon.

4.5.6 Momentum Equation Applied to Propellers

The application of the momentum equation to propellers is also of sufficient interest that this section will be devoted to illustrating the procedure. Consider the propeller of Fig. 4.17 with the streamlines shown forming the surface of a control volume in which the fluid enters with a uniform velocity V_1 and exits with a uniform velocity V_2. This flow situation would be identical to that of a propeller moving with velocity V_1 in a stagnant fluid. The momentum equation, applied to the large control volume shown, gives

$$F = \dot{m}(V_2 - V_1) \tag{4.5.20}$$

If a control volume is drawn close to the propeller such that $V_3 = V_4$, the momentum equation would give

$$F + p_3 A - p_4 A = 0 \tag{4.5.21}$$

or

$$F = (p_4 - p_3)A \tag{4.5.22}$$

Now, since viscous effects would be quite small in this flow situation the energy equation up to the propeller and then downstream from the propeller is used to obtain

$$\frac{V_1^2 - V_3^2}{2} + \frac{p_1 - p_3}{\rho} = 0 \qquad \frac{V_4^2 - V_2^2}{2} + \frac{p_4 - p_2}{\rho} = 0 \quad (4.5.23)$$

Adding these equations together, recognizing that $p_1 = p_2 = p_{\text{atm}}$, we have

$$(V_2^2 - V_1^2)\frac{\rho}{2} = p_4 - p_3 \tag{4.5.24}$$

Inserting this and Eq. 4.5.22 into Eq. 4.5.20 results in

$$V_3 = \tfrac{1}{2}(V_2 + V_1) \tag{4.5.25}$$

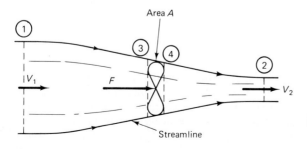

Area A

V_1

F

V_2

Streamline

Figure 4.17 Propeller in a fluid flow.

where we have used $\dot{m} = \rho A V_3$. This result shows that the velocity of the fluid moving through the propeller is the average of the upstream and downstream velocities. The input power needed to produce this effect is found by applying the energy equation between sections 1 and 2, where the pressures are atmospheric; neglecting losses, there results

$$\dot{W}_{\text{fluid}} = \frac{V_2^2 - V_1^2}{2} \dot{m} \qquad (4.5.26)$$

The moving propeller requires power given by

$$\dot{W}_{\text{prop}} = F \times V_1$$
$$= \dot{m} V_1 (V_2 - V_1) \qquad (4.5.27)$$

The theoretical propellor efficiency is then

$$\eta_P = \frac{\dot{W}_{\text{prop}}}{\dot{W}_{\text{fluid}}} = \frac{V_1}{V_3} \qquad (4.5.28)$$

In contrast to the propeller, a windmill extracts energy from the air flow; the downstream velocity is reduced and the diameter is increased.

EXAMPLE 4.11

Water flows through a horizontal pipe bend and exits into the atmosphere. The flow rate is 0.01 m³/s. Calculate the force in each of the rods holding the pipe bend in position. Neglect body forces and viscous effects.

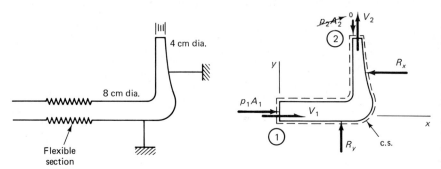

Solution

We have selected a control volume that surrounds the bend as shown. Since the rods have been cut, the forces that the rods exert on the control volume are included. The pressure forces at the entrance and exit of the control volume are also shown. The flexible section is capable of resisting the interior pressure but it transmits no axial force or moment. The body force (weight of the control volume) does not act in the x- or y-direction but normal to it. Therefore, no other forces are shown. The average velocities are found to be

$$V_1 = \frac{Q}{A_1} = \frac{0.01}{\pi(0.08)^2/4} = 1.99 \text{ m/s}$$

$$V_2 = \frac{Q}{A_2} = \frac{0.01}{\pi(0.04)^2/4} = 7.96 \text{ m/s}$$

Before we can calculate the forces R_x and R_y we need to find the pressures p_1 and p_2. The pressure p_2 is zero because the flow exits into the atmosphere. The pressure at section 1 can be determined using the energy equation or the Bernoulli equation. Neglecting losses between sections 1 and 2, the energy equation gives

$$\frac{V_1^2}{2g} + \frac{p_1}{\gamma} = \frac{V_2^2}{2g} + \frac{p_2}{\gamma}^{0}$$

$$\therefore \ p_1 = \frac{\gamma}{2g}(V_2^2 - V_1^2) = \frac{9810}{2 \times 9.81}(7.96^2 - 1.99^2) = 29\ 700 \text{ Pa}$$

Now we can apply the momentum equation (4.5.6) in the x-direction to find R_x and in the y-direction to find R_y:

x-direction: $\qquad\qquad p_1 A_1 - R_x = \dot{m}(V_{2x}^{\ 0} - V_{1x})$

$$29\ 700 \times \frac{\pi}{4} \times (0.08)^2 - R_x = 1000 \times 0.01 \times (-1.99)$$

$$\therefore \ R_x = 169 \text{ N}$$

y-direction: $\qquad\qquad + R_y = \dot{m}(V_{2y} - V_{1y}^{\ 0})$

$$\therefore \ R_y = 1000 \times 0.01 \times 7.96 = 79.6 \text{ N}$$

Note that we have assumed uniform profiles and steady flow. These are the usual assumptions if information is not given otherwise.

EXAMPLE 4.11 (English)

Water flows through a horizontal pipe bend and exits into the atmosphere. The flow rate is 0.3 ft³/sec. Calculate the force in each of the rods holding the pipe bend in position. Neglect body forces and viscous effects.

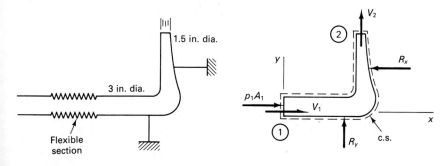

Solution

We have selected a control volume that surrounds the bend, as shown. Since the rods have been cut, the forces that the rods exert on the control volume are included. The pressure forces at the entrance and exit of the control volume are also shown. The flexible section is capable of resisting the interior pressure but it transmits no axial force or moment. The body force (weight of the control volume) does not act in the x- or y-direction but normal to it. Therefore, no other forces are shown. The average velocities are found to be

$$V_1 = \frac{Q}{A_1} = \frac{0.3}{\pi \times (3/12)^2/4} = 6.11 \text{ ft/sec}$$

$$V_2 = \frac{Q}{A_2} = \frac{0.3}{\pi \times (1.5/12)^2/4} = 24.4 \text{ ft/sec}$$

Before we can calculate the forces R_x and R_y we need to find the pressures p_1 and p_2. The pressure p_2 is zero because the flow exits into the atmosphere. The pressure at section 1 can be determined using the energy equation or the Bernoulli equation. Neglecting losses between sections 1 and 2, the energy equation gives

$$\frac{V_1^2}{2g} + \frac{p_1}{\gamma} = \frac{V_2^2}{2g} + \cancelto{0}{\frac{p_2}{\gamma}}$$

$$\therefore \; p_1 = \frac{\gamma}{2g} (V_2^2 - V_1^2) = \frac{62.4}{2 \times 32.2} (24.4^2 - 6.11^2) = 541 \text{ psf}$$

Now we can apply the momentum equation (4.5.6) in the x-direction to find R_x and in the y-direction to find R_y:

x-direction:
$$p_1 A_1 - R_x = \dot{m}(\cancelto{0}{V_{2x}} - V_{1x})$$

$$541 \times \frac{\pi}{4} \left(\frac{3}{12}\right)^2 - R_x = 1.94 \times 0.3 \times (-6.11)$$

$$\therefore \; R_x = 30.1 \text{ lb}$$

y-direction:
$$R_y = \dot{m}(V_{2y} - \cancelto{0}{V_{1y}})$$

$$\therefore \; R_y = 1.94 \times 0.3 \times 24.4 = 14.2 \text{ lb}$$

Note that we have assumed uniform profiles and steady flow. These are the usual assumptions if information is not given otherwise.

EXAMPLE 4.12

A deflector turns a sheet of water through an angle of 30° as shown. What force per unit width is necessary to hold the deflector in place if $\dot{m} = 32$ kg/s?

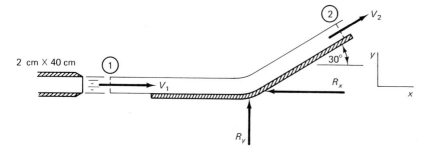

Solution

The control volume we have selected includes the deflector and the water adjacent to it. The only force that is acting on the control volume is due to the support. This force has been decomposed into R_x and R_y.

The velocity V_1 is found to be

$$V_1 = \frac{\dot{m}}{\rho A_1}$$

$$= \frac{32}{1000 \times 0.02 \times 0.4} = 4.0 \text{ m/s}$$

Bernoulli's equation (3.4.8) shows that if the pressure does not change, then the magnitude of the velocity does not change, provided that there is no significant change in elevation and that viscous effects are negligible; thus we can conclude that $V_2 = V_1$ since $p_2 = p_1$. Next, the momentum equation is applied in the x-direction to find R_x and then in the y-direction for R_y:

$$x\text{-direction:} \quad -R_x = \dot{m}(V_{2x} - V_{1x})$$

$$= 32 \,(4 \cos 30° - 4)$$

$$\therefore \quad R_x = 17.2 \text{ N}$$

$$y\text{-direction:} \quad R_y = \dot{m}(V_{2y} - \cancelto{0}{V_{1y}})$$

$$= 32 \,(4 \sin 30°) = 64 \text{ N}$$

EXAMPLE 4.13

The deflector shown moves to the right at 3 m/s while the nozzle remains stationary. Determine (a) the force components needed to move the deflector, (b) V_2 as observed from a fixed observer, and (c) the power generated by the vane. The jet velocity is 8 m/s.

Solution

(a) To solve the problem of a moving deflector, we will observe the flow from a reference frame attached to the deflector. In this moving reference frame the

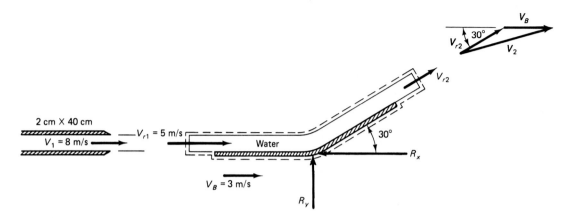

flow is steady and Bernoulli's equation can then be used to show that $V_{r1} = V_{r2} = 5$ m/s, the velocity of the sheet of water as observed from the deflector. Note that we cannot apply Bernoulli's equation in a fixed reference frame since the flow would not be steady. Applying the momentum equation to the moving control volume, which is indicated again by the dashed line, we obtain the following:

$$x\text{-direction:} \quad -R_x = \dot{m}_r[(V_{r2})_x - (V_{r1})_x]$$

$$= 1000 \times 5 \times 0.02 \times 0.4(5 \cos 30° - 5)$$

$$\therefore \quad R_x = 26.8 \text{ N}$$

$$y\text{-direction:} \quad R_y = \dot{m}_r[(V_{r2})_y - \overset{0}{\cancel{(V_{r1})_y}}]$$

$$= 1000 \times 5 \times 0.02 \times 0.4(5 \sin 30°) = 100 \text{ N}$$

When calculating $\dot{m}_r$ we must use only that water which has its momentum changed; hence the velocity used is 5 m/s.

(b) Observed from a fixed observer the velocity $\mathbf{V}_2$ of the fluid after the deflection is $\mathbf{V}_2 = \mathbf{V}_{r2} + \mathbf{V}_B$, where $\mathbf{V}_{r2}$ is directed tangential to the deflector at the exit and has a magnitude equal to V_{r1} (see the velocity diagram above). Thus

$$(V_2)_x = V_{r2} \cos 30° + V_B$$

$$= 5 \times 0.866 + 3 = 7.33 \text{ m/s}$$

$$(V_2)_y = V_{r2} \sin 30°$$

$$= 5 \times 0.5 = 2.5 \text{ m/s}$$

Finally,

$$\mathbf{V}_2 = 7.33\hat{i} + 2.5\hat{j} \text{ m/s}$$

(c) The power generated by the moving vane is equal to the velocity of the vane times the force the vane exerts in the direction of the motion. Therefore,

$$\dot{W} = V_B \times R_x$$

$$= 3 \times 26.8 = 80.4 \text{ W}$$

EXAMPLE 4.14

High-speed air jets strike the blades of a turbine rotor tangentially while the 1.5-m-diameter rotor rotates at 140 rad/s. There are 10 such jets. Calculate the maximum power output. The air density is 2.4 kg/m³.

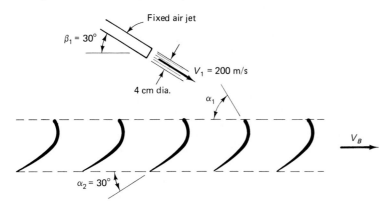

Solution

The blade angle α_1 is set by demanding that the air jet enter the blades tangentially, as observed from the moving blade; that is, the relative velocity vector $\mathbf{V}_r$ must make the angle α_1 with respect to the blade velocity $\mathbf{V}_B$. This is shown in Figure (a) below.

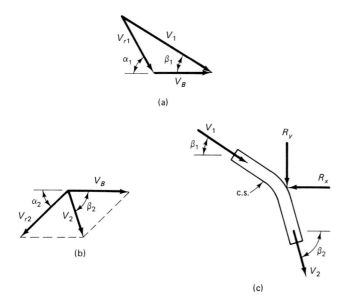

The relative entrance velocity is $\mathbf{V}_{r1}$ [Figure (a)] and the relative exit velocity is $\mathbf{V}_{r2}$ [Figure (b)]. Both velocity polygons are presented by the vector equation

$$\mathbf{V} = \mathbf{V}_r + \mathbf{V}_B$$

which states that the absolute velocity equals the relative velocity plus the blade velocity. From the polygon at the entrance we have

$$V_1 \sin \beta_1 = V_{r1} \sin \alpha_1$$

$$V_1 \cos \beta_1 = V_{r1} \cos \alpha_1 + V_B$$

$$\therefore \quad 200 \sin 30° = V_{r1} \sin \alpha_1$$

$$200 \cos 30° = V_{r1} \cos \alpha_1 + 0.75 \times 140$$

where V_B is the radius multiplied by the angular velocity. A simultaneous solution yields

$$V_{r1} = 121 \text{ m/s} \qquad \alpha_1 = 55.71°$$

The friction between the air and the blade is quite small and can be neglected when calculating the maximum output. This allows us to assume $V_{r2} = V_{r1}$. From the exiting velocity polygon we can write

$$V_B - V_{r2} \cos \alpha_2 = V_2 \cos \beta_2$$

$$V_{r2} \sin \alpha_2 = V_2 \sin \beta_2$$

$$\therefore \quad 0.75 \times 140 - 121 \cos 30° = V_2 \cos \beta_2$$

$$121 \sin 30° = V_2 \sin \beta_2$$

A simultaneous solution results in

$$V_2 = 60.5 \text{ m/s} \qquad \beta_2 = 90.0°$$

The momentum equation applied to the control volume, shown in Figure (c), gives

$$-R_x = \dot{m}(\cancel{V_{2x}}^{0} - V_{1x})$$

$$= 2.4 \times \pi \times 0.02^2 \times 200 \,(0 - 200 \cos 30°)$$

$$\therefore \quad R_x = 104.5 \text{ N}$$

There are 10 jets, each producing the force above. The maximum power output is then

$$\text{power} = 10 \times R_x \times V_B$$

$$= 10 \times 104.5 \times (0.75 \times 140) = 109\ 700 \text{ W} \quad \text{or} \quad 109.7 \text{ kW}$$

EXAMPLE 4.15

When the velocity of a flow in an open rectangular channel of width w is relatively large, it is possible for the flow to "jump" from a depth y_1 to a depth y_2 over a relatively short distance, as shown in the sketch; this is referred to as a **hydraulic jump**. Express y_2 in terms of y_1 and V_1; assume a horizontal flow.

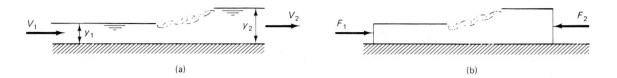

(a)   (b)

Solution

A control volume is selected with inlet and exit areas upstream and downstream of the "jump" sufficiently far that the streamlines are parallel to the wall with hydrostatic pressure distributions. Neglecting the drag that is present on the walls (if the distance between sections is relatively small, the drag force should be negligible), the momentum equation can be manipulated as follows:

$$\Sigma F_x = \dot{m}(V_{2x} - V_{1x})$$

$$F_1 - F_2 = \rho A_1 V_1 (V_2 - V_1)$$

$$\gamma \frac{y_1}{2} (y_1 w) - \gamma \frac{y_2}{2} (y_2 w) = \rho y_1 w V_1 \left(V_1 \frac{y_1}{y_2} - V_1\right)$$

where we have expressed F_1 and F_2 using Eq. 2.4.24, and continuity in the form of Eq. 4.3.6, so that

$$V_2 = \frac{y_1}{y_2} V_1$$

The momentum equation can be simplified to

$$\frac{\gamma}{2} (y_1^2 - y_2^2) = \rho y_1 V_1^2 \frac{y_1 - y_2}{y_2}$$

or

$$\frac{g}{2} (y_1 - y_2)(y_1 + y_2) = \frac{y_1}{y_2} V_1^2 (y_1 - y_2)$$

The factor $(y_1 - y_2)$ is divided out and y_2 is found as follows:

$$\frac{g}{2} (y_1 + y_2) = \frac{y_1}{y_2} V_1^2$$

$$y_2^2 + y_1 y_2 - \frac{2}{g} y_1 V_1^2 = 0$$

$$\therefore \quad y_2 = \frac{1}{2} \left(-y_1 + \sqrt{y_1^2 + \frac{8}{g} y_1 V_1^2}\right)$$

where the quadratic formula has been used. The energy equation could now be used to provide an expression for the losses in the hydraulic jump.

EXAMPLE 4.16

Consider the symmetrical flow of air around the cylinder. The control volume, excluding the cylinder, is shown. The velocity distribution downstream of the cylinder is measured to be as shown. Determine the drag force per meter of length acting on the cylinder. Use $\rho = 1.23$ kg/m³.

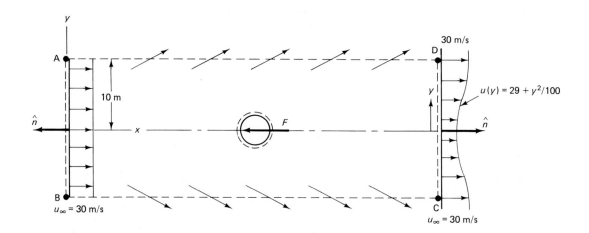

Solution

First, we must recognize that not all of the mass flux entering through AB exits through CD; consequently, some mass flux must exit AD and BC, as shown. The momentum equation for the steady flow, applied to the control volume ABCD, takes the form

$$-F = \int_{c.s.} \rho V_x \mathbf{V} \cdot \hat{n} \, dA$$

$$= \int_{A_{CD}} \rho u^2 \, dA + u_\infty \dot{m}_{AD} + u_\infty \dot{m}_{BC} - \int_{A_{AB}} \rho u^2 \, dA$$

$$= 2 \int_0^{10} 1.23 \left(29 + \frac{y^2}{100} \right)^2 dy + 2 \times 30 \dot{m}_{AD} - 1.23 \times 30^2 \times 20$$

where $\dot{m}_{BC} = \dot{m}_{AD}$ is the mass flux crossing BC and AD with the x-component velocity equal to 30 m/s. Use continuity to find $\dot{m}_{AD}$:

$$0 = \dot{m}_{AD} + \int_0^{10} \rho u(y) \, dy - \rho \times 10 \times 30$$

$$= \dot{m}_{AD} + \int_0^{10} 1.23 \times \left(29 + \frac{y^2}{100} \right) dy - 1.23 \times 10 \times 30$$

$$\therefore \quad \dot{m}_{AD} = 8.2 \text{ kg/s per meter}$$

Evaluating the terms in the momentum equation above gives us

$$F = -21\ 170 - 492 + 22\ 140$$

$$= 480 \text{ N per meter}$$

EXAMPLE 4.16 (English)

Consider the symmetrical flow of air around a cylinder. The control volume, excluding the cylinder, is shown. The velocity distribution downstream of the cylinder is measured to be as shown. Determine the force F per foot of length acting on the cylinder. Use $\rho = 0.0024$ slug/ft^3.

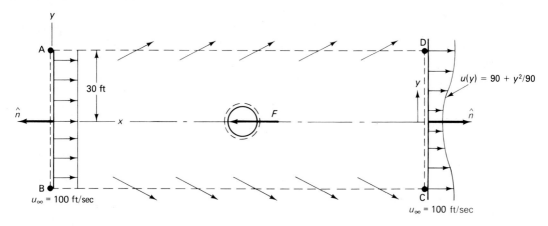

Solution

First, we must recognize that not all of the mass flux entering through AB exits through CD; consequently, some mass flux must exit AD and BC, as shown. The momentum equation for the steady flow, applied to the control volume ABCD, takes the form

$$-F = \int_{\text{c.s.}} \rho V_x \mathbf{V} \cdot \hat{n} \, dA$$

$$= \int_{A_{\text{CD}}} \rho u^2 \, dA + u_\infty \dot{m}_{\text{AD}} + u_\infty \dot{m}_{\text{BC}} - \int_{A_{\text{AB}}} \rho u^2 \, dA$$

$$= 2 \int_0^{30} 0.0024 \left(90 + \frac{y^2}{90}\right)^2 dy + 2 \times 100 \dot{m}_{\text{AD}} - 0.0024 \times 100^2 \times 60$$

where $\dot{m}_{\text{BC}} = \dot{m}_{\text{AD}}$ is the mass flux crossing BC and AD with the x-component velocity equal to 100 ft/sec. Use continuity to find $\dot{m}_{\text{AD}}$:

$$0 = \dot{m}_{\text{AD}} + \int_0^{30} \rho u(y) \, dy - \rho \times 30 \times 100$$

$$= \dot{m}_{\text{AD}} + \int_0^{30} 0.0024 \left(90 + \frac{y^2}{90}\right) dy - 0.0024 \times 3000$$

$$\therefore \quad \dot{m}_{\text{AD}} = 0.48 \text{ slug/sec per foot}$$

Evaluating the terms in the momentum equation above gives us

$$F = -1256 - 96 + 1440$$

$$= 88 \text{ lb per ft}$$

EXAMPLE 4.17

Find an expression for the head loss in a sudden expansion in a pipe in terms of V_1 and the area ratio. Assume uniform velocity profiles and assume that the pressure at the sudden enlargement is p_1.

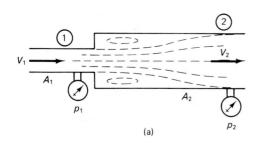

(a)

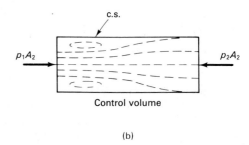

c.s.

Control volume

(b)

Solution

A sketch is shown of a sudden expansion with the diameter changing from d_1 to d_2. The pressure at the sudden enlargement is p_1, so that the force acting on the left end shown is $p_1 A_2$. Newton's second law applied to the control volume yields, assuming uniform profiles,

$$\Sigma F_x = \dot{m}(V_2 - V_1)$$

$$(p_1 - p_2)A_2 = \rho A_2 V_2(V_2 - V_1)$$

$$\therefore \quad \frac{p_1 - p_2}{\rho} = V_2(V_2 - V_1)$$

The energy equation provides

$$0 = \frac{V_2^2 - V_1^2}{2g} + \frac{p_2 - p_1}{\gamma} + z_2 \!\!\!\!\!\!\!\!\cancel{}^{\,0}\!\!\!\!\! - z_1 + h_L$$

$$\therefore \quad h_L = \frac{p_1 - p_2}{\gamma} - \frac{V_2^2 - V_1^2}{2g}$$

$$= \frac{2V_2(V_2 - V_1)}{2g} - \frac{(V_2 + V_1)(V_2 - V_1)}{2g} = \frac{(V_1 - V_2)^2}{2g}$$

To express this in terms of only V_1, we can use continuity and relate

$$V_2 = \frac{A_1}{A_2} V_1$$

Then the expression above for the head loss becomes

$$h_L = \left(1 - \frac{A_1}{A_2}\right)^2 \frac{V_1^2}{2g}$$

EXAMPLE 4.18

Calculate the momentum correction factor for a parabolic profile (a) between parallel plates and (b) in a circular pipe. The parabolic profiles are shown in the figure.

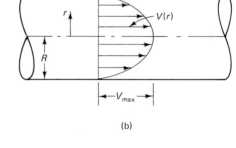

(a)

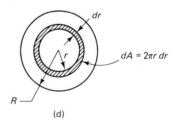

(b)

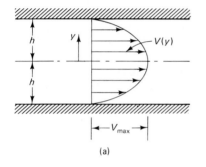

(c)

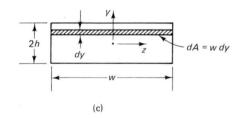

(d)

Solution

(a) A parabolic profile between parallel plates can be expressed as

$$V(y) = V_{max}\left(1 - \frac{y^2}{h^2}\right)$$

where y is measured from the centerline, the walls are at $y = \pm h$, and V_{max} is the centerline velocity at $y = 0$. First, let us find the average velocity. It is

$$\bar{V} = \frac{1}{A} \int V \, dA$$

$$= \frac{1}{2hw} \int_{-h}^{h} V_{max}\left(1 - \frac{y^2}{h^2}\right) w \, dy = \frac{V_{max}}{2h}\left(2h - \frac{2}{3}h\right) = \frac{2}{3}V_{max}$$

Then

$$\beta = \frac{\int V^2 \, dA}{\bar{V}^2 A}$$

$$= \frac{1}{\frac{4}{9} V_{\text{max}}^2 \times 2hw} \int_{-h}^{h} V_{\text{max}}^2 \left(1 - \frac{y^2}{h^2}\right)^2 w \, dy = \frac{6}{5}$$

(b) For a circular pipe a parabolic profile can be written as

$$V(r) = V_{\text{max}}\left(1 - \frac{r^2}{R^2}\right)$$

where R is the pipe radius. The average velocity is found to be

$$\bar{V} = \frac{1}{A} \int V \, dA$$

$$= \frac{1}{\pi R^2} \int_0^R V_{\text{max}}\left(1 - \frac{r^2}{R^2}\right) 2\pi r \, dr = \frac{1}{2} V_{\text{max}}$$

The momentum correction factor is then

$$\beta = \frac{\int V^2 \, dA}{\bar{V}^2 A}$$

$$= \frac{1}{\frac{1}{4} V_{\text{max}}^2 \pi R^2} \int_0^R V_{\text{max}}^2 \left(1 - \frac{r^2}{R^2}\right)^2 2\pi r \, dr = \frac{4}{3}$$

The correction factors above can be used to express the momentum flux across a cross-sectional area as $\beta \rho A \bar{V}^2$.

EXAMPLE 4.19

The rocket shown, with an initial mass of 150 kg, burns fuel at the rate of 3 kg/s with a constant exhaust velocity of 700 m/s. What is the initial acceleration of the rocket and the velocity after 4 s? Neglect the drag on the rocket.

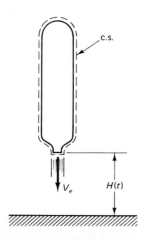

Solution

The control volume is sketched and includes the entire rocket. The reference frame attached to the rocket is accelerating upward at d^2H/dt^2. Newton's second law is written as, using z as upward,

$$\Sigma F_z - (F_I)_z = \frac{d}{dt} \int_{\text{c.v.}} \rho V_z \, d\Psi + \int_{\text{c.s.}} \rho V_z \mathbf{V} \cdot \hat{n} \, dA$$

$$\therefore \quad -W - \frac{d^2H}{dt^2} M_{\text{c.v.}} = \rho_e(-V_e)V_eA_e$$

where

$$\frac{d}{dt} \int_{\text{c.v.}} \rho V_z \, d\Psi \approx 0$$

since V_z is the velocity of each mass element $\rho \, d\Psi$ relative to the reference frame attached to the control volume; the only vertical force is the weight W; and $M_{\text{c.v.}}$ is the mass of the control volume. From continuity we see that

$$M_{\text{c.v.}} = 150 - \dot{m}t$$

$$= 150 - 3t$$

$$\therefore \quad W = (150 - 3t) \times 9.81$$

The momentum equation becomes

$$-(150 - 3t) \times 9.81 - \frac{d^2H}{dt^2}(150 - 3t) = -\dot{m}_e V_e$$

$$= -3 \times 700 = -2100$$

This is written as

$$\frac{d^2H}{dt^2} = \frac{2100}{150 - 3t} - 9.81$$

The initial acceleration is found by letting $t = 0$:

$$\frac{d^2H}{dt^2}\bigg|_{t=0} = \frac{2100}{150} - 9.81 = 4.2 \text{ m/s}^2$$

Integrate the expression for d^2H/dt^2 and obtain

$$\frac{dH}{dt} = -700 \ln(50 - t) - 9.81t + C$$

The constant $C = 700 \ln 50$ since $dH/dt = 0$ at $t = 0$. Thus at $t = 4$ s the velocity is

$$\frac{dH}{dt} = -700 \ln \frac{46}{50} - 9.81 \times 4 = 19.13 \text{ m/s}$$

4.6 MOMENT-OF-MOMENTUM EQUATION

In the preceding section we determined the magnitude of force components in a variety of flow situations. To determine the line of action of a given force component, it is often necessary to apply the moment-of-momentum equation. Also, in analyzing the flow situation in devices that have rotating components the moment-of-momentum equation is needed to relate the rotational speed to the other flow parameters. Since it may be advisable to attach the reference frame to the rotating component, we will write the general equation with the inertial forces included. It is (see Eq. 4.1.4)

$$\Sigma \mathbf{M} - \mathbf{M}_I = \frac{D}{Dt} \int_{\text{sys}} \mathbf{r} \times \mathbf{V} \rho \, d\Psi \tag{4.6.1}$$

where

$$\mathbf{M}_I = \int \mathbf{r} \times \left[\frac{d^2\mathbf{S}}{dt^2} + 2\boldsymbol{\omega} \times \mathbf{V} + \boldsymbol{\omega} \times (\boldsymbol{\omega} \times \mathbf{r}) + \frac{d\boldsymbol{\omega}}{dt} \times \mathbf{r} \right] \rho \, d\Psi \tag{4.6.2}$$

This inertial moment $\mathbf{M}_I$ accounts for the fact that a noninertial reference frame was selected; it is simply the moment of $\mathbf{F}_I$ (see Eq. 4.5.14). Applying the system-to-control volume transformation, the moment-of-momentum equation for a control volume becomes

$$\Sigma \mathbf{M} - \mathbf{M}_I = \frac{d}{dt} \int_{\text{c.v.}} \mathbf{r} \times \mathbf{V} \rho \, d\Psi + \int_{\text{c.s.}} \mathbf{r} \times \mathbf{V}(\mathbf{V} \cdot \hat{n}) \rho \, dA \tag{4.6.3}$$

Examples will illustrate the application of this equation.

EXAMPLE 4.20

A sprinkler has four 50-cm-long arms with nozzles at right angles with the arms and 45° with the ground. If the total flow rate is 0.001 m³/s and a nozzle exit diameter is 12 mm, find the rotational speed of the sprinkler. Neglect friction.

Solution

The velocity exiting a nozzle as shown is

$$V_e = \frac{Q}{A}$$

$$= \frac{0.001/4}{\pi \times 0.006^2} = 2.21 \text{ m/s}$$

Attach the reference frame to the rotating arms as shown. Then, recognizing that $\mathbf{r} \times [\boldsymbol{\omega} \times (\boldsymbol{\omega} \times \mathbf{r})] = 0$ and assuming a stationary sprinkler so that $d^2\mathbf{S}/dt^2 = 0$ and

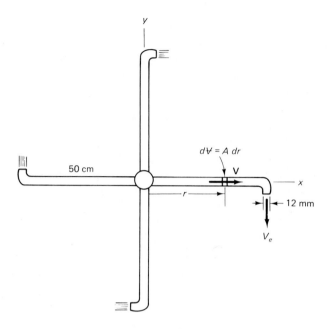

constant angular velocity so that $d\omega/dt = 0$, we have

$$\mathbf{M}_I = \int_{\text{c.v.}} \mathbf{r} \times (2\boldsymbol{\omega} \times \mathbf{V})\rho \, d\Psi$$

$$= 4 \int_0^{0.5} r\hat{i} \times (2\omega\hat{k} \times V\hat{i})\rho A \, dr$$

$$= 8\rho A V\omega\hat{k} \int_0^{0.5} r \, dr = \rho A V\omega\hat{k}$$

where $4VA = 0.001 \text{ m}^3/\text{s}$. The factor 4 accounts for the four arms. Now, there are no external moments applied to the sprinkler about the vertical z-axis; hence $\Sigma \, M_z = 0$. For the steady flow

$$-(\mathbf{M}_I)_z = \int_{\text{c.s.}} (\mathbf{r} \times \mathbf{V})_z \mathbf{V} \cdot \hat{n} \, \rho \, dA$$

$$-\rho A V\omega = 4 \int_{A_{\text{exit}}} [0.5\hat{i} \times (0.707V_e\hat{k} - 0.707V_e\hat{j})]_z V_e\rho \, dA$$

$$-VA\omega = -4 \times 0.5 \times 0.707V_e^2 A_e$$

$$\therefore \quad \omega = 4 \times 0.5 \times 0.707V_e = 3.125 \text{ rad/s}$$

where we have used $AV = A_e V_e$ from continuity considerations. Note that we have neglected the small mass in the short nozzles.

EXAMPLE 4.21

The nozzles of Example 4.20 make an angle of $0°$ with the ground and $90°$ with the arms. The water is suddenly turned on at $t = 0$ with the sprinkler motionless. Determine the resulting $\omega(t)$ if the arm diameter is 24 mm. Neglect friction.

Solution

The reference frame is again attached to the rotating arms, as sketched in Example 4.20. Referring to the control volume integral of Eq. 4.6.3, we observe that $\mathbf{r} \times \mathbf{V} = 0$ since $\mathbf{r}$ is in the same direction as $\mathbf{V}$ along an arm. Thus Eq. 4.6.3 takes the form

$$\Sigma \mathbf{M}^{0} - 4 \int_0^{0.5} r\hat{i} \times \left[(2\omega\hat{k} \times V\hat{i}) + \frac{d\omega}{dt}\hat{k} \times r\hat{i} \right] \rho A \, dr$$
$$= 4 \int_{A_{\text{exit}}} 0.5\hat{i} \times V_e(-\hat{j}) V_e \rho \, dA$$

or, performing the vector operations and dividing by 4ρ,

$$-2A\omega\bar{V} \int_0^{0.5} r \, dr - \frac{d\omega}{dt} A \int_0^{0.5} r^2 \, dr = -0.5 V_e^2 A_e$$

Performing the required integration and using $AV = A_e V_e = 0.001$ m³/s, $V_e = 2.210$ m/s, and $V = 0.5525$ m/s, we have

$$\frac{d\omega}{dt} + 1.105\omega = 4.884$$

This linear, first-order differential equation is solved by adding the homogeneous solution (suppress the right-hand side) to the particular solution to obtain

$$\omega(t) = Ce^{-1.105t} + 4.42$$

Using the initial condition $\omega(0) = 0$, we find that $C = -4.42$. Then

$$\omega(t) = 4.42(1 - e^{-1.105t}) \quad \text{rad/s}$$

Observe that as time becomes large, the angular velocity is limited to 4.42 rad/s. If friction were included, this value would be reduced.

4.7 CLOSURE

In this chapter we have presented the control-volume formulation of the fundamental laws. This formulation is useful when the intregrands (the velocities and pressure) are known or can be approximated with an acceptable degree of accuracy. If this is not the case, the differential equations must be solved (numerically or analytically) or experimental methods must be used to obtain the desired information; much of the remainder of this book is devoted to this task. After the velocities and pressures are determined, we often return to the control-volume formulation and calculate integral quantities of interest. Examples would include the lift and drag on an airfoil, the torque on a row of turbine blades, and the oscillating force on a suspension cable of a bridge.

As we have observed in the examples and problems of this chapter, the task of applying the control-volume equations is highly dependent on the appropriate selection of the boundaries of the control volume. Such boundaries are selected at locations where either information is known or where the unknowns appear. Experience is often required in the selection of a control volume, as it is in the selection of a free-body diagram in dynamics and solid mechanics. The student has undoubtedly gained some of this experience while working through the sections of this chapter. Table 4.1 presents the various forms of the fundamental laws to aid the user in selecting an appropriate form for a particular problem.

TABLE 4.1   INTEGRAL FORMS OF THE FUNDAMENTAL LAWS

Continuity	Energy	Momentum
	General Form	
$0 = \dfrac{d}{dt}\displaystyle\int_{c.v.} \rho \, d\Psi + \int_{c.s.} \rho \mathbf{V} \cdot \hat{n} \, dA$	$-\Sigma \dot{W} = \dfrac{d}{dt}\displaystyle\int_{c.v.} \left(\dfrac{V^2}{2} + gz\right)\rho \, d\Psi$ $+ \displaystyle\int_{c.s.} \left(\dfrac{V^2}{2} + \dfrac{p}{\rho} + gz\right)\rho \mathbf{V} \cdot \hat{n} \, dA + \text{losses}$	$\Sigma \mathbf{F} = \dfrac{d}{dt}\displaystyle\int_{c.v.} \rho \mathbf{V} \, d\Psi + \int_{c.s.} \rho \mathbf{V}(\mathbf{V} \cdot \hat{n}) \, dA$
	Steady Flow	
$0 = \displaystyle\int_{c.s.} \rho \mathbf{V} \cdot \hat{n} \, dA$	$-\Sigma \dot{W} = \displaystyle\int_{c.s.} \left(\dfrac{V^2}{2} + \dfrac{p}{\rho} + gz\right)\rho \mathbf{V} \cdot \hat{n} \, dA + \text{losses}$	$\Sigma \mathbf{F} = \displaystyle\int_{c.s.} \rho \mathbf{V}(\mathbf{V} \cdot \hat{n}) \, dA$
	Steady, Nonuniform Flow	
$\dot{m} = \rho_1 A_1 \bar{V}_1 = \rho_2 A_2 \bar{V}_2$	$\dfrac{-\Sigma \dot{W}}{\dot{m}g} = \alpha_2 \dfrac{\bar{V}_2^2}{2g} + \dfrac{p_2}{\gamma_2} + z_2 - \alpha_1 \dfrac{\bar{V}_1^2}{2g} - \dfrac{p_1}{\gamma_1} - z_1 + h_L$	$\Sigma F_x = \dot{m}(\beta_2 \bar{V}_{2x} - \beta_1 \bar{V}_{1x})$ $\Sigma F_y = \dot{m}(\beta_2 \bar{V}_{2y} - \beta_1 \bar{V}_{1y})$
	Steady, Uniform Flow	
$\dot{m} = \rho_1 A_1 V_1 = \rho_2 A_2 V_2$	$-\dfrac{\Sigma \dot{W}}{\dot{m}g} = \dfrac{V_2^2}{2g} + \dfrac{p_2}{\gamma_2} + z_2 - \dfrac{V_1^2}{2g} - \dfrac{p_1}{\gamma_1} - z_1 + h_L$	$\Sigma \mathbf{F} = \dot{m}(\mathbf{V}_2 - \mathbf{V}_1)$
	Steady, Uniform, Incompressible Flow	
$Q = A_1 V_1 = A_2 V_2$	$-\dfrac{\Sigma \dot{W}}{\dot{m}g} = \dfrac{V_2^2}{2g} + \dfrac{p_2}{\gamma} + z_2 - \dfrac{V_1^2}{2g} - \dfrac{p_1}{\gamma} - z_1 + h_L$	$\Sigma \mathbf{F} = \dot{m}(\mathbf{V}_2 - \mathbf{V}_1)$
	or	
	$H_P + \dfrac{V_1^2}{2g} + \dfrac{p_1}{\gamma} + z_1 = H_T + \dfrac{V_2^2}{2g} + \dfrac{p_2}{\gamma} + z_2 + h_L$	

$\dot{m}$ = mass flux

Q = flow rate

$\bar{V}$ = average velocity

$\quad = \dfrac{\int V \, dA}{A}$

α = kinetic energy correction factor

$\quad = \dfrac{\int V^3 \, dA}{\bar{V}^3 A}$

β = momentum correction factor

$\quad = \dfrac{\int V^2 \, dA}{\bar{V}^2 A}$

h_L = head loss

$\Sigma \dot{W} = \dot{W}_S + \dot{W}_{\text{shear}} + \dot{W}_I$

H_P = pump head $= -\dfrac{\dot{W}_P}{\dot{m}g}$

H_T = turbine head $= \dfrac{\dot{W}_T}{\dot{m}g}$

PROBLEMS

Basic Laws

4.1. (a) State the necessary conditions for momentum of a system to remain constant.
(b) State the necessary conditions for energy of a system to remain constant.
(c) Show the detailed steps and state the assumptions that allow Eq. 4.1.3 to be reduced to $\Sigma \mathbf{F} = m\mathbf{a}$.

4.2. Make a list of five extensive properties that are of interest in fluid mechanics. Also, list their associated intensive properties. In addition, list five other intensive properties.

4.3. A control volume is identified as the volume interior to a balloon. At an instant the system is also identified as the air inside the balloon. Air escapes during a short time increment Δt. Sketch the system and the control volume at t and at $t + \Delta t$.

4.4. At an instant in time the control volume and the system occupy the volume inside the pump and a few diameters of the pipe at the outlet side. Sketch the system and the control volume at times t and $t + \Delta t$.

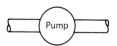

4.5. Indicate which fundamental equation would be most useful in determining the following quantity:
(a) The horsepower output of a pump.
(b) The mass flux from closing baffles.
(c) The drag force on an airfoil.
(d) The head loss in a pipeline.
(e) The rotational speed of a wind machine.

4.6. Sketch the unit vector $\hat{n}$ and the velocity vector $\mathbf{V}$ on each of the areas listed.
(a) The exit area of a fireman's nozzle.
b) The inlet area of a pump.
(c) The wall area of a pipe.
(d) The porous bottom area of a river into which a small amount of water flows.
(e) The circular exit area of a rotating impeller.

4.7. Sketch the unit vector $\hat{n}$ and the velocity vector $\mathbf{V}$ at several positions on a rectangular box surrounding the turbine blade shown.

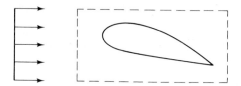

System-to-Control-Volume Transformation

4.8. We recognize that

$$\frac{d}{dt} \int_{c.v.} \rho\eta \, d\Psi = \int_{c.v.} \frac{\partial}{\partial t} \rho\eta \, d\Psi$$

What condition allows this equivalence? Why is an ordinary derivative used on the left and a partial derivative on the right?

Conservation of Mass

4.9. Water is flowing in a 6-cm-diameter pipe at 20 m/s. If the pipe enlarges to a diameter of 12 cm, calculate the reduced velocity. Also, calculate the mass flux and the flow rate.

4.9E. Water is flowing in a 2.5-in.-diameter pipe at 60 ft/sec. If the pipe enlarges to a diameter of 5 in., calculate the reduced velocity. Also, calculate the mass flux and the flow rate.

4.10. Water flows in the 5-cm-diameter pipe with an average velocity of 10 m/s. It turns a 90° angle and flows radially between two parallel plates. What is the velocity at a radius of 60 cm? What are the mass flux and the flow rate?

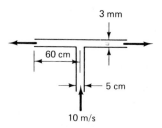

4.11. A pipe transports 200 kg/s of water. The pipe tees into a 5-cm-diameter pipe and a 7-cm-diameter pipe. If the average velocity in the smaller-diameter pipe is 25 m/s, calculate the flow rate in the larger pipe.

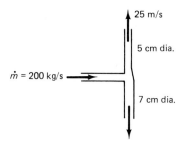

4.12. Air at 20°C and 200 kPa absolute flows in a 10-cm-diameter pipe with a mass flux of 2 kg/s. The pipe undergoes a conversion to a 5 cm by 7 cm rectangular duct in which $T = 80°C$ and $p = 50$ kPa absolute. Calculate the velocity in each section.

4.12E. Air at 60°F and 40 psia is flowing in a 4-in.-diameter pipe with a mass flux of 0.2 slug/sec. The pipe undergoes a conversion to a 2 in. by 3 in. rectangular duct in which $T = 150°F$ and $p = 7$ psia. Calculate the velocity in each section.

4.13. Air at 120°C and 500 kPa absolute is flowing in a pipe at 600 m/s and suddenly undergoes an abrupt change to 249°C and 1246 kPa absolute at a location where the diameter is 10 cm. Calculate the velocity after the sudden change (a shock wave). Also, calculate the mass flux, and flow rates before and after the shock wave.

4.14. Water flows with a velocity of 3 m/s at a depth of 1.5 m in a trapezoidal channel with a 2-m base and sides sloping at 45°. It empties into a circular pipe and flows at 2 m/s. What is the diameter if:
(a) The pipe flows full?
(b) The pipe flows half full?
(c) The water in the pipe flows at a depth of one-half the radius?

4.15. Water flows in a 8-cm-diameter pipe with the profiles shown. Find the average velocity, the mass flux, and the flow rate.

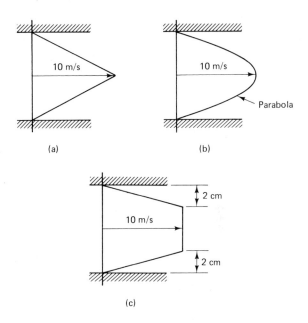

4.16. The profiles above are assumed to exist in a wide rectangular channel, 8 cm high and 80 cm wide. Find the average velocity, the mass flux, and the flow rate.

4.17. A constant-density fluid flows as shown. Find the equation of the parabola if the conduit is:
(a) A pipe.
(b) A wide rectangular channel.

4.17E. A constant-density fluid flows as shown. Find the equation of the parabola if the conduit is:
(a) A pipe.
(b) A wide rectangular channel.

4.18. A parabolic profile exists in a 10-mm-diameter pipe. The pipe contracts to a 5-mm-diameter pipe in which the velocity profile is essentially uniform at 2 m/s. Write the equation for the parabola. Assume an incompressible flow.

4.19. As air flows as shown over a flat plate, the velocity is reduced to zero at the wall. If $u(y) = 10[20y - 100y^2]$ m/s, find the mass flux through a surface parallel to the plate and 0.2 m above the plate. The plate is 2 m wide and $\rho = 1.23$ kg/m³.

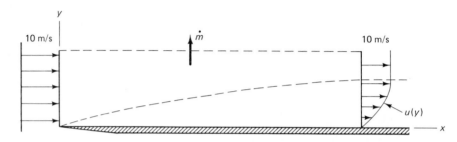

4.20. A streamline is 5 cm above the plate shown above at the leading edge. How far from the plate is that same streamline at the location of the profile $u(y) = 10(20y - 100y^2)$?

4.21. Stratified salt water flows at a depth of 10 cm in a channel with a velocity distribution $0.5(20y - 100y^2)$ m/s, where y is measured in meters. If the density varies linearly from 1200 kg/m³ at the bottom to clear water at the top, find $\dot{m}$. Also, show that $\dot{m} \neq \rho \bar{V} A$. The channel is 1.5 m wide.

4.21E. Stratified salt water flows at a depth of 4 in. in a channel with a velocity distribution $2(6y - 9y^2)$ ft/sec, where y is measured in feet. If the density varies linearly from 2.2 slug/ft³ at the bottom to clear water at the top, find $\dot{m}$. Also, show that $\dot{m} \neq \rho \bar{V} A$. The channel is 5 ft wide.

4.22. Water flows as shown. Calculate V_2.

4.23. Rain is falling vertically downward at 5.0 m/s onto a 9000-m² parking lot. All of the water flows from the lot in an open rectangular ditch with an average velocity of 1.5 m/s. Estimate the depth of flow in the 1.5-m-wide ditch if two thousand, 3-mm-diameter drops of water are contained in each cubic meter of rain.

4.24. Air at 250 kPa gage and 20°C is being forced into a tire, which has a volume of 0.5 m³, at a velocity of 60 m/s through a 6-mm-diameter opening. Determine the rate of change of density in the tire.

4.24E. Air at 37 psi gage and 60°F is being forced into a tire, which has a volume of 17 ft³, at a velocity of 180 ft/sec through a $\frac{1}{4}$ in.-diameter opening. Determine the rate of change of density in the tire.

4.25. If the mass of the control volume is not changing, find $\bar{V}_3$.

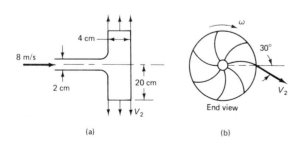

(a)

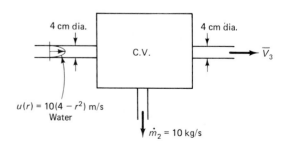

(b)

4.26. The average velocity $\bar{V}_3 = 10$ m/s in Problem 4.25. Find the rate at which the mass of the control volume is changing.

4.27. Find the velocity of the gas–fuel interface. Use $R_{gas} = 0.28$ kJ/kg · K and $d_e = 30$ cm.

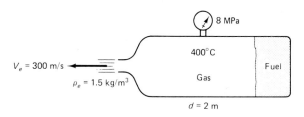

Energy Equation

4.28. Water flows in an open rectangular channel at a depth of 1 m with a velocity of 4 m/s. The bottom of the channel drops over a short length a distance of 1 m. Calculate the two possible depths of flow after the drop. Neglect all losses.

4.28E. Water flows in an open rectangular channel at a depth of 3 ft with a velocity of 12 ft/sec. The bottom of the channel drops over a short length a distance of 3 ft. Calculate the two possible depths of flow after the drop. Neglect all losses.

4.29. If the head loss in Problem 4.28 across the channel drop is 0.2 m, determine the two possible depths of flow.

4.29E. If the head loss in Problem 4.28E across the channel drop is 0.6 ft, determine the two possible depths of flow.

4.30. Find the velocity V_1 of the water in the vertical pipe. Assume no losses.

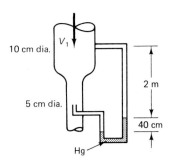

4.31. If the head loss coefficient (based on V_2) between sections 1 and 2 of Problem 4.30 is 0.05, determine V_1 of the water.

4.32. The flow rate of water in a 5-cm-diameter horizontal pipeline at a pressure of 400 kPa is 600 L/min. If the pipeline increases to 8 cm diameter, calculate the increased pressure if the loss coefficient (based on V_1) is 0.37.

4.32E. The flow rate of water in a 2-in.-diameter horizontal pipeline at a pressure of 60 psi is 120 gal/min. If the pipeline increases to 3 in. diameter, calculate the increased pressure if the loss coefficient (based on V_1) is 0.37.

4.33. Water flows at the rate of 600 L/min in a 4-cm-diameter horizontal pipeline with a pressure of 690 kPa. If the pressure after an enlargement to 6 cm diameter is measured to be 700 kPa, calculate the head loss across the enlargement.

4.34. Calculate the pressure at the position shown needed to maintain a flow rate of 0.08 m³/s of water in a 6-cm-diameter horizontal pipe leading to a nozzle if a loss coefficient, based on V_1, is 0.2 between the pressure gage and the exit.

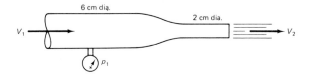

4.35. Neglect losses through the entrance, contraction and nozzle and predict the value of H and p if:
(a) $h = 15$ cm.
(b) $h = 20$ cm.

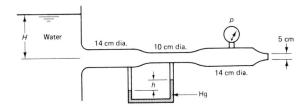

4.36. Water flows from the rectangular outlets shown. Estimate the flow rate per unit width for each if:
(a) $h = 80$ cm, $H = 2$ m.
(b) $h = 2$ ft, $H = 6$ ft.
Neglect all losses.

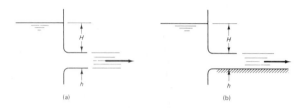

(a)   (b)

4.37. The overall loss coefficient for this pipe is 5; up to A it is 0.8, from A to B it is 1.2, from B to C it is 0.8, from C to D it is 2.2. Estimate the flow rate and the pressures at A, B, C, and D.

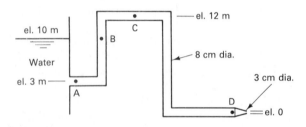

4.38. Water exits from a pressurized reservoir as shown. Calculate the flow rate if on section A we:
(a) Attach a nozzle with exit diameter 5 cm.
(b) Attach a diffuser with exit diameter 18 cm.
(c) Leave as an open pipe as shown.
Neglect losses for all cases.

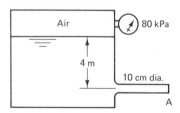

4.39. Rework Problem 4.38 assuming that $K_{pipe} = 1.5$, $K_{nozzle} = 0.04$ (based on V_1), and $K_{diffuser} = 0.8$ (based on V_1).

4.40. Relate the flow rate of the water through this venturi meter to the diameter and the manometer reading. Assume no losses.

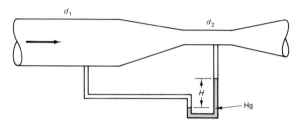

4.41. In the venturi meter of Problem 4.40, calculate the flow rate if:
(a) $H = 20$ cm, $d_1 = 2d_2 = 16$ cm.
(b) $H = 40$ cm, $d_1 = 3d_2 = 24$ cm.
(c) $H = 10$ in., $d_1 = 2d_2 = 6$ in.
(d) $H = 15$ in., $d_1 = 3d_2 = 12$ in.

4.42. Determine the maximum possible height H if cavitation is to be avoided. Let:
(a) $d = 10$ cm and $T = 20°C$.
(b) $d = 4$ in. and $T = 70°F$.
Neglect all losses and assume $p_{atm} = 100$ kPa (14.7 psi).

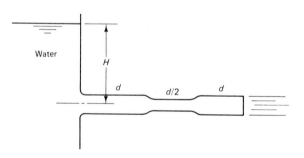

4.43. Cavitation is observed in the small section of pipe of Problem 4.42 when $H = 65$ cm. Estimate the temperature of the water. Neglect all losses and assume $p_{atm} = 100$ kPa (14.7 psi). Use d of Problem 4.42.

4.44. What is the maximum possible depth H if cavitation is to be avoided? Assume a vapor pressure of 6 kPa absolute and an overall loss coefficient of 2 based on V_2. Neglect losses up to the enlargement.

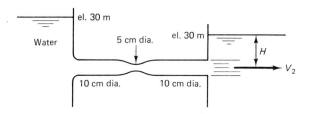

4.45. A contraction occurs in a 10-cm-diameter pipe to 6 cm followed by an enlargement back to 10 cm. The pressure upstream is measured to be 200 kPa when cavitation is first observed in the 20°C water. Calculate the flow rate. Neglect losses. Use $p_{atm} = 100$ kPa.

4.46. Calculate the maximum diameter D such that cavitation will be avoided if:
(a) $d = 20$ cm, $H = 5$ m, and $T_{water} = 20°C$.
(b) $d = 8$ in., $H = 15$ ft, and $T_{water} = 70°F$.
Neglect all losses and use $p_{atm} = 100$ kPa (14.7 psi).

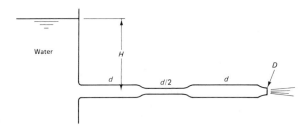

4.47. The overall loss coefficient in this siphon is 4; up to section A it is 1.5. At what height H will the siphon cease to function?

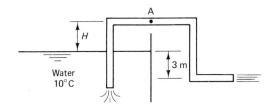

4.48. The pump shown is 85% efficient. If the pressure rise is 800 kPa, calculate the required energy input in watts.

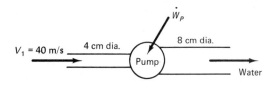

4.48E. The pump shown is 85% efficient. If the pressure rise is 120 psi, calculate the required energy input in horsepower.

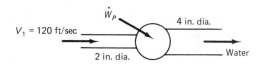

4.49. The pump shown in Problem 4.48 is powered by a 20-kW motor. If the pump is 82% efficient, determine the pressure rise.

4.50. A turbine, which is 87% efficient, accepts 2 m³/s of water from a 50-cm-diameter pipe. The pressure drop is 600 kPa and the exit velocity is small. What is the turbine output?

4.51. A turbine receives 15 m³/s of water from a 2-m-diameter pipe at a pressure of 900 kPa and delivers 10 000 kW. The pressure in the 2.5-m-diameter exit pipe is 120 kPa. Calculate the turbine efficiency.

4.51E. A turbine receives 450 ft³/sec of water from a 6-ft-diameter pipe at a pressure of 120 psi and delivers 10 000 kW. The pressure in the $7\frac{1}{2}$ ft-diameter exit pipe is 18 psi. Calculate the turbine efficiency.

4.52. Air enters a compressor with negligible velocity at 85 kPa absolute and 20° C. It leaves with a velocity of 200 m/s at 600 kPa absolute. For a mass flux of 5 kg/s, calculate the exit temperature if the power required is 1500 kW and:
(a) No heat transfer occurs.
(b) The heat transfer rate is 60 kW.

4.53. Air enters a compressor at standard conditions with negligible velocity. At the 2-cm-diameter exit the pressure, temperature, and velocity are 400 kPa absolute, 150°C, and 200 m/s, respectively. If the heat transfer is 20 kJ/kg of air, find the power required by the compressor.

4.53E. Air enters a compressor at standard conditions with negligible velocity. At the 1-in.-diameter exit the pressure, temperature, and velocity are 60 psia, 300°F, and 600 ft/sec, respectively. If the heat transfer is 10 Btu/lb of air, find the power required by the compressor.

4.54. A small river with a flow rate of 5 m³/s feeds the reservoir shown. Calculate the energy that is available continuously if the turbine is 80% efficient. The loss coefficient for the overall piping system is $K = 4.5$.

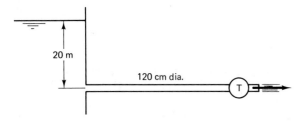

4.55. For the system shown, the average velocity in the pipe is 10 m/s. Up to point A, $K = 1.5$, from B to C, $K = 6.2$, and the pump is 80% efficient. If $p_C = 200$ kPa find p_A and p_B and the power required by the pump.

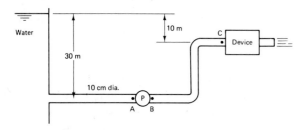

4.56. Determine the power output of the turbine for a water flow rate of 0.6 m³/s. The turbine is 90% efficient.

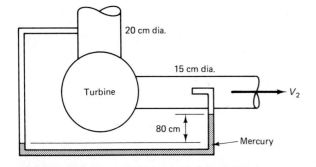

4.56E. Determine the power output of the turbine for a water flow rate of 18 ft³/sec. The turbine is 90% efficient.

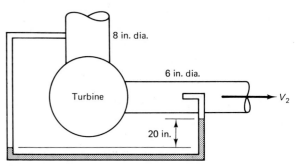

4.57. Find the power requirement of the 85%-efficient pump if the loss coefficient up to A is 3.2, and from B to C, $K = 1.5$. Neglect the losses through the exit nozzle. Also, calculate p_A and p_B.

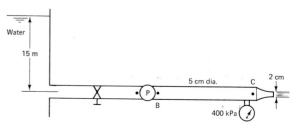

4.58. A 75%-efficient pump delivers 0.1 m³/s of water from a reservoir to a device at an elevation of 50 m above the reservoir. The pressure at the 8-cm-diameter entrance to the device is 180 kPa. If the piping loss coefficient is 5.6, find the necessary power input to the pump.

4.59. Water leaves a reservoir with a head of 10 m and flows through a horizontal 10-cm-diameter pipe, then exits to the atmosphere. At the end of the pipe a short length of smaller-diameter pipe is added. The loss coefficient, including the reduction, is 2.2, and after the reduction the losses are negligible. Calculate the minimum diameter of the smaller pipe if the flow rate is 0.02 m³/s.

4.59E. Water leaves a reservoir with a head of 30 ft and flows through a 4-in.-diameter pipe, then exits to the atmosphere. At the end of the pipe a short length of smaller-diameter pipe is added. The loss coefficient, including the reduction, is 2.2, and after the reduction the losses are negligible. Calculate the minimum diameter of the smaller pipe if the flow rate is 0.6 ft³/sec.

4.60. Neglect losses and find the depth of water on the raised section of the rectangular channel. Assume uniform velocity profiles.

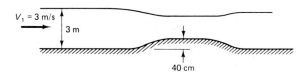

4.61. Determine the rate of kinetic energy loss, in watts, due to the 10-m-long cylinder. Assume a plane flow with $\rho = 1.23$ kg/m³.

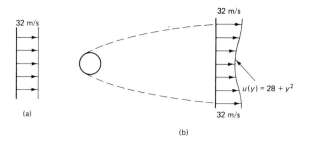

4.62. Calculate the head loss between the two sections shown.

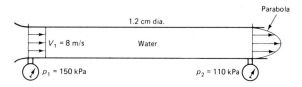

Momentum Equation

4.63. Water flows in a 6-cm-diameter pipe at a pressure of:
(a) 200 kPa.
(b) 400 kPa.
(c) 600 kPa.
It flows out a 3-cm-diameter nozzle to the atmosphere. Calculate the force of the water on the nozzle.

4.63E. Water flows in a 2.5-in.-diameter pipe at a pressure of:
(a) 30 lb/in².
(b) 60 lb/in².
(c) 90 lb/in².
It flows out a 1.2-in.-diameter nozzle to the atmosphere. Calculate the force of the water on the nozzle.

4.64. A nozzle and hose is attached to the ladder of a fire truck. What force is needed to hold a nozzle supplied by a 9-cm-diameter hose, with a pressure of 2000 kPa? The nozzle outlet diameter is 3 cm.

4.65. Water flows in a 10-cm-diameter pipe at a pressure of 400 kPa out a straight nozzle. Calculate the force of the water on the nozzle if the outlet diameter is:
(a) 8 cm.
(b) 6 cm.
(c) 4 cm.
(d) 2 cm.

4.66. Find the horizontal force of the water on the horizontal bend shown.

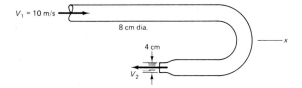

4.66E. Find the horizontal force of the water on the horizontal bend shown.

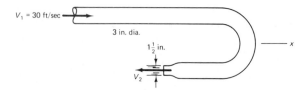

4.67. Find the horizontal force components of the water on the horizontal bend shown if p_1 is:
(a) 200 kPa.
(b) 400 kPa.
(c) 800 kPa.

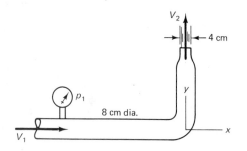

4.68. What is the net force needed to hold this orifice plate onto the pipe?

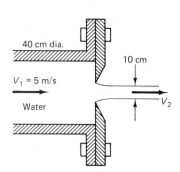

4.69. Assuming uniform velocity profiles, find F needed to hold the plug in the pipe. Neglect viscous effects.

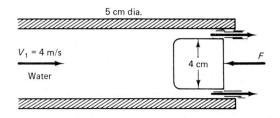

4.70. Water flows at 15 m/s in a 10-cm-diameter stem of a horizontal T-section that branches into 5-cm-diameter pipes. Find the force of the water on the T-section if the branches exit to the atmosphere. Neglect viscous effects.

4.70E. Water flows at 45 ft/sec in a 4-in.-diameter stem of a horizontal T-section that branches into 2-in.-diameter pipes. Find the force of the water on the T-section if the branches exit to the atmosphere. Neglect viscous effects.

4.71. Find the horizontal force components on the horizontal T-section shown. Neglect viscous effects.

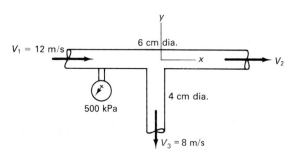

4.72. A horizontal 10-cm-diameter jet of water with $\dot{m} = 300$ kg/s strikes a vertical plate. Calculate:
(a) The force needed to hold the plate stationary.
(b) The force needed to move the plate away from the jet at 10 m/s.
(c) The force needed to move the plate into the jet at 10 m/s.

4.73. A horizontal 6-cm-diameter jet of water strikes a vertical plate. Determine the velocity issuing from the jet if a force of 900 N is needed to:
(a) Hold the plate stationary.
(b) Move the plate away from the jet at 10 m/s.
(c) Move the plate into the jet at 10 m/s.

4.73E. A horizontal $2\frac{1}{2}$ in.-diameter jet of water strikes a vertical plate. Determine the velocity issuing from the jet if a force of 200 lb is needed to:
(a) Hold the plate stationary.
(b) Move the plate away from the jet at 30 ft/sec.
(c) Move the plate into the jet at 30 ft/sec.

4.74. Determine the mass flux issuing from the jet if a force of 700 N is needed to:
(a) Hold the cone stationary.
(b) Move the cone away from the jet at 8 m/s.
(c) Move the cone into the jet at 8 m/s.

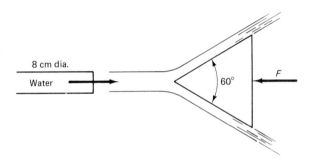

4.75. Calculate the force components of the water acting on the deflecting blade shown if:
(a) The blade is stationary.
(b) The blade moves to the right at 20 m/s.
(c) The blade moves to the left at 20 m/s.

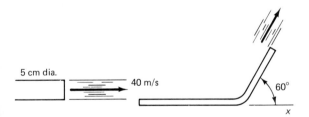

4.75E. Calculate the force components of the water acting on the deflecting blade shown if:
(a) The blade is stationary.
(b) The blade moves to the right at 60 ft/sec.
(c) The blade moves to left at 60 ft/sec.

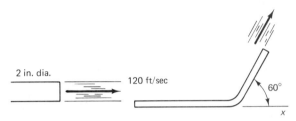

4.76. The blade of Problem 4.75 is one of a series of blades that are attached to a 50-cm-radius rotor that has a rotational speed of 30 rad/s. If there are 10 such water jets, find the power output.

4.76E. The blade of Problem 4.75E is one of a series of blades that are attached to a 20-in.-radius rotor that has a rotational speed of 30 rad/s. If there are 10 such water jets, find the power output.

4.77. Determine the force components of superheated steam acting on the blade shown if:
(a) The blade is stationary.
(b) The blade moves to the right at 100 m/s.
(c) The blade moves to the left at 100 m/s.

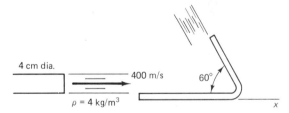

4.78. The blade of Problem 4.77 is one of a series of blades that are attached to a 1.2-m-radius rotor that rotates at 150 rad/s. Calculate the power output if there are 15 such steam jets.

4.79. Superheated steam jets strike the turbine blades shown. Find the power output of the turbine if 15 jets are involved and α_1 is:
(a) 45°.
(b) 60°.
(c) 90°.

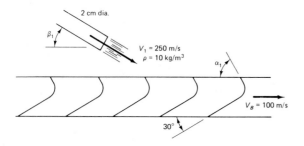

4.79E. Superheated steam jets strike the turbine blades shown. Find the power output if there are 15 jets and α_1 is:
(a) 45°.
(b) 60°.
(c) 90°.

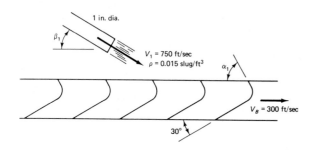

4.80. Twelve high-speed water jets strike the blades as shown. Find the power output and the blade angles if V_B is:
(a) 20 m/s.
(b) 40 m/s.
(c) 50 m/s.

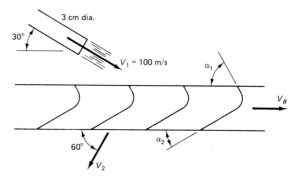

4.81. Fifteen water jets strike the blades of a turbine as shown. Calculate the power output and the blade angles if β_2 is:
(a) 60°,
(b) 70°.
(c) 80°.

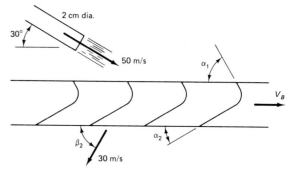

4.82. Water flows from the rectangular jet as shown. Find the force F and the mass fluxes $\dot{m}_2$ and $\dot{m}_3$ if θ is:
(a) 60°.
(b) 45°.
(c) 30°.

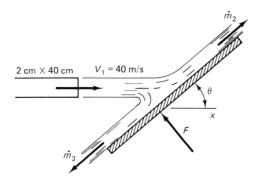

4.82E. Water flows from the rectangular jet as shown. Find the force F and the mass fluxes $\dot{m}_2$ and $\dot{m}_3$ if θ is:
(a) 60°.
(b) 45°.
(c) 30°.

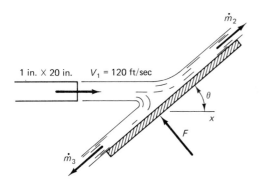

4.83. The plate of Problem 4.82(a) moves to the left at 20 m/s. Find the power required.

4.84. Calculate the velocity that the plate of Problem 4.82(a) must move with (in the x-direction) to produce the maximum power output.

4.85. A large vehicle with a mass of 100 000 kg is slowed by inserting a 180° deflector in a trough of water. If the 60-cm-wide deflector scoops off 10 cm of water, calculate the initial deceleration if the vehicle is traveling at 120 km/h. Also, find the time necessary time to reach a speed of 60 km/h.

4.85E. A large vehicle with a mass of 6000 slugs is slowed by inserting a 180° deflector in a trough of water. If the 24-in.-wide deflector scoops off 4 in. of water, calculate the initial deceleration if the vehicle is traveling at 90 mph. Also, find the necessary time to reach a speed of 40 mph.

4.86. A 2.5-m-wide snowplow travels at 50 km/h plowing snow at a depth of 0.8 m. The snow leaves the blade normal to the direction of motion of the snowplow. What power does the plowing operation require if the density of the snow is 90 kg/m³?

4.87. A vehicle with a mass of 5000 kg is traveling at 900 km/h. It is decelerated by lowering a 20-cm-wide scoop into water a depth of 6 cm. If the water is deflected through 180°, calculate the distance the vehicle must travel for the speed to be reduced to 100 km/hr.

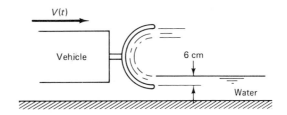

4.88. A 300-kg vehicle is required to have an initial acceleration of 2 m/s². It is proposed that a 6-cm-diameter water jet impinge a vane built into the rear of the vehicle that will divert the water through an angle of 180°. What jet velocity is necessary? What speed will be achieved after 2 s?

4.88E. A 20-slug vehicle is required to have an initial acceleration of 6 ft/sec². It is proposed that a $2\frac{1}{2}$-in.-diameter water jet strike a vane built into the rear of the vehicle that will divert the water through an angle of 180°. What jet velocity is necessary? What speed will be achieved after 2 sec?

4.89. A swamp boat is propelled at 50 km/h by a 2-m-diameter propeller that requires a 20-kW engine. Calculate the thrust on the boat, the flow rate of air through the propeller, and the propeller efficiency.

4.90. An aircraft is propelled by a 2.2-m-diameter propeller at a speed of 200 km/h. The air velocity downstream of the propeller is 320 km/h relative to the aircraft. Determine the pressure difference across the propeller blades and the required power. Use $\rho = 1.2$ kg/m^3.

4.91. The 50-cm-diameter propeller on a boat moves at 30 km/h by causing a velocity of 60 km/h relative to the boat. Calculate the required power and the mass flux of water through the propeller.

4.91E. The 20-in.-diameter propeller on a boat moves at 20 mph by causing a velocity of 40 mph relative to the boat. Calculate the required power and the mass flux of water through the propeller.

4.92. A jet boat takes in 0.2 m^3/s of water and discharges it at a velocity of 20 m/s relative to the boat. If the boat travels at 10 m/s, calculate the thrust produced and the power required.

4.93. Neglect viscous effects, assume uniform velocity profiles, and find the horizontal force component acting on the obstruction shown.

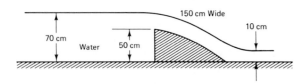

4.94. Assuming hydrostatic pressure distributions, uniform velocity profiles, and negligible viscous effects, find the force on the sluice gate shown.

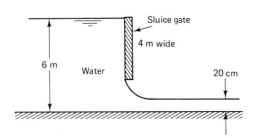

4.95. A sudden jump (a hydraulic jump) occurs in a rectangular channel as shown. Find y_2 and V_2 if
(a) $V_1 = 8$ m/s, $y_1 = 60$ cm.
(b) $V_1 = 12$ m/s, $y_1 = 40$ cm.
(c) $V_1 = 20$ ft/sec, $y_1 = 2$ ft.
(d) $V_1 = 30$ ft/sec, $y_1 = 3$ ft.

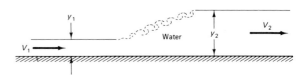

4.96. A hydraulic jump, as shown above, occurs so that $V_2 = \frac{1}{4}V_1$. Find V_1 and y_2 if
(a) $y_1 = 80$ cm.
(b) $y_1 = 2$ ft.

4.97. For a flow rate of 9 m^3/s, find V_2 and y_2 for the hydraulic jump shown. The channel is 3 m wide. Neglect losses up to the jump.

4.98. The velocity is measured to be 3 m/s downstream of a hydraulic jump where the depth is 2 m. Calculate the velocity and depth prior to the jump.

4.98E. The velocity is measured to be 10 ft/sec downstream of a hydraulic jump where the depth is 6 ft. Calculate the velocity and depth prior to the jump.

4.99. Find the pressure p_2 if $p_1 = 60$ kPa and $V_1 = 20$ m/s. *Note:* The pressure immediately after the pipe expansion is p_1.

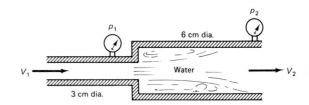

4.100. Calculate the change in the momentum flux of the water flowing through the plane contraction shown if the flow rate is 0.2 m³/s. The slope of the two profiles is the same.

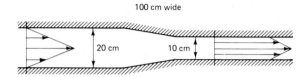

100 cm wide

20 cm   10 cm

4.101. Determine the momentum correction factor for the following profile shown in Problem 4.100:
(a) The entering profile.
(b) The exiting profile.

4.102. Water at 20°C flows in a 4-cm-diameter, horizontal pipe and experiences a pressure drop of 200 Pa over a 10-m length of pipe. Calculate the velocity gradient at the wall. Recall that $\tau = \mu |du/dr|$.

4.102E. Water at 60°F flows in a $1\frac{1}{2}$-in.-diameter, horizontal pipe and experiences a pressure drop of 0.03 psi over a 30-ft length of pipe. Calculate the velocity gradient at the wall. Recall that $\tau = \mu |du/dr|$.

4.103. Find the drag force on the walls between the two sections of the horizontal pipe shown.

1.2 cm dia.

Parabola

$V_1 = 8$ m/s

$p_1 = 150$ kPa $p_2 = 110$ kPa

4.104. The velocity distribution downstream of a 10-m-long circular cylinder is as shown. Determine the force of the air on the cylinder. Use $\rho = 1.23$ kg/m³.

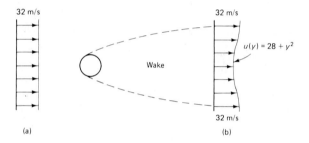

32 m/s   32 m/s

$u(y) = 28 + y^2$

Wake

32 m/s

(a)   (b)

4.104E. The velocity distribution downstream of a 30-ft-long cylinder is given by $u(y) = 96 + y^2/25$. See the figure in Problem 4.104. If the free-stream velocity upstream of the cylinder is 100 ft/sec, find the force of the air on the cylinder. Use $\rho = 0.0024$ slug/ft³.

4.105. Calculate the drag force acting on the flat 2-m-wide plate shown. Outside the viscous region the velocity is uniform. Select (use $\rho = 1.23$ kg/m³):
(a) A rectangular control volume extending outside the viscous region (mass flux crosses the top).
(b) A control volume with the upper boundary a streamline (no mass flux crosses a streamline).

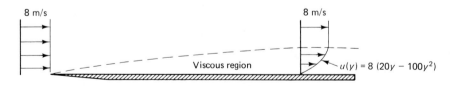

8 m/s   8 m/s

Viscous region $u(y) = 8 (20y - 100y^2)$

Momentum and Energy

4.106. Find a relationship between the acceleration of the cylindrical cart and the variables shown. Neglect friction. The initial mass of the cart and water is m_0.

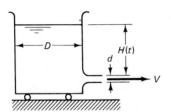

4.107. Determine the power lost in the hydraulic jump of:

(a) Problem 4.95(a).
(b) Problem 4.97.
(c) Problem 4.98.
(d) Problem 4.98E.

4.108. Calculate the loss coefficient for the expansion of Problem 4.99. Base the coefficient on the velocity V_1.

Moment of Momentum

4.109. A four-armed water sprinkler has nozzles at right angles to the 30-cm-long arms and at 45° angles with the ground. If the outlet diameters are 8 mm and 4 kg/s of water exits the four nozzles, find the rotational speed.

4.110. A four-armed rotor has 10-mm-diameter nozzles that exit water at 70 m/s relative to the arm. The nozzles are at right angles to the 25-cm-long arms and are parallel to the ground. If the rotational speed is 30 rad/s, find the power output. The arms are 3 cm in diameter.

4.110E. A four-armed rotor has $\frac{1}{2}$-in.-diameter nozzles that exit water at 200 ft/sec relative to the arm. The nozzles are at right angles to the 10-in.-long arms and are parallel to the ground. If the rotational speed is 30 rad/sec, find the power output. The arms are 1.5 in. in diameter.

4.111. Water flows out the 6-mm slots as shown. Calculate ω if 20 kg/s is delivered by the two arms.

4.112. A 1-kW motor, drives the rotor shown at 500 rad/s. Determine the flow rate neglecting all losses. Use $\rho = 1.23$ kg/m³.

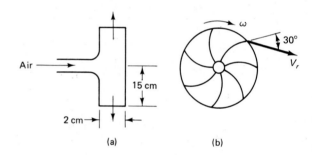

4.113. Find an expression for $\omega(t)$ if the sprinkler of Problem 4.109 is suddenly turned on at $t = 0$. Assume the arms to be 2 cm in diameter.

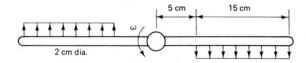

FIVE

THE DIFFERENTIAL FORMS
OF THE
FUNDAMENTAL LAWS

5.1 INTRODUCTION

In Chapter 4 the basic laws were expressed in terms of a fixed control volume, a finite volume in space. This is often described as the global approach to the mechanics of fluids. To find a solution using the control volume it was necessary either to assume an approximation to the integrands or expressions for the integrands were given. Suppose that we wish to find an integral quantity, such as the flow rate under a dam or the lift on an airfoil, and we cannot make a reasonable assumption for the velocity or pressure distribution. It is then necessary to determine the distributions that enter the integrands before the integral quantity can be found. This is done by solving the differential equations that express the basic laws.

Not only do solutions of the differential forms of the basic laws allow us to determine integral quantities of interest, they often contain information in themselves. For example, we may wish to know the exact location of the minimum pressure on a body, or the region of separated flow from a surface may be of interest. So we often solve the differential equations to answer a specific question raised about a special flow.

There are two primary methods used in deriving the differential forms of the fundamental laws. One method involves the application of Gauss's theorem, which allows the area integrals to be transformed to volume integrals; the integrands are then collected under one integral, which can be set equal to zero. The integration is valid over any arbitrary control volume, and thus the integrand can be set equal to zero, providing us with the differential form of the basic law. The other approach, the one used in this book, is to identify an infinitesimal element in space and apply the basic laws directly to that element. Both methods result in the differential forms of the basic laws; the first method, however, demands the use of vector and tensor calculus, mathematics that is usually considered unnecessary in a first course in fluids. We introduce some vector calculus but it will not be used at the operational level demanded by Gauss's theorem.

The conservation of mass, applied to an infinitesimal element, leads to the differential continuity equation; it relates the density and velocity fields.[1] Newton's second law (a vector relationship) results in partial differential equations known as the Navier–Stokes equations; they relate the velocity, pressure, and density fields and introduce the viscosity and the gravity vector in a fluid flow. The first law of thermodynamics provides us with the differential energy equation, which relates the temperature field to the velocity, density, and pressure fields and introduces the specific heat and thermal conductivity. Most problems considered in an introduc-

---

[1]When a dependent variable depends on more than one independent variable, it is referred to as a field, that is, $\mathbf{V}(x, y)$ is a velocity field and $\rho(x, y, z, t)$ is a density field. The partial differential equations describing the field quantities are often called **field equations.**

tory course are for isothermal, incompressible flows in which the temperature field does not play a role; for such flows the three Navier–Stokes
equations and the continuity equation provide four partial differential
equations that relate the three velocity components and the pressure.
Thus the energy equation would not be needed. We will, however, derive
the differential energy equation for use in a limited number of situations.

Partial differential equations require conditions that specify certain
values for the dependent variables at particular values of the independent
variables. If the independent variable is time, the conditions are called
initial conditions; if the independent variable is a space coordinate, the
conditions are **boundary conditions**. The total problem is referred to as an
initial-value problem or a **boundary value problem**.

In fluid mechanics the boundary conditions result from:

The no-slip conditions for a viscous flow. The viscosity causes the
fluid to stick to the wall and thus the velocity of the fluid at the wall
assumes the velocity of the wall.

The normal component of velocity in an inviscid flow, a flow in
which viscous effects are negligible. Near a wall the velocity vector
must be tangent to the wall, demanding that the normal component
be zero.

The pressure in a flow involving a free surface. For problems with a
free surface, such as a flow with a liquid–gas interface, the pressure
is known on the interface. This would be the situation involving
wave motion.

The temperature of the boundary or the temperature gradient at the
boundary. If the temperature of the boundary is held constant, the
temperature of the fluid next to the boundary will equal the boundary temperature. If the boundary is insulated, the temperature gradient will be zero at the boundary.

In an unsteady flow the initial condition demands that the three velocity
components be specified at a particular instant in time, usually taken as
$t = 0$.

The material in this chapter is often omitted in an introductory
course. This book has been designed to allow two possible routes: The
general equations presented in this chapter may be used, or equations
unique to a particular geometry may be derived without referring to these
general equations.

5.2 DIFFERENTIAL CONTINUITY EQUATION

Differential equations take very different forms depending on the coordinate system chosen. We derive the equations using rectangular coordi-

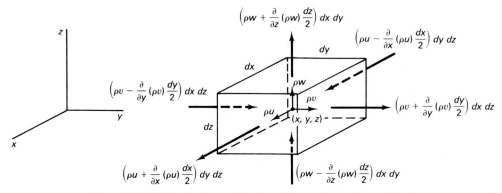

Figure 5.1 Infinitesimal control volume using rectangular coordinates.

nates and then express the equations for both cylindrical and spherical coordinates. Consider the mass flux through each face of the infinitesimal control volume shown in Fig. 5.1. We set the net flux of mass entering the element equal to the rate of change of the mass of the element; that is,

$$\dot{m}_{\text{in}} - \dot{m}_{\text{out}} = \frac{\partial}{\partial t} m_{\text{element}}$$

This takes the form

$$\left[\rho u - \frac{\partial(\rho u)}{\partial x} \frac{dx}{2}\right] dy\, dz - \left[\rho u + \frac{\partial(\rho u)}{\partial x} \frac{dx}{2}\right] dy\, dz$$

$$+ \left[\rho v - \frac{\partial(\rho v)}{\partial y} \frac{dy}{2}\right] dx\, dz - \left[\rho v + \frac{\partial(\rho v)}{\partial y} \frac{dy}{2}\right] dx\, dz$$

$$+ \left[\rho w - \frac{\partial(\rho w)}{\partial z} \frac{dz}{2}\right] dx\, dy - \left[\rho w + \frac{\partial(\rho w)}{\partial z} \frac{dz}{2}\right] dx\, dy$$

$$= \frac{\partial}{\partial t}(\rho\, dx\, dy\, dz) \tag{5.2.1}$$

Subtracting the appropriate terms and dividing by $dx\, dy\, dz$ yields

$$\frac{\partial}{\partial x}(\rho u) + \frac{\partial}{\partial y}(\rho v) + \frac{\partial}{\partial z}(\rho w) = -\frac{\partial \rho}{\partial t} \tag{5.2.2}$$

This can be put in the form

$$\frac{\partial \rho}{\partial t} + u\frac{\partial \rho}{\partial x} + v\frac{\partial \rho}{\partial y} + w\frac{\partial \rho}{\partial z} + \rho\left(\frac{\partial u}{\partial x} + \frac{\partial v}{\partial y} + \frac{\partial w}{\partial z}\right) = 0 \tag{5.2.3}$$

or, in terms of the substantial derivative (see Eq. 3.2.11),

$$\frac{D\rho}{Dt} + \rho\left(\frac{\partial u}{\partial x} + \frac{\partial v}{\partial y} + \frac{\partial w}{\partial z}\right) = 0 \qquad (5.2.4)$$

This is the preferred form of the **differential continuity equation**.

We can introduce the gradient operator, which, in rectangular coordinates, is

$$\nabla = \frac{\partial}{\partial x}\,\hat{i} + \frac{\partial}{\partial y}\,\hat{j} + \frac{\partial}{\partial z}\,\hat{k} \qquad (5.2.5)$$

The continuity equation can then be written in the form

$$\frac{D\rho}{Dt} + \rho\,\nabla \cdot \mathbf{V} = 0 \qquad (5.2.6)$$

where $\mathbf{V} = u\hat{i} + v\hat{j} + w\hat{k}$.

For the case of **incompressible flow,** a flow in which the density of a fluid particle does not change as it travels along, we see that

$$\frac{D\rho}{Dt} = \frac{\partial\rho}{\partial t} + u\,\frac{\partial\rho}{\partial x} + v\,\frac{\partial\rho}{\partial y} + w\,\frac{\partial\rho}{\partial z} = 0 \qquad (5.2.7)$$

Note that this is less restrictive than the assumption of constant density, which would require each term in Eq. 5.2.7 to be zero. Incompressible flows that have density gradients are sometimes referred to as **stratified flows**; atmospheric flow is an example of a stratified flow. Using Eq. 5.2.4, the continuity equation, for an incompressible flow, takes the form

$$\frac{\partial u}{\partial x} + \frac{\partial v}{\partial y} + \frac{\partial w}{\partial z} = 0 \qquad (5.2.8)$$

or, in vector form,

$$\nabla \cdot \mathbf{V} = 0 \qquad (5.2.9)$$

In cylindrical and spherical coordinates the continuity equation for an incompressible flow is presented in Table 5.1. The expression for D/Dt in cylindrical and spherical coordinates can also be found in Table 5.1.

EXAMPLE 5.1

Air flows in a pipe and the velocity at three neighboring points A, B, and C, 10 cm apart, is measured to be 92, 96, and 98 m/s, respectively, as shown. The tempera-

A   92 m/s   B   96 m/s   C   98 m/s

————— x

ture and pressure are 40°C and 350 kPa absolute, respectively, at point B. Find dp/dx at that point, assuming steady, uniform flow.

Solution

The continuity equation (5.2.3) for steady, uniform flow is written as

$$u \frac{d\rho}{dx} + \rho \frac{du}{dx} = 0$$

since $v = w = 0$ and $\partial \rho / \partial t = 0$. The velocity derivative is found to be approximated by

$$\frac{du}{dx} \simeq \frac{\Delta u}{\Delta x} = \frac{98 - 92}{0.20} = 30 \frac{m/s}{m}$$

where the more accurate central difference has been used.[2] The density is

$$\rho = \frac{p}{RT}$$

$$= \frac{350}{0.287 \times 313} = 3.90 \ kg/m^3$$

where absolute pressure and temperature are used. The density derivative is then

$$\frac{d\rho}{dx} = -\frac{\rho}{u} \frac{du}{dx}$$

$$= -\frac{3.90}{96} \times 30 = -1.22 \ kg/m^4$$

[2]Using a forward difference $\Delta u / \Delta x = (98 - 96)/0.1 = 20$; a backward difference provides $\Delta u / \Delta x = (96 - 92)/0.1 = 40$. These two approximations are less accurate than the central difference used in the example. A sketch of a curve $u(x)$ could graphically show this by displaying three points at $x - \Delta x$, x, and $x + \Delta x$ and sketching the slopes.

EXAMPLE 5.1 (English)

Air flows in a pipe and the velocity at three neighboring points A, B, and C, 4 in. apart, is measured to be 274, 285, and 291 ft/sec, respectively, as shown. The temperature and pressure are 100°F and 50 psia, respectively, at point B. Find dp/dx at that point, assuming steady, uniform flow.

274 ft/sec   285 ft/sec   291 ft/sec

————— x

Solution

The continuity equation (5.2.3) for steady, uniform flow is written as

$$u \frac{d\rho}{dx} + \rho \frac{du}{dx} = 0$$

since $v = w = 0$ and $\partial\rho/\partial t = 0$. The velocity derivative is found to be approximated by

$$\frac{du}{dx} \cong \frac{\Delta u}{\Delta x}$$

$$= \frac{291 - 274}{8/12} = 25.5 \frac{\text{ft/sec}}{\text{ft}}$$

where the more accurate central difference has been used.[3] The density is

$$\rho = \frac{p}{RT}$$

$$= \frac{50 \times 144}{1716 \times 510} = 0.00823 \text{ slug/ft}^3$$

where absolute pressure and temperature are used. The density derivative is then

$$\frac{d\rho}{dx} = -\frac{\rho}{u} \frac{du}{dx}$$

$$= -\frac{0.00823}{285} \times 25.5 = -0.000736 \text{ slug/ft}^4$$

---

[3]Using a forward difference $\Delta u/\Delta x = (291 - 285)/0.333 = 18$; a backward difference provides $\Delta u/\Delta x = (285 - 274)/0.333 = 33$. These two approximations are less accurate than the central difference used in the example. A sketch of a curve $u(x)$ could graphically show this by displaying three points at $x - \Delta x$, x, and $x + \Delta x$ and sketching the slopes.

EXAMPLE 5.2

The x-component of velocity at points A, B, C, and D, which are 10 mm apart, is measured to be 5.76, 6.72, 7.61, and 8.47 m/s, respectively, in the plane steady, incompressible flow shown in which $w = 0$. Approximate the x-component acceleration at C and the y-component of velocity 6 mm above B.

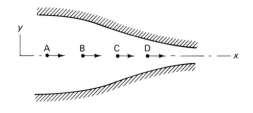

Solution

The desired acceleration component is found from Eq. 3.2.11 to be

$$a_x = u \frac{\partial u}{\partial x} + \cancel{v \frac{\partial u}{\partial y}}^{\,0}$$

$$= 761 \frac{8.47 - 6.72}{0.02} = 666 \text{ m/s}^2$$

where we have assumed a symmetrical flow so that v along the centerline is zero.

The y-component of velocity 6 mm above B is found using the continuity equation as follows:

$$\frac{\partial v}{\partial y} = -\frac{\partial u}{\partial x}$$

$$\frac{\Delta v}{\Delta y} \cong -\frac{\Delta u}{\Delta x} = -\frac{7.61 - 5.76}{0.02} = -92.5$$

$$\therefore \quad \Delta v = -92.5 \, \Delta y$$

$$= -92.5 \times 0.06 = -5.55 \text{ m/s}$$

We know that $v = 0$ at B; hence at the desired location,

$$v = -5.55 \text{ m/s}$$

EXAMPLE 5.3

Write the expression $\rho \, D\bar{u}/Dt + p \, \boldsymbol{\nabla} \cdot \mathbf{V}$ in terms of enthalpy h rather than internal energy $\bar{u}$. Recall that $h = \bar{u} + p/\rho$ (see Eq. 1.7.14).

Solution

Using the definition of enthalpy,

$$h = \bar{u} + \frac{p}{\rho}$$

we can write

$$\frac{D\bar{u}}{Dt} = \frac{Dh}{Dt} - \frac{1}{\rho} \frac{Dp}{Dt} + \frac{p}{\rho^2} \frac{D\rho}{Dt}$$

The desired expression is then

$$\rho\,\frac{D\bar{u}}{Dt} + p\,\boldsymbol{\nabla}\cdot\mathbf{V} = \rho\,\frac{Dh}{Dt} - \frac{Dp}{Dt} + \frac{p}{\rho}\frac{D\rho}{Dt} + p\,\boldsymbol{\nabla}\cdot\mathbf{V}$$

The continuity equation (5.2.6) is introduced resulting in

$$\rho\,\frac{D\bar{u}}{Dt} + p\,\boldsymbol{\nabla}\cdot\mathbf{V} = \rho\,\frac{Dh}{Dt} - \frac{Dp}{Dt} + \frac{p}{\rho}\frac{D\rho}{Dt} + p\left(-\frac{1}{\rho}\frac{D\rho}{Dt}\right)$$

$$= \rho\,\frac{Dh}{Dt} - \frac{Dp}{Dt}$$

and enthalpy has been introduced.

5.3 DIFFERENTIAL MOMENTUM EQUATION

5.3.1 General Formulation

Suppose that we do not know the velocity field or the pressure field in an incompressible[4] flow of interest and we wish to solve differential equations to provide us with that information. The differential continuity equation is one differential equation to help us toward this end; however, it has three unknowns, the three velocity components. The differential momentum equation is a vector equation and thus provides us with three scalar equations. These component equations will aid us in our attempt to determine the velocity and pressure fields. There is a difficulty in deriving these equations, however, since we must use the stress components to determine the forces required in the momentum equation. Let us identify these stress components.

There are nine stress components that act at a particular point in a fluid flow. They are the nine components of the stress tensor τ_{ij}. We will not study the properties of a stress tensor in detail in this introductory course since we do not have to maximize or minimize the stress (as would be required in a solid mechanics course); we must, however, use the nine stress components in our derivations, then relate the stress components to the velocity and pressure fields with the appropriate equations. The stress components that act at a point are displayed on two- and three-dimensional rectangular elements in Fig. 5.2. These elements are considered to be an exaggerated point, a cubical point; the stress components act in the positive direction on a positive face (a normal vector points in the positive coordinate direction) and in the negative direction on a negative face (a normal vector points in the negative coordinate direction).

---

[4] An incompressible flow, when referred to in a general discussion such as in this section, will generally refer to a constant-density flow. This is true in most fluid mechanics literature.

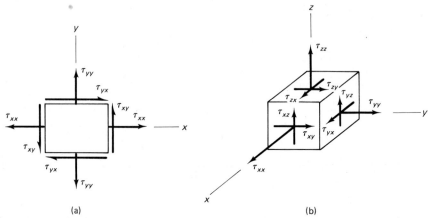

Figure 5.2 Stress components in rectangular coordinates: (a) two-dimensional stress components; (b) three-dimensional stress components.

The first subscript on a stress component denotes the face upon which the component acts, and the second subscript denotes the direction in which it acts; the component τ_{xy} acts in the positive y-direction on a positive x-face and in the negative y-direction on a negative x-face, as displayed in Fig. 5.2a. A stress component that acts perpendicular to a face is referred to as a **normal stress; the components** τ_{xx}, τ_{yy}, and τ_{zz} are normal stresses. A stress component that acts tangential to a face is called a **shear stress;** the components τ_{xy}, τ_{yx}, τ_{xz}, τ_{zx}, τ_{yz}, and τ_{zy} are the shear stress components.

There are nine stress components that act at a particular point in a fluid. To derive the differential momentum equation, consider the forces acting on the fluid particle that occupies the infinitesimal fluid element shown in Fig. 5.3. Only the forces acting on positive faces are shown. The stress components are assumed to be function of x, y, z, and t, and hence the values of the stress components change from face to face since the location of each face is slightly different.

Newton's second law, for the x-component direction, is $\Sigma F_x = ma_x$. For the fluid particle occupying the element shown, this takes the form

$$\left(\tau_{xx} + \frac{\partial \tau_{xx}}{\partial x}\frac{dx}{2}\right) dy\,dz + \left(\tau_{yx} + \frac{\partial \tau_{yx}}{\partial y}\frac{dy}{2}\right) dx\,dz + \left(\tau_{zx} + \frac{\partial \tau_{zx}}{\partial z}\frac{dz}{2}\right) dx\,dy$$

$$- \left(\tau_{xx} - \frac{\partial \tau_{xx}}{\partial x}\frac{dx}{2}\right) dy\,dz - \left(\tau_{yx} - \frac{\partial \tau_{yx}}{\partial y}\frac{dy}{2}\right) dx\,dz \qquad (5.3.1)$$

$$- \left(\tau_{zx} - \frac{\partial \tau_{zx}}{\partial y}\frac{dz}{2}\right) dx\,dy + \rho g_x\,dx\,dy\,dz = \rho\,dx\,dy\,dz\,\frac{Du}{Dt}$$

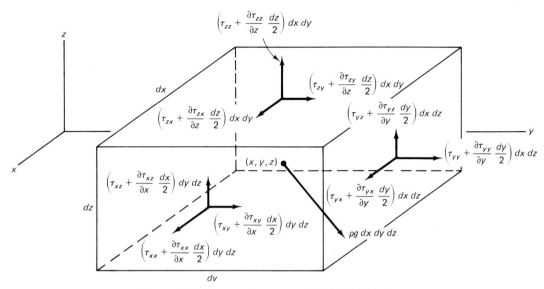

Figure 5.3   Forces acting on an infinitesimal fluid element.

where the component of the gravity vector **g** acting in the x-direction is g_x. After we divide by the volume $dx\,dy\,dz$, the equation above simplifies to

$$\rho\frac{Du}{Dt} = \frac{\partial\tau_{xx}}{\partial x} + \frac{\partial\tau_{yx}}{\partial y} + \frac{\partial\tau_{zx}}{\partial z} + \rho g_x \tag{5.3.2}$$

Similarly, for the y- and z-directions we would have

$$\rho\frac{Dv}{Dt} = \frac{\partial\tau_{xy}}{\partial x} + \frac{\partial\tau_{yy}}{\partial y} + \frac{\partial\tau_{zy}}{\partial z} + \rho g_y$$

$$\rho\frac{Dw}{Dt} = \frac{\partial\tau_{xz}}{\partial x} + \frac{\partial\tau_{yz}}{\partial y} + \frac{\partial\tau_{zz}}{\partial z} + \rho g_z \tag{5.3.3}$$

We can show, by taking moments about axes passing through the center of the infinitesimal element, that

$$\tau_{yx} = \tau_{xy} \qquad \tau_{yz} = \tau_{zy} \qquad \tau_{xz} = \tau_{zx} \tag{5.3.4}$$

That is, the stress tensor is symmetric; so there are actually six independent stress components.

The stress tensor may be displayed in the usual way as

$$\tau_{ij} = \begin{pmatrix} \tau_{xx} & \tau_{xy} & \tau_{xz} \\ \tau_{yx} & \tau_{yy} & \tau_{yz} \\ \tau_{zx} & \tau_{zy} & \tau_{zz} \end{pmatrix} \tag{5.3.5}$$

The subscripts i and j take on numerical values 1, 2, or 3. Then τ_{12} represents the element τ_{xy} in the first row, second column.

5.3.2 Euler's Equations

Good approximations to the components of the stress tensor for many flows, especially for flow away from a boundary (flow around an airfoil) or in regions of sudden change (flow through a contraction) are displayed by the array

$$\tau_{ij} = \begin{pmatrix} -p & 0 & 0 \\ 0 & -p & 0 \\ 0 & 0 & -p \end{pmatrix} \tag{5.3.6}$$

For such flows, we have assumed the shear stress components that result from viscous effects to be negligibly small and the normal stress components to be equal to the negative of the pressure; this is precisely what we did in Fig. 3.14 when deriving Bernoulli's equation. If these stress components are introduced back into Eqs. 5.3.2 and 5.3.3 there results, for this frictionless flow,

$$\rho \frac{Du}{Dt} = -\frac{\partial p}{\partial x} + \rho g_x$$

$$\rho \frac{Dv}{Dt} = -\frac{\partial p}{\partial y} + \rho g_y \tag{5.3.7}$$

$$\rho \frac{Dw}{Dt} = -\frac{\partial p}{\partial z} + \rho g_z$$

Let us assume that the z-axis is vertical so that $g_x = g_y = 0$ and $g_z = -g$. The equations above can then be written as

$$\rho \frac{D}{Dt} (u\hat{i} + v\hat{j} + w\hat{k}) = -\left(\frac{\partial p}{\partial x} \hat{i} + \frac{\partial p}{\partial y} \hat{j} + \frac{\partial p}{\partial z} \hat{k}\right) - \rho g\hat{k} \tag{5.3.8}$$

In vector form, we have the well-known **Euler's equation**

$$\rho \frac{D\mathbf{V}}{Dt} = -\nabla p - \rho g\hat{k} \tag{5.3.9}$$

If we assume a constant-density, steady flow, Eq. 5.3.9 can be integrated along a streamline to yield Bernoulli's equation, a result that does not surprise us since the same assumptions were imposed when deriving Bernoulli's equation; this will be illustrated in Example 5.5.

With the differential momentum equations in the form of Eqs. 5.3.7, we have added three additional equations to the continuity equation to give four equations and four unknowns, u, v, w, and p. With the appropriate boundary and initial conditions, a solution, yielding the velocity and pressure fields for this inviscid, incompressible flow, would be possible.

EXAMPLE 5.4

A velocity field is proposed to be

$$u = \frac{10y}{x^2 + y^2} \qquad v = -\frac{10x}{x^2 + y^2} \qquad w = 0$$

(a) Is this a possible incompressible flow? (b) If so, find the pressure gradient ∇p assuming a frictionless air flow with the z-axis vertical. Use $\rho = 1.23 \text{ kg/m}^3$.

Solution

(a) The continuity equation is used to determine if the velocity field is possible. For an incompressible flow we have

$$\frac{\partial u}{\partial x} + \frac{\partial v}{\partial y} + \frac{\partial w}{\partial z} = 0$$

Substituting in the velocity components, we have

$$\frac{\partial}{\partial x}\left(\frac{10y}{x^2 + y^2}\right) + \frac{\partial}{\partial y}\left(-\frac{10x}{x^2 + y^2}\right) = \frac{-10y(2x)}{(x^2 + y^2)^2} - \frac{-10x(2y)}{(x^2 + y^2)^2}$$

$$= \frac{1}{(x^2 + y^2)^2}[-20xy + 20xy] = 0$$

The quantity in brackets is obviously zero; hence the velocity field given is a possible incompressible flow.

(b) The pressure gradient is found using Euler's equation. In component form we have the following:

$$\rho \frac{Du}{Dt} = -\frac{\partial p}{\partial x} + \cancel{\rho g_x}^{\,0}$$

$$\therefore \quad \frac{\partial p}{\partial x} = -\rho\left[u\frac{\partial u}{\partial x} + v\frac{\partial u}{\partial y} + \cancel{\frac{\partial u}{\partial t}}^{\,0}\right]$$

$$= -1.23\left[\frac{10y}{x^2 + y^2}\frac{-20xy}{(x^2 + y^2)^2} + \frac{-10x}{x^2 + y^2}\frac{(x^2 + y^2)10 - 20y^2}{(x^2 + y^2)^2}\right]$$

$$= \frac{123x}{(x^2 + y^2)^2}$$

$$\rho \frac{Dv}{Dt} = -\frac{\partial p}{\partial y} + \cancel{\rho g_y}^{0}$$

$$\therefore \quad \frac{\partial p}{\partial y} = -\rho \left[u \frac{\partial v}{\partial x} + v \frac{\partial v}{\partial y} + \cancel{\frac{\partial v}{\partial t}}^{0} \right]$$

$$= -1.23 \left[\frac{10y}{x^2 + y^2} \frac{(x^2 + y^2)(-10) + 20x^2}{(x^2 + y^2)^2} + \frac{-10x}{x^2 + y^2} \frac{-20xy}{(x^2 + y^2)^2} \right]$$

$$= \frac{123y}{(x^2 + y^2)^2}$$

$$\rho \cancel{\frac{Dw}{Dt}}^{0} = -\frac{\partial p}{\partial z} + \rho g_z$$

$$\therefore \quad \frac{\partial p}{\partial z} = \rho g_z = 1.23 \times (-9.81) = -12.07 \text{ Pa/m}$$

Thus

$$\nabla p = \frac{123}{(x^2 + y^2)^2} (x\hat{i} + y\hat{j}) - 12.07\hat{k} \quad \text{Pa/m}$$

EXAMPLE 5.5

Assume a steady, constant-density flow and integrate Euler's equation along a streamline.

Solution

First, let us express the substantial derivative in streamline coordinates. Since the velocity vector is tangent to the streamline, we can write

$$\mathbf{V} = V\hat{s}$$

where $\hat{s}$ is the unit vector tangent to the streamline and V is the magnitude of the velocity, as shown in the sketch below. The substantial derivative is then

$$\frac{D\vec{\mathbf{V}}}{Dt} = \frac{\partial \mathbf{V}}{\partial t} + V\frac{\partial(V\hat{s})}{\partial s} + \cancel{(V)_n}^{0} \frac{\partial \mathbf{V}}{\partial n}$$

$$= \frac{\partial \mathbf{V}}{\partial t} + V\frac{\partial V}{\partial s}\hat{s} + V^2 \frac{\partial \hat{s}}{\partial s}$$

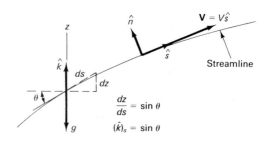

The quantity $\partial \hat{s}/\partial s$ results from the change of the unit vector $\hat{s}$; the unit vector cannot change magnitude (it must always have a magnitude of 1), it can only change direction. Hence the derivative $\partial \hat{s}/\partial s$ is in a direction normal to the streamline and does not enter the streamwise component equation. For a steady flow $\partial \mathbf{V}/\partial t = 0$. Consequently, in the streamwise direction, Euler's equation (5.3.9) takes the form

$$\rho V \frac{\partial V}{\partial s} = - \frac{\partial p}{\partial s} - \rho g \frac{\partial z}{\partial s}$$

recognizing that the component of $\hat{k}$ along the streamline can be expressed as $(\hat{k})_s = \partial z/\partial s$ (see the sketch above). Note that we use partial derivatives in this equation since velocity and pressure also vary with the normal coordinate.

The equation above can be written, assuming constant density so that $\partial \rho/\partial s = 0$, as

$$\frac{\partial}{\partial s} \left(\rho \frac{V^2}{2} + p + \rho g z \right) = 0$$

Integrating along the streamline results in

$$\rho \frac{V^2}{2} + p + \rho g z = \text{const.}$$

or

$$\frac{V^2}{2} + \frac{p}{\rho} + g z = \text{const.}$$

This is, of course, Bernoulli's equation. We have integrated along a streamline assuming constant density, steady flow, negligible viscous effects, and an inertial reference frame, so it is to be expected that Bernoulli's equation will emerge.

5.3.3 Navier–Stokes Equations

Many fluids exhibit a linear relationship between the stress components and the velocity gradients. Such fluids are called **Newtonian fluids** and include common fluids such as water, oil, and air. If in addition to linearity, we require that the fluid be **isotropic**,[5] it is possible to relate the stress components and the velocity gradients using only two fluid properties, the **viscosity** μ and the **second coefficient of viscosity** λ. The stress–velocity–gradient relations, often referred to as the **constitutive equations**,[6] are

---

[5] The condition of isotropy exists if the fluid properties are independent of direction. Polymers are examples of anisotropic fluids.

[6] Details of the development of the constitutive equations can be found in any textbook on the subject of continuum mechanics.

stated as follows:

$$\tau_{xx} = -p + 2\mu \frac{\partial u}{\partial x} + \lambda \nabla \cdot \mathbf{V} \qquad \tau_{xy} = \mu \left(\frac{\partial u}{\partial y} + \frac{\partial v}{\partial x} \right)$$

$$\tau_{yy} = -p + 2\mu \frac{\partial v}{\partial y} + \lambda \nabla \cdot \mathbf{V} \qquad \tau_{xz} = \mu \left(\frac{\partial u}{\partial z} + \frac{\partial w}{\partial x} \right) \qquad (5.3.10)$$

$$\tau_{zz} = -p + 2\mu \frac{\partial w}{\partial z} + \lambda \nabla \cdot \mathbf{V} \qquad \tau_{yz} = \mu \left(\frac{\partial v}{\partial z} + \frac{\partial w}{\partial y} \right)$$

For most gases, and for monatomic gases exactly, the second coefficient of viscosity is related to the viscosity by

$$\lambda = -\tfrac{2}{3}\mu \qquad (5.3.11)$$

a condition that is known as **Stokes' hypothesis**. With this relationship the negative average of the three normal stresses is equal to the pressure, that is,

$$-\tfrac{1}{3}(\tau_{xx} + \tau_{yy} + \tau_{zz}) = p \qquad (5.3.12)$$

Using Eqs. 5.3.10, this can be shown to always be true for a liquid in which $\nabla \cdot \mathbf{V} = 0$, and with Stokes' hypothesis it is also true for a gas.

    If we substitute the constitutive equations into the differential momentum equations (5.3.2) and (5.3.3), there results, using Stokes' hypothesis,

$$\rho \frac{Du}{Dt} = -\frac{\partial p}{\partial x} + \rho g_x + \mu \left(\frac{\partial^2 u}{\partial x^2} + \frac{\partial^2 u}{\partial y^2} + \frac{\partial^2 u}{\partial z^2} \right) + \frac{\mu}{3} \frac{\partial}{\partial x} \left(\frac{\partial u}{\partial x} + \frac{\partial v}{\partial y} + \frac{\partial w}{\partial z} \right)$$

$$\rho \frac{Dv}{Dt} = -\frac{\partial p}{\partial y} + \rho g_y + \mu \left(\frac{\partial^2 v}{\partial x^2} + \frac{\partial^2 v}{\partial y^2} + \frac{\partial^2 v}{\partial z^2} \right) + \frac{\mu}{3} \frac{\partial}{\partial y} \left(\frac{\partial u}{\partial x} + \frac{\partial v}{\partial y} + \frac{\partial w}{\partial z} \right)$$

$$\rho \frac{Dw}{Dt} = -\frac{\partial p}{\partial z} + \rho g_z + \mu \left(\frac{\partial^2 w}{\partial x^2} + \frac{\partial^2 w}{\partial y^2} + \frac{\partial^2 w}{\partial z^2} \right) + \frac{\mu}{3} \frac{\partial}{\partial z} \left(\frac{\partial u}{\partial x} + \frac{\partial v}{\partial y} + \frac{\partial w}{\partial z} \right)$$

$$(5.3.13)$$

where we have assumed a **homogeneous fluid,** that is, fluid properties (this includes the viscosity) are independent of position.

    For an incompressible flow the continuity equation allows the equations above to be reduced to

$$\rho \frac{Du}{Dt} = -\frac{\partial p}{\partial x} + \rho g_x + \mu \left(\frac{\partial^2 u}{\partial x^2} + \frac{\partial^2 u}{\partial y^2} + \frac{\partial^2 u}{\partial z^2} \right)$$

$$\rho \frac{Dv}{Dt} = -\frac{\partial p}{\partial y} + \rho g_y + \mu \left(\frac{\partial^2 v}{\partial x^2} + \frac{\partial^2 v}{\partial y^2} + \frac{\partial^2 v}{\partial z^2} \right) \qquad (5.3.14)$$

$$\rho \frac{Dw}{Dt} = -\frac{\partial p}{\partial z} + \rho g_z + \mu \left(\frac{\partial^2 w}{\partial x^2} + \frac{\partial^2 w}{\partial y^2} + \frac{\partial^2 w}{\partial z^2} \right)$$

These are called the **Navier–Stokes equations;** with these three differential equations and the differential continuity equation we have four equations and four unknowns, u, v, w, and p. The viscosity and density are fluid properties that are assumed to be known. With the appropriate boundary and initial conditions the equations can hopefully be solved. Several relatively simple geometries allow for analytical solutions; some of the solutions will be presented in Chapter 7. Numerical solutions have also been determined for many flows of interest. Because the equations are nonlinear partial differential equations, we cannot be assured that the solution we find will actually be realized in the laboratory; that is, the solutions are not unique. For example, a laminar flow and a turbulent flow may have the identical initial and boundary conditions, yet the two flows (the two solutions) are very different.

We can express the Navier–Stokes equations in vector form by multiplying equations 5.3.14 by $\hat{i}$, $\hat{j}$, and $\hat{k}$, respectively, and adding. We recognize that

$$\frac{Du}{Dt}\hat{i} + \frac{Dv}{Dt}\hat{j} + \frac{Dw}{Dt}\hat{k} = \frac{D\mathbf{V}}{Dt}$$

$$\frac{\partial p}{\partial x}\hat{i} + \frac{\partial p}{\partial y}\hat{j} + \frac{\partial p}{\partial z}\hat{k} = \nabla p \tag{5.3.15}$$

$$\nabla^2 u\hat{i} + \nabla^2 v\hat{j} + \nabla^2 w\hat{k} = \nabla^2 \mathbf{V}$$

where we have used the leplacion

$$\nabla^2 = \frac{\partial^2}{\partial x^2} + \frac{\partial^2}{\partial y^2} + \frac{\partial^2}{\partial z^2} \tag{5.3.16}$$

Combining the above, the Navier–Stokes equations (5.3.14) take the form

$$\rho \frac{D\mathbf{V}}{Dt} = -\nabla p + \rho \mathbf{g} + \mu \nabla^2 \mathbf{V} \tag{5.3.17}$$

Using this vector form we can express the Navier-Stokes equations using other coordinate systems. In Table 5.1 the equations will be listed for cylindrical and spherical coordinates.

EXAMPLE 5.6

Simplify the Navier–Stokes equations for flow in a horizontal, rectangular channel assuming all streamlines parallel to the walls. Let the x-direction be in the direction of flow.

Solution

If the streamlines are parallel to the walls, only the x-component of velocity will be nonzero. Letting $v = w = 0$ the continuity equation (5.2.8) for an incompressi-

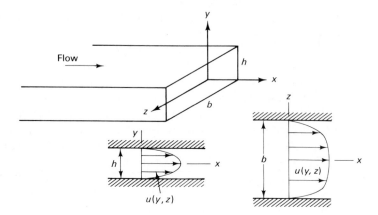

ble flow becomes

$$\frac{\partial u}{\partial x} = 0$$

showing that $u = u(y, z)$. The acceleration is then

$$\frac{Du}{Dt} = \frac{\partial u}{\partial t} + u\,\overset{0}{\cancel{\frac{\partial u}{\partial x}}} + \overset{0}{\cancel{v}}\,\frac{\partial u}{\partial y} + \overset{0}{\cancel{w}}\,\frac{\partial u}{\partial z}$$

$$= \frac{\partial u}{\partial t}$$

The x-component momentum equation then simplifies to

$$\rho\frac{\partial u}{\partial t} = -\frac{\partial p}{\partial x} + \overset{0}{\cancel{\rho g_x}} + \mu\left(\overset{0}{\cancel{\frac{\partial^2 u}{\partial x^2}}} + \frac{\partial^2 u}{\partial y^2} + \frac{\partial^2 u}{\partial z^2}\right)$$

or

$$\rho\frac{\partial u}{\partial t} = -\frac{\partial p}{\partial x} + \mu\left(\frac{\partial^2 u}{\partial y^2} + \frac{\partial^2 u}{\partial z^2}\right)$$

If the flow were a steady flow, the time-derivative term would be zero. With the appropriate boundary and initial conditions, a solution to the foregoing equation could be sought.

5.4 DIFFERENTIAL ENERGY EQUATION

Most problems of interest in fluid mechanics do not involve temperature gradients; they involve flows in which the temperature is everywhere constant. For such flows it is not necessary to introduce the differential energy equation. There are situations, however, for both compressible

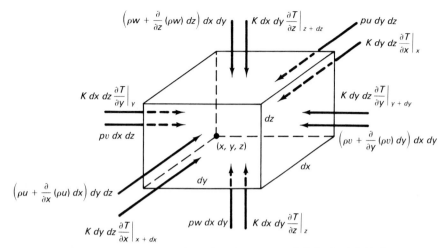

Figure 5.4   Rate of heat transfer and work rate on an infinitesimal fluid element.

and incompressible flows, in which temperature gradients are important, and for such flows the differential energy equation may be needed. We will derive the differential energy equation assuming negligible viscous effects, an assumption that significantly simplifies the derivation. Since the shear stresses that result from viscosity are quite small for many applications, this assumption may be acceptable. These shear stresses do, however, account for the high temperatures that burn up satellites on reentry to the atmosphere; if they are significant, they must be included in any analysis.

Consider the infinitesimal fluid element, shown in Fig. 5.4. The heat transfer rate through an area A is given by **Fourier's law of heat transfer,** namely,

$$\dot{Q} = -KA\frac{\partial T}{\partial n} \tag{5.4.1}$$

where n is the direction normal to the area and K is the **thermal conductivity,** assumed constant. The work rate done by a force is the magnitude of the force multiplied by the velocity in the direction of the force, that is,

$$\dot{W} = pAV \tag{5.4.2}$$

where V is the velocity in the direction of the pressure force pA. The first law of thermodynamics applied to a fluid particle is

$$\dot{Q} - \dot{W} = \frac{DE}{Dt} \tag{5.4.3}$$

where D/Dt is used since we are following a fluid particle at the instant shown. For the particle occupying the infinitesimal element of Fig. 5.4, the relationships above allow us to write

$$K \, dy \, dz \left(\frac{\partial T}{\partial x} \bigg|_{x+dx} - \frac{\partial T}{\partial x} \bigg|_x \right) - \frac{\partial}{\partial x}(pu) \, dx \, dy \, dz$$

$$+ \, K \, dx \, dz \left(\frac{\partial T}{\partial y} \bigg|_{y+dy} - \frac{\partial T}{\partial y} \bigg|_y \right) - \frac{\partial}{\partial y}(pv) \, dx \, dy \, dz$$

$$+ \, K \, dx \, dy \left(\frac{\partial T}{\partial z} \bigg|_{z+dz} - \frac{\partial T}{\partial z} \bigg|_z \right) - \frac{\partial}{\partial z}(pw) \, dx \, dy \, dz \qquad (5.4.4)$$

$$= \rho \, dx \, dy \, dz \, \frac{D}{Dt} \left[\frac{u^2 + v^2 + w^2}{2} + gz + \tilde{u} \right]$$

where E has included kinetic, potential, and internal energy, and the z-axis is assumed vertical. Also, since the mass of a fluid particle is constant $\rho \, dx \, dy \, dz$ is outside the D/Dt-operator. Divide both sides by $dx \, dy \, dz$. The result is

$$K \left(\frac{\partial^2 T}{\partial x^2} + \frac{\partial^2 T}{\partial y^2} + \frac{\partial^2 T}{\partial z^2} \right) - \frac{\partial}{\partial x}(pu) - \frac{\partial}{\partial y}(pv) - \frac{\partial}{\partial z}(pw)$$

$$= \rho \, \frac{D}{Dt} \left[\frac{u^2 + v^2 + w^2}{2} + gz + \tilde{u} \right) \qquad (5.4.5)$$

This can be rearranged as follows:

$$K \left(\frac{\partial^2 T}{\partial x^2} + \frac{\partial^2 T}{\partial y^2} + \frac{\partial^2 T}{\partial z^2} \right) - p \left(\frac{\partial u}{\partial x} + \frac{\partial v}{\partial y} + \frac{\partial w}{\partial z} \right) - u \frac{\partial p}{\partial x} - v \frac{\partial p}{\partial y} - w \frac{\partial p}{\partial z}$$

$$= \rho u \frac{Du}{Dt} + \rho v \frac{Dv}{Dt} + \rho w \frac{Dw}{Dt} + \rho g \frac{Dz}{Dt} + \rho \frac{D\tilde{u}}{Dt} \qquad (5.4.6)$$

The Euler's equations (5.3.7) are applicable for this inviscid flow; hence the last three terms on the left equal the first four terms on the right if we recognize that

$$\frac{Dz}{Dt} = \cancel{\frac{\partial z}{\partial t}}^{0} + u \cancel{\frac{\partial z}{\partial x}}^{0} + v \cancel{\frac{\partial z}{\partial y}}^{0} + w \frac{\partial z}{\partial z} = w \qquad (5.4.7)$$

since x, y, z, t are all independent variables. The simplified energy equation then takes the form

$$K \left(\frac{\partial^2 T}{\partial x^2} + \frac{\partial^2 T}{\partial y^2} + \frac{\partial^2 T}{\partial z^2} \right) - p \left(\frac{\partial u}{\partial x} + \frac{\partial v}{\partial y} + \frac{\partial w}{\partial z} \right) = \rho \frac{D\tilde{u}}{Dt} \qquad (5.4.8)$$

In vector form this is expressed as

$$\rho \frac{D\bar{u}}{Dt} = K\nabla^2 T - p \, \nabla \cdot \mathbf{V} \qquad (5.4.9)$$

Before we simplify this equation for incompressible gas flow, let us write it in terms of enthalpy rather than internal energy. Using

$$\bar{u} = h - \frac{p}{\rho} \qquad (5.4.10)$$

the energy equation becomes, using Eq. 5.2.7,

$$\rho \frac{Dh}{Dt} = K\nabla^2 T + \frac{Dp}{Dt} \qquad (5.4.11)$$

See Example 5.3 for the details of this conversion.

We have two special cases to be considered. First, for a liquid flow we can use $\nabla \cdot \mathbf{V} = 0$ and with $\bar{u} = c_v T$, c_v being the specific heat,[7] Eq. 5.4.8 simplifies to

$$\frac{DT}{Dt} = \alpha\nabla^2 T \qquad (5.4.12)$$

where we have introduced the **thermal diffusivity** α defined by

$$\alpha = \frac{K}{\rho c_v} \qquad (5.4.13)$$

For an incompressible gas flow, an interesting result occurs. In Example 5.8 we will show that

$$\left| \frac{Dp}{Dt} \right| \ll |p \, \nabla \cdot \mathbf{V}| \qquad (5.4.14)$$

Thus, when comparing Eqs. 5.4.9 and 5.4.11, it is Eq. 5.4.11 that simplifies to

$$\rho c_p \frac{DT}{Dt} = K\nabla^2 T \qquad (5.4.15)$$

---

[7]The specific heat for a liquid is listed in tables as c_p. The specific heat at constant volume for a liquid is approximately equal to c_p. Hence we often simply drop the subscript and let $c_p = c_v$ for a liquid. It is assumed to be a constant, but it does depend on temperature.

for an incompressible gas flow if we make the ideal-gas assumption that

$$dh = c_p\, dT \qquad (5.4.16)$$

If viscous effects are not negligible, the derivation would include the work input due to the shear stress components. This would add a term to the right-hand side of all of the differential energy equations above; this term is called the **dissipation function** Φ, which, in rectangular coordinates, is

$$\Phi = 2\mu\left[\left(\frac{\partial u}{\partial x}\right)^2 + \left(\frac{\partial v}{\partial y}\right)^2 + \left(\frac{\partial w}{\partial z}\right)^2 + \frac{1}{2}\left(\frac{\partial u}{\partial y} + \frac{\partial v}{\partial x}\right)^2 + \frac{1}{2}\left(\frac{\partial v}{\partial z} + \frac{\partial w}{\partial y}\right)^2 \right.$$
$$\left. + \frac{1}{2}\left(\frac{\partial u}{\partial z} + \frac{\partial w}{\partial x}\right)^2\right] \qquad (5.4.17)$$

With the addition of the energy equation, problems involving temperature variations in a flow can now be considered. Such problems are obviously present in compressible flows in which pressure and density are related to temperature by an equation of state. In incompressible gas flows and liquid flows, temperature variations are often negligible so that the differential energy equation is not of interest. If, however, a temperature field does exist in a liquid flow or an incompressible gas flow (heat exchangers, atmospheric flows, lake inversions, lubrication flows, free convection flows), the energy equation provides an additional equation relating the quantities of interest. For liquid flows involving temperature gradients it is often necessary to assume that $\mu = \mu(T)$; in free convection flows we must assume that $\rho = \rho(T)$. In gas flows we can usually assume viscosity to be constant since the temperature variation is quite small.

EXAMPLE 5.7

A constant-density liquid flows into a wide, rectangular horizontal channel, the walls of which are maintained at a higher temperature than the liquid, as shown. Assume a variable μ, include viscous dissipation, and write the describing differential equations.

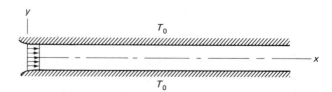

Solution

Let the x-axis coincide with centerline of the channel and the y-axis be vertical. The continuity equation would take the form

$$\frac{\partial u}{\partial x} + \frac{\partial v}{\partial y} = 0$$

The flow will be primarily in the x-direction, but we must allow for variation of the y-component v. There will be no variation in the z-direction. The accelerations for this steady flow will be

$$\frac{Du}{Dt} = u\frac{\partial u}{\partial x} + v\frac{\partial u}{\partial y}$$

$$\frac{Dv}{Dt} = u\frac{\partial v}{\partial x} + v\frac{\partial v}{\partial y}$$

The stress terms of Eqs. 5.3.2 and 5.3.3, assuming a variable μ, become

$$\frac{\partial\tau_{xx}}{\partial x} + \frac{\partial\tau_{xy}}{\partial y} = -\frac{\partial p}{\partial x} + \mu\left(\frac{\partial^2 u}{\partial x^2} + \frac{\partial^2 u}{\partial y^2}\right) + 2\frac{\partial\mu}{\partial x}\frac{\partial u}{\partial x} + \frac{\partial\mu}{\partial y}\left(\frac{\partial u}{\partial y} + \frac{\partial v}{\partial x}\right)$$

$$\frac{\partial\tau_{xy}}{\partial x} + \frac{\partial\tau_{yy}}{\partial y} = -\frac{\partial p}{\partial y} + \mu\left(\frac{\partial^2 v}{\partial x^2} + \frac{\partial^2 v}{\partial y^2}\right) + 2\frac{\partial\mu}{\partial y}\frac{\partial v}{\partial y} + \frac{\partial\mu}{\partial x}\left(\frac{\partial u}{\partial y} + \frac{\partial v}{\partial x}\right)$$

where we have used $\nabla \cdot \mathbf{V} = 0$. The momentum equations are then

$$\rho\left(u\frac{\partial u}{\partial x} + v\frac{\partial u}{\partial y}\right) = -\frac{\partial p}{\partial x} + \mu\left(\frac{\partial^2 u}{\partial x^2} + \frac{\partial^2 u}{\partial y^2}\right) + 2\frac{\partial\mu}{\partial x}\frac{\partial u}{\partial x} + \frac{\partial\mu}{\partial y}\left(\frac{\partial u}{\partial y} + \frac{\partial v}{\partial x}\right)$$

$$\rho\left(u\frac{\partial v}{\partial x} + v\frac{\partial v}{\partial y}\right) = -\frac{\partial p}{\partial y} + \mu\left(\frac{\partial^2 v}{\partial x^2} + \frac{\partial^2 v}{\partial y^2}\right) + 2\frac{\partial\mu}{\partial y}\frac{\partial v}{\partial y} + \frac{\partial\mu}{\partial x}\left(\frac{\partial u}{\partial y} + \frac{\partial v}{\partial x}\right)$$

The energy equation simplifies to

$$u\frac{\partial T}{\partial x} + v\frac{\partial T}{\partial y} = \alpha\left(\frac{\partial^2 T}{\partial x^2} + \frac{\partial^2 T}{\partial y^2}\right) + \frac{2\mu}{c_p}\left[\left(\frac{\partial u}{\partial x}\right)^2 + \left(\frac{\partial v}{\partial y}\right)^2 + \frac{1}{2}\left(\frac{\partial u}{\partial y} + \frac{\partial v}{\partial x}\right)^2\right]$$

where we have assumed K to be constant. The nonlinear, partial differential equations above, although quite formidible when attempting an analytic solution, could be solved numerically with the appropriate boundary conditions, and for a sufficiently low flow rate so that laminar flow exists.

EXAMPLE 5.8

Show that for an ideal gas, $|Dp/Dt| \ll |p\,\nabla \cdot \mathbf{V}|$ in a low-speed flow, thereby concluding that Eq. 5.4.15 is the appropriate equation.

Solution

Let us consider a steady, uniform flow in a pipe so that $|\mathbf{V}| = u$ and $Dp/Dt = u\,\partial p/\partial x$. Then the problem becomes: Show that

$$\left| u\,\frac{\partial p}{\partial x} \right| \ll \left| p\,\frac{\partial u}{\partial x} \right|$$

Viscous effects are small and would not change the conclusion, so we can ignore any possible viscous effects. Then Euler's equation (5.3.7) allows us to use

$$\frac{\partial p}{\partial x} = -\rho u\,\frac{\partial u}{\partial x}$$

Using the definition of the speed of sound (Eq. 1.7.20) and the equation of state, we see that

$$c = \sqrt{\frac{kp}{\rho}} \qquad \text{or} \qquad p = c^2\,\frac{\rho}{k}$$

Thus

$$p\,\frac{\partial u}{\partial x} = \frac{c^2}{k}\,\rho\,\frac{\partial u}{\partial x}$$

Our problem can now be stated: Show that

$$\left| \rho u^2\,\frac{\partial u}{\partial x} \right| \ll \left| \frac{c^2}{k}\,\rho\,\frac{\partial u}{\partial x} \right|$$

Or, more simply, is it true that

$$u^2 \ll \frac{c^2}{k}?$$

This can be seen to be true since we have assumed for a low-speed gas flow that the speed of the gas is much less than the speed of sound (e.g., $u < 0.3c$ or $M < 0.3$). We know that k is of order unity ($k = 1.4$ for air), so it will not affect our conclusion that

$$\left| \frac{Dp}{Dt} \right| \ll |p\,\nabla \cdot \mathbf{V}|$$

5.5 CLOSURE

We have now completed our derivation of the partial differential equations that are used in describing flows of interest. For convenience the equations are listed in Table 5.1 for rectangular, cylindrical, and spherical

Continuity

Cartesian
$$\frac{\partial u}{\partial x} + \frac{\partial v}{\partial y} + \frac{\partial w}{\partial z} = 0$$

Cylindrical
$$\frac{1}{r}\frac{\partial}{\partial r}(rv_r) + \frac{1}{r}\frac{\partial v_\theta}{\partial \theta} + \frac{\partial v_z}{\partial z} = 0$$

Spherical
$$\frac{1}{r^2}\frac{\partial}{\partial r}(r^2 v_r) + \frac{1}{r\sin\theta}\frac{\partial}{\partial \theta}(v_\theta \sin\theta) + \frac{1}{r\sin\theta}\frac{\partial v_\phi}{\partial \phi} = 0$$

Momentum

Cartesian
$$\frac{Du}{Dt} = -\frac{1}{\rho}\frac{\partial p}{\partial x} + g_x + \nu\nabla^2 u$$
$$\frac{Dv}{Dt} = -\frac{1}{\rho}\frac{\partial p}{\partial y} + g_y + \nu\nabla^2 v$$
$$\frac{Dw}{Dt} = -\frac{1}{\rho}\frac{\partial p}{\partial z} + g_z + \nu\nabla^2 w$$
$$\frac{D}{Dt} = \frac{\partial}{\partial t} + u\frac{\partial}{\partial x} + v\frac{\partial}{\partial y} + w\frac{\partial}{\partial z}$$
$$\nabla^2 = \frac{\partial^2}{\partial x^2} + \frac{\partial^2}{\partial y^2} + \frac{\partial^2}{\partial z^2}$$

Cylindrical
$$\frac{Dv_r}{Dt} - \frac{v_\theta^2}{r} = -\frac{1}{\rho}\frac{\partial p}{\partial r} + g_r + \nu\left(\nabla^2 v_r - \frac{v_r}{r^2} - \frac{2}{r^2}\frac{\partial v_\theta}{\partial \theta}\right)$$
$$\frac{Dv_\theta}{Dt} + \frac{v_r v_\theta}{r} = -\frac{1}{\rho r}\frac{\partial p}{\partial \theta} + g_\theta + \nu\left(\nabla^2 v_\theta + \frac{2}{r^2}\frac{\partial v_r}{\partial \theta} - \frac{v_\theta}{r^2}\right)$$
$$\frac{Dv_z}{Dt} = -\frac{1}{\rho}\frac{\partial p}{\partial z} + g_z + \nu\nabla^2 v_z$$
$$\frac{D}{Dt} = \frac{\partial}{\partial t} + v_r\frac{\partial}{\partial r} + \frac{v_\theta}{r}\frac{\partial}{\partial \theta} + v_z\frac{\partial}{\partial z}$$
$$\nabla^2 = \frac{\partial^2}{\partial r^2} + \frac{1}{r}\frac{\partial}{\partial r} + \frac{1}{r^2}\frac{\partial^2}{\partial \theta^2} + \frac{\partial^2}{\partial z^2}$$

Spherical
$$\frac{Dv_r}{Dt} - \frac{v_\theta^2 + v_\phi^2}{r}$$
$$= -\frac{1}{\rho}\frac{\partial p}{\partial r} + g_r + \nu\left(\nabla^2 v_r - \frac{2v_r}{r^2} - \frac{2}{r^2}\frac{\partial v_\theta}{\partial \theta}\right.$$
$$\left. - \frac{2v_\theta \cot\theta}{r^2} - \frac{2}{r^2\sin\theta}\frac{\partial v_\phi}{\partial \phi}\right)$$
$$\frac{Dv_\theta}{Dt} + \frac{v_r v_\theta - v_\phi^2\cot\theta}{r}$$
$$= -\frac{1}{\rho r}\frac{\partial p}{\partial \theta} + g_\theta + \nu\left(\nabla^2 v_\theta + \frac{2}{r^2}\frac{\partial v_r}{\partial \theta}\right.$$
$$\left. - \frac{v_\theta}{r^2\sin^2\theta} - \frac{2\cos\theta}{r^2\sin^2\theta}\frac{\partial v_\phi}{\partial \phi}\right)$$
$$\frac{Dv_\phi}{Dt} + \frac{v_\phi v_r}{r} + \frac{v_\theta v_\phi\cot\theta}{r}$$
$$= -\frac{1}{\rho r\sin\theta}\frac{\partial p}{\partial \phi} + g_\phi + \nu\left(\nabla^2 v_\phi - \frac{v_\phi}{r^2\sin^2\theta}\right.$$
$$\left. + \frac{2}{r^2\sin^2\theta}\frac{\partial v_r}{\partial \phi} + \frac{2\cos\theta}{r^2\sin^2\theta}\frac{\partial v_\theta}{\partial \phi}\right)$$
$$\frac{D}{Dt} = \frac{\partial}{\partial t} + v_r\frac{\partial}{\partial r} + \frac{v_\theta}{r}\frac{\partial}{\partial \theta} + \frac{v_\phi}{r\sin\theta}\frac{\partial}{\partial \phi}$$
$$\nabla^2 = \frac{1}{r^2}\frac{\partial}{\partial r}\left(r^2\frac{\partial}{\partial r}\right) + \frac{1}{r^2\sin\theta}\frac{\partial}{\partial \theta}\left(\sin\theta\frac{\partial}{\partial \theta}\right) + \frac{1}{r^2\sin^2\theta}\frac{\partial^2}{\partial \phi^2}$$

Energy

Cartesian
$$\rho\frac{Dh}{Dt} = K\nabla^2 T + 2\mu\left[\left(\frac{\partial u}{\partial x}\right)^2 + \left(\frac{\partial v}{\partial y}\right)^2 + \left(\frac{\partial w}{\partial z}\right)^2\right.$$
$$+ \frac{1}{2}\left(\frac{\partial u}{\partial y} + \frac{\partial v}{\partial x}\right)^2 + \frac{1}{2}\left(\frac{\partial v}{\partial z} + \frac{\partial w}{\partial y}\right)^2$$
$$\left. + \frac{1}{2}\left(\frac{\partial u}{\partial z} + \frac{\partial w}{\partial x}\right)^2\right]$$

Cylindrical
$$\rho\frac{Dh}{Dt} = K\nabla^2 T + 2\mu\left[\left(\frac{\partial v_r}{\partial r}\right)^2 + \left(\frac{1}{r}\frac{\partial v_\theta}{\partial \theta} + \frac{v_r}{r}\right)^2 + \left(\frac{\partial v_z}{\partial z}\right)^2\right.$$
$$+ \frac{1}{2}\left(\frac{1}{r}\frac{\partial v_z}{\partial \theta} + \frac{\partial v_\theta}{\partial z}\right)^2 + \frac{1}{2}\left(\frac{\partial v_r}{\partial z} + \frac{\partial v_z}{\partial r}\right)^2$$
$$\left. + \frac{1}{2}\left(\frac{1}{r}\frac{\partial v_r}{\partial \theta} + \frac{\partial v_\theta}{\partial r} - \frac{v_\theta}{r}\right)^2\right]$$

Spherical
$$\rho\frac{Dh}{Dt} = K\nabla^2 T + 2\mu\left[\left(\frac{\partial v_r}{\partial r}\right)^2 + \left(\frac{1}{r}\frac{\partial v_\theta}{\partial \theta} + \frac{v_r}{r}\right)^2 + \left(\frac{1}{r\sin\theta}\frac{\partial v_\phi}{\partial \phi} + \frac{v_r}{r} + \frac{v_\theta\cot\theta}{r}\right)^2\right.$$
$$+ \mu\left[\left(\frac{1}{r\sin\theta}\frac{\partial v_\theta}{\partial \phi} + \frac{\sin\theta}{r}\frac{\partial}{\partial \theta}\left(\frac{v_\phi}{\sin\theta}\right)\right)^2\right.$$
$$+ \left(\frac{1}{r\sin\theta}\frac{\partial v_r}{\partial \phi} + r\frac{\partial}{\partial r}\left(\frac{v_\phi}{r}\right)\right)^2$$
$$\left. + \left(r\frac{\partial}{\partial r}\left(\frac{v_\theta}{r}\right) + \frac{1}{r}\frac{\partial v_r}{\partial \theta}\right)^2\right]$$

Stresses

Cartesian
$$\tau_{xx} = -p + 2\mu\frac{\partial u}{\partial x} \qquad \tau_{xy} = \mu\left(\frac{\partial u}{\partial y} + \frac{\partial v}{\partial x}\right)$$
$$\tau_{yy} = -p + 2\mu\frac{\partial v}{\partial y} \qquad \tau_{yz} = \mu\left(\frac{\partial v}{\partial z} + \frac{\partial w}{\partial y}\right)$$
$$\tau_{zz} = -p + 2\mu\frac{\partial w}{\partial z} \qquad \tau_{xz} = \mu\left(\frac{\partial u}{\partial z} + \frac{\partial w}{\partial x}\right)$$

Cylindrical
$$\tau_{rr} = -p + 2\mu\frac{\partial v_r}{\partial r} \qquad \tau_{r\theta} = \mu\left[r\frac{\partial}{\partial r}\left(\frac{v_\theta}{r}\right) + \frac{1}{r}\frac{\partial v_r}{\partial \theta}\right]$$
$$\tau_{\theta\theta} = -p + 2\mu\left(\frac{1}{r}\frac{\partial v_\theta}{\partial \theta} + \frac{v_r}{r}\right) \qquad \tau_{\theta z} = \mu\left[\frac{\partial v_\theta}{\partial z} + \frac{1}{r}\frac{\partial v_z}{\partial \theta}\right]$$
$$\tau_{zz} = -p + 2\mu\frac{\partial v_z}{\partial z} \qquad \tau_{rz} = \mu\left[\frac{\partial v_r}{\partial z} + \frac{\partial v_z}{\partial r}\right]$$

Spherical
$$\tau_{rr} = -p + 2\mu\frac{\partial v_r}{\partial r}$$
$$\tau_{\theta\theta} = -p + 2\mu\left(\frac{1}{r}\frac{\partial v_\theta}{\partial \theta} + \frac{v_r}{r}\right)$$
$$\tau_{\phi\phi} = -p + 2\mu\left(\frac{1}{r\sin\theta}\frac{\partial v_\phi}{\partial \phi} + \frac{v_r}{r} + \frac{v_\theta\cot\theta}{r}\right)$$
$$\tau_{r\theta} = \mu\left[r\frac{\partial}{\partial r}\left(\frac{v_\theta}{r}\right) + \frac{1}{r}\frac{\partial v_r}{\partial \theta}\right]$$
$$\tau_{\theta\phi} = \mu\left[\frac{\sin\theta}{r}\frac{\partial}{\partial \theta}\left(\frac{v_\phi}{\sin\theta}\right) + \frac{1}{r\sin\theta}\frac{\partial v_\theta}{\partial \phi}\right]$$
$$\tau_{r\phi} = \mu\left[\frac{1}{r\sin\theta}\frac{\partial v_r}{\partial \phi} + r\frac{\partial}{\partial r}\left(\frac{v_\phi}{r}\right)\right]$$

coordinates. In vector form the equations, for an incompressible flow, are:

Continuity:
$$\boldsymbol{\nabla} \cdot \mathbf{V} = 0$$

Momentum:
$$\rho \left[\frac{\partial \mathbf{V}}{\partial t} + u \frac{\partial \mathbf{V}}{\partial x} + v \frac{\partial \mathbf{V}}{\partial y} + w \frac{\partial \mathbf{V}}{\partial z} \right] = -\boldsymbol{\nabla} p - \rho \mathbf{g} + \mu \nabla^2 \mathbf{V}$$

Energy:
$$\rho c \left[\frac{\partial T}{\partial t} + u \frac{\partial T}{\partial x} + v \frac{\partial T}{\partial y} + w \frac{\partial T}{\partial z} \right] = K \nabla^2 T + \Phi \quad \text{Liquids}$$

Energy:
$$\rho c_p \left[\frac{\partial T}{\partial t} + u \frac{\partial T}{\partial x} + v \frac{\partial T}{\partial y} + w \frac{\partial T}{\partial z} \right] = K \nabla^2 T + \Phi \quad \begin{array}{c} \text{Incompressible} \\ \text{gases} \end{array}$$

$$(5.5.1)$$

Note: To express these equations in the forms above, we have assumed:

A Newtonian fluid (a linear relationship between the stress components and the velocity gradients).
An isotropic fluid (the fluid properties are independent of direction).
A homogeneous fluid (the fluid properties do not depend on position).
An incompressible flow (The density of a particle is constant, that is, $D\rho/Dt = 0$; we do not demand that $\rho =$ constant. For a gas flow we require that $\mathbf{M} < 0.3$.)
An inertial reference frame.

In the derivations of the differential equations in this chapter we have made no mention of laminar or turbulent flow. The equations are applicable to either kind of flow. Some laminar flows in relatively simple geometries have been solved analytically and many others have been solved numerically. Turbulent flows, however, have not been solved even for the simplest geometry. A turbulent flow is always an unsteady flow and the presence of the time-derivative terms demand initial conditions; that is, at time $t = 0$ we must specify u, v, and w at all points in the region of interest, information that is difficult to obtain even in a simple pipe flow. Even if it were possible to solve for the turbulent flow velocity field, we are seldom interested in such detailed information. Thus experimental evidence is relied upon to provide information of interest in turbulent flows. This will be discussed in subsequent chapters.

PROBLEMS

Differential Continuity Equation

5.1. The divergence theorem (also known as Gauss's theorem) states that

$$\int_A \mathbf{V} \cdot \hat{n} \, dA = \int_V \nabla \cdot \mathbf{V} \, d\mathcal{V}$$

where A completely surrounds the volume $\mathcal{V}$. Apply this theorem to the integral continuity equation and derive the differential continuity equation.

5.2. Use the infinitesimal element shown and derive the differential continuity equation in cylindrical coordinates. The velocity vector is $\mathbf{V} = (v_r, v_\theta, v_z)$.

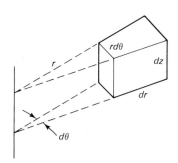

5.3. A uniform compressible flow occurs in a constant-diameter pipe. Write the simplified differential continuity equation for the steady flow.

5.4. An incompressible flow of air over the mountain range shown can be approximated by a plane, steady flow. If the z-axis is vertical and density is allowed to vary, write the differential equations that result from conservation of mass considerations.

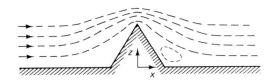

5.5. A stratified flow of salt water, in which the density increases with depth, occurs over an obstruction in the bottom of a channel. Assuming a plane, steady flow with the z-axis vertical, write the differential continuity equations.

5.6. Show that for an isothermal flow,

$$\frac{1}{p}\frac{Dp}{Dt} = -\nabla \cdot \mathbf{V}$$

5.7. A compressible flow occurs such that

$$u = 200xy \qquad v = 200(x^2 + y^2) \qquad w = 0 \text{ m/s}$$

Find the rate at which the density is changing at the point $(2, 1)$ where $\rho = 2.3$ kg/m^3.

5.8. If, in an incompressible plane flow, the velocity component $u = \text{const}$, what can we say about the y-component of velocity?

5.9. In an incompressible plane flow $u = Ax$. Find $v(x, y)$ if $v(x, 0) = 0$.

5.10. If the velocity component u is given by

$$u(x, y) = 10 + 5x/(x^2 + y^2)$$

in an incompressible plane flow, determine $v(x, y)$. Let $v(x, 0) = 0$.

5.11. The θ-component of velocity is given by

$$v_\theta = -(10 + 0.4/r^2) \cos u$$

Find the r-component of velocity for the incompressible plane flow if $v_r(0.2, \theta) = 0$.

5.12. In an incompressible plane flow

$$v_\theta = 20\left(1 + \frac{1}{r^2}\right)\sin\theta - \frac{40}{r}$$

Find $v_r(r, \theta)$ if $v_r(1, \theta) = 0$.

5.13. In an incompressible axisymmetric flow ($v_\phi = 0$) the velocity component v_θ is given by

$$v_\theta = -\left(10 + \frac{40}{r^3}\right)\sin\theta$$

Find $v_r(r, \theta)$ if $v_r(2, \theta) = 0$.

5.14. The velocity of air in a pipe is measured at points 5 cm apart to be 151, 162, and 175 m/s, respectively. At the middle point the temperature is 10°C and the pressure is 120 kPa absolute. Find dp/dx at the middle point of this steady flow.

5.14E. The velocity of air in a pipe is measured at points 2 in. apart to be 453, 486, and 526 ft/sec, respectively. At the middle point the temperature is 40°F and the pressure is 18 psia. Find dp/dx at the middle point of this steady flow.

5.15. The x-component of the velocity vector is measured, at points 5 mm apart, as 11.3, 12.6, and 13.5 m/s, respectively, in the incompressible, steady plane flow shown. Estimate:
(a) The y-component of velocity 4 mm above point B.
(b) The acceleration at point B.

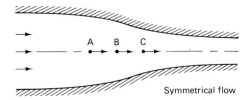

Symmetrical flow

Differential Momentum Equation

5.16. Sum forces on the element of Fig. 5.3 in the y-direction and show that Eq. 5.3.3a results.

5.17. Does the velocity field

$$u = \frac{10x}{x^2 + y^2} \qquad v = \frac{10y}{x^2 + y^2} \qquad w = 0$$

represent a possible incompressible flow? If so, find the pressure gradient ∇p assuming a frictionless flow with negligible body forces.

5.18. Does the velocity field

$$v_r = 10\left(1 - \frac{1}{r^2}\right) \cos \theta$$

$$v_\theta = -10\left(1 + \frac{1}{r^2}\right) \sin \theta$$

$$v_z = 0$$

represent a possible incompressible flow? If so, find the pressure gradient ∇p assuming a frictionless flow with negligible body forces.

5.19. Consider the velocity field

$$v_r = 10\left(1 - \frac{8}{r^3}\right) \cos \theta$$

$$v_\theta = -10\left(1 + \frac{4}{r^3}\right) \sin \theta$$

$$v_\phi = 0$$

Does it represent an incompressible flow? If so, find the pressure gradient ∇p, assuming a frictionless flow neglecting the body force.

5.20. For a steady plane flow find an expression for DV/Dt in terms of coordinates (s, n) tangential and normal to a streamline. Let R be the radius of curvature.

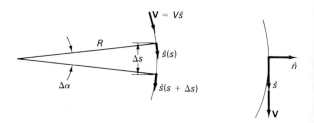

5.21. Write Euler's equation if the velocity is referred to a reference frame that is rotating with constant angular velocity.

5.22. A velocity field is given by $u = 10(y - 100y^2)$ m/s, $v = 0$, and $w = 0$. Display the stress components at $y = 2$ mm using $\mu = 10^{-3}$ N · s/m^2 and $p = 200$ kPa. Find the ratio τ_{xy}/τ_{xx}.

5.22E. A velocity field is given by $u = 30(y - 24y^2)$ ft/sec, $v = 0$, and $w = 0$. Display the stress components at $y = 0.1$ in. using $\mu = 10^{-5}$ lb-sec/ft^2 and $p = 30$ psi. Find the ratio τ_{xy}/τ_{xx}.

5.23. The velocity field near a surface is approximated by $u = 10(2y/\delta - y^2/\delta^2)$ where $\delta = Cx^{4/5}$. If $\delta = 8$ m at $x = 1000$ m, find $v(x, y)$ assuming that $w = 0$ and $v(x, 0) = 0$. Also, display the stress components at $(1000, 0)$ using $\mu = 2 \times 10^{-5}$ N · s/m^2 and $p = 100$ kPa. Assume an incompressible flow.

5.24. Show that for a steady flow Du/Dt can be written as $(\mathbf{V} \cdot \nabla)u$, and that $D\mathbf{V}/Dt = (\mathbf{V} \cdot \nabla)\mathbf{V}$.

5.25. Write the compressible flow differential momentum equations (5.3.13) as one equation in vector form.

5.26. Simplify the Navier–Stokes equations for incompressible, steady flow between horizontal, parallel plates assuming that $u = u(y)$, $w = 0$.

5.27. Simplify the Navier–Stokes equations for incompressible, steady flow in a horizontal pipe assuming that $v_z = v_z(r)$, $v_\theta = 0$, neglecting gravity.

5.28. Fluid flows in the small gap between concentrically rotating spheres such that $v_\theta = v_\theta(r)$ and $v_\phi = 0$. Simplify the Navier–Stokes equations neglecting gravity for a steady, incompressible flow.

5.29. Substitute the constitutive equations (5.3.10) into the differential momentum equations (5.3.2) and (5.3.3) and derive the Navier–Stokes equations (5.3.14).

Differential Energy Equation

5.30. For a gas flow in which Stokes' hypothesis is not applicable, the negative average of the three normal stresses, denoted $\bar{p}$, may be different from the pressure p. Find an expression for $(\bar{p} - p)$.

5.31. Derive the incompressible differential energy equation by applying Gauss's theorem (see Problem 5.1) to the integral energy equation (4.4.13) assuming that $\dot{W}_s = \dot{W}_{shear} = \dot{W}_I = 0$ and using $\dot{Q} = \int_{c.s.} K\nabla T \cdot \hat{n}\ dA$ assuming no viscous effects.

5.32. Simplify the differential energy equation for a liquid flow in which the temperature gradients are quite large and the velocity components are very small, such as in a lake heated from above.

5.33. Explain which term in the differential energy equation accounts for the extremely high temperatures that exist on satellites during reentry.

5.34. The velocity distribution in a 2.0-cm-diameter pipe is given by $u(r) = 10(1 - 10\ 000r^2)$ m/s. Find the magnitude of the dissipation function at the wall, at the centerline, and halfway between for air at 20°C.

SIX

Dimensional Analysis and Similitude

6.1 INTRODUCTION

There are very few problems of interest in the field of fluid mechanics that are solved using the differential and integral equations only. Most often it is necessary to resort to experimental methods to establish relationships between the variables of interest. Since experimental studies are usually quite expensive, it is necessary to keep the required experimentation to a minimum. This is done using a technique called **dimensional analysis,** which is based on the notion of **dimensional homogeneity**—that all terms in an equation must have the same dimensions. For example, if we write Bernoulli's equation in the form

$$\frac{V_1^2}{2g} + \frac{p_1}{\gamma} + z_1 = \frac{V_2^2}{2g} + \frac{p_2}{\gamma} + z_2 \tag{6.1.1}$$

we note that the dimension of each term is length. Furthermore, if we factored out z_1 from the left-hand side and z_2 from the right-hand side, we would have

$$\frac{V_1^2}{2gz_1} + \frac{p_1}{\gamma z_1} + 1 = \left(\frac{V_2^2}{2gz_2} + \frac{p_2}{\gamma z_2} + 1\right)\frac{z_2}{z_1} \tag{6.1.2}$$

In this form of Bernoulli's equation the terms are all dimensionless and we have written the equation as a combination of dimensionless parameters, the basic idea in dimensional analysis that will be presented in the next section.

Often in experimental work we are required to perform experiments on objects that are quite large, too large to experiment with for a reasonable cost. This would include flows over weirs and dams; wave interactions with piers and breakwaters; flows around submarines and ships; subsonic and supersonic flows around aircraft; flows around stadiums and buildings, as shown in Fig. 6.1; flows through large pumps and turbines; and flows around automobiles and trucks. Such flows are usually studied in laboratories using models that are smaller than the prototype, the actual device. This substantially reduces the costs when compared with full-scale studies and allows for the study of various configurations or flow conditions.

There are also flows of interest that involve rather small dimensions, such as flow around a turbine blade, flow into a capillary tube, flow around a microorganism, flow through a small control valve, and flow around and inside a falling droplet. These flows would require that the model be larger than the prototype so that observations could be made with an acceptable degree of accuracy.

Similitude is the study of predicting prototype conditions from model observations. This will be presented following dimensional analysis. Similitude involves the use of the dimensionless parameters obtained in dimensional analysis.

Figure 6.1 Scale model of the large buildings in a city. Air flow around the buildings is studied. Note the roughening elements on the floor to generate the desired wall turbulence. (Courtesy of Fluid Mechanics and Diffusion Laboratory, Colorado State University)

There are two approaches that can be used in the study of dimensional analysis and similitude. We will present both approaches. First, we use the **Buckingham π-theorem,** which organizes the steps of ensuring dimensional homogeneity; it requires some knowledge of the phenomenon being studied in order that the appropriate quantities of interest are included. Second, we extract the dimensionless parameters that influence a particular flow situation from the differential equations and boundary conditions that are needed to describe the phenomenon being investigated.

6.2 DIMENSIONAL ANALYSIS

6.2.1 Motivation

In the study of phenomena involving fluid flows, either analytically or experimentally, there are invariably many flow and geometric parameters involved. In the interest of saving time and money the fewest possible combinations of parameters should be utilized. For example, consider the pressure drop across the slider valve of Fig. 6.2. We may suspect that the pressure drop depends on such parameters as pipe mean velocity V, the density ρ of the fluid, the fluid viscosity[1] μ, the pipe diameter d, and the

---

[1]In this chapter we deal with isotropic Newtonian fluids only. Non-Newtonian fluids would possess additional parameters related to the appropriate stress–strain equations.

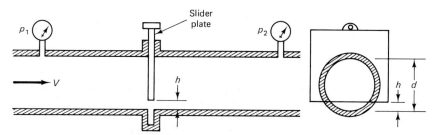

Figure 6.2 Flow around a slider valve.

gap height h. This could be expressed as

$$\Delta p = f(V, \rho, \mu, d, h) \qquad (6.2.1)$$

Now, if we attempt an experimental study of this problem, consider the strategy for finding the dependence of the pressure drop on the parameters involved. We could fix all parameters except the velocity and investigate the dependence of the pressure drop on the average velocity. Then the diameter could be changed and the experiment repeated. This would lead to the set of results shown in Fig. 6.3a. Following that set of experiments the gap height h could be changed, leading to the curves of Fig. 6.3b. Again, different fluids could be studied, leading to curves with ρ and μ changing values.

Consider next the notion that any equation that relates a certain set of variables, such as Eq. 6.2.1, can be written in terms of dimensionless parameters, as was done with the Bernoulli equation (6.1.2). We can organize the variables of Eq. 6.2.1 into dimensionless parameters (the steps needed to do this will be presented in a subsequent section) as follows:

$$\frac{\Delta p}{\rho V^2} = f\left(\frac{V\rho d}{\mu}, \frac{h}{d}\right) \qquad (6.2.2)$$

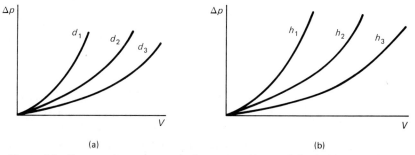

Figure 6.3 Pressure drop versus velocity curves: (a) ρ, μ, h fixed; (b) ρ, μ, d fixed.

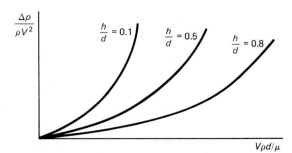

Figure 6.4 Dimensionless pressure drop versus dimensionless velocity.

Obviously, this is a much simpler relationship. We could perform an experiment with a fixed h/d (say, $h/d = 0.1$) by varying $V\rho d/\mu$ (this is done by simply varying V), resulting in a curve as shown in Fig 6.4. The quantity h is changed so that $h/d = 0.5$ and the testing is repeated. Finally, the entire experiment is presented in one figure, as in Fig. 6.4. This has greatly reduced the effort and the cost in determining the actual form of $f(V\rho d/\mu, h/d)$; we would use only one pipe and one valve, and we would use only one fluid.

It is not always clear, however, which parameters should be included in an equation such as (6.2.1). The selection of these parameters requires a detailed understanding of the physics involved. In the selection of the parameters that affect the pressure drop across the slider valve, it was assumed that density and viscosity are important parameters, while parameters such as pipe pressure and fluid compressibility are not. It should be kept in mind that the selection of the proper parameters is a first crucial step in the application of dimensional analysis.

6.2.2 Review of Dimensions

Before we present the dimensional analysis technique we shall review the dimensions of the quantities of interest in an introductory course in fluid mechanics. All of the quantities have some combination of dimensions of length, time, mass, and force which are related by Newton's second law,

$$\Sigma \mathbf{F} = m\mathbf{a} \tag{6.2.3}$$

In terms of dimensions, it is written as

$$F = \frac{ML}{T^2} \tag{6.2.4}$$

where F, M, L, and T are the dimensions of force, mass, length, and time, respectively. Thus we see that it is sufficient to use only three basic dimensions. We will choose the M-L-T system because we can eliminate the force dimension with Eq. 6.2.4.

If we were considering more complicated flow situations such as those involving electromagnetic field interactions, or those involving temperature gradients, we would need to include the appropriate additional dimensions. However, in this book such phenomena will not be introduced, except for the compressible flow of an ideal gas; for that case an equation of state relates the thermal effects to the dimensions above. That is,

$$p = \rho RT \qquad (6.2.5)$$

where T represents temperature. This allows us to write

$$[RT] = [p/\rho]$$
$$= \frac{F}{L^2} \cdot \frac{L^3}{M} = \frac{ML/T^2}{L^2} \cdot \frac{L^3}{M} = \frac{L^2}{T^2} \qquad (6.2.6)$$

where the brackets represent "the dimensions of." Notice that the equation of state does not introduce additional dimensions.

The quantities of interest in fluid mechanics are listed with their respective dimensions in Table 6.1. Reference to this table will simplify writing the dimensions of the quantities introduced in the problems.

TABLE 6.1 SYMBOLS AND DIMENSIONS OF QUANTITIES USED IN FLUID MECHANICS

Quantity	Symbol	Dimensions
Length	l	L
Time	t	T
Mass	m	M
Force	F	ML/T^2
Velocity	V	L/T
Acceleration	a	L/T^2
Frequency	ω	T^{-1}
Gravity	g	L/T^2
Area	A	L^2
Flow rate	Q	L^3/T
Mass flux	$\dot{m}$	M/T
Pressure	p	M/LT^2
Stress	τ	M/LT^2
Density	ρ	M/L^3
Specific weight	γ	M/L^2T^2
Viscosity	μ	M/LT
Kinematic viscosity	ν	L^2/T
Work	W	ML^2/T^2
Power, heat flux	$\dot{W}, \dot{Q}$	ML^2/T^3
Surface tension	σ	M/T^2
Bulk modulus	B	M/LT^2

6.2.3 Buckingham π-Theorem

In a given physical problem the dependent variable x_1 can be expressed in terms of the independent variables as

$$x_1 = f(x_2, x_3, x_4, \ldots, x_n) \tag{6.2.7}$$

where n represents the total number of variables. Referring to Eq. 6.2.1, Δp is the dependent variable and V, ρ, μ, d, and h are the independent variables. The **Buckingham π-theorem** states that $(n - m)$ dimensionless groups of variables, called π-parameters, where m is the number[2] of basic dimensions included in the variables, can be related by

$$\pi_1 = f_1(\pi_2, \pi_3, \ldots, \pi_{n\text{-}m}) \tag{6.2.8}$$

where π_1 includes the dependent variable and the remaining π-parameters include only independent variables, as in Eq. 6.2.2.

Further, it is noted that a requirement for a successful application of dimensional analysis is that a dimension should occur at least twice or not at all. For example, the equation $\Delta p = f(V, l, d)$ is ill-stated since pressure involves the dimensions of force and V, l, and d do not contain such a dimension.

The procedure used in applying the π-theorem is summarized as follows:

1. Write the functional form of the dependent variable depending on the $(n - 1)$ independent variables.
2. Identify m **repeating variables,** variables that will be combined with each remaining variable to form the π-parameters. The repeating variables selected from the independent variables must include all of the basic dimensions, but they must not form a π-parameter by themselves.
3. Form the π-parameters by combining the repeating variables with each of the remaining variables.
4. Write the functional form of the $(n - m)$ dimensionless π-parameters.

Step 3 can be accomplished by a relatively simple algebraic procedure; we will also illustrate a procedure in the examples that utilizes simple observation.

The algebraic procedure will be illustrated with an example. Suppose that we desire to combine the variables surface tension σ, velocity

---

[2]There are situations where m is less than the number of basic dimensions. Example 6.2 will illustrate.

V, density ρ, and length l into a π-parameter; this can be written as

$$\pi = \sigma^a V^b \rho^c l^d \tag{6.2.9}$$

The objective is to determine a, b, c, and d so that the grouping is dimensionless. In terms of dimensions, Eq. 6.2.9 is

$$M^0 L^0 T^0 = \left(\frac{M}{T^2}\right)^a \left(\frac{L}{T}\right)^b \left(\frac{M}{L^3}\right)^c L^d \tag{6.2.10}$$

Equating exponents on each of the basic dimensions:

$$
\begin{aligned}
M: \quad & 0 = a + c \\
L: \quad & 0 = b - 3c + d \\
T: \quad & 0 = -2a - b
\end{aligned}
\tag{6.2.11}
$$

The three algebraic equations are solved simultaneously to yield

$$a = -c \qquad b = 2c \qquad d = c \tag{6.2.12}$$

so that the π-parameter becomes

$$\pi = \left(\frac{\rho l V^2}{\sigma}\right)^c \tag{6.2.13}$$

A dimensionless parameter raised to any power remains dimensionless; consequently, we can select c to be any number other than zero. It is usually selected as $c = 1$, depending on the ratio desired. Selecting $c = 1$, the π-parameter is

$$\pi = \frac{\rho l V^2}{\sigma} \tag{6.2.14}$$

Actually, we could have selected $c = 1$ in Eq. 6.2.9 and proceeded with only three unknowns. Or if it were desired to have σ in the numerator to the first power, we could have set $a = 1$ and let b, c, and d be unknowns.

A final note: If only one π-term results, the functional form would state that the π-term must be a constant since the right-hand side of Eq. 6.2.8 would contain no additional π-terms. This would result in an expression that includes an arbitrary constant that could be determined through analysis or experimentation.

EXAMPLE 6.1

The drag force F_D on a cylinder of diameter d and length l is to be studied. What functional form relates the dimensionless variables?

Solution

First, we must determine the variables that have some influence on the drag force. If we include variables that do not, in fact, influence the drag force, we would have additional π-parameters that experimentation would show to be unimportant; if we do not include a variable that does influence the drag force, experimentation would also reveal that problem. Experience is essential in choosing the correct variables; in this example we will include as influential variables the free stream velocity V, the viscosity μ, the density ρ of the fluid, in addition to the diameter d and the length l of the cylinder. This is written as

$$F_D = f(d, l, V, \mu, \rho)$$

The variables are observed to include $m = 3$ dimensions:

$$[F_D] = \frac{ML}{T^2} \qquad [V] = \frac{L}{T} \qquad [\mu] = \frac{M}{LT}$$

$$[d] = L \qquad [l] = L \qquad [\rho] = \frac{M}{L^3}$$

We choose repeating variables with the simplest combinations of dimensions such that they do not form a π-parameter by themselves (we could not include d and l as repeating variables); the repeating variables are chosen to be d, V, and ρ. These three variables are combined with each of the remaining variables to form the π-parameters. When combined with F_D we observe that only F_D and ρ have the mass dimension; hence F_D must be divided by ρ. Only F_D and V have the time dimension; thus F_D divided by ρ has L^4 in the numerator; when divided by V^2 this results in L^2 remaining in the numerator. Hence we must have d^2 in the denominator resulting in

$$\pi_1 = \frac{F_D}{\rho V^2 d^2}$$

When d, V, and ρ are combined with l there obviously results

$$\pi_2 = \frac{l}{d}$$

The last π-parameter results from combining μ with d, V, and ρ. The mass dimension disappears if we divide μ by ρ. The time dimension disappears if we divide μ by V. This leaves one length dimension in the numerator; hence d is needed in the denominator resulting in

$$\pi_3 = \frac{\mu}{\rho V d}$$

The final dimensionless, functional relationship is

$$\pi_1 = f_1(\pi_2, \pi_3)$$

or

$$\frac{F_D}{\rho V^2 d^2} = f_1\left(\frac{l}{d}, \frac{\mu}{\rho V d}\right)$$

Rather than the original relationship of six variables we have reduced the problem to one involving three variables, a much simpler problem. To determine the particular form of the functional relationship above, we would actually have to solve the problem; experimentation would be needed if analytical or numerical methods were not available. This is often the case in fluid mechanics.

Note that we could have included several additional variables in our original list, such as gravity g, free-stream fluctuation level $(\overline{u'^2})^{1/2}$ (u' is the velocity fluctuation), and the roughness e of the cylinder surface. To not include variables that are significant, or to include variables that are not significant is a matter of experience. The novice must learn how to identify significant variables; however, even the experienced researcher is often at a loss to correlate certain phenomena; much experimentation is often needed to discover the appropriate parameters.

EXAMPLE 6.2

The rise of liquid in a capillary tube is to be studied. It is anticipated that the rise h will depend on surface tension σ, tube diameter d, liquid specific weight γ, and angle β of attachment between the liquid and tube. Write the functional form of the dimensionless variables.

Solution

The expression relating the variables is

$$h = f(\sigma, d, \gamma, \beta)$$

The dimensions of the variables are

$$[h] = L \qquad [\gamma] = \frac{M}{L^2 T^2} \qquad [\beta] = 1 \text{ (dimensionless)}$$

$$[\sigma] = \frac{M}{T^2} \qquad [d] = L$$

By observation we see that M/T^2 occurs as that combination in both σ and γ; hence M and T are not independent dimensions in this problem. There are only two independent groupings of basic dimensions, L and M/T^2. Thus $m = 2$ and we choose σ and d as the repeating variables. When combined with h, the first π-parameter is

$$\pi_1 = \frac{h}{d}$$

When σ and d are combined with γ, the second π-parameter is

$$\pi_2 = \frac{\gamma d^2}{\sigma}$$

Finally, since the angle β is dimensionless, it forms a π-parameter by itself; that is,

$$\pi_3 = \beta$$

The final functional form is

$$\pi_1 = f_1(\pi_2, \pi_3)$$

or

$$\frac{h}{d} = f_1\left(\frac{\gamma d^2}{\sigma}, \beta\right)$$

In this example we could not have chosen the angle β as a repeating variable since it already is a π-parameter. Also, we could not have chosen three repeating variables since M and T were not independent.

Also, note that we may have thought that gravity should have been included in the problem. If it had been included above, it would have not appeared in any of the π-parameters, indicating that it should not have been included. If density and gravity, rather than specific weight, had been included the relationship above would have resulted since $\gamma = \rho g$; this, by the way, would have avoided the necessity of observing that M/T^2 was a dimensional grouping.

A final note regarding the functional form of the π-parameters. The relationship above could equally have been written as

$$\frac{h}{d} = f_1\left(\frac{\sigma}{\gamma d^2}, \beta\right)$$

Also, occasionally a different set of repeating variables could be selected. This simply expresses the final functional equation in a different but equivalent form. Actually, a second form can be shown to be a combination of the π-parameters from an initial form.

6.2.4 Common Dimensionless Parameters

Consider a relatively general relationship between the pressure drop Δp, a characteristic length l, a characteristic velocity V, the density ρ, the viscosity μ, the gravity g, the surface tension σ, the speed of sound c, and an angular frequency ω, written as

$$\Delta p = f(l, V, \rho, \mu, g, \sigma, c, \omega) \tag{6.2.15}$$

The π-theorem applied to this problem, with l, V, and ρ as repeating variables, results in

$$\frac{\rho V^2}{\Delta p} = f_1\left(\frac{V\rho l}{\mu}, \frac{V^2}{lg}, \frac{V^2\rho l}{\sigma}, \frac{V}{c}, \frac{l\omega}{V}\right) \qquad (6.2.16)$$

Each of the π-parameters in this expression is a common dimensionless parameter that appears in numerous fluid flow situations. They are identified as follows:

$$
\begin{aligned}
&\text{Euler number, Eu} = \frac{\rho V^2}{\Delta p}\\[6pt]
&\text{Reynolds number, Re} = \frac{V\rho l}{\mu}\\[6pt]
&\text{Froude number, Fr} = \frac{V}{\sqrt{lg}}\\[6pt]
&\text{Mach number, M} = \frac{V}{c}\\[6pt]
&\text{Weber number, We} = \frac{V^2 l\rho}{\sigma}\\[6pt]
&\text{Strouhal number, St} = \frac{l\omega}{V}
\end{aligned}
\qquad (6.2.17)
$$

The physical significance of each parameter can be determined by observing that each dimensionless number can be written as the ratio of two forces. The forces are observed to be

$$F_p = \text{pressure force} = \Delta p A \sim \Delta p l^2$$

$$F_I = \text{inertial force} = mV\frac{dV}{ds} \sim \rho l^3 V\frac{V}{l} = \rho l^2 V^2$$

$$F_\mu = \text{viscous force} = \tau A = \mu\frac{du}{dy}A \sim \mu\frac{V}{l}l^2 = \mu l V$$

$$F_g = \text{gravity force} = mg \sim \rho l^3 g \qquad (6.2.18)$$

$$F_B = \text{compressibility force} = BA \sim \rho\frac{dp}{d\rho}l^2 = \rho c^2 l^2$$

$$F_\sigma = \text{surface tension force} = \sigma l$$

$$F_\omega = \text{centrifugal force} = mr\omega^2 \sim \rho l^3 l\omega^2 = \rho l^4 \omega^2$$

Thus we see that

$$\text{Eu} \propto \frac{\text{inertial force}}{\text{pressure force}}$$

$$\text{Re} \propto \frac{\text{inertial force}}{\text{viscous force}}$$

$$\text{Fr} \propto \frac{\text{inertial force}}{\text{gravity force}}$$

$$\text{M} \propto \frac{\text{inertial force}}{\text{compressibility force}} \qquad (6.2.19)$$

$$\text{We} \propto \frac{\text{inertial force}}{\text{surface tension force}}$$

$$\text{St} \propto \frac{\text{centrifugal force}}{\text{inertial force}}$$

Thinking of the dimensionless parameters in terms of the ratios of forces allows us to anticipate the significant parameters in a particular flow of interest. If viscous forces are important, we know that the Reynolds number is a significant dimensionless parameter. If surface tension forces are instrumental in affecting the flow, as a droplet formation or flow over a weir with a small head, we expect the Weber number to be important. Similar analysis can be applied to other fluid flow situations.

Obviously, all of the effects included in the general relationship (6.2.16) would not be of interest in any one situation. It would be very unlikely that both compressibility effects and surface tension effects would influence a flow simultaneously. In addition, there is often more than one length of importance, thereby introducing additional geometric, dimensionless ratios. We have, however, introduced the more common dimensionless flow parameters of interest in fluid mechanics.

6.3 SIMILITUDE

6.3.1 General Information

As stated in the introduction, **similitude** is the study of predicting proto-type conditions from model observations. When an analytical or numeri-cal solution is not practical, or when calculations are based on a simplified model so that uncertainty is introduced, it is usually advisable to perform tests on a model if testing is not practical on a full-scale prototype, be it too large or too small.

If it is decided that a model study is to be performed, it is necessary to develop the means whereby a quantity measured on the model can be used to predict the associated quantity on the prototype. We can develop such a means if we have **dynamic similarity** between model and prototype, that is, if the forces which act on corresponding masses in the model flow and the prototype flow are in the same ratio throughout the entire flow field. Suppose that pressure forces, inertial forces, viscous forces, and gravity forces are present; then dynamic similarity requires that, at corresponding points in the flow fields,

$$\frac{(F_I)_m}{(F_I)_p} = \frac{(F_p)_m}{(F_p)_p} = \frac{(F_\mu)_m}{(F_\mu)_p} = \frac{(F_g)_m}{(F_g)_p} = \text{const.} \tag{6.3.1}$$

These can be rearranged to read

$$\left(\frac{F_I}{F_p}\right)_m = \left(\frac{F_I}{F_p}\right)_p \qquad \left(\frac{F_I}{F_\mu}\right)_m = \left(\frac{F_I}{F_\mu}\right)_p \qquad \left(\frac{F_I}{F_g}\right)_m = \left(\frac{F_I}{F_g}\right)_p \tag{6.3.2}$$

which, in the preceding section have been shown to be

$$Eu_m = Eu_p \qquad \text{Re}_m = \text{Re}_p \qquad \text{Fr}_m = \text{Fr}_p \tag{6.3.3}$$

If the forces above were the only ones present, we could write

$$F_I = f(F_p, F_\mu, F_g) \tag{6.3.4}$$

Recognizing that there is only one basic dimension, namely force, dimensional analysis would allow us to write (see Eq. 6.2.8) the equation above in terms of force ratios or

$$Eu = f(\text{Re}, \text{Fr}) \tag{6.3.5}$$

Hence we could conclude that if the Reynolds number and the Froude number are the same on the model and prototype, the Euler number must also be the same. Thus dynamic similarity between model and prototype is guaranteed by equating the Reynolds number and the Froude number of the model to those on the prototype, respectively. If compressibility forces were included here, the analysis above would result in the Mach number being included in Eq. 6.3.5.

We can write the inertial force ratio as

$$\frac{(F_I)_m}{(F_I)_p} = \frac{a_m m_m}{a_p m_p} = \text{const.} \tag{6.3.6}$$

showing that the acceleration ratio between corresponding points on the model and prototype is a constant provided that the mass ratio of corresponding fluid elements is a constant. We can write the acceleration ratio as

$$\frac{a_m}{a_p} = \frac{V_m^2/l_m}{V_p^2/l_p} = \text{const.} \tag{6.3.7}$$

showing that the velocity ratio between corresponding points is a constant providing the length ratio is a constant. The velocity ratio being a constant between all corresponding points in the flow fields is the statement of **kinematic similarity.** This would result in the streamline pattern around the model being the same as that around the prototype except for a scale factor. The length ratio being constant between all corresponding points in the flow fields is the demand of **geometric similarity** which results in the model having the same shape as the prototype. Hence, to ensure complete similarity between model and prototype, we demand that:

Geometric similarity be satisfied.

The mass ratio of corresponding fluid elements be a constant.

The appropriate dimensionless parameters of Eq. 6.2.17 be equal.

Assuming that complete similarity between model and prototype exists, we can now predict quantities of interest on a prototype from measurements on a model. If we measure a drag force F_D on a model and wish to predict the corresponding drag on the prototype, we could equate the ratio of the drag forces to the ratio of the inertial forces (see Eq. 6.2.18) as

$$\frac{(F_D)_m}{(F_D)_p} = \frac{(F_I)_m}{(F_I)_p} = \frac{\rho_m V_m^2 l_m^2}{\rho_p V_p^2 l_p^2} \tag{6.3.8}$$

If we measure the power input to a model and wish to predict the power requirement of the prototype, we would recognize that power is force times velocity and write

$$\frac{\dot{W}_m}{\dot{W}_p} = \frac{(F_I)_m V_m}{(F_I)_p V_p} = \frac{\rho_m V_m^2 l_m^2 V_m}{\rho_p V_p^2 l_p^2 V_p} \tag{6.3.9}$$

Hence we can predict a prototype quantity if we select the model fluid (this provides ρ_m/ρ_p), the scale ratio (this provides l_m/l_p), and the appropriate dimensionless number from Eq. 6.2.17 (this provides V_m/V_p). Examples will illustrate.

6.3.2 Confined Flows

A confined flow is a flow that has no free surfaces (a liquid–gas surface) or interfaces (two different liquids forming an interface). It is confined to move within a specified region; such flows include external flows around objects, such as airfoils, buildings, and submarines, as well as internal flows in pipes and conduits.

Gravity does not influence the flow pattern in confined flows; that is, if gravity could be changed in magnitude, the flow pattern and associated flow quantities would not change. The dominant effect is that of viscosity in incompressible flows (all liquid flows and gas flows in which M < 0.3). Surface tension is obviously not a factor, as it would be in bubble formation, and for steady flows there would be no unsteady effects due to oscillations in the flow. The three relevant forces are pressure forces, inertial forces, and viscous forces. Therefore, in confined flows dynamic similarity is achieved if the ratios of viscous forces, inertial forces, and pressure forces between model and prototype are the same. This leads to the conclusion (see Eq. 6.3.5) that Eu = f(Re), so that it is only necessary to consider the Reynolds number as the dominant dimensionless parameter in a confined incompressible flow. If compressibility effects are significant, the Mach number would also be important.

EXAMPLE 6.3

A test is to be performed on a proposed design for a large pump that is to deliver 1.5 m³/s of water from a 40-cm-diameter impeller with a pressure rise of 400 kPa. A model with an 8-cm-diameter impeller is to be used. What flow rate should be used and what pressure rise is to be expected? The model fluid is water at the same temperature as the water in the prototype.

Solution

For similarity to exist in this confined flow problem, the Reynolds numbers must be equal, that is,

$$\text{Re}_m = \text{Re}_p$$

$$\frac{V_m d_m}{\nu_m} = \frac{V_p d_p}{\nu_p}$$

Recognizing that $\nu_m = \nu_p$, we see that

$$\frac{V_m}{V_p} = \frac{d_p}{d_m}$$

$$= \frac{0.4}{0.08} = 5$$

The ratio of flow rates is found recognizing that $Q = VA$:

$$\frac{Q_m}{Q_p} = \frac{V_m d_m^2}{V_p d_p^2}$$

$$= 5 \times \left(\frac{1}{5}\right)^2 = \frac{1}{5}$$

Thus we find that

$$Q_m = \frac{Q_p}{5} = \frac{1.5}{5} = 0.3 \text{ m}^3/\text{s}$$

The dimensionless pressure rise is found using the Euler number:

$$\left(\frac{\Delta p}{\rho V^2}\right)_m = \left(\frac{\Delta p}{\rho V^2}\right)_p$$

Hence the pressure rise for the model is

$$\Delta p_m = \Delta p_p \frac{\rho_m}{\rho_p} \frac{V_m^2}{V_p^2}$$

$$= 400 \times 1 \times 5^2 = 10\ 000 \text{ kPa}$$

Note that in this example we see that the velocity in the model is equal to the velocity in the prototype multiplied by the length ratio, and the pressure rise in the model is equal to the pressure rise in the prototype multiplied by the length ratio squared. If the length ratio were very large, it is obvious that to maintain Reynolds number equivalence would be quite difficult, indeed. This observation is discussed in more detail in section 6.3.4.

6.3.3 Free-Surface Flows

A free-surface flow is one in which part of the boundary involves a pressure boundary condition. This includes flows over weirs and dams, as shown in Fig. 6.5, flows in channels and spillways, flows involving two fluids separated by an interface, and flows around floating objects with waves and around submerged objects with cavitation present. In all of these flows the location of the free surface is unknown and the velocity at the free surface is unknown; it is the pressure that must be the same[3] on either side of the interface. In free-surface flows, gravity controls both the location and the motion of the free surface. This introduces the Froude number because of the influence of the gravity forces. If we consider flows that do not exhibit periodic motions, which have negligible surface tension and compressibility effects, we can ignore the influence of St, We, and M. That leaves only the viscous effects to be considered. There are many free-surface flows in which viscous effects are significant. Consider, however, that in most model studies water is the only economical fluid to use; if the prototype fluid is also water, as it often is, we would find from the Froude numbers

$$\frac{V_m^2}{l_m g_m} = \frac{V_p^2}{l_p g_p} \qquad \therefore \quad \frac{V_m}{V_p} = \left(\frac{l_m}{l_p}\right)^{1/2} \qquad (6.3.10)$$

[3]The surface tension, if significant, results in a pressure difference across the interface (see Eq. 1.3.20).

Figure 6.5 Model of the Bonneville lock and dam on the Columbia River. (Courtesy of the
U.S. Army Corps of Engineers Waterways Experiment Station.)

assuming that $g_m = g_p$. Matching the Reynolds numbers (using $\nu_m = \nu_p$):

$$\frac{V_m l_m}{\nu_m} = \frac{V_p l_p}{\nu_p} \qquad \therefore \quad \frac{V_m}{V_p} = \frac{l_p}{l_m} \qquad (6.3.11)$$

Thus we have a conflict. If we use the same fluid in the model study as in
the prototype flow, we cannot satisfy both the Froude number criterion
and the Reynolds number criterion. If we demand that both criteria be
satisfied by using different fluids for model and prototype ($\nu_m \neq \nu_p$), we
must select a model fluid with a viscosity $\nu_m = \nu_p (l_m/l_p)^{3/2}$ (this results
from a matching of the Froude and Reynolds numbers). A fluid with this
viscosity is probably either an impossibility or an impracticality. Hence,
when modeling free-surface flows in which viscous effects are important,
we equate the Froude numbers and include viscous effects by some other
technique. For example, if we measured the total drag on the model of a
ship, we would approximate the viscous drag (using some technique not
included here) and subtract it from the total drag, leaving the drag due to
wave resistance. The wave drag on the prototype would then be predicted
using similitude, and the approximated viscous drag would be added to
the wave drag, yielding the expected drag on the ship. For the better
designs the viscous drag on shiphulls can be of the same order of magnitude as the wave drag.

EXAMPLE 6.4

A 1:20 scale model of a surface vessel is used to test the influence of a proposed design on the wave drag. A drag of 24 N is measured at a model speed of 2.6 m/s. What speed does this correspond to on the prototype, and what wave drag is predicted for the prototype? Neglect viscous effects, and assume the same fluid for model and prototype.

Solution

The Froude number must be equated for both model and prototype. Thus

$$\text{Fr}_m = \text{Fr}_p \qquad \frac{V_m}{\sqrt{l_m g}} = \frac{V_p}{\sqrt{l_p g}}$$

This yields

$$V_p = V_m \left(\frac{l_p}{l_m}\right)^{1/2}$$

$$= 2.6\sqrt{20} = 11.63 \text{ m/s}$$

To find the wave drag on the prototype, we equate the drag ratio to the inertia force ratio:

$$\frac{(F_D)_m}{(F_D)_p} = \frac{\rho_m V_m^2 l_m^2}{\rho_p V_p^2 l_p^2}$$

This allows us to calculate the wave drag on the prototype as, using $\rho_p = \rho_m$,

$$(F_D)_p = (F_D)_m \frac{\rho_p V_p^2 l_p^2}{\rho_m V_m^2 l_m^2}$$

$$= 24 \times \frac{11.63^2}{2.6^2} \times 20^2 = 192\ 000 \text{ N} \quad \text{or} \quad 192 \text{ kN}$$

EXAMPLE 6.4 (English)

A 1:20 scale model of a surface vessel is used to test the influence of a proposed design on the wave drag. A drag of 6.2 lb is measured at a model speed of 8.0 ft/sec. What speed does this correspond to on the prototype, and what wave drag is predicted for the prototype? Neglect viscous effects, and assume the same fluid for model and prototype.

Solution

The Froude number must be equated for both model and prototype. Thus

$$\text{Fr}_m = \text{Fr}_p \qquad \frac{V_m}{\sqrt{l_m g}} = \frac{V_p}{\sqrt{l_p g}}$$

This yields

$$V_p = V_m \left(\frac{l_p}{l_m}\right)^{1/2}$$

$$= 8.0\sqrt{20} = 35.8 \text{ ft/sec}$$

To find the wave drag on the prototype, we equate the drag ratio to the inertia force ratio:

$$\frac{(F_D)_m}{(F_D)_p} = \frac{\rho_m V_m^2 l_m^2}{\rho_p V_p^2 l_p^2}$$

This allows us to calculate the wave drag on the prototype as, using $\rho_p = \rho_m$,

$$(F_D)_p = (F_D)_m \frac{\rho_p V_p^2 l_p^2}{\rho_m V_m^2 l_m^2}$$

$$= 6.2 \times \frac{35.8^2}{8^2} \times 20^2 = 49{,}700 \text{ lb}$$

6.3.4 High-Reynolds-Number Flows

In a confined flow in which the Reynolds number is the dimensionless parameter that guarantees dynamic similarity, we note that if the same fluid is used in model and prototype, the velocity in the model study is $V_m = V_p l_p / l_m$; the velocity in the model is the velocity in the prototype multiplied by the scale factor. This often results in velocities that are prohibitively large in the model study. Also, the pressures encountered in the model study are large, as shown in Example 6.3, and the energy consumption is also very large. Because of these problems the Reynolds numbers may not be matched in studies involving large Reynolds numbers.

There is, however, some justification for not matching the Reynolds number in model studies. Consider a typical drag coefficient C_D versus Reynolds number curve, as shown in Fig. 6.6 (the complete curve is

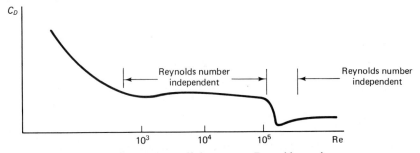

Figure 6.6 Drag coefficient versus Reynolds number.

presented in Fig. 8.8). The drag coefficient is a dimensionless drag, defined as $C_D = \text{drag}/\frac{1}{2}\rho V^2 A$. At a sufficiently high Reynolds number, between 10^3 and 10^5, the flow is insensitive to changes in the Reynolds number; note that the drag coefficient is essentially constant and independent of Re. That implies that the flow field is similar at $\text{Re} = 10^3$ to that $\text{Re} = 10^5$. Thus if $\text{Re}_p = 10^5$, it is only necessary that $10^3 < \text{Re}_m < 10^5$ for the viscous effects to have the same effect on model and prototype. This often allows another parameter of interest to be matched, such as the Froude number or the Mach number. There are, however, high-Reynolds-number flows in which compressibility effects and free-surface effects are negligible, so that neither the Froude number nor the Mach number are applicable. Examples include flow around automobiles, large smoke stacks, and dirigibles. For such flows we must only ensure that the Reynolds number be within the range where the drag coefficient is constant. We should note that for Reynolds numbers quite large ($\text{Re} > 5 \times 10^5$ in Fig. 6.6) the flow may also become Reynolds number independent. If that is the case, it is only necessary that Re_m be sufficiently large.

EXAMPLE 6.5

A 1:10 scale model of an automobile is used to measure the drag on a proposed design. It is to simulate a prototype speed of 90 km/h. What speed should be used in the wind tunnel if Reynolds numbers are equated? For this condition, what is the ratio of drag forces?

Solution

The same fluid exists on model and prototype, thus equating the Reynolds numbers results in

$$\frac{V_m l_m}{\nu_m} = \frac{V_p l_p}{\nu_p} \qquad \therefore \quad V_m = V_p \frac{l_p}{l_m}$$

$$= 90 \times 10 = 900 \text{ km/h}$$

This speed would, of course, introduce compressibility effects, effects that do not exist in the prototype. Hence the model study would be inappropriate.

     If we did use this velocity in the model, the drag force ratio would be

$$\frac{(F_D)_p}{(F_D)_m} = \frac{\rho_p V_p^2 l_p^2}{\rho_m V_m^2 l_m^2} \qquad \therefore \quad \frac{(F_D)_p}{(F_D)_m} = 1$$

Thus we see that the drag force on the model is the same as the drag force on the prototype if the same fluids are used when we equate Reynolds numbers.

EXAMPLE 6.6

In Example 6.5, if the Reynolds numbers were equated, the velocity in the model study was observed to be in the compressible flow regime (i.e., $V_m > 360$ km/h).

To conduct an acceptable model study, could we use a velocity of 90 km/h on a model with a characteristic length of 10 cm? Assume that the drag coefficient $(C_D = F_D/\frac{1}{2}\rho V^2 A$, where A is the projected area), is independent of Re for Re > 10^5. If so, what drag force on the prototype would correspond to a drag force of 1.2 N measured on the model?

Solution

The proposed model study in a wind tunnel is to be conducted with $V_m = 90$ km/h and $l_m = 0.1$ m. Using $\nu = 1.6 \times 10^{-5}$ m²/s, the Reynolds number is

$$\text{Re}_m = \frac{V_m l_m}{\nu_m}$$

$$= \frac{\dfrac{90 \times 1000}{3600} \times 0.1}{1.6 \times 10^{-5}} = 1.56 \times 10^5$$

This Reynolds number is greater than 10^5, so we will assume that similarity exists between model and prototype. The velocity of 90 km/h is sufficiently high.

The drag force on the prototype traveling at 90 km/h corresponding to 1.2 N on the model is found from

$$\frac{(F_D)_p}{(F_D)_m} = \frac{\rho_p V_p^2 l_p^2}{\rho_m V_m^2 l_m^2}$$

$$\therefore \quad (F_D)_p = (F_D)_m \frac{\rho_p^{\,1}}{\rho_m} \frac{V_p^2}{V_m^2} \frac{l_p^2}{l_m^2}$$

$$= 1.2 \times 10^2 = 120 \text{ N}$$

Note that in this example we have assumed that the drag coefficient is independent of Re for Re > 10^5. If the drag coefficient continued to vary above Re = 10^5 (this would be evident from experimental data), the foregoing analysis would have to be modified accordingly.

6.3.5 Compressible Flows

For most compressible flow situations the Reynolds number is so large (refer to Fig. 6.6) that it is not a parameter of significance; the compressibility effects lead to the Mach number as the primary dimensionless parameter for model studies. Thus for a particular model study we require

$$M_m = M_p \quad \text{or} \quad \frac{V_m}{c_m} = \frac{V_p}{c_p} \qquad (6.3.12)$$

If the model study is carried out in a wind tunnel and the prototype fluid is air, we can assume that $c_m \simeq c_p$. This shows that the velocity in the model study is equal to the velocity associated with the prototype. Of course, if the speeds of sound are different, the velocity ratio will be different from unity, accordingly.

EXAMPLE 6.7

The pressure rise from free stream to the nose of a fusilage section of an aircraft is measured in a wind tunnel at 20°C to be 34 kPa with a wind-tunnel airspeed of 900 km/h. If the test is to simulate flight at an elevation of 12 km, what is the prototype velocity and the expected nose pressure rise?

Solution

To find the prototype velocity corresponding to a wind-tunnel airspeed of 900 km/h, we equate the Mach numbers

$$\mathbf{M}_m = \mathbf{M}_p \qquad \text{or} \qquad \frac{V_m}{\sqrt{kRT_m}} = \frac{V_p}{\sqrt{kRT_p}}$$

Thus

$$V_p = V_m \left(\frac{kRT_p}{kRT_m}\right)^{1/2}$$

$$= 900 \left(\frac{216.7}{293}\right)^{1/2} = 774 \text{ km/h}$$

The pressure at the nose of the prototype fusilage is found as follows:

$$\frac{\Delta p_m}{\rho_m V_m^2} = \frac{\Delta p_p}{\rho_p V_p^2}$$

$$\therefore \quad \Delta p_p = \Delta p_m \frac{\rho_p}{\rho_m} \frac{V_p^2}{V_m^2}$$

$$= 34 \times 0.2546 \times \frac{774^2}{900^2} = 6.4 \text{ kPa}$$

where the density ratio and temperature T_p come from Appendix B.

EXAMPLE 6.7 (English)

The pressure rise from free stream to the nose of a fusilage section of an aircraft is measured in a wind tunnel at 60°F to be 5 psi with a wind-tunnel airspeed of 600 mph. If the test is to simulate flight at an elevation of 40,000 ft, what is the prototype velocity and the expected nose pressure rise?

Solution

To find the prototype velocity corresponding to a wind-tunnel airspeed of 600 mph we equate the Mach numbers

$$M_m = M_p \quad \text{or} \quad \frac{V_m}{\sqrt{kRT_m}} = \frac{V_p}{\sqrt{kRT_p}}$$

Thus

$$V_p = V_m \left(\frac{kRT_p}{kRT_m}\right)^{1/2}$$

$$= 600 \left(\frac{392}{520}\right)^{1/2} = 521 \text{ mph}$$

The pressure at the nose of the prototype fusilage is found as follows:

$$\frac{\Delta p_m}{\rho_m V_m^2} = \frac{\Delta p_p}{\rho_p V_p^2}$$

$$\therefore \quad \Delta p_p = \Delta p_m \frac{\rho_p}{\rho_m} \frac{V_p^2}{V_m^2}$$

$$= 5 \times 0.247 \times \frac{521^2}{600^2} = 0.932 \text{ psi}$$

where the density ratio and temperature T_p come from Appendix B.

___

6.3.6 Periodic Flows

In many flow situations there are regions of the flows in which periodic motions occur. Such flows include the periodic vortex shedding that takes place when a fluid flows past a cylindrical object such as a bridge, a TV tower, a cable, or a tall building; the flow past a wind machine; and the flow through turbomachinery. In flows such as these it is necessary to equate the Strouhal numbers, which can be written as

$$\frac{V_m}{\omega_m l_m} = \frac{V_p}{\omega_p l_p} \tag{6.3.13}$$

in order that the periodic motion be properly modeled.

In addition to the Strouhal number, there may be additional dimensionless parameters that must be equated: in viscous flows, the Reynolds number; in free-surface flows, the Froude number; and in compressible flows, the Mach number.

EXAMPLE 6.8

A large wind turbine, designed to operate at 50 km/h, is to be tested in a laboratory by constructing a 1:15 scale model. What airspeed should be used in the wind tunnel, what angular velocity should be used to simulate a prototype speed of 5 rpm, and what power output is expected from the model if the prototype output is designed to be 500 kW?

Solution

The speed in the wind tunnel can be any speed above that needed to provide a sufficiently large Reynolds number. Let us select the same speed with which the prototype is to operate, namely, 50 km/h, and calculate the minimum characteristic length that a Reynolds number of 10^5 would demand; this gives

$$\text{Re} = \frac{Vl}{\nu} \qquad 10^5 = \frac{\dfrac{50 \times 1000}{3600} \times l}{1.6 \times 10^{-5}} \qquad \therefore \quad l = 0.12 \text{ m}$$

Obviously, in a reasonably large wind tunnel we can maintain a characteristic length (e.g., the blade length) that large.

The angular velocity is found by equating the Strouhal numbers. There results

$$\frac{V_m}{\omega_m l_m} = \frac{V_p}{\omega_p l_p}$$

$$\therefore \quad \omega_m = \omega_p \frac{V_m}{V_p}^{1} \frac{l_p}{l_m}$$

$$= 5 \times 1 \times 15 = 75 \text{ rpm}$$

assuming that the wind velocities are equal.

The power is found by observing that power is force times velocity.

$$\frac{\dot{W}_m}{\dot{W}_p} = \frac{\rho_m V_m^3 l_m^2}{\rho_p V_p^3 l_p^2}$$

or

$$\dot{W}_m = \dot{W}_p \frac{\rho_m}{\rho_p} \frac{V_m^3}{V_p^3}^{1} \frac{l_m^2}{l_p^2}$$

$$= 500 \times \left(\frac{1}{15}\right)^2 = 2.22 \text{ kW}$$

6.4 NORMALIZED DIFFERENTIAL EQUATIONS

In Chapter 5 we derived the partial differential equations used to describe any flow of interest that involves an isotropic, homogeneous, Newtonian fluid. Such a flow could be laminar or turbulent, steady or unsteady, compressible or incompressible, a confined flow or a free-surface flow, a flow with surface tension effects or one in which surface tension is negligible. Most often, when utilizing the differential equations, we express them in dimensionless, or normalized, form. This form of the equations provides us with information not contained in the dimensional form, information similar to that provided by dimensional analysis. Let us normalize the differential equations that would describe the motion of an incompressible, homogeneous flow. The energy equation will not be needed.

Before we normalize the equations, let's review them. In vector form the continuity equation and the Navier-Stokes equation are

$$\nabla \cdot \mathbf{V} = 0$$
$$\rho \frac{D\mathbf{V}}{Dt} = -\nabla p - \rho g \nabla h + \mu \nabla^2 \mathbf{V} \qquad (6.4.1)$$

where we have assumed h to be vertical. Because the first two terms on the right are of the same form (the gradient of a scalar function) we can combine them as follows:

$$\nabla p + \rho g \nabla h = \nabla p_k \qquad (6.4.2)$$

where the **kinetic pressure** p_k is defined to be

$$p_k = p + \rho g h \qquad (6.4.3)$$

It is the pressure that results from motion alone. As the fluid motion ceases, p_k goes to zero. In terms of the kinetic pressure the Navier–Stokes equation becomes

$$\rho \frac{D\mathbf{V}}{Dt} = -\nabla p_k + \mu \nabla^2 \mathbf{V} \qquad (6.4.4)$$

If the pressure p is never used in a boundary condition, we can retain the kinetic pressure in the equations, thereby "hiding" the gravity effects. If the pressure enters a boundary condition, we must return to Eq. 6.4.3 and gravity becomes important.

To normalize the differential equations and boundary conditions, we must select characteristic quantities that best describe the problem of interest. For example, consider a flow that has an average velocity V and a primary dimension l. The dimensionless velocities and coordinate variables are

$$u^* = \frac{u}{V} \quad v^* = \frac{v}{V} \quad w^* = \frac{w}{V} \quad x^* = \frac{x}{l} \quad y^* = \frac{y}{l} \quad z^* = \frac{z}{l} \qquad (6.4.5)$$

where asterisks are used to denote dimensionless quantities. The characteristic pressure will be twice the inviscid pressure rise between free stream and the stagnation point (i.e., ρV^2); the characteristic time, provided that no characteristic time such as a period of oscillation exists in the problem, will be the time it takes a fluid particle to travel the distance l at the velocity V (i.e., l/V). Hence

$$p^* = \frac{p}{\rho V^2} \qquad t^* = \frac{t}{l/V} \tag{6.4.6}$$

Using the dimensionless quantities of Eq. 6.4.5, we see that

$$\mathbf{V}^* = u^*\hat{i} + v^*\hat{j} + w^*\hat{k}$$

$$= \frac{u}{V}\hat{i} + \frac{v}{V}\hat{j} + \frac{w}{V}\hat{k} = \frac{\mathbf{V}}{V}$$

$$\boldsymbol{\nabla}^* = \frac{\partial}{\partial x^*}\hat{i} + \frac{\partial}{\partial y^*}\hat{j} + \frac{\partial}{\partial z^*}\hat{k} \tag{6.4.7}$$

$$= l\frac{\partial}{\partial x}\hat{i} + l\frac{\partial}{\partial y}\hat{j} + l\frac{\partial}{\partial z}\hat{k} = l\boldsymbol{\nabla}$$

Introducing the above into the Navier–Stokes equation and the continuity equation, we have

$$\rho\frac{V}{l/V}\frac{D^*\mathbf{V}^*}{Dt^*} = \frac{\rho V^2}{l}\boldsymbol{\nabla}^*p_k^* + \mu\frac{V}{l^2}\nabla^{*2}\mathbf{V}^*$$

$$\frac{V}{l}\boldsymbol{\nabla}^* \cdot \mathbf{V}^* = 0 \tag{6.4.8}$$

where D^*/Dt^* represents the dimensionless material derivative. Finally, in dimensionless form,

$$\frac{D^*\mathbf{V}^*}{Dt^*} = -\boldsymbol{\nabla}^*p_k^* + \frac{\mu}{\rho Vl}\nabla^{*2}\mathbf{V}^*$$

$$\boldsymbol{\nabla}^* \cdot \mathbf{V}^* = 0 \tag{6.4.9}$$

Note that we have introduced the Reynolds number

$$\text{Re} = \frac{\rho Vl}{\mu} \tag{6.4.10}$$

as a coefficient in the normalized Navier–Stokes equation.

Consider now the boundary conditions. The no-slip condition on a fixed boundary $\mathbf{V}^* = 0$ introduces geometric parameters that are necessary to specify the geometry of the fixed boundary. These could include, for example, a roughness parameter e/l, where e is the average height of the roughness elements; a thickness parameter t/l, where t is the maximum thickness of a turbine blade; and a nose radius parameter r_0/l, where r_0 is the nose radius of an airfoil.

Another boundary condition would be the entering velocity distribution, for example, $u(y)/V$. This may introduce parameters having to do with the profile and the turbulent eddy structure of the entering flow. Such features could be extremely significant for particular flows.

Also, a portion of the boundary may be oscillating, like that of a rotating component. On such a boundary we would require that the velocity of the fluid be the same as the velocity of the rotating part, that is, $v = r\omega$. This would introduce the Strouhal number St,

$$v^* = \text{St } r^* \qquad (6.4.11)$$

where

$$\text{St} = \frac{\omega l}{V} \qquad (6.4.12)$$

Another boundary condition that introduces an additional parameter is that of a free surface. It is required that, neglecting surface tension effects, the pressure be constant across a free surface. On an air–water surface this requires that $p = 0$ on the free surface. Returning to Eq. 6.4.3, and normalizing, there results

$$\rho V^2 p_k^* = \rho V^2 p^* + \rho g l h^* \qquad \text{or} \qquad p_k^* = p^* + \frac{gl}{V^2} h^* \quad (6.4.13)$$

where we have introduced the Froude number

$$\text{Fr} = \frac{V}{\sqrt{lg}} \qquad (6.4.14)$$

Finally, a boundary condition may involve surface tension, as in droplet formation problems. Such a boundary condition would involve the normalized equation

$$\Delta p^* = \frac{\sigma}{l\rho V^2} \left(\frac{1}{r_1^*} + \frac{1}{r_2^*} \right) \qquad (6.4.15)$$

where we have introduced the Weber number

$$\text{We} = \frac{V^2 l \rho}{\sigma} \qquad (6.4.16)$$

Note that we have introduced all of the dimensionless parameters discussed in Section 6.3 with the exception of the Mach number. It could be introduced in a normalized energy equation, a task that is left to the home problems.

Obviously, if we can write the differential equations that describe a particular flow, those equations and the boundary conditions contain all the parameters of interest; hence the Buckingham π-theorem is not actually necessary for flow situations in which the equations and boundary conditions are known.

In addition, we can obtain the similitude conditions from the normalized equations and boundary conditions. The idea is to make the normalized differential equations and boundary conditions the same for both model and prototype. If they are the same they should possess the same solution.[4] The normalized boundary conditions being identical would demand that geometric similarity be satisfied. In order that the equations and boundary conditions be identical on both model and prototype, it is necessary that the dimensionless parameters be the same for both model and prototype. Thus, similitude conditions are also included in the normalized problem.

[4]Questions of uniqueness, which we will not consider, enter here. For example, it is posssible to have identical conditions and equations, yet in one case a laminar flow could result and in the other it could be a turbulent flow; that is, different solutions would exist.

PROBLEMS

6.1. Write Bernoulli's equation in dimensionless form by dividing Eq. 6.1.1 by V_1^2 and multiplying by g. Express the equation in a form similar to that of Eq. 6.1.2.

6.2. If the *F-L-T* system of units were used, what would be the dimensions on each of the following?
 (a) Mass flux.   (e) Work.
 (b) Pressure.   (f) Power.
 (c) Density.   (g) Surface tension.
 (d) Viscosity.

Dimensional Analysis

6.3. If the velocity V in a fluid flow depends on a dimension l, the fluid density ρ, and the viscosity μ, show that this implies that the Reynolds number is constant.

6.4. If the velocity V in a fluid depends on the surface tension σ, the density ρ, and a diameter d, show that this implies that the Weber number is constant.

6.5. Assume that the velocity V of fall of an object depends on the height H through which it falls, the gravity g, and the mass m of the object. Find an expression for V.

6.6. Include the density ρ and viscosity μ of the surrounding fluid and repeat Problem 6.5. This would account for the resistance (drag) of the fluid.

6.7. Select l, V, and ρ as the repeating variables in Example 6.1 and find an expression for F_D. Show that this is an equivalent form to that of Example 6.1.

6.8. Select d, μ, and V as the repeating variables in Example 6.1 and find an expression for F_D. Show that this is equivalent to the expression of Example 6.1.

6.9. Include gravity g in the list of variables of Example 6.2 and determine the final expression that results for h.

6.10. Find an expression for the centrifugal force F_c if it depends on the mass m, the angular velocity ω, and the radius R of an impeller.

6.11. The normal stress σ in a beam depends on the bending moment M, the distance y from the neutral axis, and the moment of inertia I. Relate σ to the other variables if we know that σ varies linearly with y.

6.12. Find an expression for the average velocity in a smooth pipe if it depends on the viscosity, the diameter, and the pressure gradient $\partial p/\partial x$.

6.13. It is suggested that the velocity of the water flowing from a hole in the side of an open tank depends on the height H of the hole from the surface, the gravity, and the density of the water. What expression relates the variables?

6.14. Derive an expression for the velocity of liquid issuing from a hole in the side of an open tank if the velocity depends on the height H of the hole from the free surface, the fluid viscosity and density, gravity, and the hole diameter.

6.15. The pressure drop Δp in a pipe depends on the average velocity, the pipe diameter, the kinematic viscosity, the length L of pipe, the wall roughness height e, and the fluid density. Find an expression for Δp.

6.16. The flow rate Q in an open channel depends on the hydraulic radius R, the cross-sectional area A, the wall roughness height e, gravity g, and the slope S. Relate Q to the other variables.

6.17. The velocity of propagation of ripples in a shallow liquid depends on the liquid depth h, gravity g, surface tension σ, and the liquid density. Find an expression for the propagation velocity.

6.18. The drag force F_D on a sphere depends on the velocity V, the viscosity, the density, the surface roughness height e, the freestream fluctuation intensity I (a dimensionless quantity), and the diameter. Find an expression for F_D.

6.19. The drag force F_D on a smooth sphere falling in a liquid depends on the sphere speed V, the solid density ρ_s, the liquid density ρ and viscosity, the sphere diameter D, and gravity g. Find an expression for F_D.

6.20. The drag force F_D on a golf ball depends on the velocity, the viscosity, the density, the diameter, the dimple depth, the dimple radius, and the concentration C of dimples measured by the number of dimples per unit area. What expression relates F_D to the other variables?

6.21. The frequency f with which vortices are shed from a cylinder depends on the viscosity, density, velocity, and diameter. Derive an expression for f.

6.22. The lift F_L on an airfoil depends on the aircraft speed V, the speed of sound c, the fluid density, the airfoil chord length l_c, the airfoil thickness t, and the angle of attack α. Find an expression for F_L.

6.23. Derive an expression for the torque T needed to rotate a disk of diameter d at the angular velocity ω in a fluid with density ρ and viscosity μ if the disk is a distance t from a wall. Also, find an expression for the power requirement.

6.24. A vacuum cleaner creates a pressure drop Δp across its fan. Relate this pressure drop to the impeller diameter D and width h, its rotational speed ω, the air density ρ, and the inlet and outlet diameters d_i and d_o. Also, find an expression for the power requirement of the fan.

6.25. Derive an expression for the maximum torque needed to rotate an agitator if it depends on the frequency of oscillation, the angular velocity with which the agitator rotates during an oscillation, the diameter of the agitator, the nominal height of the paddles, the length of the paddles, the number of paddles, the depth of liquid, and the liquid density.

6.26. The flow rate of water over a weir depends on the head of water, the width of the weir, gravity, viscosity, density, and surface tension. Relate the flow rate to the other variables.

6.27. The size of droplets in the spray of a fruit sprayer depends on the air velocity, the spray jet velocity, the spray jet exit diameter, the liquid spray surface tension, density, and viscosity, and the air density. Relate the droplet diameter to the other variables.

6.28. Relate the torque T to the other variables shown.

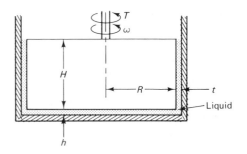

6.29. A viscometer is composed of an open tank in which the liquid level is held constant. A small-diameter tube empties into a calibrated volume. Find an expression for the viscosity using the relevant parameters.

6.30. Derive an expression for the torque T necessary to rotate the shaft.

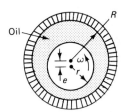

6.31. Derive an expression for the depth y_2 in this hydraulic jump.

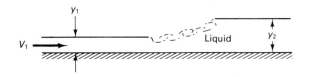

6.32. Derive an expression for the frequency with which a cylinder, suspended by a string in a liquid flow, oscillates.

Similitude

6.33. After a model study has been performed, it is necessary to predict quantities of interest that are to be expected on the prototype. Write expressions, in terms of density, velocity, and length, for the ratio of model to prototype of each of the following: flow rate Q, pressure drop Δp, pressure force F_p, shear stress τ_0, torque T, and heat transfer rate $\dot{Q}$.

6.34. A $1:7$ scale model simulates the operation of a large turbine that is to generate 200 kW with a flow rate of 1.5 m³/s. What flow rate should be used in the model, and what power output is expected?
(a) Water at the same temperature is used in both model and prototype.
(b) The model water is at 25°C and the prototype water is at 10°C.

6.35. A $1:5$ scale model of a large pump is used to test a proposed change. The prototype pump produces a pressure rise of 600 kPa at a mass flux of 800 kg/s. Determine the mass flux to be used in the model and the expected pressure rise.
(a) Water at the same temperature is used in both model and prototype.
(b) The water in the model study is at 30°C and the water in the prototype is at 15°C.

6.36. A force on a component of a $1:10$ scale model of a large pump is measured to be 45 N. What force is expected on the prototype component if water is used for both model and prototype:
(a) With the water at the same temperature.
(b) With the prototype water at 12°C and the model water at 25°C.

6.36E. A force on a component of a $1:10$ scale model of a large pump is measured to be 10 lb. What force is expected on the prototype component if water is used for both model and prototype:
(a) With the water at the same temperature.
(b) With the prototype water at 50°F and the model water at 70°F.

6.37. A $1:10$ scale model study of an automobile is proposed. A prototype speed of 100 km/h is desired. What wind-tunnel speed should be selected for the model study? Comment on the advisability of such a speed selection.

6.38. A $1:10$ scale model study of a torpedo is proposed. Prototype speeds of 90 km/h are to be studied. Should a wind tunnel or a water facility be used?

6.39. It is proposed that a model study be carried out on a proposed low-speed airfoil that is to fly at low altitudes at a speed of 50 m/s. If a $1:10$ scale model is to be constructed, what velocity should be used in a wind tunnel? Comment as to the advisability of such a test. Would it be more advisable to perform the test in a 20°C water channel? If a water channel study were undertaken, calculate the drag ratio between the model and prototype.

6.40. A model study of oil (SAE-10W) at 5°C flowing in a 0.8-m-diameter pipe is performed by using water at 20°C. Pipe of what diameter should be used if the average velocities are to be the same? What pressure drop ratio is expected? $S_{\text{oil}} = 0.9$.

6.40E. A model study of oil (SAE-10W) at 30°F flowing in a 2.5-ft-diameter pipe is performed by using water at 70°F. Pipe of what diameter should be used if the average velocities are to be the same? What pressure drop ratio is expected? $S_{oil} = 0.9$.

6.41. A 0.025-mm-long microorganism moves through water at the rate of 0.1 body length per second. Could a model study be performed for such a prototype in a water channel or a wind tunnel?

6.42. A 1:30 scale model of a ship is to be tested with complete similarity. What should be the viscosity of the model fluid? Is such a liquid possible? What conclusion can be stated?

6.43. A 1:60 scale model of a ship is used in a water tank to simulate a ship speed of 10 m/s. What should be the model speed? If a towing force of 10 N is measured on the model, what force is expected on the prototype? Neglect viscous effects.

6.43E. A 1:60 scale model of a ship is used in a water tank to simulate a ship speed of 30 ft/sec. What should be the model speed? If a towing force of 3 lb is measured on the model, what force is expected on the prototype? Neglect viscous effects.

6.44. The flow rate over a weir is 2 m³/s of water. A 1:10 scale model of the weir is tested in a water channel.
(a) What flow rate should be used?
(b) If a force of 12 N is measured on the model, what force would be expected on the prototype?

6.45. A 1:10 scale model of a hydrofoil is tested in a water channel with a force of 2.5 N measured at a speed of 6 m/s. Determine the speed and force predicted for the prototype. Neglect viscous effects.

6.45E. A 1:10 scale model of a hydrofoil is tested in a water channel with a force of 0.8 lb measured at a speed of 20 ft/sec. Determine the speed and force predicted for the prototype. Neglect viscous effects.

6.46. The propeller of a ship is to be studied with a 1:10 scale model.
(a) Assuming that the propeller operates close to the surface, select the propeller speed of the model if the prototype speed is 600 rpm.
(b) What torque would be expected if 1.2 N · m is measured on the model?

6.47. The pontoon of a seaplane is to be studied in a water channel that has a channel speed capability of 6 m/s. If the seaplane is to lift off at 100 m/s, what model scale should be selected?

6.48. A 1:30 scale model study of a submarine is proposed in an attempt to study the influence of a suggested shape modification. The prototype is 2 m in diameter and it is designed to travel at 15 m/s. The model is towed in a water tank at 2 m/s and a drag force of 2.15 N is measured. Does similitude exist for this test? If so, predict the power needed for the prototype.

6.48E. A 1:30 scale model study of a submarine is proposed in an attempt to study the influence of a suggested shape modification. The prototype is 6 ft in diameter and it is designed to travel at 45 ft/sec. The model is towed in a water tank at 6 ft/sec and a drag force of 0.5 lb is measured. Does similitude exist for this test? If so, predict the power needed for the prototype.

6.49. The smoke emanating from the stacks on a ship has a tendency to make its way to the deck down the outside of the stacks causing discomfort to the passengers. This problem is studied with a 1:20 scale model of a 4-m-diameter stack. The ship is traveling at 10 m/s. What range of wind-tunnel airspeeds could be used in this study?

6.50. A model study of a dirigible (a large balloon that travels through the air) is to be performed. The 10-m-diameter dirigible is to travel at 20 m/s. If a 40-cm-diameter model is proposed for use in a wind tunnel, or a 10-cm-diameter model for use in a 20°C water channel, which should be selected? Suppose that the wind-tunnel model is used with a speed of 15 m/s and a drag force of 3.2 N is measured. What force would be expected on the water-channel model with a speed of 2.4 m/s in the water channel? What horsepower would be predicted to overcome drag on the prototype? Assume that the flow is Reynolds number independent for Re > 10⁵.

6.51. A model study is to be performed of a 300-m-tall, 15-m-diameter smokestack of a power plant. It is known that the smokestack is submerged in a ground boundary layer that is 400 m thick. Could the study be performed in a wind tunnel that produces a 1.2-m-thick boundary layer?

6.51E. A model study is to be performed of a 1000-ft-tall, 45-ft-diameter smokestack of a power plant. It is known that the smokestack is submerged in a ground boundary layer that is 1200 ft thick. Could the study be performed in a wind tunnel that produces a 4-ft-thick boundary layer?

6.52. A 1:20 scale model of an aircraft is tested in a 23°C wind tunnel. A speed of 200 m/s is used in the model study. A drag force of 10 N is measured. What prototype speed and drag force does this simulate if the elevation is:
(a) Sea level?
(b) 5000 m?
(c) 10 000 m?

6.53. A 1:10 scale model of an airfoil is tested in a wind tunnel that uses outside air at 0°C. The test is to simulate an aircraft speed of 250 m/s at an elevation of 10 000 m. What wind-tunnel speed should be used? If a velocity of 290 m/s and a pressure of 80 kPa absolute are measured on the model at a particular location at a 5° angle of attack, what speed and pressure are expected on the prototype at the corresponding location, and what is the angle of attack?

6.54. A 1:10 scale model of a propeller on a ship is to be tested in a water channel. What should be the rotational speed of the model if the rotational speed of the propeller is 2000 rpm and:
(a) The Froude number governs the study?
(b) The Reynolds number governs the study?

6.55. A torque of 12 N · m is measured on a 1:10 scale model of a large wind machine with a wind tunnel speed of 60 m/s. This is to simulate a wind velocity of 15 m/s since viscous effects are considered negligible. What torque is expected on the prototype? If the model rotates at 500 rpm, what prototype angular velocity is simulated?

6.55E. A torque of 10 ft-lb is measured on a 1:10 scale model of a large wind machine with a wind-tunnel speed of 180 ft/sec. This is to simulate a wind velocity of 45 ft/sec. What torque is expected on the prototype? If the model rotates at 500 rpm, what prototype angular velocity is simulated?

6.56. An underwater study of a porpoise is to be performed by using a 1:10 scale model. A porpoise swimming at 10 m/s and making one swimming motion each second is to be simulated. What speed could be used in the water channel, and for that speed, how many swimming motions per second should be used?

Normalized Diferential Equations

6.57. Normalize the continuity equation

$$\frac{\partial \rho}{\partial t} + \frac{\partial}{\partial x}(\rho u) + \frac{\partial}{\partial y}(\rho v) = 0$$

using a characteristic velocity V, length l, density ρ_0, and time f^{-1}, where f is the frequency. What dimensionless parameter is introduced?

6.58. Normalize Euler's equation

$$\frac{\partial \mathbf{V}}{\partial t} + u\frac{\partial \mathbf{V}}{\partial x} + v\frac{\partial \mathbf{V}}{\partial y} + w\frac{\partial \mathbf{V}}{\partial z} = -\frac{\nabla p}{\rho}$$

using a characteristic velocity U, length l, pressure ρU^2, and time f^{-1}, where f is the frequency. Find any dimensionless parameters that are introduced.

6.59. Normalize Euler's equation

$$\rho\frac{D\mathbf{V}}{Dt} = -\nabla p - \rho g\nabla h$$

using a characteristic velocity U, length l, pressure ρU^2, and time l/U. Determine any dimensionless parameters introduced in the normalized equation.

6.60. Nondimensionalize the energy equation

$$\rho c_p\left(u\frac{\partial T}{\partial x} + v\frac{\partial T}{\partial y}\right) = K\nabla^2 T$$

using a characteristic velocity U, length l, and temperature T_0. Express the dimensionless parameter that results in terms of the Prandtl number

$$\text{Pr} = \mu\frac{c_p}{K}$$

6.61. Nondimensionalize the compressible flow differential momentum equation

$$\rho\frac{D\mathbf{V}}{Dt} = -\nabla p + \mu\nabla^2\mathbf{V} + \frac{\mu}{3}\nabla(\nabla \cdot \mathbf{V})$$

and the energy equation

$$\rho c_v\frac{DT}{Dt} = K\nabla^2 T - p\nabla \cdot \mathbf{V}$$

using characteristic quantities ρ_0, p_0, T_0, U, and l. The characteristic time is l/U. The speed of sound is $c = \sqrt{kRT_0}$. Find any dimensionless parameters that result if $\text{Pr} = \mu c_p/K$.

SEVEN

Internal Flows

7.1 INTRODUCTION

In this chapter the effects of viscosity on an incompressible, internal flow will be studied. Such flows are of particular importance to engineers. Flow in a circular pipe is undoubtedly the most common internal fluid flow. It is encountered in the veins and arteries in a body, in a city's water system, in a farmer's irrigation system, in the piping systems transporting fluids in a factory, in the hydraulic lines of an aircraft, and in the ink jet of a computer's printer. Flows in noncircular ducts and open channels must also be included in our study. In Chapter 6 we discovered that the viscous effects in a flow result in the introduction of the Reynolds number,

$$\text{Re} = \frac{V\rho l}{\mu} \tag{7.1.1}$$

The Reynolds number was observed to be a ratio of the inertial force to the viscous force. Hence, when this ratio becomes large, it is expected that the inertial forces may dominate the viscous forces. This is usually true when short, sudden geometric changes occur; for long reaches of pipe or open channels, this is not the situation. When the surface areas, such as the wall area of a pipe, are relatively large, viscous effects become quite important and must be included in our study.

Internal flow between parallel plates, in a pipe, between rotating cylinders, and in an open channel will be considered in detail. For a sufficiently low Reynolds number (Re < 2000 in a pipe and Re < 1500 in a wide channel) a laminar flow results, and at sufficiently high Reynolds number a turbulent flow occurs. See Section 3.3.3 for a more detailed discussion. We consider laminar flow first and then turbulent flow.

7.2 ENTRANCE FLOW AND DEVELOPED FLOW

When considering internal flows we are interested primarily in developed flows. A **developed flow** results when the velocity profile ceases to change in the flow direction. In the **entrance region** the velocity profile changes in the flow direction, as shown in Fig. 7.1. The idealized flow from a reser-

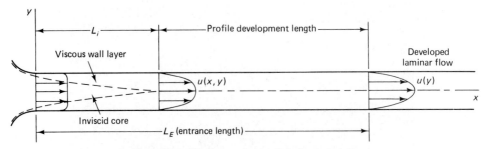

Figure 7.1 Laminar entrance flow in a conduit.

voir begins at the inlet as a uniform flow (there is a thin boundary layer on the wall, as shown); a wall viscous layer then grows over the **inviscid core length** L_i until the viscous stresses dominate the entire cross section; the profile then continues to change in the **profile development region** due to viscous effects until a developed flow is achieved. The inviscid core length is approximately one-fourth of the **entrance length** L_E, depending on the conduit geometry, shape of the inlet, and the Reynolds number.

For a laminar flow in a circular pipe with a uniform profile at the inlet, the entrance length is given by

$$\frac{L_E}{D} = 0.065 \mathrm{Re} \qquad\qquad \mathrm{Re} = \frac{VD}{\nu} \qquad\qquad (7.2.1)$$

where the Reynolds number is based on the average velocity and the diameter. Laminar flow in a pipe has been observed at Reynolds numbers in excess of 40 000 for carefully controlled conditions. However, for engineering applications a value of 2000 is the highest Reynolds number for which laminar flow is assured; this is due to vibrations of the pipe, fluctuations in the flow, or roughness elements on the pipe wall.

For a laminar flow between two parallel plates and a uniform profile at the inlet the entrance length is

$$\frac{L_E}{h} = 0.04 \mathrm{Re} \qquad\qquad \mathrm{Re} = \frac{Vh}{\nu} \qquad\qquad (7.2.2)$$

where the Reynolds number is based on the average velocity and the distance h between the plates. The inviscid core length is approximately one-third of the entrance length. In a high-aspect-ratio channel (the aspect ratio is the width divided by the distance between the plates) a laminar flow cannot exist over $\mathrm{Re} = 7700$; for engineering situations a value of 1500 is often used as the upper limit for laminar flow.

For a turbulent flow the situation is slightly different, as shown in Fig. 7.2. The inviscid core exists followed by the velocity profile develop-

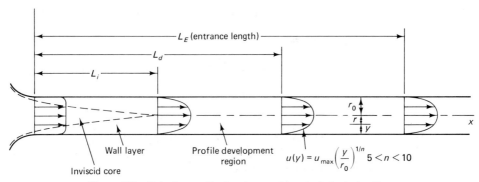

Figure 7.2   Velocity profile development in a turbulent pipe flow.

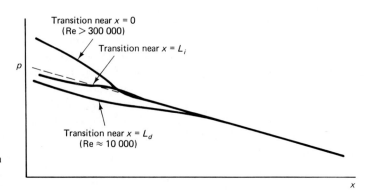

Figure 7.3 Pressure variation in a turbulent flow in a horizontal pipe.

ment region, which terminates at $x = L_d$. An additional length is needed, however, for the turbulence structure (e.g., the eddy size and frequency) to develop. For large Reynolds number flow (Re $> 10^5$) in a pipe, where the turbulence fluctuations initiate near $x = 0$, tests have yielded

$$\frac{L_i}{D} \simeq 10 \qquad \frac{L_d}{D} \simeq 40 \qquad \frac{L_E}{D} \simeq 120 \qquad (7.2.3)$$

For turbulent flow with Re $= 4000$ the foregoing developmental lengths would be substantially higher,[1] perhaps five times the values listed; experimental data are not available for low-Reynolds-number turbulent flow.

In Fig. 7.3 the pressure variation is sketched. In the flow beyond a sufficiently large x it is noted that the pressure variation decreases linearly with x. If transition to turbulent flow occurs near the origin, the linear pressure variation begins near L_i and the pressure gradient in the inlet region is higher than in the developed flow region; if transition occurs near L_d, as it does for low Re, the linear variation begins at the end of the transition process and the pressure gradient in the inlet region is less than that of developed flow. For a laminar flow, the pressure variation qualitatively resembles that associated with a large Reynolds number.

7.3 LAMINAR FLOW IN A PIPE

In this section we investigate incompressible, steady, developed laminar flow in a pipe, as sketched in Fig. 7.4. Two methods will be used: an elemental approach and a direct solution of the x-component Navier–Stokes equation. Either can be used.

---

[1]This is true since, for low-Re turbulent flows, transition to a turbulent flow occurs in the profile development region; hence much of the entry region is laminar with relatively low wall shear.

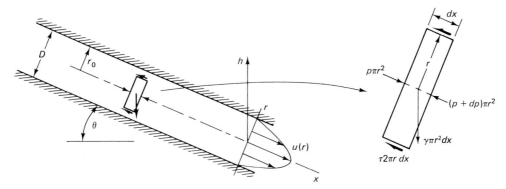

Figure 7.4   Developed flow in a circular pipe.

7.3.1 Elemental Approach

An elemental volume of the fluid is shown in Fig. 7.4. This volume can be considered an infinitesimal control volume into which and from which fluid flows, or it can be considered an infinitesimal fluid mass upon which forces are acting. If it is considered a control volume, we would apply the momentum equation (4.5.6); if it is a fluid mass, we would apply Newton's second law. Since the velocity profile does not change in the x-direction, the momentum flux in equals the momentum flux out and the resultant force is zero; since there is no acceleration of the mass element, the resultant force must also be zero. Consequently,

$$p\pi r^2 - (p + dp)\pi r^2 - \tau 2\pi r\, dx + \gamma\pi r^2\, dx \sin\theta = 0 \qquad (7.3.1)$$

which can be simplified to

$$\tau = -\frac{r}{2}\frac{d}{dx}(p + \gamma h) \qquad (7.3.2)$$

where we have used $\sin\theta = -dh/dx$, the vertical direction being denoted by h. The shear stress[2] in this flow is related to the velocity gradient and the viscosity, giving

$$-\mu\frac{du}{dr} = -\frac{r}{2}\frac{d}{dx}(p + \gamma h) \qquad (7.3.3)$$

which can be integrated to give the velocity distribution,

$$u(r) = \frac{r^2}{4\mu}\frac{d}{dx}(p + \gamma h) + A \qquad (7.3.4)$$

[2]The shearing stress is a positive quantity, as displayed in Fig. 7.4. The minus sign in $\tau = -\mu\, du/dr$ is necessary since du/dr is negative.

where A is a constant of integration. Using $u = 0$ at $r = r_0$, we can evaluate A and find the velocity distribution to be

$$u(r) = \frac{1}{4\mu} \frac{d(p + \gamma h)}{dx} (r^2 - r_0^2) \tag{7.3.5}$$

a parabolic profile. It is often referred to as **Poiseuille flow.**

The foregoing result can also be obtained by integrating the appropriate Navier–Stokes equations, as illustrated in the following section. If such an exercise is not of interest, go to Section 7.3.3.

7.3.2 Solving the Navier–Stokes Equations

For developed flow in a circular pipe the streamlines are parallel to the wall, so that $v_r = v_\theta = 0$ and $u = u(r)$ only. Referring to the momentum equations in cylindrical coordinates of Table 5.1 (note that $u = v_z$ and the z-coordinate has been replaced by x) the x-component Navier–Stokes equation is

$$\rho \left(\underset{\text{steady}}{\cancel{\frac{\partial u}{\partial t}}} + \underset{\substack{\text{streamlines} \\ \parallel \text{ wall}}}{\cancel{v_r \frac{\partial u}{\partial r}}} + \underset{\text{no swirl}}{\cancel{\frac{v_\theta}{r} \frac{\partial u}{\partial \theta}}} + \underset{\substack{\text{developed} \\ \text{flow}}}{\cancel{u \frac{\partial u}{\partial x}}} \right)$$

$$= -\frac{\partial p}{\partial x} + \gamma \sin \theta + \mu \left(\frac{\partial^2 u}{\partial r^2} + \frac{1}{r} \frac{\partial u}{\partial r} + \underset{\substack{\text{symmetric} \\ \text{flow}}}{\frac{1}{r^2} \cancel{\frac{\partial^2 u}{\partial \theta^2}}} + \underset{\substack{\text{developed} \\ \text{flow}}}{\cancel{\frac{\partial^2 u}{\partial x^2}}} \right) \tag{7.3.6}$$

This equation simplifies to, using $\sin \theta = -dh/dx$ (see Fig. 7.4),

$$\frac{1}{\mu} \frac{\partial}{\partial x} (p + \gamma h) = \frac{1}{r} \frac{\partial}{\partial r} \left(r \frac{\partial u}{\partial r} \right) \tag{7.3.7}$$

where the first two terms in the parentheses on the right-hand side of Eq. 7.3.6 have been combined. We see that the left-hand side is at most a function of x and the right-hand side is at most a function of r. Since x and r can be varied independently, we must have

$$\frac{1}{r} \frac{d}{dr} \left(r \frac{du}{dr} \right) = \lambda \tag{7.3.8}$$

where λ is a constant and we have used ordinary derivatives since u depends on only one variable r. This can be integrated twice as follows:

$$r \frac{du}{dr} = \frac{\lambda}{2} r^2 + A \tag{7.3.9}$$

$$u(r) = \frac{\lambda}{4} r^2 + A \ln r + B \tag{7.3.10}$$

The velocity must remain finite at $r = 0$; hence $A = 0$. Also, at $r = r_0$, $u = 0$; thus B can be evaluated and we have

$$u(r) = \frac{\lambda}{4} (r^2 - r_0^2)$$
$$= \frac{1}{4\mu} \frac{d}{dx} (p + \gamma h) (r^2 - r_0^2) \tag{7.3.11}$$

where we have used λ as the left-hand side of Eq. 7.3.7. This is the parabolic velocity distribution for flow in a pipe, often referred to as **Poiseuille flow.**

7.3.3 Pipe Flow Quantities

For steady, laminar, developed flow in a circular pipe, the velocity distribution has been shown to be

$$u(r) = \frac{1}{4\mu} \frac{d(p + \gamma h)}{dx} (r^2 - r_0^2) \tag{7.3.12}$$

The average velocity V is found to be

$$V = \frac{Q}{A} = \frac{\int_0^{r_0} u(r) \, 2\pi r \, dr}{\pi r_0^2}$$
$$= \frac{2}{r_0^2} \int_0^{r_0} \frac{1}{4\mu} \frac{d(p + \gamma h)}{dx} (r^2 - r_0^2) r \, dr = -\frac{r_0^2}{8\mu} \frac{d(p + \gamma h)}{dx} \tag{7.3.13}$$

Or, expressing the pressure drop Δp in terms of the average velocity, we have, for a horizontal[3] pipe,

$$\Delta p = \frac{8\mu V L}{r_0^2} \tag{7.3.14}$$

where we have used $\Delta p / L = -dp/dx$ since dp/dx is a constant for developed flow. Note that the pressure drop is a positive quantity, whereas the pressure gradient is negative.

[3]For a pipe on an incline, simply replace p with $(p + \gamma h)$.

The maximum velocity at $r = 0$ is

$$u_{max} = -\frac{r_0^2}{4\mu}\frac{d(p + \gamma h)}{dx} \qquad (7.3.15)$$

so that (see Eq. 7.3.13)

$$V = \tfrac{1}{2}u_{max} \qquad (7.3.16)$$

The shearing stress is determined to be

$$\tau = -\mu\frac{du}{dr}$$
$$= -\frac{r}{2}\frac{d(p + \gamma h)}{dx} \qquad (7.3.17)$$

Letting $\tau = \tau_0$ at $r = r_0$, we see that the pressure drop Δp over a length L of a horizontal section of pipe is

$$\Delta p = \frac{2\tau_0 L}{r_0} \qquad (7.3.18)$$

where we have again used $dp/dx = -\Delta p/L$. If we introduce the **friction factor** f, a dimensionless wall shear, defined by

$$f = \frac{\tau_0}{\tfrac{1}{8}\rho V^2} \qquad (7.3.19)$$

we see that

$$\frac{\Delta p}{\gamma} = h_L = f\frac{L}{D}\frac{V^2}{2g} \qquad (7.3.20)$$

where h_L is the head loss (see Eq. 4.4.17) with dimension of length. This equation is often referred to as the **Darcy–Weisbach equation.**[4] Combining Eqs. 7.3.14, 7.3.18, and 7.3.19, we find that

$$f = \frac{64}{Re} \qquad (7.3.21)$$

for laminar flow in a pipe. Substituting this back into Eq. 7.3.20, we see that the head loss is directly proportional to the velocity to the first power, a result that generally is applied to developed, laminar flows in conduits.

[4]This equation will be shown to be valid for both laminar and turbulent flows.

EXAMPLE 7.1

A small-diameter horizontal tube is connected to a supply reservoir as shown. If 6600 mm³ is captured at the outlet in 10 s, calculate the viscosity of the water.

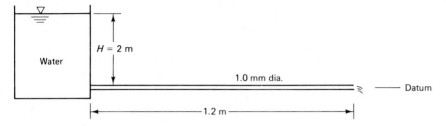

Solution

The tube is very small, so we expect viscous effects to limit the velocity to a small value. Using Bernoulli's equation from the surface to the entrance to the tube, and neglecting the velocity head, we have, letting 0 be a point on the surface,

$$\cancel{\frac{p_0}{\gamma}}^{\,0} + H = \cancel{\frac{V^2}{2g}}^{\text{negligible}} + \frac{p}{\gamma}$$

where we have used gage pressure with $p_0 = 0$. This becomes, assuming $V \cong 0$,

$$p = \gamma H$$
$$= 9800 \times 2 = 19\,600 \text{ Pa}$$

At the exit of the tube the pressure is zero; hence

$$\frac{\Delta p}{L} = \frac{19\,600}{1.2} = 16\,300 \text{ Pa/m}$$

The average velocity is found to be

$$V = \frac{Q}{A}$$
$$= \frac{6600 \times 10^{-9}/10}{\pi \times 0.001^2/4} = 0.840 \text{ m/s}$$

This is quite small ($V^2/2g = 0.036$ m compared with $p/\gamma = 2$ m), so the assumption of negligible velocity head is valid. Our pressure calculation is acceptable. Using Eq. 7.3.13, we can find the viscosity to be

$$\mu = -\frac{r_0^2}{8V}\frac{dp}{dx}$$
$$= \frac{0.0005^2}{8 \times 0.84}(16\,300) = 6.06 \times 10^{-4} \text{ N} \cdot \text{s/m}^2$$

using $\Delta p/L = -dp/dx$. We should check the Reynolds number to determine if our assumption of a laminar flow was acceptable. It is

$$\text{Re} = \frac{\rho VD}{\mu}$$

$$= \frac{1000 \times 0.84 \times 0.001}{6.06 \times 10^{-4}} = 1390$$

This is obviously a laminar flow since $\text{Re} < 2000$, so the calculations are valid providing the inviscid core length is not too long. It is

$$L_i = 0.065 \text{ Re } \frac{D}{4}$$

$$= 0.065 \times 1390 \times \frac{0.001}{4} = 0.022 \text{ m}$$

This is approximately 2% of the total length, a rather small quantity; hence the calculations are assumed reliable.

EXAMPLE 7.2

Derive an expression for the velocity distribution between horizontal, concentric pipes for a steady, incompressible developed flow.

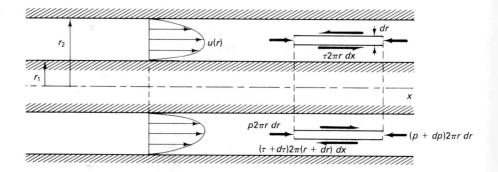

Solution

Let us use an elemental approach. The element is a hollow cylindrical shell as sketched in the figure. If we sum forces, we obtain

$$p2\pi r \, dr - (p + dp)2\pi r \, dr + \tau 2\pi r \, dx - (\tau + d\tau)2\pi(r + dr) \, dx = 0$$

Simplifying, there results

$$\frac{dp}{dx} = -\frac{\tau}{r} - \frac{d\tau}{dr}$$

Substituting $\tau = -\mu\, du/dr$ we have

$$\frac{dp}{dx} = \mu\left(\frac{1}{r}\frac{du}{dr} + \frac{d^2u}{dr^2}\right)$$

$$= \frac{\mu}{r}\frac{d}{dr}\left(r\frac{du}{dr}\right)$$

This is integrated to yield

$$r\frac{du}{dr} = \frac{1}{2\mu}\frac{dp}{dx}r^2 + A$$

Integration again gives

$$u(r) = \frac{1}{4\mu}\frac{dp}{dx}r^2 + A\ln r + B$$

where A and B are arbitrary constants. They are found by setting $u = 0$ at $r = r_1$ and at $r = r_2$; that is,

$$0 = \frac{1}{4\mu}\frac{dp}{dx}r_1^2 + A\ln r_1 + B$$

$$0 = \frac{1}{4\mu}\frac{dp}{dx}r_2^2 + A\ln r_2 + B$$

The solution is

$$A = \frac{1}{4\mu}\frac{dp}{dx}\frac{r_1^2 - r_2^2}{\ln(r_2/r_1)}$$

$$B = -A\ln r_2 - \frac{r_2^2}{4\mu}\frac{dp}{dx}$$

Thus

$$u(r) = \frac{1}{4\mu}\frac{dp}{dx}\left[r^2 - r_2^2 + \frac{r_2^2 - r_1^2}{\ln(r_1/r_2)}\ln(r/r_2)\right]$$

This is integrated to give the flow rate:

$$Q = \int_{r_1}^{r_2} u(r)2\pi r\, dr$$

$$= -\frac{\pi}{8\mu}\frac{dp}{dx}\left[r_2^4 - r_1^4 - \frac{(r_2^2 - r_1^2)^2}{\ln(r_2/r_1)}\right]$$

As $r_1 \to 0$ the velocity distribution approaches the parabolic distribution of pipe flow. As $r_1 \to r_2$ this distribution approaches that of parallel-plate flow. These two conclusions are presented as a problem at the end of this chapter.

7.4 LAMINAR FLOW BETWEEN PARALLEL PLATES

Consider the incompressible, steady, developed flow of a fluid between parallel plates, with the upper plate moving with velocity U as shown in Fig. 7.5. We will derive the velocity distribution by two methods; either can be used.

7.4.1 Elemental Approach

Consider an elemental volume of unit depth, as sketched in Fig. 7.5. If we sum forces in the x-direction, we can write

$$p \, dy - (p + dp) \, dy - \tau \, dx + (\tau + d\tau) \, dx + \gamma \, dx \, dy \sin \theta = 0 \quad (7.4.1)$$

since there is no acceleration. The pressure is assumed to depend only on x; variation with y is assumed to be negligible since the dimension a is quite small for most applications. After dividing by $dx \, dy$, the above is simplified to

$$\frac{d\tau}{dy} = \frac{dp}{dx} - \gamma \sin \theta \quad (7.4.2)$$

Since this is one-dimensional flow, the shear stress is

$$\tau = \mu \frac{du}{dy} \quad (7.4.3)$$

Using this and $\sin \theta = -dh/dx$, there results

$$\mu \frac{d^2 u}{dy^2} = \frac{dp}{dx} + \gamma \frac{dh}{dx} = \frac{d}{dx}(p + \gamma h) \quad (7.4.4)$$

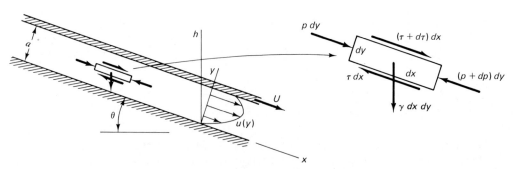

Figure 7.5 Developed flow between parallel plates.

This can be integrated twice to yield (first, divide by μ)

$$u(y) = \frac{y^2}{2\mu} \frac{d}{dx} (p + \gamma h) + Ay + B \qquad (7.4.5)$$

where A and B are integration constants. If we require $u = 0$ at $y = 0$ and $u = U$ at $y = a$, we have

$$A = \frac{U}{a} - \frac{a}{2\mu} \frac{d}{dx} (p + \gamma h) \qquad B = 0 \qquad (7.4.6)$$

Thus the velocity distribution is the parabola

$$u(y) = \frac{1}{2\mu} \frac{d}{dx} (p + \gamma h)(y^2 - ay) + \frac{U}{a} y \qquad (7.4.7)$$

If the motion is due to the motion of the plate only (a linear profile), the flow is called a **Couette flow.**

Rather than summing forces on an element we can integrate the appropriate Navier–Stokes equation, as follows. If such an exercise is not of interest, go to Section 7.4.3.

7.4.2 Integrating the Navier–Stokes Equations

For developed flow between parallel plates the streamlines are parallel to the plates so that $u = u(y)$ only and $v = w = 0$. The Navier–Stokes equation for the x-direction is (see Eq. 5.3.14)

$$\rho \left(\frac{\partial u}{\partial t} + u \frac{\partial u}{\partial x} + v \frac{\partial u}{\partial y} + w \frac{\partial u}{\partial z} \right)$$

steady   developed   0   0
flow

$$(7.4.8)$$

developed   wide
flow   channel

$$= -\frac{\partial p}{\partial x} + \mu \left(\frac{\partial^2 u}{\partial x^2} + \frac{\partial^2 u}{\partial y^2} + \frac{\partial^2 u}{\partial z^2} \right) + \gamma \sin \theta$$

If the parallel plates are the top and bottom of a channel, the channel must be very wide; the analysis then applies to the midsection away from the side walls. Equation 7.4.8 reduces to, using $\sin \theta = -dh/dx$,

$$\frac{\partial^2 u}{\partial y^2} = \frac{1}{\mu} \frac{d}{dx} (p + \gamma h) \qquad (7.4.9)$$

The left-hand side is at most a function of y and the right-hand side is at most a function of x. Hence we must have

$$\frac{\partial^2 u}{\partial y^2} = \lambda \tag{7.4.10}$$

where λ is a constant, since x and y are independent variables. This can be integrated twice to yield

$$u(y) = \frac{\lambda}{2} y^2 + Ay + B \tag{7.4.11}$$

where A and B are arbitrary constants of integration. We demand that $u = 0$ at $y = 0$ and $u = U$ at $y = a$. This gives

$$A = \frac{U}{a} - \frac{\lambda a}{2} \qquad B = 0 \tag{7.4.12}$$

Using these constants, we can write Eq. 7.4.11 as

$$\begin{aligned}
u(y) &= \frac{\lambda}{2}(y^2 - ay) + \frac{U}{a} y \\
&= \frac{1}{2\mu} \frac{d}{dx}(p + \gamma h)(y^2 - ay) + \frac{U}{a} y
\end{aligned} \tag{7.4.13}$$

where we have used λ as the right-hand side of Eq. 7.4.9. If the flow is due to the motion of the plate only (a linear profile), the flow is called a **Couette flow.**

7.4.3 Simplified Flow Situation

Using the results derived above, we can write the expression for the velocity distribution between fixed plates (let $U = 0$) as

$$u(y) = \frac{1}{2\mu} \frac{d(p + \gamma h)}{dx}(y^2 - ay) \tag{7.4.14}$$

Using this distribution, we can find the flow rate per unit width to be

$$\begin{aligned}
Q &= \int u \, dA \\
&= \int_0^a \frac{1}{2\mu} \frac{d(p + \gamma h)}{dx}(y^2 - ay) \, dy = -\frac{a^3}{12\mu} \frac{d(p + \gamma h)}{dx}
\end{aligned} \tag{7.4.15}$$

The average velocity V is found to be

$$V = \frac{Q}{a \times 1}$$
$$= -\frac{a^2}{12\mu} \frac{d(p + \gamma h)}{dx}$$
(7.4.16)

This can be expressed as the pressure drop in terms of the average velocity; for a horizontal[5] channel we have

$$\Delta p = \frac{12\mu VL}{a^2}$$
(7.4.17)

where we have used $\Delta p/L = -dp/dx$ since dp/dx is constant for developed flow. Observe that the maximum velocity occurs at $y = a/2$ and is

$$u_{max} = -\frac{a^2}{8\mu} \frac{dp}{dx}$$
(7.4.18)

Thus the average velocity is related to the maximum velocity by

$$V = \tfrac{2}{3} u_{max}$$
(7.4.19)

The shearing stress is found to be

$$\tau = \mu \frac{du}{dy}$$
$$= \frac{1}{2} \frac{dp}{dx} (2y - a)$$
(7.4.20)

At the wall where $y = 0$ there results

$$\tau_0 = -\frac{a}{2} \frac{dp}{dx}$$
(7.4.21)

The pressure drop Δp over a length L of horizontal channel is found to be

$$\Delta p = \frac{2\tau_0}{a} L$$
(7.4.22)

[5] For plates on an incline, simply replace p with $(p + \gamma h)$.

recognizing that $dp/dx = -\Delta p/L$. This can be expressed in a more convenient form as

$$\frac{\Delta p}{\gamma} = f\,\frac{L}{2a}\,\frac{V^2}{2g} \tag{7.4.23}$$

if we introduce the friction factor f, defined by

$$f = \frac{\tau_0}{\frac{1}{8}\rho V^2} \tag{7.4.24}$$

In terms of the head loss (see Eq. 4.4.17), Eq. 7.4.23 becomes

$$h_L = f\,\frac{L}{2a}\,\frac{V^2}{2g} \tag{7.4.25}$$

If we combine Eqs. 7.4.16, 7.4.21, and 7.4.24, we find that

$$
\begin{aligned}
f &= \frac{8}{\rho V^2}\left(-\frac{a}{2}\frac{dp}{dx}\right)\\[1mm]
&= \frac{8}{\rho V^2}\left(-\frac{a}{2}\right)\left(-\frac{12\mu V}{a^2}\right) = \frac{48\mu}{\rho a V} = \frac{48}{\mathrm{Re}}
\end{aligned}
\tag{7.4.26}
$$

where the Reynolds number $\mathrm{Re} = \rho a V/\mu$ has been introduced. Observe that if this is substituted into Eq. 7.4.25, we see that the head loss is directly proportional to the average velocity, a conclusion that is true for laminar flows, in general.

We have calculated most of the quantities of interest for laminar flow of an incompressible, steady flow between parallel plates. This would, of course, be an acceptable approximation for a wide channel, one for which the width exceeds the height by at least a factor of 8. For channels with smaller aspect ratios, the edge effects become significant and must be accounted for by adding an additional shear stress to the sides of the element in Fig. 7.5, or by maintaining $\partial^2 u/\partial z^2$ in Eq. 7.4.8. The solution is beyond the scope of this book.

EXAMPLE 7.3

Water at 20°C flows between horizontal plates 1.5 cm apart and 50 cm wide with a Reynolds number of 1500. Calculate (a) the flow rate, (b) the wall shear stress, (c) the pressure drop over 3 m, and (d) the velocity at $y = 0.5$ cm.

Solution

Since the Reynolds number is 1500, the laminar flow equations are assumed applicable.

(a) Using the definition of the Reynolds number, the average velocity is found as follows:

$$1500 = \frac{Va}{\nu}$$

$$\therefore \quad V = \frac{1500\nu}{a} = \frac{1500 \times 10^{-6}}{0.015} = 0.1 \text{ m/s}$$

Thus

$$Q = VA$$

$$= 0.1 \times 0.015 \times 0.5 = 7.5 \times 10^{-4} \text{ m}^3/\text{s}$$

(b) Using Eq. 7.4.26, the friction factor is

$$f = \frac{48}{Re} = \frac{48}{1500} = 0.032$$

The shearing stress at the wall is found, using Eq. 7.4.24, to be

$$\tau_0 = \tfrac{1}{8}\rho V^2 f$$

$$= \tfrac{1}{8} \times 1000 \times 0.1^2 \times 0.032 = 0.04 \text{ Pa}$$

(c) The pressure drop over 3 m is found, using Eq. 7.4.22, to have the value

$$\Delta p = \frac{2\tau_0 L}{a}$$

$$= \frac{2 \times 0.04 \times 3}{0.015} = 16 \text{ Pa}$$

(d) The velocity distribution of Eq. 7.4.14 is

$$u(y) = \frac{1}{2\mu}\frac{dp}{dx}(y^2 - ay)$$

$$= \frac{1}{2 \times 10^{-3}}\frac{-16}{3}(y^2 - 0.015y) = -2667(y^2 - 0.015y)$$

where we have used $dp/dx = -\Delta p/L$ because dp/dx is constant. At $y = 0.005$ m the velocity is

$$u = -2667(0.005^2 - 0.015 \times 0.005) = 0.133 \text{ m/s}$$

EXAMPLE 7.3 (English)

Water at 60°F flows between horizontal plates 0.5 in. apart and 20 in. wide with a Reynolds number of 1500. Calculate (a) the flow rate, (b) the wall shear stress, (c) the pressure drop over 10 ft, and (d) the velocity at $y = 0.2$ in.

Solution

Since the Reynolds number is 1500, the laminar flow equations are assumed applicable.

(a) Using the definition of the Reynolds number the average velocity is found as
follows:

$$1500 = \frac{Va}{\nu}$$

$$\therefore \quad V = \frac{1500\nu}{a} = \frac{1500 \times 1.22 \times 10^{-5}}{0.5/12} = 0.439 \text{ ft/sec}$$

Thus

$$Q = VA$$

$$= 0.439 \times \frac{0.5 \times 20}{144} = 0.0305 \text{ ft}^3/\text{sec}$$

(b) Using Eq. 7.4.26, the friction factor is

$$f = \frac{48}{\text{Re}}$$

$$= \frac{48}{1500} = 0.032$$

The shearing stress at the wall is found, using Eq. 7.4.24, to be

$$\tau_0 = \tfrac{1}{8}\rho V^2 f$$

$$= \tfrac{1}{8} \times 1.94 \times 0.439^2 \times 0.032 = 0.00150 \text{ psf}$$

(c) The pressure drop over 10 ft is found, using Eq. 7.4.22, to have the value

$$\Delta p = \frac{2\tau_0 L}{a}$$

$$= \frac{2 \times 0.0015 \times 10}{0.5/12} = 0.72 \text{ psf}$$

(d) The velocity distribution of Eq. 7.4.14 is

$$u(y) = \frac{1}{2\mu} \frac{dp}{dx} (y^2 - ay)$$

$$= \frac{1}{2 \times 2.37 \times 10^{-5}} \frac{-0.72}{10} \left(y^2 - \frac{0.5}{12} y \right) = -1520(y^2 - 0.0417y)$$

where we have used $dp/dx = -\Delta p/L$ because dp/dx is constant. At $y = 0.2$ in., the
velocity is

$$u = -1520\left[\left(\frac{0.2}{12}\right)^2 - 0.417 \times \frac{0.2}{12} \right] = 0.634 \text{ ft/sec}$$

EXAMPLE 7.4

Find the pressure gradient that results in a zero shear stress at the lower wall, where $y = 0$, of two parallel plates; also, sketch the velocity profiles for a top plate speed of U with various pressure gradients. Assume horizontal plates.

Solution

The velocity distribution for plates with the top plate moving with velocity U is given by Eq. 7.4.13. Letting $\partial h/\partial x = 0$, we have

$$u(y) = \frac{1}{2\mu}\frac{dp}{dx}(y^2 - ay) + \frac{U}{a}y$$

The shear stress is

$$\tau = \mu\frac{du}{dy}$$

$$= \frac{1}{2}\frac{dp}{dx}(2y - a) + \mu\frac{U}{a}$$

If $\tau = 0$ at $y = 0$, then $du/dy = 0$ at $y = 0$ and the pressure gradient is

$$\frac{dp}{dx} = \frac{2\mu U}{a^2}$$

If dp/dx is greater than this value, the slope du/dy at $y = 0$ is negative and thus the velocity u will be negative near $y = 0$. If $dp/dx = 0$, we observe that a linear velocity distribution results, namely,

$$u(y) = \frac{U}{a}y$$

If dp/dx is negative, $u(y)$ is greater at each y-location than the linear distribution since $(y^2 - ay)$ is a negative quantity for all y's of interest.

All of the results above can be qualitatively displayed on a sketch of $u(y)$ for several dp/dx as shown.

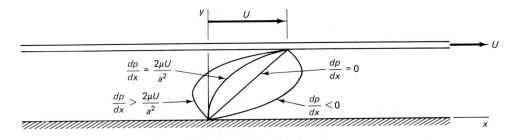

7.5 LAMINAR FLOW BETWEEN ROTATING CYLINDERS

Fully developed, steady flow between concentric, rotating cylinders, as shown in Fig. 7.6, is another flow that yields a rather simple solution. It has particular application in the field of lubrication, where the fluid may be oil and the inner cylinder a rotating shaft. We will again use two approaches to find the velocity distribution. The laminar flow solution we will find will be valid up to a Reynolds number[6] of 1700, providing the angular velocity of the outer cylinder $\omega_2 = 0$, as is often the case. Above Re = 1700 a secondary laminar flow (a flow with a different velocity distribution) may develop and eventually a turbulent flow develops.

7.5.1 Elemental Approach

Body forces will be neglected in this derivation, or the cylinders will be assumed to be vertical. Since pressure does not vary with θ, an element in the form of a thin cylindrical shell will be used. The resultant torque acting on this element is zero because it has no angular acceleration; this is expressed as

$$\tau 2\pi r L \times r - (\tau + d\tau) 2\pi (r + dr) L \times (r + dr) = 0 \qquad (7.5.1)$$

where L, the length of the cylinders, should be large with respect to the gap width $(r_2 - r_1)$ to avoid three-dimensional end effects. Equation 7.5.1 reduces to

$$2\tau + r \frac{d\tau}{dr} = 0 \qquad (7.5.2)$$

Using the one-dimensional constitutive equation (see Table 5.1), and recognizing that $\tau = -\tau_{r\theta}$, we have

$$\tau = -\mu r \frac{d}{dr}\left(\frac{v_\theta}{r}\right) \qquad (7.5.3)$$

There results

$$-2\mu r \frac{d}{dr}\left(\frac{v_\theta}{r}\right) - r\mu \frac{d}{dr}\left[r \frac{d}{dr}\left(\frac{v_\theta}{r}\right)\right] = 0 \qquad (7.5.4)$$

Divide by μr, multiply by dr, and integrate to find

$$2 \frac{v_\theta}{r} + r \frac{d}{dr}\left(\frac{v_\theta}{r}\right) = A \qquad (7.5.5)$$

---

[6]The Reynolds number is defined as Re $= \omega_1 r_1 \delta/\nu$, where $\delta = r_2 - r_1$.

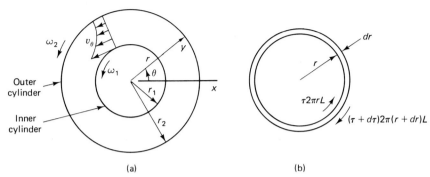

Figure 7.6 Flow between concentric cylinders: (a) basic flow variables; (b) an element from between the cylinders.

This can be rearranged as

$$\frac{dv_\theta}{dr} + \frac{v_\theta}{r} = A \qquad (7.5.6)$$

or

$$\frac{1}{r}\frac{d}{dr}(rv_\theta) = A \qquad (7.5.7)$$

Multiply by $r\,dr$ and integrate again to find that

$$v_\theta(r) = \frac{A}{2}r + \frac{B}{r} \qquad (7.5.8)$$

The boundary conditions are $v_\theta = r_1\omega_1$ at $r = r_1$, and $v_\theta = r_2\omega_2$ at $r = r_2$. These conditions allow the constants to be evaluated as

$$A = 2\,\frac{\omega_2 r_2^2 - \omega_1 r_1^2}{r_2^2 - r_1^2} \qquad B = \frac{r_1^2 r_2^2(\omega_1 - \omega_2)}{r_2^2 - r_1^2} \qquad (7.5.9)$$

We can obtain the same result by integrating the appropriate Navier–Stokes equation, or if that is not of interest, go directly to Section 7.5.3.

7.5.2 Solving the Navier–Stokes Equations

For steady, laminar flow between concentric cylinders we assume the streamines to be circular so that $v_r = v_z = 0$, $v_\theta(r)$ only, and $\partial p/\partial\theta = 0$. The

θ-component Navier–Stokes equation from Table 5.1 is

$$
\cancelto{0}{\left(\frac{\partial v_\theta}{\partial t}\right)}_{\text{steady}} + \cancel{v_r} \frac{\partial v_\theta}{\partial r} + \underbrace{\frac{v_\theta}{r}\cancel{\frac{\partial v_\theta}{\partial \theta}}}_{\substack{\text{symmetric}\\ \text{flow}}} + \cancelto{0}{v_z \frac{\partial v_\theta}{\partial z}} + \cancelto{0}{\frac{v_r v_\theta}{r}}
$$

steady   0   symmetric flow

(7.5.10)

$$
= -\frac{1}{r}\cancel{\frac{\partial p}{\partial \theta}} + \mu\left[\frac{\partial^2 v_\theta}{\partial r^2} + \frac{1}{r}\frac{\partial v_\theta}{\partial r} + \frac{1}{r^2}\cancel{\frac{\partial^2 v_\theta}{\partial \theta^2}} + \cancel{\frac{\partial^2 v_\theta}{\partial z^2}} + \frac{2}{r^2}\cancel{\frac{\partial v_r}{\partial \theta}} - \frac{v_\theta}{r^2}\right]
$$

symmetric   long   symmetric flow   cylinders   flow

This reduces to

$$
0 = \frac{\partial^2 v_\theta}{\partial r^2} + \frac{1}{r}\frac{\partial v_\theta}{\partial r} - \frac{v_\theta}{r^2} \tag{7.5.11}
$$

which can be written as

$$
\frac{d^2 v_\theta}{dr^2} + \frac{d}{dr}\left(\frac{v_\theta}{r}\right) = 0 \tag{7.5.12}
$$

Integrating once, there results

$$
\frac{dv_\theta}{dr} + \frac{v_\theta}{r} = A \tag{7.5.13}
$$

or

$$
\frac{1}{r}\frac{d}{dr}(rv_\theta) = A \tag{7.5.14}
$$

A second integration yields

$$
v_\theta = \frac{A}{2}r + \frac{B}{r} \tag{7.5.15}
$$

Applying the boundary conditions $v_\theta = r_1\omega_1$ at $r = r_1$ and $v_\theta = r_2\omega_2$ at $r = r_2$, we can find that

$$
A = 2\frac{r_2^2\omega_2 - r_1^2\omega_1}{r_2^2 - r_1^2} \qquad B = \frac{r_1^2 r_2^2(\omega_1 - \omega_2)}{r_2^2 - r_1^2} \tag{7.5.16}
$$

7.5.3 Flow with the Outer Cylinder Fixed

For a number of situations the outer cylinder is fixed. Setting $\omega_2 = 0$, the velocity distribution is

$$v_\theta = \frac{r_1^2 \omega_1}{r_2^2 - r_1^2} \left(\frac{r_2^2}{r} - r \right) \tag{7.5.17}$$

The shearing stress τ_1 on the inner cylinder (see Table 5.1 and let $\tau = -\tau_{r\theta}$) is found to be

$$
\begin{aligned}
\tau_1 &= -\left[\mu r \frac{d}{dr} \left(\frac{v_\theta}{r} \right) \right]_{r=r_1} \\
&= \mu \frac{2}{r_1^2} \frac{r_1^2 r_2^2 \omega_1}{r_2^2 - r_1^2} = \frac{2\mu r_2^2 \omega_1}{r_2^2 - r_1^2}
\end{aligned} \tag{7.5.18}
$$

The torque T necessary to rotate the inner cylinder of length L is

$$
\begin{aligned}
T_1 &= \tau_1 A_1 r_1 \\
&= \frac{2\mu r_2^2 \omega_1}{r_2^2 - r_1^2} 2\pi r_1 L r_1 = \frac{4\pi \mu r_1^2 r_2^2 L \omega_1}{r_2^2 - r_1^2}
\end{aligned} \tag{7.5.19}
$$

The power $\dot{W}$ necessary to rotate the shaft is found by multiplying the torque by the rotational speed ω_1; it is

$$
\begin{aligned}
\dot{W} &= T\omega_1 \\
&= \frac{4\pi \mu r_1^2 r_2^2 L \omega_1^2}{r_2^2 - r_1^2}
\end{aligned} \tag{7.5.20}
$$

This power is necessary to overcome the resistance of viscosity and results in an increase in the internal energy and thus the temperature of the fluid. The removal of this energy from the fluid often requires special heat exchangers.

EXAMPLE 7.5

Show that as the inner cylinder radius approaches the outer cylinder radius the velocity distribution approaches the linear distribution between parallel plates with one plate moving and a zero pressure gradient. This is Couette flow.

Solution

For this problem we will let $\omega_2 = 0$; the velocity distribution (7.5.17) is

$$v_\theta(r) = \frac{r_1^2\omega_1}{r_2^2 - r_1^2}\left(\frac{r_2^2}{r} - r\right)$$

$$= \frac{r_1^2\omega_1}{r_2^2 - r_1^2}\frac{r_2^2 - r^2}{r} = \frac{r_1^2\omega_1}{r_2^2 - r_1^2}(r_2 - r)\frac{r_2 + r}{r}$$

Introduce the independent variable y, measured from the outer cylinder defined by $r + y = r_2$ (see Fig. 7.6); let $\delta = r_2 - r_1$. Then the above can be written as

$$v_\theta(r) = \frac{r_1^2\omega_1(r_2 - r)}{(r_2 - r_1)(r_2 + r_1)}\frac{r_2 + r}{r}$$

$$= \frac{r_1^2\omega_1 y}{\delta(r_2 + r_1)}\frac{2r_2 - y}{r_2 - y}$$

As the inner radius approaches the outer radius we can write $r_1 \simeq r_2$. Letting $r_2 \simeq r_1 \simeq R$ we have $r_2 + r_1 \simeq 2R$ and

$$\frac{2R - y}{R - y} \simeq 2$$

since $y \ll R$. The velocity distribution then simplifies to

$$v_\theta(y) = \frac{R^2\omega_1 y}{\delta\ 2R} \times 2 = \frac{R\omega_1}{\delta}y$$

this is a linear distribution and is a good approximation to the flow whenever $\delta \ll R$.

7.6 TURBULENT FLOW IN A PIPE

7.6.1 General Information

The study of developed turbulent flow in a circular pipe is of substantial interest in actual flows since most flows encountered in practical applications are turbulent flows. Turbulent flow is assumed to occur whenever the Reynolds number in a pipe

$$\text{Re} = \frac{VD}{\nu}$$

exceeds 2000. Consider, for example, water at 20°C flowing in a rather small 5-mm-diameter pipe; the average velocity need only be 0.4 m/s (1 ft/sec in a $\frac{1}{4}$-in. pipe) for turbulent flow to occur. For larger-diameter

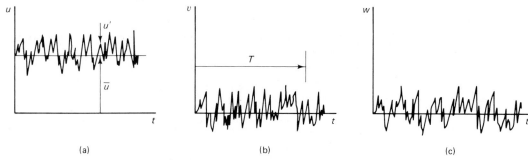

Figure 7.7   Velocity components in a turbulent pipe flow: (a) x-component velocity; (b) r-component velocity; (c) θ-component velocity.

pipes the average velocity needed to produce a turbulent flow will occur in most engineering situations.

In a turbulent flow all three velocity components are nonzero. If we measure the components as a function of time, graphs similar to those shown in Fig. 7.7 would result. There is seldom any interest (to the engineer) in the details of the randomly fluctuating velocity components; hence we introduce the notion of a time-average quantity. The velocity components (u, v, w) are written as

$$u = \bar{u} + u' \qquad v = \bar{v} + v' \qquad w = \bar{w} + w' \tag{7.6.1}$$

where a bar over a quantity denotes a time average and a prime denotes the fluctuating part. Using the component u as an example, the **time average** is defined as,

$$\bar{u} = \frac{1}{T} \int_0^T u(t)\, dt \tag{7.6.2}$$

where T is a time increment large enough[7] to eliminate all time dependence from $\bar{u}$. In a developed pipe flow $\bar{u}$ would be nonzero and $\bar{v} = \bar{w} = 0$, as is obvious from Fig. 7.7.

EXAMPLE 7.6

Show that $\overline{u'} = 0$ and $\dfrac{\overline{\partial u}}{\partial y} = \dfrac{\partial \bar{u}}{\partial y}$ for a turbulent flow.

[7]At a given location in a pipe this could be checked experimentally as follows: start with a relatively large value of T and then decrease T for subsequent runs. If $\bar{u}$ remains unchanged as T decreases, T is sufficiently large. If T is too small, $\bar{u}$ will be different for each run.

Solution

To show that $\overline{u'} = 0$ we simply substitute the expression (7.6.1) for $u(t)$ into Eq. 7.6.2 and obtain

$$\bar{u} = \frac{1}{T} \int_0^T (\bar{u} + u') \, dt$$

$$= \frac{1}{T} \int_0^T \bar{u} \, dt + \frac{1}{T} \int_0^T u' \, dt$$

$$= \bar{u} \frac{1}{T} \int_0^T dt + \overline{u'}$$

$$= \bar{u} + \overline{u'}$$

Subtracting $\bar{u}$ from both sides results in

$$\overline{u'} = 0$$

Now, let us time average the derivative $\partial u / \partial y$. We have

$$\overline{\frac{\partial u}{\partial y}} = \frac{1}{T} \int_0^T \frac{\partial u}{\partial y} \, dt$$

$$= \frac{\partial}{\partial y} \left(\frac{1}{T} \int_0^T u \, dt \right) = \frac{\partial}{\partial y} \bar{u}$$

Thus

$$\overline{\frac{\partial u}{\partial y}} = \frac{\partial \bar{u}}{\partial y}$$

7.6.2 Differential Equation

The differential equation that must be solved to provide us with the time-average velocity distribution is derived by using a particle approach. The Navier–Stokes equations could be time averaged, arriving at the same differential equation, but that approach is not attempted in this book.

Consider the situation for a turbulent flow in a horizontal pipe, as shown in Fig. 7.8. Fluid particles[8] move randomly throughout the flow. At an instant in time a fluid particle moves through the incremental area dA due to the velocity fluctuation v'; it enters a neighboring layer of fluid moving with a higher x-component velocity, providing a retarding effect on the neighboring layer. A fluid particle that moves to a neighboring layer that is traveling with a lower x-component velocity would accelerate

---

[8] A particle is a relatively small mass of the fluid that tends to move as a unit.

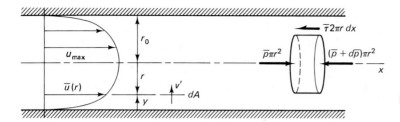

Figure 7.8 Turbulent flow in a pipe.

the slower moving fluid. The x-component force that results due to the random motion of a fluid particle passing through the incremental area dA would be (see Eq. 4.5.6)

$$dF = -\rho v' \, dA \, u' \qquad (7.6.3)$$

where u' is the negative change in x-component velocity due to the momentum exchange. The turbulent shear stress that results is

$$\tau_{\text{turb}} = \frac{dF}{dA} = -\rho u'v' \qquad (7.6.4)$$

where we know that $(u'v')$ is, on the average, a negative quantity since a positive v' produces a negative u'. The negative sign is introduced in the equation so that the desired positive shear stress results. The time average turbulent shear stress, often called the **apparent shear stress,** which is of primary interest, is

$$\bar{\tau}_{\text{turb}} = -\rho\overline{u'v'} \qquad (7.6.5)$$

Note that $\overline{u'w'}$ would be zero since a w'-component (in the θ-direction) would not move a fluid particle into a layer of higher or lower x-component velocity.

The total shear stress at a particular location would be due to both the viscosity and the momentum exchange described above; that is,

$$\bar{\tau} = \bar{\tau}_{\text{viscous}} + \bar{\tau}_{\text{turb}}$$
$$= \mu \frac{\partial \bar{u}}{\partial y} - \rho\overline{u'v'} \qquad (7.6.6)$$

The shear stress can be related to the pressure gradient by summing forces on the horizontal cylindrical element shown in Fig. 7.8. There results

$$\bar{\tau} = -\frac{r}{2}\frac{d\bar{p}}{dx} \qquad (7.6.7)$$

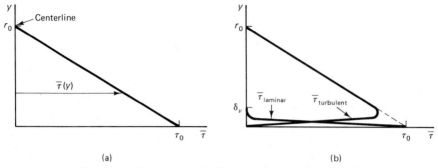

Figure 7.9   Shear stress distributions in a developed pipe flow.

which shows that the shear stress distribution is linear for a turbulent flow as well as a laminar flow (see Eq. 7.3.17). The turbulent shear obviously goes to zero at the wall since the velocity perturbations are zero at the wall, and the total shear is zero at the centerline, as shown in Fig. 7.9. The viscous shear is nonzero only in a very thin viscous wall layer, of thickness δ_ν, near the wall, as shown in part (b). Note that the turbulent shear reaches a maximum near the wall in the viscous wall layer.

The differential equation that must be solved if the time-average velocity distribution is to be determined is found by combining the two preceding equations; it is

$$\frac{r}{2}\frac{d\bar{p}}{dx} = \mu\frac{d\bar{u}}{dr} + \rho\overline{u'v'} \qquad (7.6.8)$$

where we have used $r + y = r_0$ so that $dy = -dr$. For developed flow we know that $d\bar{p}/dx = $ const.; hence if we know how $\overline{u'v'}$ varied with r, the differential equation could be solved. The quantity $\overline{u'v'}$ cannot be determined analytically, however, so the solution to Eq. 7.6.8 cannot be attempted until an empirical expression is found for $\overline{u'v'}$. Rather than finding the empirical expression for $\overline{u'v'}$ and then solving the differential equation for $\bar{u}$, we will simply present the empirical results obtained for the velocity profile $\bar{u}(y)$. However, before we do this we should introduce the eddy viscosity and the correlation coefficient.

Instead of using the quantity $\overline{u'v'}$ as the unknown in Eq. 7.6.8, we often introduce the **eddy viscosity** η, defined by

$$\eta = \frac{\overline{u'v'}}{d\bar{u}/dr} \qquad (7.6.9)$$

Note that it has the same dimensions as the kinematic viscosity. In terms of the eddy viscosity the differential equation becomes

$$\frac{r}{2}\frac{d\bar{p}}{dx} = \rho(\nu + \eta)\frac{d\bar{u}}{dr} \qquad (7.6.10)$$

The **correlation coefficient** K_{uv}, a normalized apparent shear stress, often used when describing turbulent motions, has limits of ± 1 and is defined by

$$K_{uv} = \frac{\overline{u'v'}}{\sqrt{\overline{u'^2}}\,\sqrt{\overline{v'^2}}} \tag{7.6.11}$$

The quantities η and K_{uv} are functions of r that simply replace the variable $\overline{u'v'}$. They do not simplify the differential equation; they allow the equation to take slightly different forms.

EXAMPLE 7.7

Note that in Fig. 7.9b there is a region near the wall where the turbulent shear is relatively constant and the viscous shear is quite small. Assume that the eddy viscosity η depends on the product of the velocity gradient $\partial \bar{u}/\partial y$ and y^2, the square of the distance from the wall. With these assumptions show that $\bar{u}(y)$ is logarithmic in this region near the wall.

Solution

If the viscous shear is negligible, we have, combining Eqs. 7.6.6 and 7.6.9,

$$\bar{\tau}_{\text{turb}} = \rho \eta \frac{d\bar{u}}{dy}$$

where we have used $dr = -dy$. Now, if $\bar{\tau}_{\text{turb}} = \text{const.} = c_1$ and

$$\eta = c_2 y^2 \frac{d\bar{u}}{dy}$$

there results

$$\rho c_2 y^2 \left(\frac{du}{dy}\right)^2 = c_1$$

or

$$y \frac{d\bar{u}}{dy} = c_3$$

where c_3 is a constant. This is integrated to yield

$$\bar{u}(y) = c_3 \ln y + c_4$$

Hence, with the foregoing assumptions we see that a logarithmic profile is predicted for the region of constant turbulent shear near the wall.

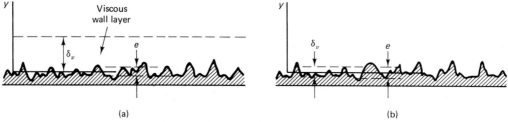

Figure 7.10   (a) A smooth wall and (b) a rough wall.

7.6.3 Velocity Profile

The time-average velocity profile in a pipe is quite sensitive to the magnitude of the average wall roughness height e, as sketched in Fig. 7.10. All materials are "rough" when viewed with sufficient magnification, although glass and plastic are assumed to be smooth with $e = 0$. (Values for e are listed in Fig. 7.13.) As noted in the preceding section, the turbulent shear $(-\rho\overline{u'v'})$ becomes negligibly small near the wall in a viscous wall layer with thickness δ_ν. If the thickness δ_ν is sufficiently large, it submerges the wall roughness elements so that they have no effect on the flow; it is as if the wall were smooth. Such a condition is often referred to as being **hydraulically smooth.** If the viscous wall layer is relatively thin, the roughness elements protrude out of this layer and the wall is rough. The **relative roughness** e/D and the Reynolds number can be used to determine if a pipe is smooth or rough. This will be observed from the friction factor data presented in the next section.

We do not solve for the velocity distribution in a developed turbulent flow since $\overline{u'v'}$ cannot be determined analytically. Thus we will present the empirical data for $\bar{u}(y)$ directly and not solve Eq. 7.6.8. The two most common expressions for turbulent flow are now presented.

The first method of empirically expressing the velocity distribution requires us to identify two regions of the flow, the **wall region** and the **outer region.** In the wall region the characteristic velocity and length are the **shear velocity** u_τ defined by[9] $u_\tau = \sqrt{\tau_0/\rho}$, and the **viscous length** ν/u_τ. The dimensionless velocity distribution in the wall region, for a smooth pipe, is

$$\frac{\bar{u}}{u_\tau} = \frac{u_\tau y}{\nu} \quad \text{(viscous wall layer)} \qquad\qquad 0 \le \frac{u_\tau y}{\nu} \le 5 \qquad (7.6.12)$$

$$\frac{\bar{u}}{u_\tau} = 2.44 \ln \frac{u_\tau y}{\nu} + 4.9 \quad \left(\begin{array}{c}\text{turbulent}\\\text{region}\end{array}\right) \qquad 30 < \frac{u_\tau y}{\nu}, \quad \frac{y}{r_0} < 0.15 \quad (7.6.13)$$

[9]The shear velocity u_τ is a fictitious velocity and has no relationship to an actual velocity in the flow. It is a quantity with dimensions of velocity that allows the experimental data to be presented in dimensionless form as universal profiles (profiles that are valid for all turbulent pipe flows).

In the interval $5 < u_\tau y/\nu < 30$, the **buffer zone,** the experimental data do not fit either of the curves above but merge the two curves as shown in Fig. 7.11a. The viscous wall layer has thickness δ_ν and it is in the viscous layer that turbulence is initiated; this layer possesses a time-average, linear velocity distribution, but instantaneously the layer is very time dependent. The outer edge of the wall region is quite dependent on the

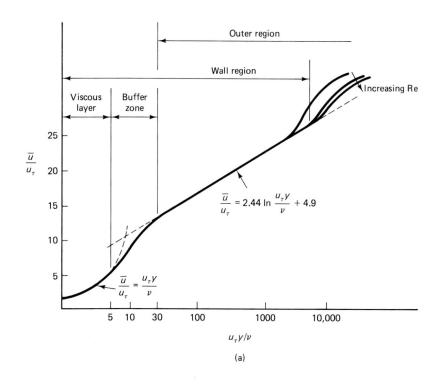

(a)

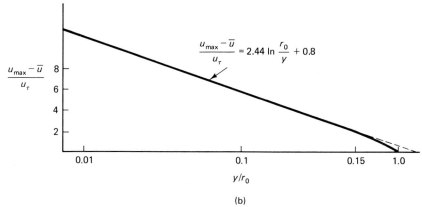

(b)

Figure 7.11   Empirical relations for turbulent flow in a smooth pipe: (a) wall region; (b) outer region. (Based on data from J. Laufer, "The Structure of Turbulence in Fully Developed Pipe Flow," NACA Report 1174, 1954.)

Reynolds number, as shown; for low Re it may be located near $u_\tau y / \nu = 3000$.

For rough pipes the wall layer does not play an important role since turbulence intitates from the protruding wall elements, so only a logarithmic profile is necessary in the wall region. The characteristic length is obviously the average roughness height e; the dimensionless velocity profile for the rough pipe is

$$\frac{\bar{u}}{u_\tau} = 2.44 \ln \frac{y}{e} + 8.5 \qquad (7.6.14)$$

where the constants 8.5 and 2.44 allow a good fit with experimental data.

In the outer region, sketched in Fig. 7.11b, the characteristic length is r_0; the velocity defect $(u_{max} - \bar{u})$ is normalized with u_τ, and the empirical relation, for both smooth and rough pipes, is

$$\frac{u_{max} - \bar{u}}{u_\tau} = 2.44 \ln \frac{r_0}{y} + 0.8 \quad \text{(outer region)} \qquad \frac{y}{r_0} \le 0.15 \qquad (7.6.15)$$

The wall region and the outer region overlap as shown in Fig. 7.11a. In this overlap region we can combine the equations above to obtain an expression for the maximum velocity; for a smooth pipe it is

$$\frac{u_{max}}{u_\tau} = 2.44 \ln \frac{u_\tau r_0}{\nu} + 5.7 \quad \text{(smooth pipes)} \qquad (7.6.16)$$

and for a rough pipe we find that

$$\frac{u_{max}}{u_\tau} = 2.44 \ln \frac{r_0}{e} + 8.5 \quad \text{(rough pipes)} \qquad (7.6.17)$$

Although we do not often desire the actual time-average velocity at a specified radial position in a pipe, the distributions above are of occasional use and are presented for completeness. Note, however, that before u_{max} can be found we must know u_τ; before u_τ can be found we must know τ_0. To find τ_0 we use the pressure gradient,

$$\tau_0 = -\frac{r_0}{2} \frac{dp}{dx} \qquad (7.6.18)$$

or the friction factor with Eq. 7.3.19.

An alternative, and simpler form that adequately describes the turbulent flow velocity distribution in a pipe is the **power-law profile,** namely,

$$\frac{\bar{u}}{u_{max}} = \left(\frac{y}{r_0}\right)^{1/n} \qquad (7.6.19)$$

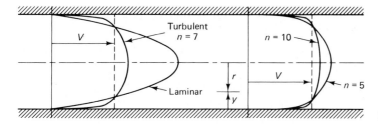

Figure 7.12   Turbulent velocity profile.

where y is measured from the pipe wall and n is an integer between 5 and 10. Using this distribution the average velocity is found to be

$$V = \frac{\int_0^{r_0} \bar{u}(r)2\pi r \, dr}{\pi r_0^2}$$

$$= \frac{2n^2}{(n+1)(2n+1)} u_{max}$$

(7.6.20)

This distribution is compared with a laminar profile in Fig. 7.12.

The value of n in the exponent is related to the friction factor f by the empirical expression

$$n = \frac{1}{\sqrt{f}}$$

(7.6.21)

The constant n varies from 5 to 10 depending on the Reynolds number and the pipe wall roughness e/D. For smooth pipes the exponent n is related to the Reynolds number as shown in Table 7.1.

TABLE 7.1   EXPONENT n FOR SMOOTH PIPES

Re = VD/ν	4×10^3	10^5	10^6	$> 2 \times 10^6$
n	6	7	9	10

The power-law profile cannot be used to obtain the slope at the wall since it will always yield $(\partial u/\partial y)_{wall} = \infty$ for all n. Thus it cannot be used to predict the wall shear stress. The wall shear is found by combining Eqs. 7.6.21 and 7.3.19.

It should be noted that the kinetic-energy-correction factor α in pipes is 1.11, 1.06, and 1.03 for $n = 5$, 7, and 10, respectively. Because it is close to unity, it is often set equal to unity in the energy equation when working problems involving turbulent flow.

EXAMPLE 7.8

A 4-cm-diameter smooth, horizontal pipe transports 0.004 m³/s of water at 20°C. Calculate (a) the friction factor, (b) the maximum velocity, (c) the radial position where $\bar{u} = V$, (d) the wall shear, (e) the pressure drop over a 10-m length, and (f) the maximum velocity using Eq. 7.6.16.

Solution

(a) The average velocity is calculated to be

$$V = \frac{Q}{A}$$

$$= \frac{0.004}{\pi} \times 0.02^2 = 3.18 \text{ m/s}$$

The Reynolds number is

$$\text{Re} = \frac{VD}{\nu}$$

$$= \frac{3.18 \times 0.04}{10^{-6}} = 1.27 \times 10^5$$

From Table 7.1 we see that $n \cong 7$ and from Eq. 7.6.21,

$$f = \frac{1}{n^2}$$

$$= \frac{1}{7^2} = 0.02$$

(b) The maximum velocity is found using Eq. 7.6.20 to be

$$u_{max} = \frac{(n + 1)(2n + 1)}{2n^2} V$$

$$= \frac{8 \times 15}{2 \times 49} \times 3.18 = 3.89 \text{ m/s}$$

(c) The distance from the wall where $\bar{u} = V = 3.18$ m/s is found using Eq. 7.6.19 as follows:

$$\frac{\bar{u}}{u_{max}} = \left(\frac{y}{r_0}\right)^{1/7}$$

$$\therefore \quad y = r_0 \left(\frac{u}{u_{max}}\right)^7$$

$$= 2 \left(\frac{3.18}{3.89}\right)^7 = 0.488 \text{ cm}$$

The radial position is thus

$$r = r_0 - y$$

$$= 2 - 0.488 = 1.51 \text{ cm}$$

(d) The wall shear is found using the definition of the friction factor

$$f = \frac{\tau_0}{\frac{1}{8}\rho V^2}$$

and is

$$\tau_0 = \tfrac{1}{8}\rho V^2 f$$
$$= \tfrac{1}{8} \times 1000 \times 3.18^2 \times 0.02 = 25.3 \text{ Pa}$$

(e) The pressure drop is calculated using Eq. 7.6.18 with $\Delta p/L = -dp/dx$ to be

$$\Delta p = \frac{2\tau_0 L}{r_0}$$
$$= \frac{2 \times 25.3 \times 10}{0.02} = 25\,300 \text{ Pa} \quad \text{or} \quad 25.3 \text{ kPa}$$

(f) To use Eq. 7.6.16 we must know the shear velocity. It is

$$u_\tau = \sqrt{\frac{\tau_0}{\rho}}$$
$$= \sqrt{\frac{25.3}{1000}} = 0.159 \text{ m/s}$$

Then we find u_{max} to be

$$u_{max} = 0.159 \left(2.44 \ln \frac{0.159 \times 0.02}{10^{-6}} + 5.7 \right) = 4.04 \text{ m/s}$$

very close to that given by the power-law formula in part (b). This answer is considered to be more accurate.

7.6.4 Losses in Developed Pipe Flow

Perhaps the most calculated quantity in pipe flow is the head loss. If the head loss is known in a developed flow, the pressure change can be calculated; for a pipe the energy equation provides us with

$$h_L = \frac{\Delta(p + \gamma h)}{\gamma} \tag{7.6.22}$$

The head loss that results from the wall shear in a developed flow is related to the friction factor (see Eq. 7.3.20) by the Darcy–Weisbach equation, namely,

$$h_L = f \frac{L}{D} \frac{V^2}{2g} \tag{7.6.23}$$

Consequently, if the friction factor were known, we could find the head loss and then the pressure drop.

The friction factor f depends on the various quantities that affect the flow, written as

$$f = f(\rho, \mu, V, D, e) \qquad (7.6.24)$$

where the average wall roughness height e accounts for the influence of the wall roughness elements. A dimensional analysis following the steps of Section 6.2 provides us with

$$f = f\left(\frac{\rho VD}{\mu}, \frac{e}{D}\right) \qquad (7.6.25)$$

where e/D is the relative roughness.

Experimental data that relate the friction factor to the Reynolds number have been obtained for fully developed pipe flow over a wide range of wall roughnesses. The results of these data are presented in Fig. 7.13, which is commonly referred to as the **Moody diagram.** There are several features of the Moody diagram that should be noted.

- For a given wall roughness, measured by the relative roughness e/D, there is a sufficiently large value of Re above which the friction factor is constant, thereby defining the **completely turbulent regime.** The average roughness element size e is substantially greater than the viscous wall layer thickness δ_ν, so that viscous effects are not significant; the resistance to the flow is produced by the drag of the roughness elements that protrude into the flow.
- For the smaller relative roughness e/D values it is observed that, as Re decreases, the friction factor increases in the **transition zone** and eventually becomes the same as that of a smooth pipe. The roughness elements become submerged in the viscous wall layer so that they produce little effect on the main flow.
- For Reynolds numbers less than 2000, the friction factor of laminar flow is shown. The **critical zone** couples the turbulent flow to the laminar flow and may represent an oscillating flow that alternately exists between turbulent and laminar flow.
- The e values in this diagram are for new pipes. With age a pipe will corrode and become fouled, changing both the roughness and the pipe diameter, with a resulting increase in the friction factor. Such factors should be included in design considerations; they will not be reviewed here.

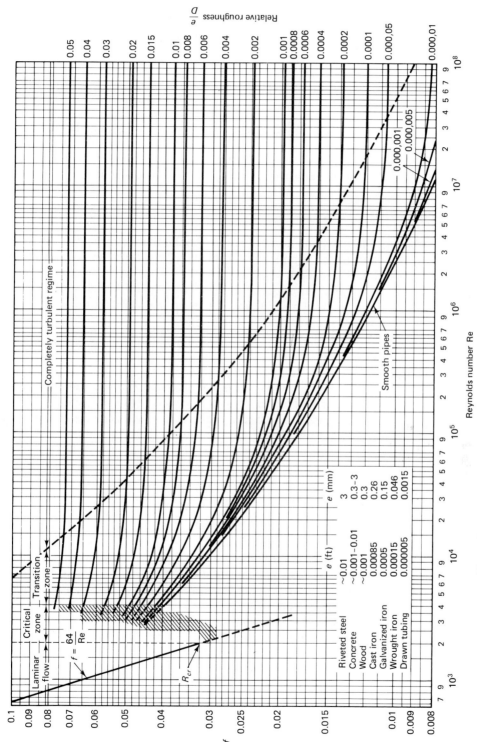

Figure 7.13 Moody diagram. (From L. F. Moody, *Trans. ASME*, Vol. 66, 1944.)

The Moody diagram is a graphical representation of the following empirical equations:

Smooth pipe flow: $\qquad\qquad \dfrac{1}{\sqrt{f}} = 0.86 \ln \text{Re}\sqrt{f} - 0.8 \qquad (7.6.26)$

Completely turbulent zone: $\quad \dfrac{1}{\sqrt{f}} = -0.86 \ln \dfrac{e}{3.7D} \qquad\qquad (7.6.27)$

Transition zone: $\qquad\qquad \dfrac{1}{\sqrt{f}} = -0.86 \ln\left(\dfrac{e}{3.7D} + \dfrac{2.51}{\text{Re}\sqrt{f}}\right)$

$$(7.6.28)$$

The transition zone equation (7.6.28) that couples the smooth pipe equation to the completely turbulent regime equation is known as the **Colebrook equation.**

Good approximations can be made for the head loss in conduits with noncircular cross sections by using the **hydraulic radius** R, defined by

$$R = \frac{A}{P} \qquad (7.6.29)$$

where A is the cross-sectional area and P is the **wetted perimeter,** that perimeter where the fluid is in contact with the solid boundary. For a circular pipe the hydraulic radius is $R = r_0/2$. Hence we simply replace the radius r_0 with $2R$ and use the Moody diagram with

$$\text{Re} = \frac{4VR}{\nu} \qquad \frac{\text{relative}}{\text{roughness}} = \frac{e}{4R}$$

$$h_L = f\frac{L}{4R}\frac{V^2}{2g} \qquad (7.6.30)$$

To use this hydraulic radius technique the cross section should be fairly "open", such as a rectangle with aspect ratio less than 4 : 1, and equilateral triangle, or an oval. For other shapes, such as an annulus, the error would be significant.

Three categories of problems can be identified for developed turbulent flow in a pipe length L:

Category	Known	Unknown
1	Q, D, e, ν	h_L
2	D, e, ν, h_L	Q
3	Q, e, ν, h_L	D

A category 1 problem is straightforward and requires no iteration procedure when using the Moody diagram. Category 2 and 3 problems are more like problems encountered in engineering design situations and require an iterative trial-and-error process when using the Moody diagram. Each of these types will be illustrated with an example.

An alternative to using the Moody diagram is made possible by empirically derived formulas. Perhaps the best of such formulas were presented by Swamee and Jain[10]; an explicit expression that provides an approximate value for the unknown in each category above is as follows:

$$h_L = 1.07 \frac{Q^2 L}{gD^5} \left\{ \ln\left[\frac{e}{3.7D} + 4.62\left(\frac{\nu D}{Q}\right)^{0.9} \right] \right\}^{-2} \qquad \begin{array}{l} 10^{-6} < e/D < 10^{-2} \\ 3000 < Re < 3 \times 10^8 \end{array}$$

$$(7.6.31)$$

$$Q = -0.965 \left(\frac{gD^5 h_L}{L}\right)^{0.5} \ln\left[\frac{e}{3.7D} + \left(\frac{3.17\nu^2 L}{gD^3 h_L}\right)^{0.5} \right] \qquad Re > 2000$$

$$(7.6.32)$$

$$D = 0.66 \left[e^{1.25} \left(\frac{LQ^2}{gh_L}\right)^{4.75} + \nu Q^{9.4} \left(\frac{L}{gh_L}\right)^{5.2} \right]^{0.04} \qquad \begin{array}{l} 10^{-6} < e/D < 10^{-2} \\ 5000 < Re < 3 \times 10^8 \end{array}$$

$$(7.6.33)$$

One may use either English or SI units in the equations above.

Equation 7.6.32 is as accurate as the Moody diagram, and Eqs. 7.6.31 and 7.6.33 are accurate to within approximately 2% of the Moody diagram. These tolerances are acceptable for engineering calculations. It is important to realize that the Moody diagram is based on experimental data that likely is accurate to within no more than 5%. Hence the foregoing three formulas of Swamee and Jain, which can easily be input on a programmable hand-held calculator, are often used by design engineers. The following examples also illustrate the use of these approximate formulas.

EXAMPLE 7.9

Water at 20°C is transported for 500 m in a 4-cm-diameter wrought iron horizontal pipe with a flow rate of 0.003 m³/s. Calculate the pressure drop over the 500-m length of pipe.

[10]P. K. Swamee and A. K. Jain, Explicit Equations for Pipe-Flow Problems, *J. Hydraulics Div., ASCE*, Vol. 102, No. HY5, May 1976.

Solution

The average velocity is

$$V = \frac{Q}{A}$$

$$= \frac{0.003}{\pi \times 0.02^2} = 2.39 \text{ m/s}$$

The Reynolds number is

$$\text{Re} = \frac{VD}{\nu}$$

$$= \frac{2.39 \times 0.04}{10^{-6}} = 9.6 \times 10^4$$

Obtaining e from Fig. 7.13, we have, using $D = 40$ mm,

$$\frac{e}{D} = \frac{0.046}{40} = 0.00115$$

The friction factor is read from the Moody diagram to be

$$f = 0.023$$

The head loss is calculated as

$$h_L = f \frac{L}{D} \frac{V^2}{2g}$$

$$= 0.023 \frac{500}{0.04} \frac{2.39^2}{2 \times 9.81} = 84 \text{ m}$$

This answer is given to two significant numbers since the friction factor is known to at most two significant numbers. The pressure drop is found to be

$$\Delta p = \gamma h_L$$

$$= 9800 \times 84 = 820\,000 \text{ Pa} \quad \text{or} \quad 820 \text{ kPa}$$

Using Eq. 7.6.31 we find

$$h_L = 1.07 \frac{0.003^2 \times 500}{9.81 \times (0.04)^5} \left\{ \ln \left[\frac{0.00115}{3.7} + 4.62 \left(\frac{10^{-6} \times 0.04}{0.003} \right)^{0.9} \right] \right\}^{-2}$$

$$= 1.07 \times 4480 \times 0.01731 = 83 \text{ m}$$

This value is within 1.2% of the value using the Moody diagram.

EXAMPLE 7.9 (English)

Water at 74°F is transported for 1500 ft in a 1½-in.-diameter wrought iron horizontal pipe with a flow rate of 0.1 ft³/sec. Calculate the pressure drop over the 1500-ft length of pipe.

Solution

The average velocity is

$$V = \frac{Q}{A}$$

$$= \frac{0.1}{\pi \times 0.75^2/144} = 8.15 \text{ ft/sec}$$

The Reynolds number is

$$\text{Re} = \frac{VD}{\nu}$$

$$= \frac{8.15 \times 1.5/12}{10^{-5}} = 1.02 \times 10^5$$

Obtaining e from Fig. 7.13, we have, using $D = 1.5/12$ ft,

$$\frac{e}{D} = \frac{0.00015}{0.125} = 0.0012$$

The friction factor is read from the Moody diagram to be

$$f = 0.023$$

The head loss is calculated as

$$h_L = f \frac{L}{D} \frac{V^2}{2g}$$

$$= 0.023 \frac{1500}{1.5/12} \frac{8.15^2}{2 \times 32.2} = 280 \text{ ft}$$

This answer is given to two significant numbers since the friction factor is known to at most two significant numbers. The pressure drop is found by Eq. 7.6.22 to be

$$\Delta p = \gamma h_L$$

$$= 62.4 \times 280 = 17,500 \text{ psf} \quad \text{or} \quad 120 \text{ psi}$$

EXAMPLE 7.10

A pressure drop of 700 kPa is measured over a 300-m length of horizontal, 10-cm-diameter wrought iron pipe that transports oil ($S = 0.9$, $\nu = 10^{-5}$ m²/s). Calculate the flow rate.

Solution

The relative roughness is

$$\frac{e}{D} = \frac{0.046}{100} = 0.00046$$

Assuming that the flow is completely turbulent (Re is not needed), the Moody diagram gives

$$f = 0.0165$$

The head loss is found to be

$$h_L = \frac{\Delta p}{\gamma}$$

$$= \frac{700\,000}{9800 \times 0.9} = 79.4 \text{ m}$$

The velocity is calculated from Eq. 7.6.23 to be

$$V = \left(\frac{2gDh_L}{fL}\right)^{1/2}$$

$$= \left(\frac{2 \times 9.8 \times 0.1 \times 79.4}{0.0165 \times 300}\right)^{1/2} = 5.61 \text{ m/s}$$

This provides us with a Reynolds number of

$$\text{Re} = \frac{VD}{\nu}$$

$$= \frac{5.61 \times 0.1}{10^{-5}} = 5.61 \times 10^4$$

Using this Reynolds number and $e/D = 0.00046$, the Moody diagram gives the friction factor as

$$f = 0.023$$

This corrects the original value for f. The velocity is recalculated to be

$$V = \left(\frac{2 \times 9.8 \times 0.1 \times 79.4}{0.023 \times 300}\right)^{1/2} = 4.75 \text{ m/s}$$

The Reynolds number is then

$$\text{Re} = \frac{4.75 \times 0.1}{10^{-5}} = 4.75 \times 10^4$$

From the Moody diagram $f = 0.023$ appears to be satisfactory. Thus the flow rate is

$$Q = VA$$
$$= 4.75 \times \pi \times 0.05^2 = 0.037 \text{ m}^3/\text{s}$$

Only two significant numbers are given since f is known to at most two significant numbers.

Using the explicit relationship (7.6.32), we can directly calculate Q to be

$$Q = -0.965 \left(\frac{9.8 \times 0.1^5 \times 79.4}{300}\right)^{0.5} \ln\left[\frac{0.00046}{3.7} + \left(\frac{3.17 \times 10^{-10} \times 300}{9.8 \times 0.1^3 \times 79.4}\right)^{0.5}\right]$$
$$= -0.965 \times 5.096 \times 10^{-3} \times (-7.655) = 0.038 \text{ m}^3/\text{s}$$

This value is within 2.5% of the value using the Moody diagram.

EXAMPLE 7.11

Drawn tubing of what diameter should be selected to transport 0.002 m³/s of 20°C water over a 400-m length so that the head loss does not exceed 30 m?

Solution

In this problem we do not know D. Thus, a trial-and-error solution is anticipated. The average velocity is related to D by

$$V = \frac{Q}{A}$$
$$= \frac{0.002}{\pi D^2/4} = \frac{0.00255}{D^2}$$

The friction factor and D are related as follows:

$$h_L = f\frac{L}{D}\frac{V^2}{2g}$$
$$30 = f\frac{400}{D}\frac{(0.00255/D^2)^2}{2 \times 9.8}$$
$$\therefore \quad D^5 = 4.42 \times 10^{-6}f$$

The Reynolds number is

$$\text{Re} = \frac{VD}{\nu}$$

$$= \frac{0.00255D}{D^2 \times 10^{-6}} = \frac{2550}{D}$$

Now, let us simply guess a value for f and check with the relations above and the Moody diagram. The first guess of $f = 0.03$ and the correction is listed in the following table.

f	D (m)	Re	e/D	f (Fig. 7.13)
0.03	0.0421	6.06×10^4	0.000036	0.02
0.02	0.0388	6.57×10^4	0.000039	0.02

The value of $f = 0.02$ is acceptable, yielding a diameter of 3.88 cm. Since this diameter would undoubtedly not be standard, a diameter of

$$D = 4 \text{ cm}$$

would be the tube size selected. This tube would have a head loss less than the limit of $h_L = 30$ m imposed in the problem statement. Any larger-diameter tube would also satisfy this criterion but would be more costly, so should not be selected.

Using the explicit relationship (7.6.33), we can directly calculate D to be

$$D = 0.66 \left[(1.5 \times 10^{-6})^{1.25} \left(\frac{400 \times 0.002^2}{9.8 \times 30} \right)^{4.75} + 10^{-6} \times 0.002^{9.4} \left(\frac{400}{9.8 \times 30} \right)^{5.2} \right]^{0.04}$$

$$= 0.66 \, [5.163 \times 10^{-33} + 2.102 \times 10^{-31}]^{0.04} = 0.039 \text{ m}$$

Hence $D = 4$ cm would be the tube size selected. This is the same tube size as that selected using the Moody diagram.

EXAMPLE 7.12

Air at standard conditions is to be transported through 500 m of a smooth, horizontal, 30 cm × 20 cm rectangular duct at a flow rate of 0.24 m³/s. Calculate the pressure drop.

Solution

The hydraulic radius is

$$R = \frac{A}{P}$$

$$= \frac{0.3 \times 0.2}{(0.3 + 0.2) \times 2} = 0.06 \text{ m}$$

The average velocity is

$$V = \frac{Q}{A}$$

$$= \frac{0.24}{0.3 \times 0.2} = 4.0 \text{ m/s}$$

This gives a Reynolds number of

$$\text{Re} = \frac{4VR}{\nu}$$

$$= \frac{4 \times 4 \times 0.06}{1.6 \times 10^{-5}} = 6 \times 10^4$$

Using the smooth pipe curve of the Moody diagram, there results

$$f = 0.0198$$

Hence,

$$h_L = f \frac{L}{4R} \frac{V^2}{2g}$$

$$= 0.0198 \frac{500}{4 \times 0.06} \frac{4^2}{2 \times 9.8} = 33.7 \text{ m}$$

The pressure drop is

$$\Delta p = \rho g h_L$$

$$= 1.23 \times 9.8 \times 33.7 = 406 \text{ Pa}$$

EXAMPLE 7.12 (English)

Air at standard conditions is to be transported through 1500 ft of a smooth, horizontal, 12 in. × 8 in. rectangular duct at a flow rate of 500 ft³/min. Calculate the pressure drop.

Solution

The hydraulic radius is

$$R = \frac{A}{P}$$

$$= \frac{12 \times 8/144}{(12 + 8) \times 2/12} = 0.2 \text{ ft}$$

The average velocity is

$$V = \frac{Q}{A}$$

$$= \frac{500/60}{12 \times 8/144} = 12.5 \text{ ft/sec}$$

This gives a Reynolds number of

$$\text{Re} = \frac{4VR}{\nu}$$

$$= \frac{4 \times 12.5 \times 0.2}{1.56 \times 10^{-4}} = 6.41 \times 10^4$$

Using the smooth pipe curve of the Moody diagram, there results

$$f = 0.0195$$

Hence,

$$h_L = f \frac{L}{4R} \frac{V^2}{2g}$$

$$= 0.0195 \frac{1500}{4 \times 0.2} \frac{12.5^2}{2 \times 32.2} = 88.7 \text{ ft}$$

The pressure drop is

$$\Delta p = \rho g h_L$$

$$= 0.00237 \times 32.2 \times 88.7 = 6.77 \text{ psf}$$

7.6.5 Minor Losses in Pipe Flow

We now know how to calculate the losses due to a developed flow in a pipe. Pipe systems do, however, include valves, elbows, enlargements, contractions, inlets, outlets, bends, and other fittings that cause additional losses, referred to as **minor losses.** Each of these devices causes a change in the magnitude and/or the direction of the velocity vectors and hence result in losses. In general, if the flow is gradually accelerated by a device, the losses are very small; relatively large losses are associated with sudden enlargements because of the separated regions that result (a separated flow occurs when the primary flow separates from the wall).

A minor loss is expressed in terms of a loss coefficient K, defined by

$$h_L = K \frac{V^2}{2g} \tag{7.6.34}$$

Values of K have been determined experimentally for the various fittings and geometry changes of interest in piping systems. One exception is the sudden expansion from area A_1 to area A_2, for which the loss can be calculated; this was done in Example 4.17, where we found that

$$h_L = \left(1 - \frac{A_1}{A_2}\right)^2 \frac{V_1^2}{2g} \qquad (7.6.35)$$

Thus, for the sudden expansion

$$K = \left(1 - \frac{A_1}{A_2}\right)^2 \qquad (7.6.36)$$

If A_2 is extremely large (e.g., a pipe exiting into a reservoir), $K = 1.0$.

A pipe fitting that has a relatively large loss coefficient with no change in cross-sectional area is the pipe bend, or the elbow. This results primarily from the secondary flow caused by the fluid flowing from the high-pressure region to the low-pressure region, as shown in Fig. 7.14; this secondary flow is eventually dissipated after the fluid leaves the long sweep bend or elbow. In addition, a separated region occurs at the sharp corner in a standard elbow. Energy is needed to maintain such a secondary flow and the flow in the separated region. This wasted energy is measured in terms of a loss coefficient.

The loss coefficients for various geometries are presented in Table 7.2 and Fig. 7.15. A globe valve may be used to control the flow rate by introducing large losses by partially closing the valve. The other types of valves should not be used to control the flow; damage could result.

Loss coefficients for sudden contractions and orifice plates can be approximated by neglecting the losses in the converging flow up to the vena contracta and calculating the losses in the diverging flow using the

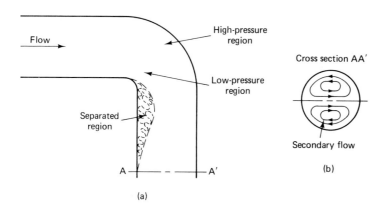

(a)

Figure 7.14 Flow in an elbow.

TABLE 7.2 NOMINAL LOSS COEFFICIENTS K (TURBULENT FLOW)[a]

Type of Fitting	Screwed			Flanged		
Diameter	1 in.	2 in.	4 in.	2 in.	4 in.	8 in.
Globe valve (fully open)	8.2	6.9	5.7	8.5	6.0	5.8
(half open)	20	17	14	21	15	14
(one-quarter open)	57	48	40	60	42	41
Angle valve (fully open)	4.7	2.0	1.0	2.4	2.0	2.0
Swing check valve (fully open)	2.9	2.1	2.0	2.0	2.0	2.0
Gate valve (fully open)	0.24	0.16	0.11	0.35	0.16	0.07
Return bend	1.5	.95	.64	0.35	0.30	0.25
Tee (branch)	1.8	1.4	1.1	0.80	0.64	0.58
Tee (line)	0.9	0.9	0.9	0.19	0.14	0.10
Standard elbow	1.5	0.95	0.64	0.39	0.30	0.26
Long sweep elbow	0.72	0.41	0.23	0.30	0.19	0.15
45° elbow	0.32	0.30	0.29			

Square-edged entrance   0.5

Reentrant entrance   0.8

Well-rounded entrance   0.03

Pipe exit   1.0

Sudden contraction[b]	Area ratio	
	2:1	0.25
	5:1	0.41
	10:1	0.46

Orifice plate	Area ratio A/A_0	
	1.5:1	0.85
	2:1	3.4
	4:1	29
	$\geq 6:1$	$2.78\left(\dfrac{A}{A_0} - 0.6\right)^2$

Sudden enlargement[c] $\left(1 - \dfrac{A_1}{A_2}\right)^2$

90° miter bend (without vanes)   1.1

(with vanes)   0.2

General contraction	(30° included angle)	0.02
	(70° included angle)	0.07

[a]Values for other geometries can be found in *Technical Paper 410*, The Crane Company, 1957.

[b]Based on exit velocity V_2.

[c]Based on entrance velocity V_1.

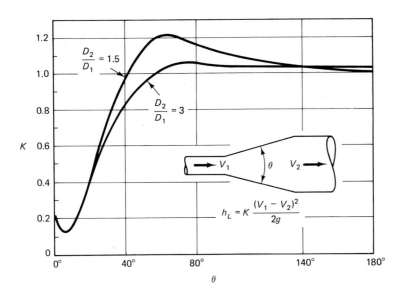

Figure 7.15 Loss coefficients in a conical expansion. (From A. H. Gibson, *Engineering*, Vol. 93, 1912.)

loss coefficient for a sudden expansion. Figure 7.16 provides the information necessary to establish the area of the **vena contracta**, the minimum area; this minimum area results when the converging streamlines begin to expand to fill the downstream area.

It is often the practice to express a loss coefficient as an **equivalent length** L_e of pipe. This is done by equating Eq. 7.6.34 to Eq. 7.6.23:

$$K \frac{V^2}{2g} = f \frac{L_e}{D} \frac{V^2}{2g} \qquad (7.6.37)$$

giving the relationship

$$L_e = K \frac{D}{f} \qquad (7.6.38)$$

Hence the pipe exit of a 20-cm-diameter pipe with a friction factor of $f = 0.02$ could be replaced by an equivalent pipe length of $L_e = 10$ m.

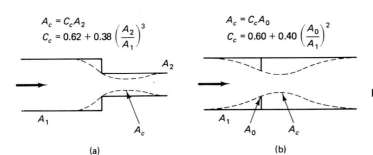

$$A_c = C_c A_2$$
$$C_c = 0.62 + 0.38 \left(\frac{A_2}{A_1}\right)^3$$

$$A_c = C_c A_0$$
$$C_c = 0.60 + 0.40 \left(\frac{A_0}{A_1}\right)^2$$

(a)

(b)

Figure 7.16 Vena contractas in contractions and orifices: (a) sudden contraction; (b) concentric orifice.

Finally, a comment should be made concerning the magnitude of the minor losses. In piping systems involving intermediate lengths (i.e., 50 diameters) of pipe between minor losses, the minor losses may be of the same order of magnitude as the frictional losses; for short lengths the minor losses may be substantially greater than the frictional losses; and for long lengths (e.g., 1000 diameters) of pipe, the minor losses are usually neglected.

EXAMPLE 7.13

If the flow rate through this pipe is 0.04 m³/s, find the difference in elevation H of the two reservoirs.

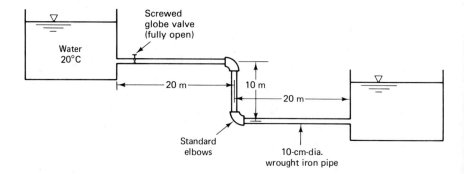

Solution

The energy equation written for a control volume that contains the two reservoir surfaces (see Eq. 4.4.17), where $V_1 = V_2 = 0$ and $p_1 = p_2 = 0$, is

$$0 = z_2 - z_1 + h_L$$

Thus, letting $z_1 - z_2 = H$, we have

$$H = (K_{entrance} + K_{valve} + 2K_{elbow} + K_{exit}) \frac{V^2}{2g} + f \frac{L}{D} \frac{V^2}{2g}$$

The average velocity, Reynolds number, and relative roughness are

$$V = \frac{0.04}{\pi \times 0.05^2} = 5.09 \text{ m/s}$$

$$Re = \frac{5.09 \times 0.1}{10^{-6}} = 5.09 \times 10^5$$

$$\frac{e}{D} = \frac{0.046}{100} = 0.00046$$

From the Moody diagram we find that

$$f = 0.0173$$

Using the loss coefficients from Table 7.2 for screwed 4-in. elements there results

$$H = (0.5 + 5.7 + 2 \times 0.64 + 1.0) \frac{5.09^2}{2 \times 9.8} + 0.0173 \frac{50}{0.1} \frac{5.09^2}{2 \times 9.8}$$

$$= 11.2 + 11.4 = 22.6 \text{ m}$$

EXAMPLE 7.14

Approximate the loss coefficient for the sudden contraction $A_1/A_2 = 2$ by neglecting the losses in the contracting portion up to the vena contracta and assuming that all the losses occur in the expansion from the vena contracta to A_2 (see Fig. 7.16). Compare with that given in Table 7.2.

Solution

The head loss from the vena contracta to area A_2 is (see Table 7.2, sudden enlargement)

$$h_L = \left(1 - \frac{A_c}{A_2}\right)^2 \frac{V_c^2}{2g}$$

Continuity allows us to write

$$V_c = \frac{A_2}{A_c} V_2$$

Thus

$$h_L = \left(1 - \frac{A_c}{A_2}\right)^2 \left(\frac{A_2}{A_c}\right)^2 \frac{V_2^2}{2g}$$

so

$$K = \left(1 - \frac{A_c}{A_2}\right)^2 \left(\frac{A_2}{A_c}\right)^2$$

Using the expression of C_c given in Fig. 7.16, we have

$$\frac{A_c}{A_2} = C_c$$

$$= 0.62 + 0.38 \left(\frac{1}{2}\right)^3 = 0.67$$

Finally,

$$K = (1 - 0.67)^2 \frac{1}{0.67^2} = 0.24$$

This compares favorably with the value of 0.25 in Table 7.2.

7.6.6 Hydraulic and Energy Grade Lines

When the energy equation is written in the form of Eq. 4.4.17 or 4.4.24, all the terms have dimensions of length. This has led to the conventional use of the hydraulic grade line and the energy grade line. The **hydraulic grade line** (HGL) in a piping system is formed by the locus of points located a distance p/γ above the center of the pipe, or $p/\gamma + z$ above a preselected datum. The **energy grade line** (EGL) is formed by the locus of points a distance $V^2/2g$ above the HGL, or the distance $V^2/2g + p/\gamma + z$ above the datum. The following points are noted relating to the HGL and the EGL:

- As the velocity goes to zero, the HGL and the EGL approach each other. Thus, in a reservoir, they are identical and lie on the surface (see Fig. 7.17).
- The EGL and, consequently, the HGL slope downward in the direction of the flow due to the head loss in the pipe. The greater the loss per unit length, the greater the slope. As the average velocity in the pipe increases, the loss per unit length increases.
- A sudden change occurs in the HGL and the EGL whenever a loss occurs due to a sudden geometry change as represented by the valve or the sudden enlargement of Fig. 7.17.
- A jump occurs in the HGL and the EGL whenever useful energy is added to the fluid as occurs with a pump, and a drop occurs if useful energy is extracted from the flow.

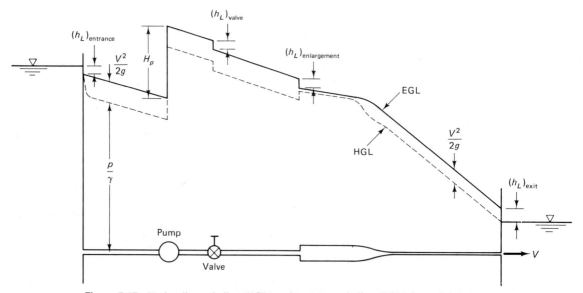

Figure 7.17   Hydraulic grade line (HGL) and energy grade line (EGL) for a piping system.

• At points where the HGL passes through the centerline of the pipe, the pressure is zero. If the pipe lies above the HGL, there is a vacuum in the pipe, a condition that is often avoided, if possible, in the design of piping systems; an exception would be in the design of a siphon.

The concepts of energy grade line and hydraulic grade line may also be applied to open-channel flows. The HGL coincides with the free surface and the EGL is a distance $V^2/2g$ above the free surface. Uniform flows in open channels will be discussed in the following section, and Chapter 10 is devoted to nonuniform open-channel flow.

EXAMPLE 7.15

Water at 20°C flows between two reservoirs at the rate of 0.06 m³/s as shown. Sketch the HGL and the EGL. What is the minimum diameter D_B allowed to avoid the occurrence of cavitation?

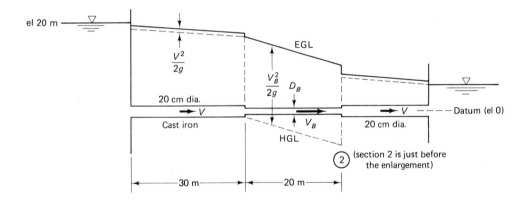

Solution

The EGL and the HGL are sketched on the figure, including sudden changes at the entrance, contraction, enlargement, and the exit. Note the large velocity head in the smaller pipe because of the high velocity. The velocity, Reynolds number, and relative roughness in the 20-cm-diameter pipe are calculated to be

$$V = \frac{Q}{A} = \frac{0.06}{\pi \times 0.20^2/4} = 1.91 \text{ m/s}$$

$$\text{Re} = \frac{VD}{\nu} = \frac{1.91 \times 0.2}{10^{-6}} = 3.8 \times 10^5$$

$$\frac{e}{D} = \frac{0.26}{200} = 0.0013$$

Thus $f = 0.022$ from Fig. 7.13. The velocity, Reynolds number, and relative roughness in the smaller pipe are

$$V_B = \frac{0.06}{\pi D_B^2/4} = \frac{0.0764}{D_B^2}$$

$$\text{Re} = \frac{0.0764 \times D_B}{D_B^2 \times 10^{-6}} = \frac{76\,400}{D_B}$$

$$\frac{e}{D_B} = \frac{0.00026}{D_B}$$

The minimum possible diameter is established by recognizing that the water vapor pressure (2450 Pa absolute) at 20°C is the minimum allowable pressure. Since the distance between the pipe and the HGL is an indication of the pressure in the pipe, we can conclude that the minimum pressure will occur at section 2. Hence the energy equation applied between section 1, the reservoir surface, and section 2 gives

$$\frac{\cancel{V_1^2}^{\,0}}{2g} + \frac{p_1}{\gamma} + z_1 = \frac{V_B^2}{2g} + \frac{p_2}{\gamma} + \cancel{z_2}^{\,0} + K_{\text{ent}} \frac{V^2}{2g} + K_{\text{cont}} \frac{V_B^2}{2g}$$

$$+ f_A \frac{L_A}{D_A} \frac{V^2}{2g} + f_B \frac{L_B}{D_B} \frac{V_B^2}{2g}$$

where the subscripts A refer to the 20-cm-diameter pipe. This simplifies to

$$\frac{101\,000}{9810} + 20 = \frac{(0.0764/D_B^2)^2}{2 \times 9.81} \left(1 + 0.25 + f_B \frac{20}{D_B}\right) + \frac{2450}{9810}$$

$$+ \left(0.5 + 0.022 \frac{30}{0.2}\right) \frac{1.91^2}{2 \times 9.81}$$

$$100\,000 = \frac{1.25}{D_B^4} + f_B \frac{20}{D_B^5}$$

where we have used $K_{\text{ent}} = 0.5$ and $K_{\text{cont}} = 0.25$. This requires a trial-and-error solution. The following illustrates the procedure.

Let $D_B = 0.1$ m. Then $e/D = 0.0026$ and $\text{Re} = 7.6 \times 10^5$. Therefore, $f = 0.026$.

$$100\,000 \stackrel{?}{=} 12\,500 + 52\,000$$

Let $D_B = 0.09$ m. Then $e/D = 0.0029$ and $\text{Re} = 8.4 \times 10^5$. Therefore, $f = 0.027$.

$$100\,000 \stackrel{?}{=} 19\,000 + 91\,000$$

Let $D_B = 0.08$ m. Then $e/D = 0.0032$ and $\text{Re} = 9.5 \times 10^5$. Therefore, $f = 0.027$.

$$100\,000 \stackrel{?}{=} 30\,500 + 165\,000.$$

We see that 0.1 m is too large and 0.08 m is too small. In fact, the value of 0.09 m is also slightly too small. Consequently, to be safe we must select the next larger pipe size, of 0.1 m diameter. If there were a pipe size of 9.5 cm diameter, that could be selected. Assuming that that size is not available, we select

$$D_B = 10 \text{ cm}$$

Note that the assumption of a $2:1$ area ratio for the contraction is too small. It is actually $4:1$. This would give $K_{\text{cont}} \approx 0.4$. This value would not influence the result.

7.6.7 Simple Pipe System with a Pump

The problems we have considered thus far have not involved a pump. If a centrifugal pump is included in the piping system and the flow rate is specified, the solution is straightforward using the techniques we have already developed. If, on the other hand, the flow rate is not specified, as is usually the case, a trial-and-error solution involving the centrifugal pump results since the head produced by a centrifugal pump and its efficiency depend on the discharge, as shown by the pump characteristic curves in Fig. 7.18. Companies furnish such characteristic curves for each centrifugal pump manufactured. Such a curve provides one equation relating the flow rate Q and pump head H_P. The other equation is provided by the energy equation. The two must be solved simultaneously to yield the desired flow rate.

    Note that for the piping system, the required pump energy head H_P, demanded by the energy equation, increases with Q and from the pump characteristic curve we see that H_P decreases with Q; hence the two curves will intersect at a point, called the **operating point** of the system. An example will illustrate the solution technique.

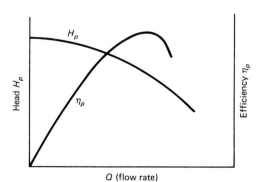

Figure 7.18 Pump characteristic curves.

EXAMPLE 7.16

Estimate the flow rate in the simple piping system if the pump characteristic curves are as shown. Also, find the pump power requirement.

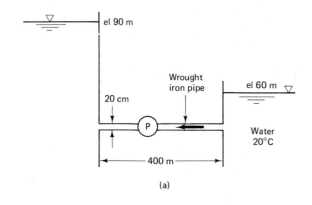

(a)

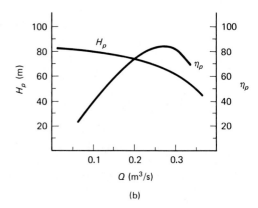

(b)

Solution

We will assume that the Reynolds number is sufficiently large that the flow is completely turbulent. So, using $e/D = 0.046/200 = 0.00023$, the friction factor from the Moody diagram is

$$f = 0.014$$

The energy equation (see Eq. 4.4.24), applied between the two surfaces, yields

$$H_P = \frac{V_2^2 - V_1^2}{2g} + z_2 - z_1 + \frac{p_2 - p_1}{\gamma} + h_L$$

or

$$H_P = 30 + \left(K_{\text{entrance}} + K_{\text{exit}} + f\frac{L}{D}\right)\frac{V^2}{2g}$$

$$= 30 + \left(0.5 + 1.0 + 0.014\,\frac{400}{0.2}\right)\frac{Q^2}{2 \times 9.8 \times [\pi \times (0.1)^2]^2}$$

$$= 30 + 1520Q^2$$

This equation, termed the **system demand curve,** and the characteristic curve $H_P(Q)$ of the pump are now solved simultaneously by trial and error. Actually, the curve above could be plotted on the same graph as the characteristic curve and the point of intersection, the operating point, would provide Q. Try $Q = 0.2$ m³/s: $(H_P)_{\text{energy}} = 91$ m, $(H_P)_{\text{char}} = 75$ m. Try $Q = 0.15$ m³/s: $(H_P)_{\text{energy}} = 64$ m, $(H_P)_{\text{char}} = 78$ m. Try $Q = 0.17$ m³/s: $(H_P)_{\text{energy}} = 74$ m, $(H_P)_{\text{char}} = 76$ m. This is our solution. We have

$$Q = 0.17 \text{ m}^3/\text{s}$$

The power requirement of the pump is given by Eq. 4.4.26:

$$\dot{W}_P = \frac{Q\gamma H_P}{\eta_P}$$

$$= 0.17 \times 9800 \times \frac{75}{0.67} = 186\,000 \text{ W} \quad \text{or} \quad 186 \text{ kW}$$

Note: Since $L/D > 1000$, minor losses could have been neglected.

7.7 UNIFORM TURBULENT FLOW IN OPEN CHANNELS

The last internal flow situation we consider is that of uniform (constant depth) flow in an open channel, shown in Fig. 7.19. We could already treat this flow by using the Darcy–Weisbach relation presented in Section 7.6.4. In fact, that technique predicts better results than does the more common method we present here. Both methods will be compared in an example. However, unless otherwise stated, flow in open, rough channels is commonly analyzed using the following method.

If the energy equation is applied between two sections of the channel sketched in Fig. 7.19, we obtain

$$0 = \frac{V_2^2 - \cancel{V_1^2}^{\,0}}{2g} + \frac{p_2 - \cancel{p_1}^{\,0}}{\gamma} + z_2 - z_1 + h_L \qquad (7.7.1)$$

which shows that the head loss is

$$h_L = z_2 - z_1$$
$$= L \sin \theta = LS \qquad (7.7.2)$$

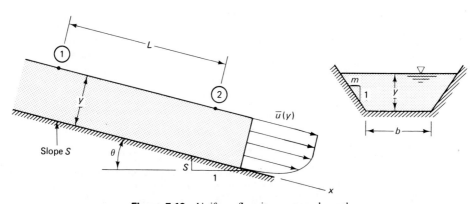

Figure 7.19   Uniform flow in an open channel.

where L is the length of the channel between the two sections and S is the slope of the channel, which is assumed small, so that $\sin \theta \simeq S$.

The Darcy–Weisbach equation (7.6.30), with $h_L = LS$ from Eq. 7.7.2, takes the form

$$RS = \frac{f}{8g} V^2 \qquad (7.7.3)$$

where R is the hydraulic radius. Since open channels are usually quite large with large Reynolds numbers, the friction factor is invariably constant (the flow is completely turbulent). Thus the equation above is written as

$$V = C\sqrt{RS} \qquad (7.7.4)$$

where the **Chezy coefficient** C is a dimensional constant; the equation above is referred to as the **Chezy equation**. The Chezy coefficient is related to the channel roughness and the hydraulic radius (much like f is in a pipe) by

$$C = \frac{c_1}{n} R^{1/6} \qquad (7.7.5)$$

where the dimensional constant c_1 has a value of 1.0 if SI units are used and 1.49 if English units are used. The dimensionless constant n is directly

TABLE 7.3   AVERAGE VALUES[a] OF THE MANNING n

Well material	Manning n
Planed wood	0.012
Unplaned wood	0.013
Finished concrete	0.012
Unfinished concrete	0.014
Sewer pipe	0.013
Brick	0.016
Cast iron, wrought iron	0.015
Concrete pipe	0.015
Rivited steel	0.017
Earth, straight	0.022
Corrugated metal flumes	0.025
Rubble	0.03
Earth with stones and weeds	0.035
Mountain streams	0.05

[a]The values in this table result in flow rates too large for hydraulic radii greater than about 3 m (10 ft). The Manning n should be increased by 10 to 15% for such large conduits.

related to the wall roughness; it is called the **Manning** n. Values for various wall materials are listed in Table 7.3.

The flow rate, which is of primary interest in open-channel flow problems, is found to be

$$Q = \frac{c_1}{n} AR^{2/3}S^{1/2} \qquad c_1 = \begin{cases} 1.0 \text{ in SI units} \\ 1.49 \text{ in English units} \end{cases} \qquad (7.7.6)$$

This is the **Chezy-Manning equation**.

For smooth-surfaced channels, the use of the Chezy–Manning equation is discouraged since it implicitly assumes a rough wall. Calculations for smooth-surfaced channels such as glass or plastic should be based on the Darcy–Weisbach relation with variable f; see Section 7.6.4.

EXAMPLE 7.17

The depth of water at 20°C flowing in a 4-m-wide rectangular, finished concrete channel is measured to be 1.2 m. The slope is measured to be 0.0016. Estimate the flow rate using (a) the Chezy–Manning equation and (b) the Darcy–Weisbach equation.

Solution

(a) The hydraulic radius is calculated to be

$$R = \frac{A}{P} = \frac{yb}{2y + b}$$

$$= \frac{1.2 \times 4}{2 \times 1.2 + 4} = 0.75 \text{ m}$$

Using the Chezy–Manning equation, with $n = 0.012$ from Table 7.3, we have

$$Q = \frac{1.0}{n} AR^{2/3}S^{1/2}$$

$$= \frac{1.0}{0.012} \times (1.2 \times 4) \times 0.75^{2/3} \times 0.0016^{1/2} = 13.2 \text{ m}^3/\text{s}$$

(b) The relative roughness is, using an intermediate value $e = 1.0$ mm shown on the Moody diagram,

$$\frac{e}{4R} = \frac{1.0}{4 \times 750} = 0.000333$$

Assuming a completely turbulent flow, the Moody diagram gives the friction factor as

$$f = 0.0152$$

The Darcy–Weisbach equation (7.6.30) then yields the velocity as follows:

$$S = \frac{h_L}{L} = \frac{f}{4R} \frac{V^2}{2g}$$

$$\therefore \quad V = \left(\frac{8RgS}{f}\right)^{1/2} = \left(\frac{8 \times 0.75 \times 9.81 \times 0.0016}{0.0152}\right)^{1/2} = 2.49 \text{ m/s}$$

This gives a Reynolds number of

$$\text{Re} = \frac{V \times 4R}{\nu} = \frac{2.49 \times 4 \times 0.75}{10^{-6}} = 7.47 \times 10^6$$

which is in the completely turbulent region. Thus the assumption above for f is acceptable. The flow rate is calculated as

$$Q = VA$$

$$= 2.49 \times 1.2 \times 4 = 12.0 \text{ m}^3/\text{s}$$

These two values are within 10%, an acceptable engineering tolerance for this type of problem. The second is considered to be more accurate, however.

EXAMPLE 7.17 (English)

The depth of water at 60°F flowing in a 12-ft-wide rectangular, finished concrete channel is measured to be 4 ft. The slope is measured to be 0.0016. Estimate the flow rate using (a) the Chezy–Manning equation and (b) the Darcy–Weisbach equation.

Solution

(a) The hydraulic radius is calculated to be

$$R = \frac{A}{P} = \frac{yb}{2y + b} = \frac{4 \times 12}{2 \times 4 + 12} = 2.4 \text{ ft}$$

Using the Chevy–Manning equation, with $n = 0.012$ from Table 7.3, we have

$$Q = \frac{1.49}{n} AR^{2/3} S^{1/2}$$

$$= \frac{1.49}{0.012} \times (4 \times 12) \times 2.4^{2/3} \times 0.0016^{1/2} = 427 \text{ ft}^3/\text{sec}$$

(b) The relative roughness is, using an intermediate value $e = 0.003$ ft shown on the Moody diagram,

$$\frac{e}{4R} = \frac{0.003}{4 \times 2.4} = 0.00031$$

Assuming a completely turbulent flow, the Moody diagram gives the friction factor as

$$f = 0.015$$

The Darcy–Weisbach equation (7.6.30) then yields the velocity as follows:

$$S = \frac{h_L}{L} = \frac{f}{4R}\frac{V^2}{2g}$$

$$\therefore \quad V = \left(\frac{8RgS}{f}\right)^{1/2}$$

$$= \left(\frac{8 \times 2.4 \times 32.2 \times 0.0016}{0.015}\right)^{1/2} = 8.12 \text{ ft/sec}$$

This gives a Reynolds number of

$$\text{Re} = \frac{V \times 4R}{\nu}$$

$$= \frac{8.12 \times 4 \times 2.4}{1.2 \times 10^{-5}} = 6.5 \times 10^6$$

which is in the completely turbulent region. Thus the assumption above for f is acceptable. The flow rate is calculated as

$$Q = VA$$

$$= 8.12 \times 4 \times 12 = 390 \text{ ft}^3/\text{sec}$$

These two values are within 10%, an acceptable engineering tolerance for this type of problem. The second is considered to be more accurate, however.

EXAMPLE 7.18

A 1.0-m-diameter concrete pipe transports 20°C water at a depth of 0.4 m. If the slope is 0.001, find the flow rate using (a) Chezy–Manning equation and (b) the Darcy–Weisbach equation.

$$\alpha = \sin^{-1}\frac{0.1}{0.5} = 11.54°$$

$$\therefore \quad \theta = 156.9°$$

$$\therefore \quad A = \pi \times 0.5^2 \times \frac{156.9}{360} - 0.49 \times 0.1$$

$$= 0.2933 \text{ m}^2$$

$$P = 2\pi \times 0.5 \times \frac{156.9}{360} = 1.369 \text{ m}$$

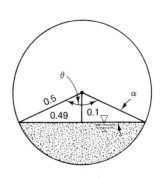

Solution

(a) The hydraulic radius is found, referring to the sketch above, to be

$$R = \frac{A}{P}$$

$$= \frac{0.2933}{1.369} = 0.2142 \text{ m}$$

The Chezy–Manning equation yields

$$Q = \frac{1.0}{n} AR^{2/3} S^{1/2}$$

$$= \frac{1.0}{0.015} \times 0.2933 \times 0.2142^{2/3} \times 0.001^{1/2} = 0.22 \text{ m}^3/\text{s}$$

(b) The relative roughness is, using a relatively rough value for concrete pipe (see Fig. 7.13) of $e = 2.0$ mm,

$$\frac{e}{4R} = \frac{2}{4 \times 214.2} = 0.0023$$

Assuming completely turbulent flow, the Moody diagram yields

$$f = 0.025$$

The Darcy–Weisbach equation (7.6.30) then gives the following:

$$S = \frac{h_L}{L} = \frac{f}{4R} \frac{V^2}{2g}$$

$$\therefore \quad V = \left(\frac{8RgS}{f}\right)^{1/2}$$

$$= \left(\frac{8 \times 0.2142 \times 9.81 \times 0.001}{0.025}\right)^{1/2} = 0.820 \text{ m/s}$$

The Reynolds number is

$$\text{Re} = \frac{V \times 4R}{\nu}$$

$$= \frac{0.82 \times 4 \times 0.2142}{10^{-6}} = 7.03 \times 10^5$$

This is in the completely turbulent region, so the calculations are acceptable. The flow rate is

$$Q = VA$$

$$= 0.820 \times 0.2933 = 0.24 \text{ m}^3/\text{s}$$

This is within 10% of the result above, an acceptable tolerance for this type of problem.

PROBLEMS

Laminar or Turbulent Flow

7.1. Calculate the maximum average velocity V with which 20°C water can flow in a pipe in the laminar state if the critical Reynolds number ($Re = VD/\nu$) at which transition occurs is 2000; the pipe diameter is:
(a) 2 m.
(b) 2 cm.
(c) 2 mm.

7.2. Water at 20°C flows in a wide river. Using a critical Reynolds number ($Re = Vh/\nu$) at which transition occurs of 1500 calculate the average velocity V which will result in a laminar flow if the depth h of the river is:
(a) 4 m.
(b) 1 m.
(c) 0.3 m.

7.3. Water at 10°C is flowing in a thin sheet in a parking lot at a depth of 5 mm with an average velocity of 0.5 m/s. Is the flow laminar or turbulent?

7.3E. Water at 50°F is flowing in a thin sheet in a parking lot at a depth of 0.2 in. with an average velocity of 1.5 ft/sec. Is the flow laminar or turbulent?

Entrance and Developed Flow

7.4. Calculate the laminar entrance length in a 4-cm-diameter pipe if 2×10^{-4} m³/s of water is flowing at:
(a) 10°C.
(b) 20°C.
(c) 40°C.
(d) 80°C.

7.5. A 6-cm-diameter pipe originates in a tank and delivers 0.025 m³/s of water at 20°C to a receiver 50 m away. Is the assumption of developed flow acceptable?

7.6. Air at 23°C is used as the working fluid in a parallel-plate research project. If the plates are 1.2 cm apart, how long is the longest possible entrance length for laminar flow? What is the shortest entrance length?

7.6E. Air at 72°F is used as the working fluid in a parallel-plate research project. If the plates are $\frac{1}{2}$ in. apart, how long is the longest possible entrance length for laminar flow? What is the shortest entrance length?

7.7. Air at 25°C can exist in either the laminar state or the turbulent state (a trip wire near the entrance is used to make it turbulent) for flow in a 10-cm-diameter pipe in a research lab. If the average velocity is 5 m/s, compare the length of the entrance region for the laminar flow with that of the turbulent flow.

7.8. Draw an incremental control volume with length Δx and radius r_0 and show that for a laminar flow $(\Delta p/\Delta x)_{\text{entrance}} > (\Delta p/\Delta x)_{\text{developed}}$. Also, explain the pressure variations observed for a turbulent flow in Fig. 7.3.

Laminar Flow in a Pipe

7.9. A pressure drop of 400 Pa occurs over a section of 2-cm-diameter pipe transporting water at 20°C. Determine the length of the horizontal section if the Reynolds number is 1600. Also, find the shear stress at the wall and the friction factor.

7.9E. A pressure drop of 0.07 psi occurs over a section of 0.8-in.-diameter pipe transporting water at

70°F. Determine the length of the horizontal section if the Reynolds number is 1600. Also, find the shear stress at the wall and the friction factor.

7.10. Find the angle of a 10-mm-diameter pipe in which water at 40°C is flowing with $Re = 1500$ such that no pressure drop occurs. Also, find the flow rate.

7.11. A liquid is pumped through a 2-cm-diameter pipe at a flow rate of 0.0002 m³/s. Calculate the pressure drop in a 10-m horizontal section if the liquid is:

(a) SAE-10W oil at 20°C.

(b) Water at 20°C.

(c) Glycerine at 40°C.

Is the assumption of laminar flow acceptable?

7.12. A liquid flows with no pressure drop in a vertical 2-cm-diameter pipe. Find the flow rate if, assuming a laminar flow, the liquid is:

(a) Water at 5°C.

(b) SAE-30W oil at 25°C.

(c) Glycerine at 20°C.

Is the assumption of laminar flow acceptable?

7.13. A laminar flow is to exist in a pipe transporting 0.004 m³/s of SAE-10W oil at 20°C. What is the maximum allowable diameter? What is the pressure drop over 10 m of horizontal pipe for this diameter?

7.13E. A laminar flow is to exist in a pipe transporting 0.12 ft³/sec of SAE-10W oil at 70°F. What is the maximum allowable diameter? What is the pressure drop over 30 ft of horizontal pipe for this diameter?

7.14. Estimate the flow rate through the smooth pipe shown. How long is the entrance region? Assume a laminar flow.

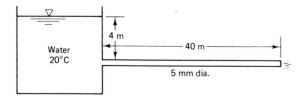

7.15. Air at 20°C flows in a horizontal 2-cm-diameter pipe. Calculate the maximum pressure drop in a 10-m section for a laminar flow. Assume $\rho = 1.2$ kg/m³.

7.15E. Air at 70°F flows in a horizontal, 0.8-in.-diameter pipe. Calculate the maximum pressure drop in a 30-ft section for a laminar flow. Assume $\rho = 0.0024$ slug/ft³.

7.16. Water at 20°C flows in a 4-mm-diameter pipe that is at an angle of 10°. The pressure rise over a 10-m section is 6 kPa. Find the Reynolds number of the flow and the wall shear stress. Assume laminar flow.

7.17. A research experiment requires a laminar flow of air at 20°C through a 10-cm-diameter pipe at a Reynolds number of 40 000. What maximum velocity is to be expected? What would be the pressure drop over a 10-m length of developed flow? How long would the entrance region be? Use $\rho = 1.2$ kg/m³.

7.18. A laminar flow of water at 20°C is obtained in a research lab at Re = 20 000 in a horizontal, 4-cm-diameter pipe. Calculate the head loss in a 10-m section of developed flow, the wall shear stress, and the length of the entrance region.

7.18E. A laminar flow of water at 60°F is obtained in a research lab at Re = 20,000 in a horizontal, 2-in.-diameter pipe. Calculate the head loss in a 30-ft section of developed flow, the wall shear stress, and the length of the entrance region.

7.19. Water at 20°C flows between two concentric, horizontal pipes with diameters of 2 cm and 3 cm. A pressure drop of 100 Pa is measured over a 10-m section of developed flow. Find the flow rate and the shear stress on the inner pipe.

7.20. Air at 20°C is to flow in the annulus between two concentric, horizontal pipes, with respective diameters of 2 cm and 3 cm, such that a pressure drop of 10 Pa occurs over a 10-m length. Find the average velocity and the shear stress at the inner pipe.

7.21. Fluid flows in the annulus between two concentric, horizontal pipes. The inner pipe is maintained at a higher temperature than the outer pipe so that the viscosity in the annulus cannot be assumed to be constant but $\mu = \mu(T)$. What differential equation would be solved to yield $u(r)$ assuming a laminar flow?

7.22. Show that the velocity distribution of Example 7.2 approaches that of pipe flow as $r_1 \to 0$, and approaches that of parallel plate flow as $r_1 \to r_2$.

Laminar Flow Between Parallel Plates

7.23. A flow occurs in a horizontal channel 1.2 cm × 50 cm with Re = 2000. Calculate the flow rate if the fluid is:

(a) Water at 20°C.

(b) Atmospheric air at 20°C.

7.23E. A flow occurs in a horizontal channel $\frac{1}{2}$ in. × 20 in. with Re = 2000. Calculate the flow rate if the fluid is:

(a) Water at 60°F.

(b) Atmospheric air at 60°F.

7.24. A 1-m² board, that weighs 40 N, moves down the incline shown with a velocity of $V = 0.2$ m/s. Estimate the viscosity of the fluid if θ is:
(a) 20°.
(b) 30°.

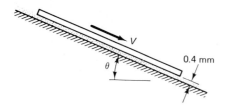

7.25. Water at 20°C exists between the plate and the surface of Problem 7.24. Calculate the velocity of the plate for an angle θ of:
(a) 20°.
(b) 30°.

7.26. Water at 20°C flows down a 20° incline at a depth of 6 mm and a width of 50 m. Calculate the flow rate and the Reynolds number assuming laminar flow. Also, find the maximum velocity and the wall shear.

7.27. Water at 20°C flows down a 100-m-wide parking lot on a slope of 0.00015 at a depth of 10 mm. Determine the flow rate and the Reynolds number assuming laminar flow. Also, calculate the friction factor and the wall shear.

7.27E. Water at 70°F flows down a 300-ft-wide parking lot on a slope of 0.00015 at a depth of 0.4 in. Determine the flow rate and the Reynolds number assuming laminar flow. Also, calculate the friction factor and the wall shear.

7.28. A pressure drop of 50 Pa is measured over a 60-m length of 90 cm × 2 cm rectangular, horizontal channel transporting 20°C air. Calculate the maximum flow rate and the associated Reynolds number. Use $\rho = 1.2$ kg/m³.

7.29. A pressure difference $p_A - p_B$ is measured to be 96 kPa. Find the friction factor for the wide channel assuming laminar flow. The flow direction is unknown.

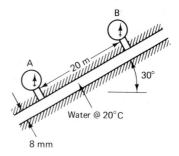

7.30. A slit with dimensions 0.5 mm × 100 mm exists in the 4-cm-thick side of a pressure vessel containing SAE-10W oil at 30°C and 4000 kPa. What is the maximum flow rate that can exist from the slit? Assume developed flow.

7.30E. A slit with dimensions 0.02 in. × 4 in. exists in the 2-in.-thick side of a pressure vessel containing SAE-10W oil at 80°F and 600 psi. What is the maximum flow rate that can exist from the slit? Assume developed flow.

7.31. Air flows between the parallel plates as shown. Find the pressure gradient such that:
(a) The shear stress at the upper surface is zero.
(b) The shear stress at the lower surface is zero.
(c) The flow rate is zero.
(d) The velocity at $y = 2$ mm is 4 m/s.

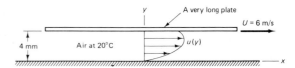

7.32. A pressure gradient of -20 Pa/m exists in 50°C air flowing between horizontal parallel plates spaced 6 mm apart. Find the velocity of the upper plate so that:
(a) The shear stress at the upper plate is zero.
(b) The shear stress at the lower plate is zero.
(c) The flow rate is zero.
(d) The velocity at $y = 2$ mm is 2 m/s.

7.33. Oil with $\mu = 0.01$ N · s/m² fills the concentric space between the rod and the surface shown. Find the force F if $V = 15$ m/s. Assume $dp/dx = 0$.

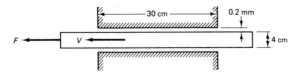

7.33E. Oil with $\mu = 10^{-4}$ lb-sec/ft² fills the concentric space between the rod and the surface shown. Find the force F if $V = 45$ ft/sec. Assume $dp/dx = 0$.

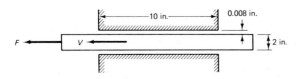

7.34. Calculate the torque T necessary to rotate the rod shown at 30 rad/s if the fluid filling the gap is SAE-10W oil at 20°C.
(a) Assume a linear velocity profile.
(b) Use the results of rotating cylinders.

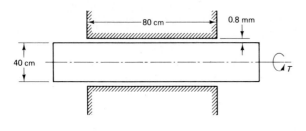

7.35. Oil with $\mu = 0.01$ N · s/m² fills the gap. Calculate the torque necessary to rotate the disk shown. Is the assumption of laminar flow valid? Use $S = 0.86$.

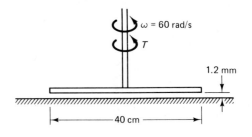

7.36. Approximate the torque necessary to rotate the inner 20-cm-diameter cylinder shown. SAE-30W oil at 20°C fills the gap.

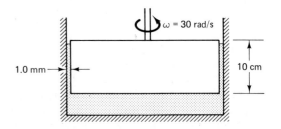

7.37. Find the torque needed to rotate the cone if oil with $\mu = 0.01$ N · s/m² fills the gap as shown.

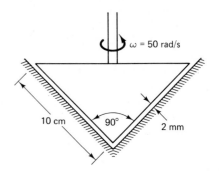

Laminar Flow between Rotating Cylinders

7.38. A long cylinder of radius R rotates in a large container of liquid. What is the velocity distribution in the liquid? Calculate the torque needed to rotate the 100-cm-long, 2-cm-radius cylinder at 1000 rpm if the liquid is water at 20°C. Assume a laminar flow.

7.38E. A long cylinder of radius R rotates in a large container of liquid. What is the velocity distribution in the liquid? Calculate the torque needed to rotate the 40-in.-long, 2-in.-diameter cylinder at 1000 rpm if the liquid is water at 60°F. Assume a laminar flow.

7.39. SAE-10W oil at 40°C fills the gap between two concentric, 40-cm-long cylinders with respective radii of 2 cm and 3 cm. What torque is necessary to rotate the inner cylinder at 3000 rpm if the outer cylinder is fixed? What power is required? Is the velocity distribution given by Eq. 7.5.15 the one to be expected?

7.40. A torque of 0.015 N · m is required to rotate a 4-cm-radius cylinder inside a fixed, 5-cm-radius cylin-

der at 40 rad/s. The concentric cylinders are 50 cm long. Calculate the viscosity of the fluid. Use $S = 0.9$. Is the velocity distribution given by Eq. 7.5.15 the one to be expected?

7.41. Find an expression for the torque necessary to rotate the outer cylinder if the inner cylinder of Fig. 7.6 is fixed.

Turbulent Flow

7.42. Time-average the differential continuity equation for an incompressible flow and show that two continuity equations result: the instantaneous continuity equation

$$\frac{\partial u'}{\partial x} + \frac{\partial v'}{\partial y} + \frac{\partial w'}{\partial z} = 0$$

and the time-average continuity equation

$$\frac{\partial \bar{u}}{\partial x} + \frac{\partial \bar{v}}{\partial y} + \frac{\partial \bar{w}}{\partial z} = 0$$

7.43. Find an expression for the difference between the time-average acceleration $\dfrac{Du}{Dt}$ and the quantity $\dfrac{D\bar{u}}{Dt}$ using the fact that

$$\overline{u'\frac{\partial u'}{\partial x}} + \overline{v'\frac{\partial u'}{\partial y}} + \overline{w'\frac{\partial u'}{\partial z}} = \frac{\partial}{\partial x}\overline{u'^2} + \frac{\partial}{\partial y}\overline{u'v'} + \frac{\partial}{\partial z}\overline{u'w'}$$

7.44. Prove that the last equation written in Problem 7.43 is indeed a fact.

7.45. The velocity components at a point in a turbulent flow are as given in the following table. Find $\bar{u}$, $\bar{v}$, $\overline{u'^2}$, $\overline{v'^2}$, and $\overline{u'v'}$.

t (s)	u (m/s)	v (m/s)	t (s)	u (m/s)	v (m/s)
0.00	16.1	1.6	0.06	17.1	−1.4
0.01	25.7	−5.4	0.07	28.6	6.7
0.02	10.6	−8.6	0.08	6.7	−5.2
0.03	17.3	3.5	0.09	19.2	−8.2
0.04	5.2	4.1	0.10	21.6	1.5
0.05	10.2	−6.0			

7.46. Over a small radial distance in a developed turbulent flow, the time-average velocity is as given in the following table. Find the pressure drop in a 10-m horizontal section if $\overline{u'v'}$ is measured at $r = 0.22$ m to be -3.2 m²/s². Air with $\rho = 1.23$ kg/m³ and $\nu = 1.6 \times 10^{-5}$ m²/s is flowing.

r (m)	0.20	0.21	0.22	0.23
ū (m/s)	27.2	25.5	23.4	20.2

7.46E. Over a small radial distance in a developed turbulent flow the time-average velocity is as given in the following table. Find the pressure drop in a 30-ft horizontal section if $\overline{u'v'}$ is measured at $r = 0.63$ ft to be -20.4 ft²/sec². Air with $\rho = 0.0035$ slug/ft³ and $\nu = 1.6 \times 10^{-4}$ ft²/sec is flowing.

r (ft)	0.60	0.63	0.69	0.72
ū (ft/sec)	81.4	76.9	70.2	60.7

7.47. If at $r = 0.22$ m for the data of Problem 7.46, we measure $\overline{u'^2} = 37$ m²/s² and $\overline{v'^2} = 17.2$ m²/s², what are the magnitudes of the eddy viscosity and the correlation coefficient?

7.47E. If at $r = 0.63$ ft. for the data of Problem 7.46E we measure $\overline{u'^2} = 316$ ft²/sec² and $\overline{v'^2} = 156$ ft²/sec², what are the magnitudes of the eddy viscosity and the correlation coefficient?

7.48. The velocity components are measured at a point in a laminar flow to be as shown. Find $\overline{u'v'}$, η, and K_{uv} if $d\bar{u}/dy = 10 \text{ s}^{-1}$ at the point.

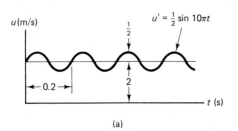

(a)

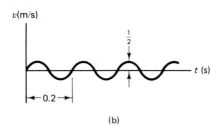

(b)

7.49. The velocity profile for water at 20°C in a turbulent flow in a 10-cm-diameter smooth pipe is approximated by $\bar{u} = 9.2y^{1/7}$ m/s. Find
(a) The wall shear.
(b) The velocity gradient $d\bar{u}/dy$ at the wall.
(c) The pressure gradient.
(d) The value of η at $r = 2.5$ cm.

7.50. Water at 20°C flows in a 12-cm-diameter horizontal pipe at the rate of 0.08 m³/s. Find the constant n in the exponent of Eq. 7.6.19. What is the maximum velocity?

7.50E. Water at 70°F flows in a 5-in.-diameter horizontal pipe at the rate of 2.5 ft³/sec. Find the constant n in the exponent of Eq. 7.6.19. What is the maximum velocity?

7.51. Show that the kinetic-energy correction factor is 1.10 for $n = 5$ and 1.03 for $n = 10$ using $u = u_{\text{max}}(y/r_0)^{1/n}$ in a circular pipe.

7.52. Water at 20°C flows in a 10-cm-diameter pipe with an average velocity of 10 m/s. Using $u = u_{\text{max}}(y/r_0)^{1/n}$ with $n = 7$, plot the viscous shear and the turbulent shear as a function of r. Also, find $d\bar{p}/dx$.

7.53. SAE-10W oil at 10°C is transported in a smooth, 80-cm-diameter pipe at the rate of 1.2 m³/s.
(a) Find the Reynolds number.
(b) Find the friction factor.
(c) Find the maximum velocity.
(d) Find the viscous wall layer thickness.
(e) Compare part (c) with the solution using the logarithmic velocity profile.

7.54. A pressure drop of 10 kPa is measured with gages placed 5 m apart on a smooth, horizontal, 10-cm-diameter pipe transporting water at 40°C. Estimate:
(a) The wall shear.
(b) The maximum velocity.
(c) The average velocity.
(d) The Reynolds number.
(e) The flow rate.

7.54E. A pressure drop of 1.5 psi is measured with gages placed 15 ft apart on a smooth, horizontal, 4-in.-diameter pipe transporting water at 100°F. Estimate:
(a) The wall shear.
(b) The maximum velocity.
(c) The average velocity.
(d) The Reynolds number.
(e) The flow rate.

7.55. Make a linear plot (not a semilog plot) of the velocity profile of the flow in Problem 7.54 using:
(a) The log profile.
(b) The power-law profile.

Turbulent Flow in Pipes and Conduits

7.56. A flow rate of 0.02 m³/s occurs in a 10-cm-diameter wrought iron pipe. Calculate the pressure drop over a 100-m horizontal section if the pipe transports:
(a) Water at 20°C.
(b) Glycerine at 60°C.
(c) SAE-30W at 30°C.
(d) Kerosene at 10°C.

7.57. Water at 20°C flows in a 4-cm-diameter pipe with a flow rate of 0.002 m³/s. Determine the head loss in a 200-m section if the pipe is:
(a) Cast iron.
(b) Galvanized iron.
(c) Wrought iron.
(d) Plastic.

7.57E. Water at 60°F flows in a 1.5-in.-diameter pipe with a flow rate of 0.06 ft³/sec. Determine the head loss in a 600-ft section if the pipe is:
(a) Cast iron.
(b) Galvanized iron.
(c) Wrought iron.
(d) Plastic.

7.58. A mass flux of 1.2 kg/s occurs in a 10-cm-diameter plastic pipe at 20°C and 500 kPa absolute. Assume an incompressible flow and find the pressure drop in a 100-m section of pipe if the fluid flowing is:
(a) Air.
(b) Carbon dioxide.
(c) Hydrogen.

7.59. SAE-30W oil flows at the rate of 0.08 m³/s in a 15-cm-diameter horizontal, galvanized iron pipe. Find the pressure drop in 100 m if the temperature of the oil is:
(a) 0°C.
(b) 30°C.
(c) 60°C.
(d) 90°C.
(e) 120°C.

7.60. Water at 20°C flows up a 30° incline in a 6-cm-diameter plastic pipe with a flow rate of 0.01 m³/s. Find the pressure change over a 100-m length of pipe.

7.60E. Water at 50°F flows up a 30° incline in a 2½-in.-diameter plastic pipe with a flow rate of 0.3 ft³/sec. Find the pressure change over a 300-ft length of pipe.

7.61. Water at 40°C flows in a horizontal section of 5-cm-diameter wrought iron pipe with a flow rate of 0.02 m³/s. Does the pipe behave as a smooth pipe, or is the roughness significant?

7.62. An 80-cm-diameter concrete pipe is to transport storm water at 20°C at a rate of 5 m³/s. What pressure drop is to be expected over a 100-m section of horizontal pipe?

7.63. A pressure drop of 500 kPa is not to be exceeded over a 200-m length of horizontal, 10-cm-diameter cast iron pipe. Calculate the maximum flow rate if the fluid is:
(a) Water at 20°C.
(b) Glycerine at 20°C.
(c) SAE-10W oil at 20°C.
(d) Kerosene at 20°C.

7.63E. A pressure drop of 75 psi is not to be exceeded over a 600-ft length of horizontal, 4-in.-diameter cast iron pipe. Calculate the maximum flow rate if the fluid is:
(a) Water at 60°F.
(b) Glycerine at 60°F.
(c) SAE-10W oil at 60°F.
(d) Kerosene at 60°F.

7.64. A pressure drop of 200 kPa is not to be exceeded over a 100-m length of horizontal, 4-cm-diameter pipe. What is the maximum flow rate if water at 20°C is transported and the pipe is:
(a) Cast iron?
(b) Wrought iron?
(c) Plastic?

7.65. A pressure drop of 400 Pa is allowable in gas flow in a 400-m horizontal section of 12-cm-diameter wrought iron pipe. If the temperature and pressure are 40°C and 200 kPa absolute, find the maximum mass flux if the gas is:
(a) Air.
(b) Carbon dioxide.
(c) Hydrogen.

7.66. A pressure drop of 200 kPa is not to be exceeded over a 200-m length of horizontal, 1.2-m-diameter concrete pipe transporting water at 20°C. What flow rate can be accommodated?

7.66E. A pressure drop of 30 psi is not to be exceeded over a 600-ft length of horizontal, 4-ft-diameter concrete pipe transporting water at 60°F. What flow rate can be accommodated?

7.67. What size plastic tubing should be selected if 0.002 m³/s of fluid is to be transported such that the pressure drop does not exceed 200 kPa in a 100-m horizontal section? The fluid is:
(a) Water at 20°C.
(b) Glycerine at 60°C.
(c) Kerosene at 20°C.
(d) SAE-10W oil at 40°C.

7.68. Select a concrete pipe size that will transport 5 m³/s of 20°C water so that the head loss does not exceed 20 m in a 300-m horizontal pipe section.

7.69. Atmospheric air at 30°C is to be transported through a square sheet metal conduit at the rate of 4 m³/s. What should be the conduit dimensions so that the head loss does not exceed 10 m over a horizontal length of 200 m?

7.70. Water at 20°C is transported through a 2 cm ×
4 cm copper conduit and experiences a pressure drop
of 80 Pa over a 2-m horizontal length. What is the flow
rate?

7.71. A 4 cm × 10 cm plastic conduit transports water
at 20°C. If a pressure drop of 100 Pa is measured by
gages spaced 5 m apart on a horizontal section, find
the flow rate.

Minor Losses

7.72. If $Q = 0.02$ m³/s of air at 20°C and $p_1 = 50$ kPa,
find p_2.

7.75. The flow rate is measured to be 0.004 m³/s from
the pipe shown. Find the loss coefficient of the valve.
Neglect wall friction.

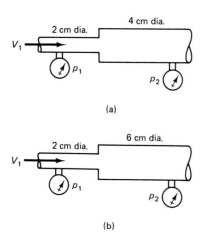

(a)

(b)

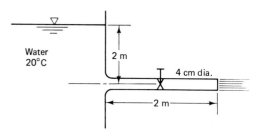

7.73. Replace the sudden enlargement above with a
20° expansion angle and rework Problem 7.72.

7.74. Estimate the loss coefficient based on V_2 using
the data of Fig. 7.16.

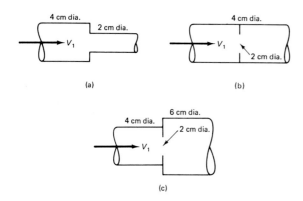

(a)

(b)

(c)

7.75E. The flow rate is measured to be 0.12 ft³/sec
from the pipe shown. Find the loss coefficient of the
valve. Neglect wall friction.

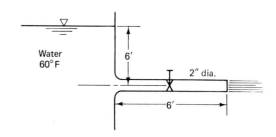

7.76. The flow rate is measured to be 0.006 m³/s in the
pipe shown. Find the loss coefficient of the valve if
H is

(a) 4 cm.

(b) 8 cm.

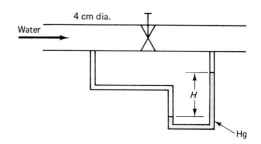

Simple Piping Systems

7.77. Find the flow rate from the pipe shown. Sketch the EGL and the HGL.

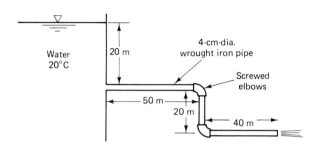

7.78. Water at 20°C flows from a 10-cm-diameter 400-m length of cast iron horizontal pipe that is attached to a reservoir with a square-edged entrance. A screwed globe valve that controls the flow is half open. Find the flow rate if the reservoir elevation above the exit is:

(a) 5 m.

(b) 10 m.

(c) 20 m.

7.78E. Water at 70°F flows from a 4-in.-diameter 1200-ft length of cast iron horizontal pipe that is attached to a reservoir with a square-edged entrance. A screwed globe valve that controls the flow is half open. Find the flow rate if the reservoir elevation above the pipe exit is:

(a) 15 ft.

(b) 30 ft.

(c) 60 ft.

7.79. Estimate the flow rate to be expected through the plastic siphon shown if the diameter is:

(a) 4 cm.

(b) 8 cm.

(c) 12 cm.

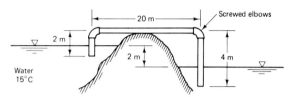

7.80. (a) What flow rate exists through the cast iron pipe shown if $H = 20$ m?

(b) What is the maximum pressure in the piping system?

(c) Sketch the HGL and the EGL.

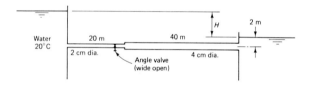

7.81. What is the maximum flow rate through the pipe shown if the elevation difference of the reservoir surfaces is:

(a) 80 m?

(b) 150 m?

(c) 200 m?

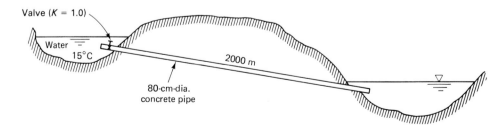

7.82. Water at 20°C is to be pumped through 300 m of cast iron pipe from a reservoir to a device that is 10 m above the reservoir surface. It is to enter the device at 200 kPa. Screwed components include two elbows, a square-edged entrance, and an angle valve. If the flow rate is to be 0.02 m³/s, what pump power is needed (assume 80% efficiency) if the pipe diameter is:
(a) 4 cm?
(b) 8 cm?
(c) 12 cm?

7.82E. Water at 60°F is to be pumped through 900 ft of cast iron pipe from a reservoir to a device that is 30 ft above the reservoir surface. It is to enter the device at 30 psi. Screwed components include two elbows, a square-edged entrance, and an angle valve. If the flow rate is to be 0.6 ft³/sec, what pump power is needed (assume 80% efficiency) if the pipe diameter is:
(a) 1.5 in.?
(b) 3 in.?
(c) 4.5 in.?

7.83. What pump power (85% efficient) is needed for a flow rate of 0.01 m³/s in the pipe shown? What is the greatest distance from the left reservoir that the pump can be located?

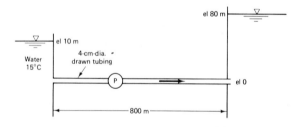

7.84. A flow rate of 2 m³/s exists in the pipe. What is the expected power output of the turbine (85% efficient) if the elevation difference of the reservoir surfaces is:
(a) 20 m?
(b) 60 m?
(c) 100 m?

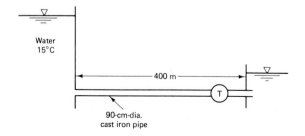

7.85. What pump power (75% efficient) is needed in the piping system shown? What is the greatest distance from the reservoir that the pump can be located?

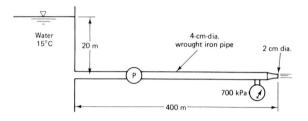

7.85E. What pump power (75% efficient) is needed in the piping system shown? What is the greatest distance from the reservoir that the pump can be located?

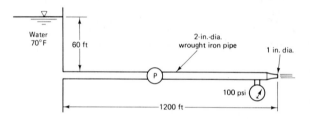

7.86. A pump has the characteristic curves shown in Example 7.16. Estimate the flow rate and the power required by the pump. Sketch the EGL and the HGL.

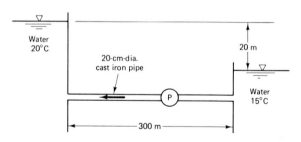

7.87. Reverse the flow direction in Problem 7.86 and redo the problem.

7.88. The pump shown has characteristic curves shown in Example 7.16. Estimate the flow rate and:
(a) Calculate the pump power requirement.
(b) Calculate the pressure at the pump inlet.
(c) Calculate the pressure at the pump outlet.
(d) Sketch the EGL and the HGL.

7.89. Reverse the flow direction in Problem 7.88 and redo the problem.

7.90. A turbine with a characteristic curve shown is inserted in the pipeline. Calculate the turbine power output. Assume that $\eta_T = 0.90$.

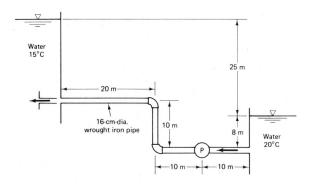

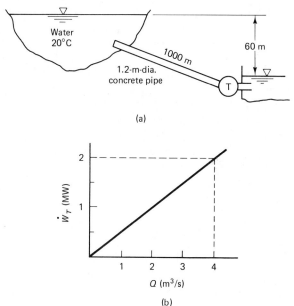

(a)

(b)

Open Channel Flow

7.91. Using a control volume surrounding a finite length of the liquid in a channel flowing at constant depth, find the average shear stress on the walls if water flows in a 3-m-wide rectangular channel at a depth of 2 m on a 0.001 slope.

7.91E. Using a control volume surrounding a finite length of the liquid in a channel flowing at constant depth, find the average shear stress on the walls if water flows in a 10-ft-wide rectangular channel at a depth of 6 ft on a 0.001 slope.

7.92. Using the approach mentioned in Problem 7.91, determine the average shear stress on that portion of the wall of a 40-cm-diameter circular conduit in contact with the water, which is flowing at a constant depth of 10 cm. The slope is 0.0016.

7.93. Calculate the flow rate in a 2-m-wide rectangular, planed wood channel on a slope of 0.001 if the depth is 60 cm.
(a) Use the Chezy–Manning equation.
(b) Use the Darcy–Weisbach equation.

7.94. Water flows in a 2-m-diameter, finished concrete conduit on a 0.0012 slope. Predict the flow rate if the depth of flow is:
(a) 2 m minus a little.
(b) 1.9 m.
(c) 1 m.
(d) 0.5 m.
(e) 0.2 m.

7.94E. Water flows in a 6-ft/diameter, finished concrete conduit on a 0.0012 slope. Predict the flow rate if the depth of flow is:
(a) 6 ft minus a little.
(b) 5.7 ft.
(c) 3 ft.
(d) 1.5 ft.
(e) 0.5 ft.

7.95. Find the flow rate and average velocity if $S = 0.001$ using:
(a) The Chezy–Manning equation.
(b) The Darcy–Weisbach equation.

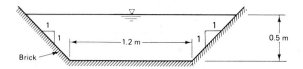

7.96. At what depth will 5 m³/s of water flow in a 2-m-wide, rectangular brick channel with $S = 0.001$?
(a) Use the Chezy–Manning equation.
(b) Use the Darcy–Weisbach equation.

7.97. The cross section of a straight river is approximated as shown. At what depth will 100 m³/s of water flow? The slope is 0.001.

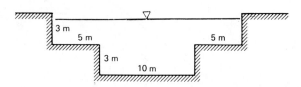

7.98. If $S = 0.0016$, find the flow depth if $Q = 10$ m³/s and the channel is constructed with:
(a) Planed wood.
(b) Brick.

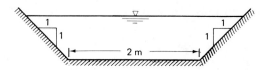

7.99. Water flows in a 120-cm-diameter sewer pipe at a flow rate of 0.8 m³/s. Estimate the depth if the slope is 0.001.

7.99E. Water flows in a 4-ft-diameter sewer pipe at a flow rate of 24 ft³/sec. Estimate the depth if the slope is 0.001.

7.100. Water flows in a 80-cm-diameter sewer pipe at a flow rate of 0.2 m³/s. Determine the depth if the slope is 0.001.

EIGHT

External Flows

8.1 INTRODUCTION

The study of external flows is of particular importance to the aeronautical engineer in the analysis of airflow around the various components of an aircraft. In fact, much of the present knowledge of external flows has been obtained from studies motivated by such aerodynamic problems. There is, however, substantial interest by other engineers in external flows; the flow of fluid around turbine blades, automobiles, buildings, athletic stadiums, smokestacks, spray droplets, bridge abutments, submarine pipelines, river sediment, and red blood cells suggest a variety of phenomena that can be understood only from the perspective of external flows.

It is a difficult task to determine the flow field external to a body and the pressure distribution on the surface of a body, even for the simplest geometry. To discuss this subject, consider low-Reynolds-number flows (Re < 5, or so) and high-Reynolds-number flows (Re > 1000, or so). Low-Reynolds-number flows, called **creeping flows** or **Stokes flows,** seldom occur in engineering applications (flow around spray droplets and red blood cells, lubrication in small gaps, and flow in porous media would be exceptions) and are not presented in this book; they are left to the specialist. We will direct our attention to high-Reynolds-number flows only.

High-Reynolds-number flows can be subdivided into three major catagories: (1) incompressible immersed flows involving such objects as automobiles, helicopters, submarines, low-speed aircraft, buildings, and turbine blades; (2) flows of liquids that involve a free surface as experienced by a ship or a bridge abutment; and (3) compressible flows involving high-speed objects ($V > 100$ m/s) such as aircraft, missiles, and bullets. We will focus our attention on the first category of flows and consider cases in which the object is far from a solid boundary or other objects.

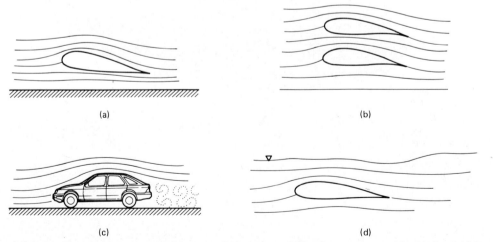

(a)

(b)

(c)

(d)

Figure 8.1   Examples of immersed flows: (a) flow near a solid boundary; (b) flow between two turbine blades; (c) flow around an automobile; (d) flow near a free surface.

The flow becomes significantly influenced by the presence of a boundary or another object, as shown in Fig. 8.1; in part (d) the slender object must be at least five body lengths below the free surface before free surface effects can be neglected. The flows shown in Fig. 8.1 are not included in an introductory presentation.

High-Reynolds-number, incompressible immersed flows are divided into two categories: flows around blunt bodies and flows around streamlined bodies, as displayed in Fig. 8.2. The boundary layer at the stagnation point is a laminar boundary layer, and for a sufficiently large Reynolds number, undergoes transition downstream to a turbulent boundary layer, as shown; the boundary layer may separate from the body and form a **separated region,** a region of recirculating flow, as shown for the blunt body, or it simply leaves the streamlined body at the trailing edge. The **wake,** which is characterized by a **velocity defect** (velocities less than the free-stream velocity) in a growing (diffusion) region, trails the body, as shown. The boundaries of the wake, the separated region, and the turbu-

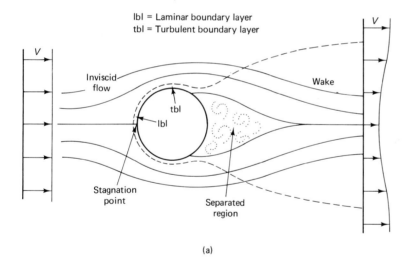

(a)

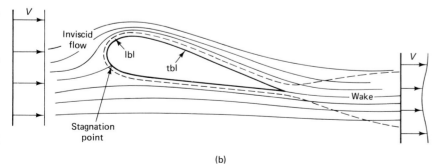

(b)

Figure 8.2   Flow around a blunt body and a streamlined body.

lent boundary layer are quite time dependent; in the sketch the time-average location of the wake is shown by the dashed lines. Shear stresses due to viscosity are concentrated in the thin boundary layer, the separated region, and the wake; outside these regions the flow is approximated by an inviscid flow.

From the figure it could be assumed that the separated region does not exchange mass with the free stream since mass does not cross a streamline. When viewed instantaneously, however, the separation streamline is highly time dependent, and due to this unsteady character, it is able to exchange mass slowly with the free stream.

Several final comments should be made regarding the separated region and the wake. The separated region eventually closes; the wake keeps diffusing into the main flow and eventually disappears as its area becomes exceedingly large (the fluid regains the free-stream velocity). Time-average streamlines do not enter a separated region; they do enter a wake. The separated region is always submerged within the wake.

Flow around a blunt object is usually treated empirically, as was done for a turbulent flow in a conduit. We follow this procedure here. We are interested primarily in the force, the **drag,** the flow exerts on the body in the direction of the flow. The **lift,** which acts normal to the direction of flow, will be of interest for airfoil shapes, as presented in Section 8.4. The actual details of the flow field are seldom of interest and are not presented in an introductory course. We present the drag F_D and lift F_L as dimensionless **drag coefficient** and **lift coefficient,** defined as

$$C_D = \frac{F_D}{\frac{1}{2}\rho V^2 A} \qquad C_L = \frac{F_L}{\frac{1}{2}\rho V^2 A} \qquad (8.1.1)$$

where A is the projected area (projected on a plane normal to the direction of the flow). Drag coefficients for several common shapes are presented in Section 8.3. Since the drag on a blunt object is due primarily to the separated region, there is little interest in studying the boundary layer growth on the front part of a blunt body and the associated viscous shear at the wall. Hence interest is focused on the empirical data that provide the drag coefficient.

Flow around a streamlined body—the separated region is insignificantly small or nonexistent—provides the motivation for detailed study of the laminar and turbulent boundary layers. A boundary layer that develops on a plane streamlined surface, such as an airfoil, is usually sufficiently thin that the curvature of the surface can be ignored and the problem can be treated as a boundary layer developing on a flat plate with a pressure gradient. We will provide a detailed study of flow on a flat plate with zero pressure gradient; once that problem is understood, the influence of a pressure gradient can be discussed. If the flow in the boundary layer on a streamlined body can be determined, the drag can be calculated, since the drag is a result of the shear stress and pressure force acting on the body surface.

Outside the boundary layer there exists an inviscid, **free-stream flow,** as shown in Fig. 8.2. Initially, we will assume that the free-stream flow is known. Before the velocity profile in the boundary layer can be determined, it is necessary that the inviscid flow solution be known. It is found by completely ignoring the boundary layer, since it is so thin, and solving the appropriate inviscid equations. The inviscid flow solution is then used to provide the lift on the body, and the pressure gradient and the velocity at the boundary, two quantities used in the boundary layer flow solution. With the inviscid flow known and the boundary layer flow determined, quantities of interest can be obtained in the flow about a streamlined body.

8.2 SEPARATION

Before we present the empirical data associated with the flow around blunt bodies, the general nature of separation will be discussed. Separation occurs when the main stream flow leaves the body, resulting in a separated region of flow, as sketched in Fig. 8.2. When separation occurs on a streamlined body near the forward portion, as it will with a sufficiently large angle of attack (the angle the oncoming flow makes with the **chord,** a line connecting the trailing edge with the nose), it is referred to as **stall,** as shown in Fig. 8.3. Stall is highly undesirable on aircraft at cruise conditions and leads to inefficiencies when it occurs on turbine blades. It is used, however, to provide the high drag needed when landing an aircraft. For blunt bodies, however, separation is unavoidable at high Reynolds numbers and its effect must be understood.

The location of the separation point is dependent primarily on the geometry of the body; if the body has an abrupt change in geometry, such as that shown in Fig. 8.4, separation will occur at, or near, the abrupt change; however, it will also occur upstream on the flat surface as shown. In addition, reattachment will occur at some location, as shown. Let us

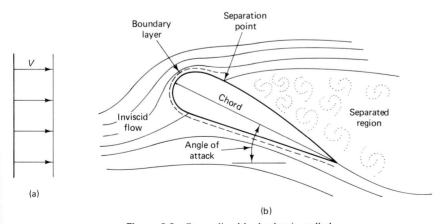

Figure 8.3   Streamlined body that is stalled.

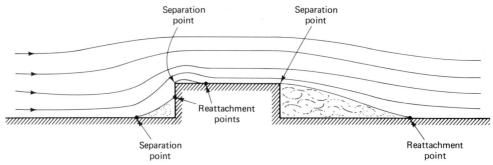

Figure 8.4 Separation due to abrupt geometry changes.

establish the criterion used in predicting the location of the separation
point on a surface with no abrupt geometry change.

Consider the flow on the flat surface just before the step of Fig. 8.4.
The region near the forward separation point is enlarged and shown in
Fig. 8.5; the y-coordinate is normal to the wall and the x-coordinate is
measured along the wall. Downstream of the separation point the x-com-
ponent velocity near the wall is in the negative x-direction and thus at the
wall $\partial u/\partial y$ must be negative. Upstream of the separation point the
x-component velocity near the wall is in the positive x-direction, demanding that $\partial u/\partial y$ at the wall be positive. Hence we conclude that the separation point is defined as that point where $(\partial u/\partial y)_{\text{wall}} = 0$.

Observe that separation on the flat surface occurs as the flow is
approaching a stagnation region where the velocity is low and the pressure is high. As the flow approaches the stagnation region the pressure
increases, that is, $\partial p/\partial x > 0$; the pressure gradient is positive. Since
separation is often undesirable, a positive pressure gradient is called an
adverse pressure gradient; a negative gradient is a **favorable pressure gradient.** In general, the effect of an adverse pressure gradient results in
decreasing velocities in the streamwise direction; if an adverse pressure

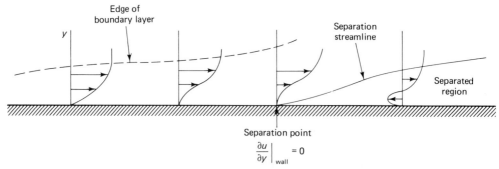

Figure 8.5 Flow separation on a flat surface due to an adverse pressure gradient.

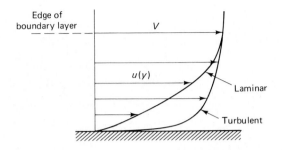

Figure 8.6 Comparison of laminar and turbulent velocity profiles.

gradient acts on a surface over a sufficient distance, separation may result. This is true even if the surface is a flat plate, such as the wall of a diffuser. See Section 8.6.7 for further discussion.

In addition to the influence of the geometry and pressure gradient, several other parameters influence separation. These include the Reynolds number as a very important parameter, with the wall roughness, the free-stream fluctuation level (the intensity of the disturbances that exist away from the boundary), and the wall temperature having less, but occasionally, significant influence.[1] Visualize, for example, flow around a sphere; at sufficiently low Reynolds numbers no separation will occur. As the Reynolds number increases to a particular value, separation will occur over a small area at the rear; this area will become larger and larger as the Reynolds number increases until at a sufficiently large Reynolds number no additional increase in separation area will be observed. The boundary layer before separation will still be laminar. An interesting phenomenon takes place as the boundary layer before separation becomes turbulent; there is a sudden movement of the separation point to the rear of the sphere, resulting in a substantial reduction in the separation area and thus the drag. This phenomenon is explained by comparing the velocity profile of a laminar boundary layer with that of a turbulent boundary, as displayed in Fig. 8.6. As was true in a pipe flow, the turbulent profile has a much larger gradient near the wall (much larger wall shear stress) and thus the momentum of the fluid near the wall is substantially larger in the turbulent boundary layer. For a given geometry a greater distance is required to reduce the velocity near the wall to zero, resulting in the movement of the separation point to the rear, as is quite obvious in Fig. 8.7. In Fig. 8.7a it is observed that separation occurs on the front half of the sphere, in a region of favorable pressure gradient. This separation is due to the centrifugal effects as the fluid moves around the sphere. This phenomenon is observed in the sudden drop in the drag coefficient curves for a sphere and a cylinder, presented in the following section.

[1] If a fluid is flowing past a body that is rigidly supported, the vibration level of the support system will also influence the separation phenomenon. External sound waves can also be significant.

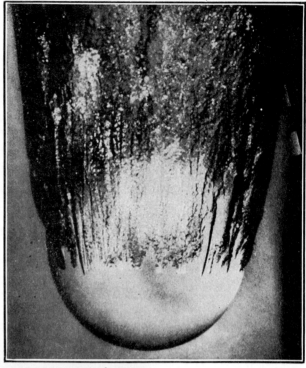

(a)

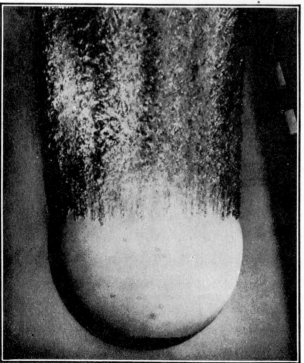

(b)

Figure 8.7 Effect of boundary layer transition on separation: (a) laminar boundary layer before separation; (b) turbulent boundary layer before separation. (U.S. Navy photographs.)

8.3 DRAG CHARACTERISTICS FOR VARIOUS IMMERSED BODIES

8.3.1 Drag Coefficients

From our study of dimensional analysis we know that for a steady, incompressible flow in which gravity, thermal, and surface tension effects are neglected, the primary flow parameter that influences the flow is the Reynolds number; other occasionally important parameters include the relative wall roughness and the free-stream velocity fluctuation intensity[2].

The drag coefficient curves for two bodies that do not exhibit sudden geometric changes will be presented; the drag coefficients for the smooth sphere and the smooth cylinder are shown in Fig. 8.8 over the full range of Reynolds numbers. At Re < 1 creeping flow with no separation results. For the sphere, this creeping flow problem has been solved, with the result that

$$C_D = \frac{24}{\text{Re}} \qquad \text{Re} < 1 \qquad (8.3.1)$$

Separation is observed at Re ≃ 10 over a very small area on the rear of the body. The separated area increases as the Reynolds number increases until Re ≃ 1000, where the separated region ceases to enlarge; during this growth of the separated region the drag coefficient decreases. At Re = 1000, 95% of the drag is due to form drag (the drag force due to the pressure acting on the body) and 5% is due to frictional drag (the drag force due to the shear stresses acting on the body).

The drag coefficient curve is relatively flat over the range $10^3 <$ Re $< 2 \times 10^5$. The boundary layer before the point of separation is laminar and the separated region is as shown in Fig. 8.7a. At Re ≃ 2×10^5, for a smooth surface and with low free-stream fluctuation intensity,[2] the boundary layer before separation undergoes transition to a turbulent state and the increased momentum in the boundary layer "pushes" the separation back, as shown in Fig. 8.7b, with a substantial decrease (a 60 to 80% drop) in the drag. If the surface is rough (dimples on a golf ball) or the free stream has high free-stream fluctuation intensity, the drop in the C_D curve may occur at Re ≃ 8×10^4. Since a lower drag is usually desirable, surface roughness is often added; the dimples on the golf ball may increase the flight distance by 50 to 100%.

After the sudden drop in drag, the C_D curve is observed to again increase with increased Reynolds number. Experimental data are not readily available for Re > 10^6 for a sphere and Re > 10^7 for a cylinder;

---

[2]Free-stream fluctuation intensity is defined as $\sqrt{\overline{u'^2}}/V$. A value of 0.001 is quite low and 0.1 is quite high.

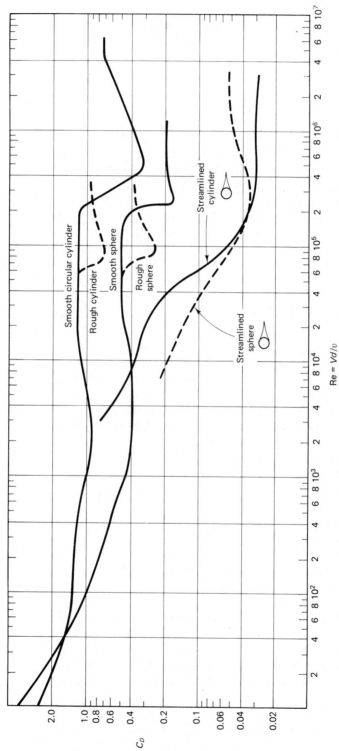

Figure 8.8 Drag coefficients for flow around a cylinder and a sphere. (See E. Achenbach, *J. of Fluid Mechanics*, V. 46, 1971, and V. 54, 1972.)

TABLE 8.1   DRAG COEFFICIENTS OF FINITE-LENGTH
CIRCULAR CYLINDERS[a] AND OF INFINITE-LENGTH
ELLIPTIC CYLINDERS

Circular cylinder		Elliptic cylinder[b]		
Length $\dfrac{\text{Length}}{\text{Diameter}}$	$\dfrac{C_D}{C_{D_x}}$	Major axis $\dfrac{\text{Major axis}}{\text{Minor axis}}$	Re	C_D
∞	1	2	4×10^4	0.6
40	0.82	4	10^5	0.46
20	0.76	4	2.5×10^4 to 10^5	0.32
10	0.68	8	2.5×10^4	0.29
5	0.62	8	2×10^5	0.20
3	0.62			
2	0.57			
1	0.53			

[a] C_{D_x} is the drag coefficient for the infinite-length circular cylinder
obtained in Fig. 8.8.

[b] Flow is in the direction of the major axis.

however, a value of $C_D = 0.2$ for a sphere at large Reynolds number
appears acceptable.

For cylinders of finite length and for elliptic cylinders, the drag
coefficients are presented in Table 8.1. Blunt objects with sudden geometry changes have separated regions that are relatively insensitive to the

TABLE 8.2   DRAG COEFFICIENTS FOR VARIOUS BLUNT
OBJECTS

Object	L/w	Re	C_D
Square cylinder → $\square w$ $\quad w$	∞	$>10^4$	2.0
Rectangular plates → $\mid w$	∞	$>10^3$	2.0
	20	$>10^3$	1.5
	5	$>10^3$	1.2
	1	$>10^3$	1.1
Automobile			
1920	—	$>10^5$	0.9
Modern	—	$>10^5$	0.35
Van	—	$>10^5$	0.42
Circular cylinder → $\square w = D$ $\quad L$	0	$>10^3$	1.10
	4	$>10^3$	0.90
	7	$>10^3$	1.0
Equilateral cylinders → $\triangleleft$	2.0	$>10^4$	2.0
→ $\triangleright$	1	$>10^4$	1.4
Cone → $\triangleleft$ $60°$		$>10^4$	0.5
Parachute		$>10^7$	1.2

Reynolds number; the drag coefficients for some common shapes are given in Table 8.2.

EXAMPLE 8.1

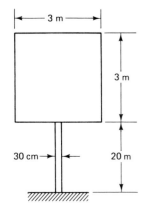

A square sign, 3 m × 3 m, is attached to the top of a 20-m-high pole that is 30 cm in diameter. Approximate the moment that must be resisted by the base for a wind speed of 30 m/s. (We use $p = 1.20$ kg/m³ unless otherwise given.)

Solution

The force F_1 acting on the sign is

$$F_1 = C_D \times \tfrac{1}{2}\rho V^2 A$$
$$= 1.1 \times \tfrac{1}{2} \times 1.20 \times 30^2 \times 3^2 = 5350 \text{ N}$$

The force F_2 acting on the cylindrical pole is (using the projected area as $A = 20 \times 0.3$ m²)

$$F_2 = C_D \times \tfrac{1}{2}\rho V^2 A$$
$$= 0.8 \times \tfrac{1}{2} \times 1.20 \times 30^2 \times (20 \times 0.3) = 2600 \text{ N}$$

where C_D is found from Fig. 8.8 with Re $= 30 \times 0.3/1.6 \times 10^{-5} = 5.63 \times 10^5$, assuming a high-intensity fluctuation level (i.e., a rough cylinder).

The resisting moment that must be supplied by the supporting base is

$$M = d_1 F_1 + d_2 F_2$$
$$= 21.5 \times 5350 + 10 \times 2600 = 141\,000 \text{ N} \cdot \text{m}$$

assuming the forces act at the centers of their respective areas.

EXAMPLE 8.1 (English)

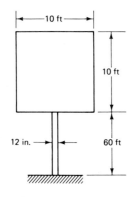

A square sign, 10 ft × 10 ft, is attached to the top of a 60-ft-high pole which is 12 in. in diameter. Approximate the moment that must be resisted by the base for a wind speed of 100 ft/sec.

Solution

The force F_1 acting on the sign is

$$F_1 = C_D \times \tfrac{1}{2}\rho V^2 A$$
$$= 1.1 \times \tfrac{1}{2} \times 0.0024 \times 100^2 \times 10^2 = 1320 \text{ lb}$$

The force F_2 acting on the cylindrical pole is (using the projected area as $A = 60 \times 1 \text{ ft}^2$)

$$F_2 = C_D \times \tfrac{1}{2}\rho V^2 A$$

$$= 0.8 \times \tfrac{1}{2} \times 0.0024 \times 100^2 \times 60 = 576 \text{ lb}$$

where C_D is found from Fig. 8.8 with Re $= 100 \times 1/1.6 \times 10^{-4} = 6.2 \times 10^5$, assuming a high-intensity fluctuation level (i.e., a rough cylinder).

The resisting moment that must be supplied by the supporting base is

$$M = d_1 F_1 + d_2 F_2$$

$$= 65 \times 1320 + 30 \times 576 = 103,000 \text{ ft-lb}$$

assuming that the forces act at the centers of their respective areas.

EXAMPLE 8.2

Determine the terminal velocity of a 30-cm-diameter smooth sphere ($S = 1.02$) if it is dropped in (a) air at 20°C, and (b) water at 20°C.

Solution

(a) When terminal velocity is reached by a falling object, the weight of the object is balanced by the drag force acting on the object. Using $\Sigma F = 0$ and Eq. 8.1.1, we have

$$W = \text{drag}$$

$$\therefore \quad \gamma_{\text{sphere}} \times \text{volume} = C_D \times \tfrac{1}{2}\rho V^2 A$$

The volume of a sphere is $\tfrac{4}{3}\pi R^3$ and $\gamma_{\text{sphere}} = S\gamma_{\text{water}}$; the projected area is $A = \pi R^2$. Thus

$$S\gamma_{\text{water}} \times \tfrac{4}{3}\pi R^3 = C_D \times \tfrac{1}{2}\rho V^2 \pi R^2$$

The velocity is expressed as

$$V = \left(\frac{8RS\gamma_{\text{water}}}{3\rho C_D}\right)^{1/2}$$

$$= \left(\frac{8 \times 0.15 \times 1.02 \times 9800}{3 \times 1.20 \times C_D}\right)^{1/2} = \frac{57.7}{\sqrt{C_D}}$$

Let us assume that the Reynolds number will be quite large and use C_D from Fig. 8.8 as $C_D = 0.2$. Then

$$V = \frac{57.7}{\sqrt{0.2}} = 129 \text{ m/s}$$

We must check the Reynolds number to verify the C_D value assumed. It is

$$\text{Re} = \frac{VD}{\nu}$$

$$= \frac{129 \times 0.3}{1.6 \times 10^{-5}} = 2.42 \times 10^6$$

This is beyond the end of the curve where data are unavailable; we will assume that the drag coefficient is unchanged at 0.2, so that the terminal velocity is 129 m/s.

(b) For the sphere falling in water, we must include the buoyancy force B acting in the same direction as the drag force. Hence the summation of forces yields

$$W = \text{drag} + B$$

$$\therefore \quad \gamma_{\text{sphere}} \times \text{volume} = C_D \times \tfrac{1}{2}\rho V^2 A + \gamma_{\text{water}} \times \text{volume}$$

This gives

$$(S - 1)\gamma_{\text{water}} \times \tfrac{4}{3}\pi R^3 = C_D \times \tfrac{1}{2}\rho V^2 A$$

Using $\rho = 1000$ kg/m³, there results

$$V = \left(\frac{8R(S - 1)\gamma_{\text{water}}}{3\rho C_D}\right)^{1/2}$$

$$= \left(\frac{8 \times 0.15 \times 0.02 \times 9800}{3 \times 1000 \times C_D}\right)^{1/2} = \frac{0.28}{\sqrt{C_D}}$$

We anticipate the Reynolds number being lower than in part (a), so let's assume that it is in the range $2 \times 10^4 < \text{Re} < 2 \times 10^5$. Then $C_D = 0.5$ and there results

$$V = 0.40 \text{ m/s}$$

This gives a Reynolds number of

$$\text{Re} = \frac{VD}{\nu}$$

$$= \frac{0.40 \times 0.3}{10^{-6}} = 1.2 \times 10^5$$

This is in the required range, so the terminal velocity is expected to be 0.40 m/s. Of course, if the sphere were roughened (sand glued to the surface), the C_D value would be 0.2 and we would have $V = 0.63$ m/s, an increase of 58%.

8.3.2 Vortex Shedding

Long blunt objects, such as circular cylinders, exhibit a particularly interesting phenomenon when placed normal to a fluid flow; vorticies or eddies (regions of circulating fluid) are shed from the object, regularly and alternately from opposite sides, as shown in Fig. 8.9. The resulting flow downstream is often referred to as a **von Kármán vortex street.** It occurs in the Reynolds number range $40 <$ Re $< 10\ 000$, and is accompanied by turbulence above Re $= 300$. Photographs of low- and high-Reynolds-number vortex shedding are presented in Fig. 8.10.

Dimensional analysis may be applied to find an expression for the shedding frequency. For high-Reynolds-number flows, that is, flows with insignificant viscous forces, the shedding frequency depends only on the

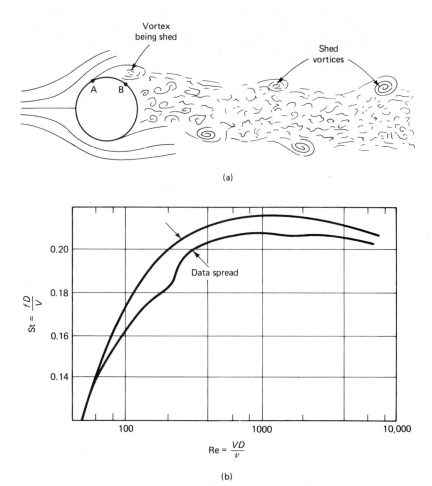

(a)

(b)

Figure 8.9   Vortex shedding from a cylinder: (a) vortex shedding; (b) Strouhal number versus Reynolds number. (From NACA Rep. 1191, by A. Roshko, 1954.)

(a)

Figure 8.10 Vortex shedding at high and low Reynolds numbers: (a) Re = 10 000 (photograph by Thomas Corke and Hassan Nagib); (b) Re = 140 (photograph by Sadatoshi Taneda).

(b)

velocity and diameter. Thus $f = f(V, D)$. Using dimensional analysis we can show that $fD/V = $ const. The shedding frequency, expressed as a dimensionless quantity, is expressed as the Strouhal number,

$$\text{St} = \frac{fD}{V}. \tag{8.3.2}$$

From the experimental results of Fig. 8.9, we observe that the Strouhal number is essentially constant (0.21) over the range $400 < \text{Re} < 10\,000$; hence, the frequency is directly proportional to the velocity over this relatively large Reynolds number range.

The engineer or architect must be very careful when designing structures, such as towers and bridges, that shed vortices. When a vortex is shed, a small force is applied to the structure; if the frequency of shedding is close to the natural frequency[3] (or one of the harmonics) of the

[3]The natural frequency is that frequency with which a structure vibrates when given a "knock."

structure, the phenomenon of resonance may occur in which the response to the applied force is multiplied by a large factor. For example, when resonance occurs on a television tower, the deflection of the tower due to the applied force may become so large that the supporting cables fail, leading to collapse of the structure. This has occurred many times, leading to severe damage and numerous deaths and injuries. The collapse of the Tacoma Narrows suspension bridge is undoubtedly the most spectacular failure due to vortex shedding. Galloping power lines, in which a power line alternates between the usual catenary and an inverted catenary, is another example that may lead to significant damage.

EXAMPLE 8.3

The velocity of a slow-moving, 30°C air stream is to be measured using a cylinder and a pressure tap located between points A and B on the cylinder in Fig. 8.9. The velocity range is expected to be $0.1 < V < 1$ m/s. What size cylinder should be selected and what frequency would be observed by the pressure-measuring device for $V = 1$ m/s?

Solution

The Reynolds number should be in the vortex shedding range, say 4000. For the maximum velocity the diameter would be found as follows:

$$4000 = \frac{VD}{\nu}$$

$$= \frac{1.0 \times D}{1.6 \times 10^{-5}}$$

$$\therefore \quad D = 0.064 \text{ m} \quad \text{or} \quad 6.4 \text{ cm}$$

At $V = 0.1$ m/s the Reynolds number is $0.1 \times 0.064/1.6 \times 10^{-5} = 400$. Vortex shedding would occur, so this is acceptable. The expected vortex shedding frequency at $V = 1.0$ m/s is found using a Strouhal number from Fig. 8.9 of 0.21. Hence

$$0.21 = \frac{fD}{V}$$

$$= \frac{f \times 0.064}{1.0}$$

$$\therefore \quad f = 3.28 \text{ hertz}$$

8.3.3 Streamlining

If the flow is to remain attached to the surface of a blunt object, such as a cylinder or a sphere, it must move into regions of higher and higher pressure as it progresses to the rear stagnation point. At sufficiently high Reynolds numbers (Re > 10) the slow-moving boundary layer flow near the surface is unable to make its way to the high-pressure region near the rear stagnation point, so it separates from the object. **Streamlining** reduces the high pressure at the rear of the object so that the slow-moving flow near the surface is able to negotiate its way into a slightly higher pressure region. The fluid may not be able to make it all the way to the trailing edge of the streamlined object, but the separation region will be reduced to only a small percentage of the initial separated region on the blunt object. The included angle at the trailing edge must not be greater than about 20° or the separation region will be too large and the effect of streamlining will be negated.

When a body is streamlined, the surface area is increased substantially. This eliminates the majority of the pressure drag but increases the shear drag on the surface. To minimize drag, the idea is to minimize the sum of the pressure drag and the shear drag. Consequently, the streamlined body cannot be so long that the shear drag is larger than the pressure drag plus the shear drag for a shorter body. An optimization procedure is required. Such a procedure would lead to a thickness-to-chord length ratio of about 0.25 for a strut.

Obviously, for a low-Reynolds-number flow (Re < 10) the drag is due primarily to shear drag and thus streamlining is unnecessary; it would undoubtedly lead to increased drag since the surface area would be increased.

Finally, it should be pointed out that another advantage of streamlining is that the periodic shedding of vortices is usually eliminated. The vibrations produced by vortex shedding are often undesirable, so streamlining not only decreases drag but can eliminate the vibrations.

EXAMPLE 8.4

A strut on a stunt plane traveling at 60 m/s is 4 cm in diameter and 24 cm long. Calculate the drag force acting on the strut as a cylinder and as a streamlined strut. Would you expect vortex shedding from the circular cylinder?

Solution

The Reynolds number associated with the cylinder and the streamlined strut is, assuming that $T = 20°C$,

$$\text{Re} = \frac{VD}{\nu}$$

$$= \frac{60 \times 0.04}{1.5 \times 10^{-5}} = 1.6 \times 10^{5}$$

Assuming a smooth surface, the drag coefficient is $C_D = 1.2$ from Fig. 8.8. The drag force is then

$$F_D = C_D \times \tfrac{1}{2}\rho V^2 A$$

$$= 1.2 \times \tfrac{1}{2} \times 1.20 \times 60^2 \times (0.24 \times 0.04) = 24.9 \text{ N}$$

For the streamlined strut, Fig. 8.8 yields $C_D = 0.04$. The drag force is

$$F_D = C_D \times \tfrac{1}{2}\rho V^2 A$$

$$= 0.04 \times \tfrac{1}{2} \times 1.20 \times 60^2 \times (0.24 \times 0.04) = 0.82 \text{ N}$$

This is a reduction of 97% in the drag, a rather substantial percentage.

Vortex shedding is not to be expected on the circular cylinder; the Reynolds number is too high.

8.3.4 Cavitation

Cavitation is a very rapid change of phase from liquid to vapor which occurs in a liquid whenever the local pressure is equal to or less than the vapor pressure. The first appearance of cavitation is at the position of lowest pressure in the field of flow. Four types of cavitation have been identified:

1. **Traveling cavitation,** which exists when vapor bubbles or cavities are formed, are swept downstream, and collapse.
2. **Fixed cavitation,** which exists when a fixed cavity of vapor exists as a separated region. The separated region may reattach to the body or the separated region may enclose the rear of the body and be closed by the main flow, in which case it is referred to as **supercavitation.**
3. **Vortex cavitation,** which is found in the high-velocity, and thus low pressure, core of a vortex, often observed in the tip vortex leaving a propeller.
4. **Vibratory cavitation,** which may exist when a pressure wave moves in a liquid. A pressure wave consists of a pressure pulse, which has a high pressure followed by a low-pressure. The low-pressure part of the wave (or vibration) can result in cavitation.

The first type of cavitation, in which vapor bubbles are formed and collapse, is associated with potential damage. The pressures resulting from the collapse can be extremely high (perhaps 1400 MPa) and may cause damage to stainless steel components.

TABLE 8.3   DRAG COEFFICIENTS FOR ZERO CAVITATION NUMBER
FOR BLUNT OBJECTS

Two-dimensional body				Axisymmetric body		
Geometry		θ	$C_D(0)$	Geometry	θ	$C_D(0)$
Flat plate		—	0.88	Disk	—	0.8
Circular cylinder		—	0.50	Sphere	—	0.30
Wedge		120	0.74	Cone	120	0.64
		90	0.64		90	0.52
		60	0.49		60	0.38
		30	0.28		30	0.20

Cavitation occurs whenever the **cavitation number** σ, defined by

$$\sigma = \frac{p_\infty - p_v}{\frac{1}{2}\rho V^2} \tag{8.3.3}$$

is less than the critical cavitation number σ_{crit}, which depends on the
geometry of the body and the Reynolds number. Here p_∞ is the absolute
pressure in the undisturbed free stream and p_v is the vapor pressure. As σ
decreases below σ_{crit}, the cavitation increases in intensity, moving from
traveling cavitation to fixed cavitation to supercavitation.

    The drag coefficient of a body is dependent on the cavitation number
and for small cavitation numbers is given by

$$C_D(\sigma) = C_D(0)(1 + \sigma) \tag{8.3.4}$$

TABLE 8.4   DRAG AND LIFT COEFFICIENTS
AND CRITICAL CAVITATION NUMBER FOR
A TYPICAL HYDROFOIL

Angle (°)	Lift coefficient C_L	Drag coefficient C_D	Critical cavitation number σ_{crit}
−2	0.2	0.014	0.5
0	0.4	0.014	0.6
2	0.6	0.015	0.7
4	0.8	0.018	0.8
6	0.95	0.022	1.2
8	1.10	0.03	1.8
10	1.22	0.04	2.5

where some values of $C_D(0)$ for common shapes are listed in Table 8.3 for Re $\simeq 10^5$.

The hydrofoil, an airfoil-type body that is used to lift a vessel out of the water, is a shape that is invariably associated with cavitation. Drag and lift coefficients and critical cavitation numbers are given in Table 8.4 for a typical hydrofoil with $10^5 < \text{Re} < 10^6$, where the Reynolds number is based on the chord length and the area used with C_D and C_L is the chord times the length.

EXAMPLE 8.5

A hydrofoil is to operate 50 cm below the surface of 20°C water at an angle of attack of 8° and travel at 15 m/s. If its chord length is 60 cm and it is 2 m long, calculate its lift and drag. Is cavitation present?

Solution

The absolute pressure p_∞ is

$$p_\infty = \gamma h + p_{atm}$$

$$= 9800 \times 0.5 + 100\ 000 = 104\ 900 \text{ Pa absolute}$$

The vapor pressure is $p_v = 2540$ Pa absolute, so

$$\sigma = \frac{p_\infty - p_v}{\frac{1}{2}\rho V^2}$$

$$= \frac{104\ 900 - 2450}{\frac{1}{2} \times 1000 \times 15^2} = 0.91$$

Answering the last question first, we see that this is less than 1.8, hence cavitation exists.

The lift force is

$$F_L = C_L \times \tfrac{1}{2}\rho V^2 A$$

$$= 1.1 \times \tfrac{1}{2} \times 1000 \times 15^2 \times (0.6 \times 2) = 148\ 500 \text{ N}$$

The drag force is

$$F_D = C_D \times \tfrac{1}{2}\rho V^2 A$$

$$= 0.03 \times \tfrac{1}{2} \times 1000 \times 15^2 \times (0.6 \times 2) = 4050 \text{ N}$$

EXAMPLE 8.5 (English)

A hydrofoil is to operate 20 in. below the surface of 60°F water at an angle of attack of 8° and travel at 45 ft/sec. If its chord length is 24 in. and it is 6 ft long, calculate its lift and drag. Is cavitation present?

Solution

The absolute pressure p_∞ is

$$p_\infty = \gamma h + p_{atm}$$

$$= 62.4 \times \frac{20}{12} + 2117 = 2221 \text{ psf absolute}$$

The vapor pressure is $p_v = 0.256$ psia, so

$$\sigma = \frac{p_\infty - p_v}{\frac{1}{2}\rho V^2}$$

$$= \frac{2221 - 0.256 \times 144}{\frac{1}{2} \times 1.94 \times 45^2} = 1.11$$

Answering the last question first, we see that this is less than 1.8, hence cavitation exists.

The lift force is

$$F_L = C_L \times \tfrac{1}{2}\rho V^2 A$$

$$= 1.1 \times \tfrac{1}{2} \times 1.94 \times 45^2 \times \left(\frac{24}{12} \times 6\right) = 25{,}900 \text{ lb}$$

The drag force is

$$F_D = C_D \times \tfrac{1}{2}\rho V^2 A$$

$$= 0.03 \times \tfrac{1}{2} \times 1.94 \times 45^2 \times (2 \times 6) = 707 \text{ lb}$$

8.3.5 Added Mass

The previous sections in this chapter have dealt with bodies moving at constant velocity. In this section we consider bodies accelerating from rest in a fluid. When a body accelerates, we say that an unbalanced force acts on the body; not only does the body accelerate but so does some of the fluid surrounding the body. The acceleration of the surrounding fluid requires an added force over and above the force required to accelerate only the body. A relatively simple way of accounting for the fluid mass being accelerated is to add a mass, called the **added mass** m_a, to the mass of the body. Summing forces in the direction of motion for a symmetrical body moving in the direction of its axis of symmetry we have, for horizontal motion,

$$F - F_D = (m + m_a)\frac{dV_B}{dt} \qquad (8.3.5)$$

where V_B is the velocity of the body and F_D is the drag force. For an initial acceleration from rest F_D would be zero.

The added mass is related to the mass of the fluid m_f displaced by the body by the relationship

$$m_a = km_f$$

where k is an added mass coefficient. For a sphere $k = 0.5$; for an ellipsoid with major axis twice the minor axis and moving in the direction of the major axis, $k = 0.2$; for a long cylinder moving normal to its axis, $k = 1.0$. These values were calculated for inviscid flows and thus are applicable for motions starting from rest so that the viscous forces are negligible.

For dense bodies accelerating in the atmosphere the added mass is negligibly small and is typically ignored. Masses accelerating from rest in a liquid are more influenced by the added mass and it must usually be accounted for. Offshore structures that are subjected to oscillatory wave motions experience periodic forces the determination of which must include the added mass effect.

EXAMPLE 8.6

A sphere with specific gravity 2.5 is released from rest in water. Calculate its initial acceleration. What is the percentage error if the added mass is ignored?

Solution

The summation of forces in the vertical direction, with zero drag, is

$$W - B = (m + m_a)\frac{dV_B}{dt}$$

where B is the buoyant force. Substituting in the appropriate quantities gives, letting V = sphere volume,

$$S\gamma_{H_2O}V - \gamma_{H_2O}V = (\rho_{H_2O}SV + 0.5\rho_{H_2O}V)\frac{dV_B}{dt}$$

This gives

$$g(S - 1) = (S + 0.5)\frac{dV_B}{dt}$$

Hence

$$\frac{dV_B}{dt} = \frac{g(S - 1)}{S + 0.5} = \frac{9.8(2.5 - 1)}{2.5 + 0.5} = 4.90 \text{ m/s}^2$$

If we ignored the added mass, the acceleration would be

$$\frac{dV_B}{dt} = \frac{g(S-1)}{S} = \frac{9.8(2.5-1)}{2.5} = 5.88 \text{ m/s}^2$$

This is an error of 20%.

8.4 LIFT AND DRAG ON AIRFOILS

Separation occurs on a blunt body, such as a cylinder, due to the strong adverse pressure gradient in the boundary layer on the rear of the body. An airfoil is a streamlined body designed to reduce the adverse pressure gradient so that separation will not occur, usually with a small angle of attack, as shown in Fig. 8.11. Without separation the drag is due primarily to the wall shear stress, which results from viscous effects in the boundary layer.

The boundary layer on an airfoil is very thin, and thus it can be ignored when solving for the flow field (the streamline pattern and the pressure distribution) surrounding the airfoil. Since the boundary layer is so thin, the pressure on the wall is not significantly influenced by the boundary layer's existence. Hence the lift on an airfoil can be approximated by integrating the pressure distribution on the wall as given by the inviscid flow solution. In the next section we will demonstrate how this is done; in this section we simply give empirical results.

The drag on an airfoil can be predicted by solving the boundary layer equations (simplified Navier–Stokes equations) for the shear stress on the wall, and performing the appropriate integration. The inviscid flow field must be known before the boundary layer equations can be solved since the pressure gradient on the wall and the inviscid flow velocity at the wall are needed as inputs in solving for the boundary layer flow. Boundary layer calculations will be presented in Section 8.6; in this section we present the empirical results for the drag.

The drag coefficient presented may seem quite low compared with the coefficients of the preceding section. For airfoils a much larger projected area is used, namely, the plan area, which is the chord c (see Fig.

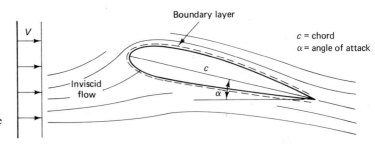

Figure 8.11 Flow around an airfoil at an angle of attack.

8.11) times the length L of the airfoil. Thus the drag and lift coefficients are defined as

$$C_D = \frac{\text{drag}}{\frac{1}{2}\rho V^2 cL} \qquad C_L = \frac{\text{lift}}{\frac{1}{2}\rho V^2 cL} \qquad (8.4.1)$$

For a typical airfoil the lift and drag coefficients are given in Fig. 8.12. For a specially designed airfoil the drag coefficient may be as low as 0.0035, but the maximum lift coefficient decreases to about 1.5. The design lift coefficient (cruise condition) is about 0.3, which is near the minimum drag coefficient condition. This corresponds to an angle of attack of about 2°, far from the stall condition of about 16°.

Conventional airfoils are not symmetric; hence there is a positive lift coefficient for zero angle of attack. The lift is directly proportional to the angle of attack but deviates from the straight-line function just before stall. The drag coefficient, however, is highly nonlinear with angle of attack.

To take off and land at relatively low speeds, it is necessary to attain significantly higher lift coefficients than the maximum of 1.7 of Fig. 8.12. Or if a relatively low lift coefficient is to be accepted, the area $c \times L$ must be enlarged. Both are actually accomplished. Flaps are moved out from a section of each airfoil, resulting in an increased chord, and the angle of

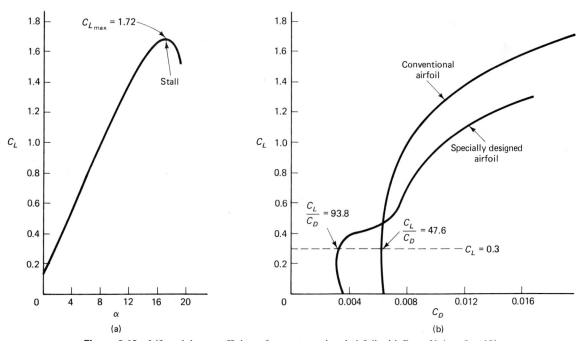

Figure 8.12   Lift and drag coefficients for a conventional airfoil with Re $= Vc/\nu \approx 9 \times 10^6$.

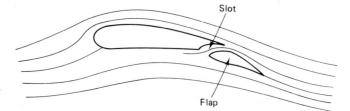

Figure 8.13 Flapped airfoil
with slot for sepa-
ration control.

attack of the flap is also increased. Slots are used to move high-pressure
air from the underside into the relatively low momentum boundary layer
flow on the top side, as shown in Fig. 8.13; this prevents separation from
the flap, thereby maintaining high lift. On some modern aircraft there may
be three flaps in series with three slots, to ensure that the boundary layer
does not separate from the upper surface of the airfoil.

The drag coefficient is essentially constant on airfoils up to a Mach
number of about 0.75. Then a sudden rise occurs until the Mach number
reaches unity; see Fig. 8.14. The drag coefficient then slowly falls. Obviously, the condition of $M = 1$ is to be avoided. Thus aircraft either fly at
$M < 0.75$ or $M > 1.5$ or so, to avoid the high drag coefficients near $M = 1$.

It is useful to use swept-back airfoils since it is the component of
velocity normal to the leading edge of the airfoil that must be used in
calculating the Mach number in Fig. 8.14. Cruise speeds at $M = 0.8$ with
swept-back wings are not uncommon. It should be pointed out, though,
that fuel consumption depends on power required, and power is drag
force times velocity; hence fuel consumption depends on the velocity
cubed, assuming all other parameters are constant. A lower velocity
results in a fuel savings even though the engines must operate longer.

A final comment on airfoils regards the influence of a finite airfoil.
To understand flow around a finite airfoil we make reference to a vortex.
Fluid particles rotate about the center of a vortex as they travel along in
the flow field. There is a high pressure on the bottom and a low pressure
on the top side of the airfoil displayed in Fig. 8.15. This results in a
movement of air from the bottom side to the top side around the ends of
the airfoil, as shown, resulting in a strong tip vortex. Distributed vortices
are also shed all along the airfoil and they all collect into two large trailing

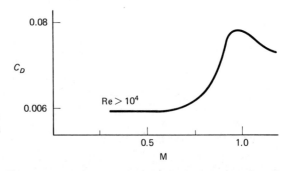

Figure 8.14 Drag coefficient as
a function of Mach
number for an
unswept airfoil.

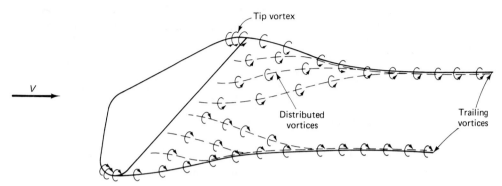

Figure 8.15   Trailing vortex.

vortices. On a clear day, the two trailing vortices may show up as visible white streaks behind a high-flying aircraft. The trailing vortices persist for a considerable distance (perhaps 15 km) behind a large aircraft, and their 90-m/s velocities can cause a small aircraft to flip over. Also, the trailing vortices induce a downwash, a downward velocity component, that must be accounted for in the design of the aircraft. The tail section is located up high to minimize the effect of this downwash.

EXAMPLE 8.7

A light airplane weighs 10 000 N, its wingspan measures 12 m, its chord measures 1.8 m, and a payload of 2000 N is anticipated. Predict (a) the takeoff speed if an angle of attack of 8° is desired, (b) the stall speed, and (c) the power required for cruise at 50 m/s.

Solution

(a) The lift on an airplane is equal to its weight. With the payload the total weight is 12 000 N; hence the lift-coefficient equation (8.4.1) gives the following:

$$V = \left(\frac{\text{lift}}{\frac{1}{2}\rho C_L c L}\right)^{1/2}$$

$$= \left(\frac{12\ 000}{\frac{1}{2} \times 1.20 \times 1.0 \times 1.8 \times 12}\right)^{1/2} = 30.4 \text{ m/s}$$

where we have used $C_L = 1.0$ at $\alpha = 8°$ from Fig. 8.12.

(b) The stall speed is found using a maximum lift coefficient of 1.72 from Fig. 8.12. It is found to be

$$V_{\text{stall}} = \left(\frac{\text{lift}}{\frac{1}{2}\rho C_L c L}\right)^{1/2}$$

$$= \left(\frac{12\ 000}{\frac{1}{2} \times 1.20 \times 1.72 \times 1.8 \times 12}\right)^{1/2} = 23.2 \text{ m/s}$$

(c) The power demanded during cruise is equal to the drag force times the velocity. The design lift coefficient is equal to 0.3 and thus from Fig. 8.12 $C_D = 0.0063$. This gives

$$\text{drag} = \tfrac{1}{2}\rho V^2 cLC_D$$

$$= \tfrac{1}{2} \times 1.20 \times 50^2 \times 1.8 \times 12 \times 0.0063 = 204 \text{ N}$$

The power is then

$$\text{power} = \text{drag} \times V$$

$$= 204 \times 50 = 10\,200 \text{ W} \qquad \text{or 13.7 hp}$$

8.5 POTENTIAL FLOW THEORY

8.5.1 Basic Flow Equations

Inviscid flow exists outside the boundary layer and the wake in high-Reynolds-number flow around bodies. For an airfoil the boundary layer is quite thin, and the inviscid flow provides a good approximation to the actual flow; it is used to predict the pressure distribution on the surface, thereby giving a good estimate of the lift. It will also provide us with the velocity to be used as a boundary condition in the boundary layer solution of Section 8.6; from such a solution we can estimate the drag and predict possible points of separation. Consequently, the inviscid flow solution is very important in our study of external flows. Obviously, if we use the empirical results of previous sections, the details of the inviscid flow solution are unnecessary. If, on the other hand, we wish to predict quantities such as the lift and drag, and locate possible points of separation, the inviscid flow solution is essential.

Consider a velocity field that is given by the gradient of a scalar function ϕ, that is,

$$\mathbf{V} = \nabla\phi \tag{8.5.1}$$

in which ϕ is called a **velocity potential function.** Such a velocity field is called a **potential flow** (or an **irrotational flow**) and possesses the property that the vorticity $\boldsymbol{\omega}$, which is the curl of the velocity vector, is zero; this is expressed by

$$\boldsymbol{\omega} = \nabla \times \mathbf{V} = 0 \tag{8.5.2}$$

The fact that the vorticity is zero for a potential flow can easily be shown by letting $\mathbf{V} = \nabla\phi$ and expanding Eq. 8.5.2 in rectangular coordinates. A fluid particle that does not possess vorticity (i.e., is not rotating) cannot obtain vorticity without the action of viscosity; normal pressure forces and body forces that act through the mass center cannot impart rotation to a fluid particle. This result is also observed from the **vorticity equation,** which is obtained by taking the curl of the Navier–Stokes equation (4.3.17) (a rather difficult task); it is

$$\frac{D\boldsymbol{\omega}}{Dt} = (\boldsymbol{\omega} \cdot \boldsymbol{\nabla})\mathbf{V} + \nu\nabla^2\boldsymbol{\omega} \tag{8.5.3}$$

Note that if $\boldsymbol{\omega} = 0$, the only way that $D\boldsymbol{\omega}/Dt$ can be nonzero is for viscous effects to act through the last term. If viscous effects are absent, as in an invisid flow, then $D\boldsymbol{\omega}/Dt = 0$ and vorticity must remain zero.

With the velocity given by the gradient of a scalar function, the differential continuity equation, for an incompressible flow, gives

$$\boldsymbol{\nabla} \cdot \boldsymbol{\nabla}\phi = V^2\phi = 0 \tag{8.5.4}$$

and is known as **Laplace's equation.** In rectangular coordinates this is

$$\frac{\partial^2\phi}{\partial x^2} + \frac{\partial^2\phi}{\partial y^2} + \frac{\partial^2\phi}{\partial z^2} = 0 \tag{8.5.5}$$

With the appropriate boundary conditions this equation could be solved. However, three-dimensional problems are quite difficult, so we focus our attention on plane flows in which the velocity components u and v depend on x and y. This is acceptable for two-dimensional airfoils and other plane flows, such as flow around cylinders.

Before we attempt a solution to Eq. 8.5.5, let us define another scalar function that will aid us in our study of plane fluid flows. The continuity equation

$$\frac{\partial u}{\partial x} + \frac{\partial v}{\partial y} = 0 \tag{8.5.6}$$

motivates the definition. If we let

$$u = \frac{\partial\psi}{\partial y} \quad \text{and} \quad v = -\frac{\partial\psi}{\partial x} \tag{8.5.7}$$

we observe that the continuity equation is automatically satisfied; the scalar function $\psi(x, y)$ is called a **stream function.** By using the mathematical description of a streamline, $\mathbf{V} \times d\mathbf{r} = 0$, we see that, for a plane flow, $u\, dy - v\, dx = 0$. Substituting from Eqs. 8.5.7, this becomes

$$\frac{\partial\psi}{\partial y}\, dy + \frac{\partial\psi}{\partial x}\, dx = 0 \tag{8.5.8}$$

This is, of course, $d\psi = 0$. Thus ψ is constant along a streamline. Example 8.8 will show that the difference $(\psi_2 - \psi_1)$ between any two streamlines is equal to the flow rate per unit depth between the two streamlines.

The vorticity vector for a plane flow has only a z-component since $w = 0$ and there is no variation with z. The vorticity is

$$\nabla \times \mathbf{V} = \left(\frac{\partial v}{\partial x} - \frac{\partial u}{\partial y}\right)\hat{k} \tag{8.5.9}$$

For our potential flow we demand that the vorticity be zero, so Eq. 8.5.9 gives, using Eqs. 8.5.7,

$$\frac{\partial^2 \psi}{\partial x^2} + \frac{\partial^2 \psi}{\partial y^2} = 0 \tag{8.5.10}$$

So we see that both the stream function ψ and the potential function ϕ satisfy Laplace's equation.

Rather than attempt a solution of Laplace's equation for a particular flow of interest, we will use a different technique; we will identify some relatively simple functions that satisfy Laplace's equation and then superimpose these simple functions to create flows of interest. It is possible to generate any plane flow desired using this technique. Hence we will not actually solve Laplace's equation.

Before we present some simple functions, some additional observations concerning ϕ and ψ will be made. Using Eqs. 8.5.1 and 8.5.7, we see that

$$u = \frac{\partial \phi}{\partial x} = \frac{\partial \psi}{\partial y} \qquad v = \frac{\partial \phi}{\partial y} = -\frac{\partial \psi}{\partial x} \tag{8.5.11}$$

These relationships between the derivatives of ϕ and ψ are the famous **Cauchy–Riemann equations** from the theory of complex variables. The functions ϕ and ψ are harmonic functions since they satisfy Laplace's equation and form an **analytic function** $(\phi + i\psi)$ called the **complex velocity potential.** The theory of complex variables with all its powerful theorems is thus applicable to this restricted class of problems: namely, plane, incompressible, potential flows. An example will show that the streamlines and constant-potential lines intersect each other at right angles.

EXAMPLE 8.8

Show that the difference in the stream function between any two streamlines is equal to the flow rate per unit depth between the two streamlines. The flow rate per unit depth is denoted by q.

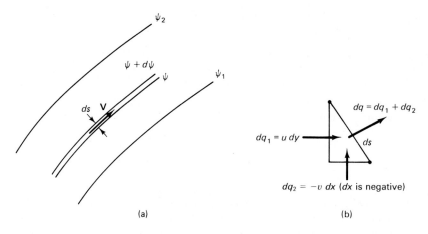

(a)   (b)

Solution

Consider the flow between two streamlines infinitesimally close, as shown in the sketch. The flow rate per unit depth through the elemental area is

$$dq = dq_1 + dq_2$$
$$= u\,dy - v\,dx$$
$$= \frac{\partial \psi}{\partial y}\,dy + \frac{\partial \psi}{\partial x}\,dx = d\psi$$

If this is integrated between two streamlines with $\psi = \psi_1$ and $\psi = \psi_2$, there results

$$q = \psi_2 - \psi_1$$

thereby proving the statement of the example.

EXAMPLE 8.9

Show that the streamlines and equipotential lines of a plane, incompressible, potential flow intersect one another at right angles.

Solution

If, at a point, the slope of a streamline is the negative reciprocal of the slope of an equipotential line, the two lines are perpendicular to each other. The slope of a streamline is given by

$$\left.\frac{dy}{dx}\right|_{\psi=\text{const.}} = \frac{v}{u}$$

The slope of an equipotential line is found from

$$d\phi = \frac{\partial \phi}{\partial x}\,dx + \frac{\partial \phi}{\partial y}\,dy = 0$$

since ϕ = const. along an equipotential line. This gives

$$\frac{dy}{dx}\bigg|_{\phi=\text{const.}} = -\frac{\partial\phi/\partial x}{\partial\phi/\partial y} = -\frac{u}{v}$$

Hence we see that the slope of the streamline is the negative reciprocal of the slope of the equipotential line, that is,

$$\frac{dy}{dx}\bigg|_{\phi=\text{const.}} = -\left(\frac{dy}{dx}\bigg|_{\psi=\text{const.}}\right)^{-1}$$

Thus, whenever the streamlines intersect the equipotential lines, they must do so at right angles. A sketch of the streamlines and equipotential lines (equally spaced at large distances from the body), known as a **flow net,** is shown for flow over a wier.

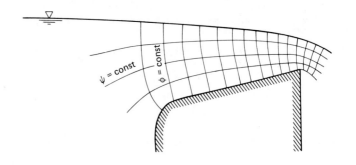

EXAMPLE 8.10

A scalar potential function is given by $\phi = A \tan^{-1}(y/x)$. Find the stream function $\psi(x, y)$.

Solution

The relationship between ϕ and ψ is given by Eq. 8.5.11. We have

$$\frac{\partial\psi}{\partial y} = \frac{\partial\phi}{\partial x} = \frac{\partial}{\partial x}\left(A \tan^{-1}\frac{y}{x}\right) = -\frac{Ay}{x^2 + y^2}$$

This can be integrated as follows:

$$\int \frac{\partial\psi}{\partial y}\, dy = -A \int \frac{y}{x^2 + y^2}\, dy$$

$$\therefore \quad \psi = -\frac{A}{2} \ln(x^2 + y^2) + f(x)$$

A function of x must be added since partial derivatives are being used. Now, let us differentiate this expression with respect to x. There results

$$\frac{\partial \psi}{\partial x} = -A \frac{x}{x^2 + y^2} + \frac{df}{dx}$$

This must equal $(-\partial \phi / \partial y)$; that is,

$$-\frac{Ax}{x^2 + y^2} + \frac{df}{dx} = -\frac{Ax}{x^2 + y^2}$$

Thus

$$\frac{df}{dx} = 0 \quad \text{or} \quad f = C$$

Since ϕ and ψ are used to find the velocity components by differentiation, the constant is of no concern; it is usually set equal to zero. Hence

$$\psi = -\frac{A}{2} \ln(x^2 + y^2)$$

8.5.2 Simple Solutions

Now, let us identify some relatively simple functions that satisfy Laplace's equation. However, before we do, it is often more convenient to use polar coordinates. Laplace's equation, the continuity equation, and the velocity components take the following forms:

$$\nabla^2 \psi = \frac{1}{r} \frac{\partial}{\partial r} \left(r \frac{\partial \psi}{\partial r} \right) + \frac{1}{r^2} \frac{\partial^2 \psi}{\partial \theta^2} = 0 \qquad (8.5.12)$$

$$\frac{1}{r} \frac{\partial}{\partial r} (r v_r) + \frac{1}{r} \frac{\partial v_\theta}{\partial \theta} = 0 \qquad (8.5.13)$$

$$v_r = \frac{1}{r} \frac{\partial \psi}{\partial \theta} = \frac{\partial \phi}{\partial r} \qquad v_\theta = -\frac{\partial \psi}{\partial r} = \frac{1}{r} \frac{\partial \phi}{\partial \theta} \qquad (8.5.14)$$

We will introduce the names of four simple flows, sketched in Fig. 8.16, and their corresponding functions. Each function obviously satisfies Laplace's equation. The names and functions are:

Uniform flow: $\quad \psi = U_\infty y, \qquad \phi = U_\infty x \qquad (8.5.15)$

Line source: $\quad \psi = \dfrac{q}{2\pi} \theta, \qquad \phi = \dfrac{q}{2\pi} \ln r \qquad (8.5.16)$

Irrotational vortex: $\quad \psi = \dfrac{\Gamma}{2\pi} \ln r, \qquad \phi = -\dfrac{\Gamma}{2\pi} \theta \qquad (8.5.17)$

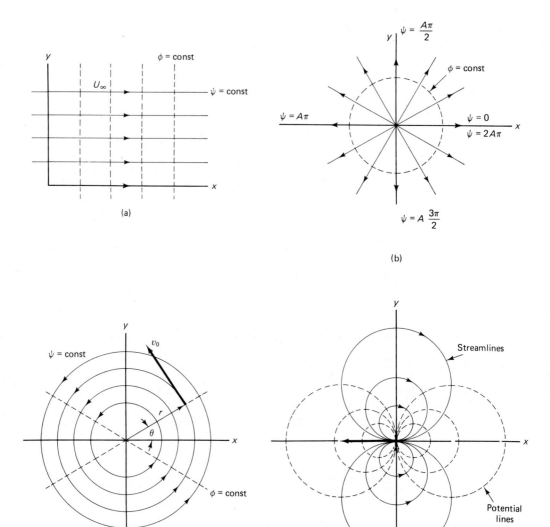

Figure 8.16   Four simple potential flows: (a) uniform flow U_∞; (b) line source q; (c) irrotational vortex Γ; (d) doublet μ.

Doublet:
$$\psi = -\frac{\mu \sin \theta}{r}, \quad \phi = \frac{\mu \cos \theta}{r} \quad (8.5.18)$$

The uniform flow velocity U_∞ is assumed to be in the x-direction; if a y-component is desired an appropriate term is simply added. The **source strength** q is the volume rate of flow per unit depth issuing from the source; a negative value would create a sink. The vortex strength Γ is the

circulation about the origin, defined by

$$\Gamma = \oint_L \mathbf{V} \cdot d\mathbf{s} \tag{8.5.19}$$

where L must be a closed curve (a circle is usually used) around the origin and clockwise is positive. The **doublet strength** μ is for a doublet oriented in the negative x-direction; note the arrow (in Fig. 8.16d) showing the direction of the doublet. Doublets oriented in other directions are seldom of interest and are not considered here.

The velocity components for the four simple flows are easily shown, using Eqs. 8.5.11 and 8.5.14 for rectangular and polar coordinates, to be as follows:

Uniform flow:
$$u = U_\infty, \qquad v = 0$$
$$v_r = U_\infty \cos\theta, \qquad v_\theta = -U_\infty \sin\theta \tag{8.5.20}$$

Line source:
$$v_r = \frac{q}{2\pi r}, \qquad v_\theta = 0$$
$$u = \frac{q}{2\pi}\frac{x}{x^2 + y^2}, \qquad v = \frac{q}{2\pi}\frac{y}{x^2 + y^2} \tag{8.5.21}$$

Irrotational vortex:
$$v_r = 0, \qquad v_\theta = -\frac{\Gamma}{2\pi r}$$
$$u = -\frac{\Gamma}{2\pi}\frac{y}{x^2 + y^2}, \qquad v = \frac{\Gamma}{2\pi}\frac{x}{x^2 + y^2} \tag{8.5.22}$$

Doublet:
$$v_r = -\frac{\mu \cos\theta}{r^2}, \qquad v_\theta = -\frac{\mu \sin\theta}{r^2}$$
$$u = -\mu\frac{x^2 - y^2}{(x^2 + y^2)^2}, \qquad v = -\mu\frac{2xy}{(x^2 + y^2)^2} \tag{8.5.23}$$

EXAMPLE 8.11

The pressure far from an irrotational vortex (a simplified tornado) in the atmosphere is zero gage. If the velocity at $r = 20$ m is 20 m/s, estimate the velocity and the pressure at $r = 5$ m. (The irrotational vortex ceases to be a good model for a tornado when r is small. In the "eye" of the tornado the motion is approximated by rigid-body motion.)

Solution

For an irrotational vortex, we know that

$$v_\theta = -\frac{\Gamma}{2\pi r}$$

Hence

$$\Gamma = -2\pi r v_\theta$$

$$= -2\pi \times 20 \times 20 = -800\pi \text{ m}^2/\text{s}$$

The velocity at $r = 5$ m is then

$$v_\theta = -\frac{-800\pi}{2\pi \times 5} = 80 \text{ m/s}$$

Bernoulli's equation for this incompressible, inviscid, steady flow then gives the pressure as follows:

$$\cancel{p_\infty}^{\,0} + \cancel{\frac{U_\infty^2}{2}}^{\,0} \rho = p + \frac{v_\theta^2}{2} \rho$$

$$\therefore \quad p = -\tfrac{1}{2}\rho v_\theta^2$$

$$= -\tfrac{1}{2} \times 1.20 \times 80^2 = -3840 \text{ Pa}$$

The negative sign denotes a vacuum.

8.5.3 Superposition

The simple flows presented in Section 8.5.2 are of particular interest because they can be superimposed with each other to form more complicated flows of engineering importance. In fact, the most complicated incompressible, plane flow situation can be constructed using these simple flows. For example, suppose that the flow around an airfoil with a slotted flap is desired. We could divide the surface of the airfoil into a relatively large number (say, 200) of panels, locate a source or a sink (or alternatively, a doublet) at the center of each panel, add a uniform flow and an irrotational vortex, and then by adjusting[4] the panel source strengths, the desired inviscid flow could be created. The development of the model and the computer routine necessary to perform the calculations are considered beyond the scope of this book.

In this section we demonstrate superposition by creating flow around a circular cylinder with and without circulation. First, superimpose a uniform flow and a doublet; there results

$$\psi = U_\infty y - \frac{\mu \sin \theta}{r} \tag{8.5.24}$$

[4]The source strengths are adjusted so that the normal component of velocity at the center of each panel is zero. The vortex strength is adjusted so that the rear stagnation point occurs at the trailing edge.

The velocity component v_r is (let $y = r \sin \theta$)

$$v_r = \frac{1}{r} \frac{\partial \psi}{\partial \theta}$$

$$= U_\infty \cos \theta - \frac{\mu}{r^2} \cos \theta \qquad (8.5.25)$$

Let us ask the question: Is there a radius r_c for which $v_r = 0$? If we set $v_r = 0$ we find that

$$r_c = \sqrt{\frac{\mu}{U_\infty}} \qquad (8.5.26)$$

At this radius v_r is identically zero for all θ and thus the circle $r = r_c$ must be a streamline.

The stagnation points are found by setting $v_\theta = 0$ on the circle $r = r_c$. There results

$$v_\theta = -\frac{\partial \psi}{\partial r}$$

$$= -U_\infty \sin \theta - \frac{\mu \sin \theta}{r_c^2} = -2U_\infty \sin \theta = 0 \qquad (8.5.27)$$

Thus we see that $v_\theta = 0$ at $\theta = 0°$ and $180°$. The flow is as sketched in Fig. 8.17a. We only have an interest in the flow external to the circular streamline $r = r_c$.

If the pressure distribution were desired on the cylinder, Bernoulli's equation could be used between the stagnation point where $\mathbf{V} = 0$ and $p = p_0$ and some arbitrary point on the cylinder to yield

$$p_c = p_0 - \rho \frac{v_\theta^2}{2}$$

$$= p_0 - 2\rho U_\infty^2 \sin^2\theta \qquad (8.5.28)$$

This provides a symmetric pressure distribution that yields zero drag and zero lift. The zero-lift prediction is acceptable for a real flow, but the zero-drag result is obviously unacceptable. This may suggest that we ignore the inviscid flow solution; however, compared with the actual flow situation of Fig. 8.17b, the measured pressure distribution on the cylinder up to the point of separation is very nearly the same as that predicted by the potential flow solution. Hence the potential flow solution is quite useful to us, even for blunt bodies that experience a separated flow. At low Reynolds numbers, viscous effects are not confined to a thin boundary layer, so potential flow theory is not useful.

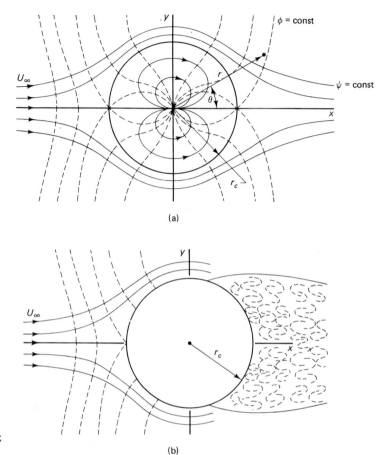

Figure 8.17 Flow around a
circular cylinder:
(a) potential flow;
(b) actual flow.

Let us now consider flow around a rotating cylinder. This is accomplished by adding an irrotational vortex to the doublet and uniform flow so that

$$\psi = U_\infty y - \frac{\mu \sin \theta}{r} + \frac{\Gamma}{2\pi} \ln r \qquad (8.5.29)$$

Since the vortex flow does not influence the velocity component v_r the cylinder $r = r_c$ remains unchanged. The stagnation points, however, are changed and are located by setting $v_\theta = 0$ on $r = r_c$; that is,

$$v_\theta = -\frac{\partial \psi}{\partial r}$$

$$= -2U_\infty \sin \theta - \frac{\Gamma}{2\pi r_c} = 0 \qquad (8.5.30)$$

This gives the location of the stagnation points as shown in Fig. 8.18.

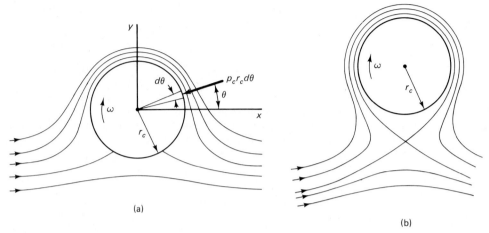

Figure 8.18   Flow around a circular cylinder with circulation: (a) $\Gamma < 4\pi U_\infty r_c$; (b) $\Gamma > 4\pi U_\infty r_c$.

Bernoulli's equation gives the pressure distribution as

$$p_c = p_0 - \rho \frac{U_\infty^2}{2}\left(2\sin\theta + \frac{\Gamma}{2\pi r_c U_\infty}\right)^2 \qquad (8.5.31)$$

This can be integrated to give drag = 0 and the lift per unit length as

$$\text{lift} = -\int_0^{2\pi} p_c \sin\theta\, r_c\, d\theta$$
$$= \rho U_\infty \Gamma \qquad (8.5.32)$$

This expression for the lift provides an excellent approximation to the lift for all cylinders, including the airfoil. Along with the zero-drag conclusion, it forms the **Kutta–Joukowsky theorem.**

Other superpositions of the simple flows are included in the problems.

EXAMPLE 8.12

A 20-cm-diameter cylinder rotates clockwise at 1000 rpm in a 20°C–atmospheric airstream flowing at 5 m/s. Locate any stagnation points and find the minimum pressure on the cylinder.

Solution

The circulation is calculated to be

$$\Gamma = \oint \mathbf{V} \cdot d\mathbf{s}$$
$$= 2\pi r_c^2 \omega = 2\pi \times 0.1^2 \times \frac{1000 \times 2\pi}{60} = 6.58 \text{ m}^2/\text{s}$$

This is greater than $4\pi U_\infty r_c = 4\pi \times 5 \times 0.1 = 6.28$ m²/s; hence the stagnation point is off the cylinder (see Fig. 8.18b) at

$$\theta = 270° \qquad r_0 = \frac{\Gamma}{4\pi U_\infty} = \frac{6.58}{4\pi \times 5} = 0.105 \text{ m}$$

Only one stagnation point exists.

     The minimum pressure is located on the top of the cylinder where $\theta = 90°$. Using Bernoulli's equation from the free stream to that point, we have, letting $p_\infty = 0$,

$$p_\infty^{\nearrow 0} + \rho \frac{U_\infty^2}{2} = p_{min} + \rho \frac{(v_\theta)^2_{max}}{2}$$

$$\therefore \quad p_{min} = \frac{\rho}{2} [U_\infty^2 - (v_\theta)^2_{max}] = \frac{\rho}{2} \left[U_\infty^2 - \left(2U_\infty + \frac{\Gamma}{2\pi r_c} \right)^2 \right]$$

$$= \frac{1.20}{2} \left[5^2 - \left(2 \times 5 + \frac{6.58}{2\pi \times 0.1} \right)^2 \right] = -236 \text{ Pa}$$

EXAMPLE 8.12 (English)

An 8-in.-diameter cylinder rotates clockwise at 1000 rpm in a 60°F–atmospheric airstream flowing at 15 ft/sec. Locate any stagnation points and find the minimum pressure on the cylinder.

Solution

The circulation is calculated to be

$$\Gamma = \oint \mathbf{V} \cdot d\mathbf{s}$$

$$= 2\pi r_c^2 \omega = 2\pi \times \left(\frac{4}{12} \right)^2 \times \frac{1000 \times 2\pi}{60} = 73.1 \text{ ft}^2/\text{sec}$$

This is greater than $4\pi U_\infty r_c = 4\pi \times 15 \times 4/12 = 62.8$ ft²/sec; hence the stagnation point is off the cylinder (see Fig. 8.18b) at

$$\theta = 270° \qquad r_0 = \frac{\Gamma}{4\pi U_\infty} = \frac{73.1}{4\pi \times 15} = 0.388 \text{ ft}$$

Only one stagnation point exists.

     The minimum pressure is located on the top of the cylinder where $\theta = 90°$. Using Bernoulli's equation from the free stream to that point, we have, letting

$p_\infty = 0,$

$$\cancel{p_\infty}^{0} + \rho\,\frac{U_\infty^2}{2} = p_{min} + \frac{\rho}{2}\,(v_\theta)_{max}^2$$

$$\therefore\ p_{min} = \frac{\rho}{2}\,[U_\infty^2 - (v_\theta)_{max}^2] = \frac{\rho}{2}\left[U_\infty^2 - \left(2U_\infty + \frac{\Gamma}{2\pi r_c}\right)^2\right]$$

$$= \frac{0.0024}{2}\left[15^2 - \left(2 \times 15 + \frac{73.1}{2\pi \times 4/12}\right)^2\right] = -4.78\ \text{psf}$$

8.6 BOUNDARY LAYER THEORY

8.6.1 General Background

In our study of high-Reynolds-number external flows we have observed that the viscous effects are confined to a thin layer of fluid, a boundary layer, next to the body and to the wake downstream of the body. For a streamlined body such as an airfoil, a good approximation to the drag can be obtained by integrating the viscous shear stress on the wall. To predict the wall shear, the velocity gradient must be known. This requires a complete solution of the flow field inside the boundary layer. Such a solution also allows us to predict locations of possible separation. In this section we derive the integral and differential equations and provide solution techniques for a boundary layer flow on a flat plate with a zero pressure gradient; this simplified flow has many applications. Nonzero-pressure-gradient flows on flat plates and curved surfaces, are beyond the scope of this book.

Let us now discuss some of the characteristics of a boundary layer. The edge of the boundary layer, with thickness designated by $\delta(x)$, cannot be observed in an actual flow; we arbitrarily define it to be that locus of points where the velocity is equal to 99% of the free-stream velocity [the free-stream velocity is the inviscid flow wall velocity $U(x)$], as shown in Fig. 8.19.

The boundary layer begins as a laminar flow with zero thickness at the leading edge of a flat plate, as shown in Fig. 8.20, or with some finite thickness at the stagnation point of a blunt object or an airfoil. After a distance x_T, which depends on the free-stream velocity, the viscosity, the pressure gradient, the wall roughness, and the free-stream fluctuation level, the laminar flow undergoes a transition process that results, after a short distance, in a turbulent flow, as sketched. For flow over a flat plate with zero pressure gradient, this transition process occurs when $U_\infty x_T/\nu = 3 \times 10^5$ for flow on rough plates or with high free-stream fluctuation intensity, or when $U_\infty x_T/\nu = 6 \times 10^5$ for flow on smooth plates with low free-stream fluctuation intensity. For extremely low fluctuation levels in

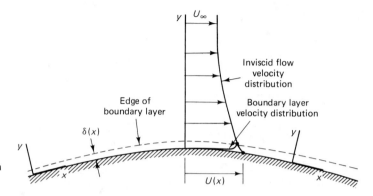

Figure 8.19 Boundary layer on an airfoil.

research labs, laminar flows have been observed on smooth rigid plates up to $U_\infty x_T/\nu = 10^6$.

The quantity $U_\infty x/\nu$ is the **local Reynolds number** and $U_\infty x_T/\nu$ is the **critical Reynolds number.** For a smooth flat plate and a very low free-stream fluctuation level, a zero-pressure-gradient flow becomes unstable (i.e., small disturbances will grow) at a local Reynolds number of about 6×10^4. The small disturbances grow initially as a two-dimensional wave, then a three-dimensional wave, and finally burst into a turbulent spot; the initial burst forms the beginning of the transition region. The transition region is relatively short and is usually ignored in calculations. The flow up to x_T is assumed to be laminar, and the flow after x_T is considered to be turbulent.

The turbulent boundary layer thickens much more rapidly than the laminar layer. It also has a substantially greater wall shear. A sketch of a turbulent boundary layer with its submerged viscous wall layer is shown in Fig. 8.21a, and an actual photograph in Fig. 8.21b. The time-average thickness is $\delta(x)$ and the time-average viscous wall layer thickness is $\delta_\nu(x)$. Both layers are time dependent. The instantaneous boundary layer thickness varies between 0.4δ and 1.2δ, as shown. The turbulent profile has a

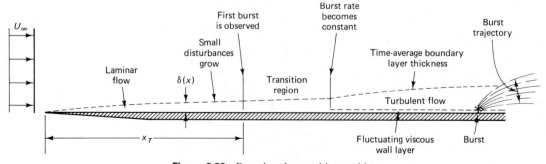

Figure 8.20 Boundary layer with transition.

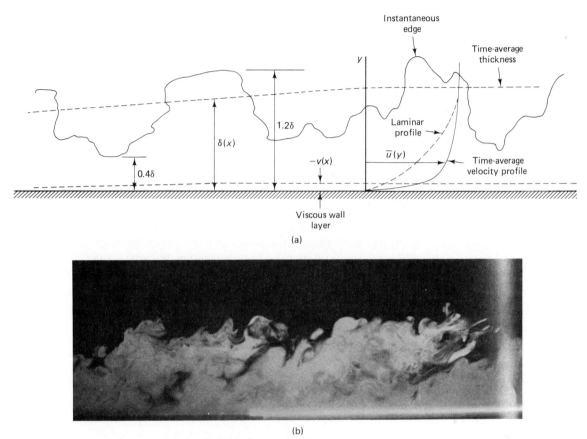

(a)

(b)

Figure 8.21   Turbulent boundary layer: (a) nomenclature sketch; (b) streamwise slice of the boundary layer (photograph by R. E. Falco).

greater slope at the wall than does a laminar profile with the same boundary layer thickness, as can easily be observed in Fig. 8.21a.

Finally, we should emphasize that the boundary layer is quite thin. A thick boundary layer is drawn to scale in Fig. 8.22. We have assumed a laminar flow up to x_T and a turbulent flow thereafter. For higher velocities

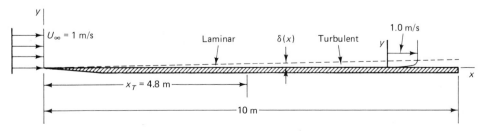

Figure 8.22   Boundary layer in air with $Re_{crit} = 3 \times 10^5$.

the boundary layer thickness decreases. If we assumed that $U_\infty = 100$ m/s, the boundary layer would hardly be noticed drawn to the same scale, yet all the viscous effects are confined in that thin layer; the velocity is brought to rest with very large gradients. Viscous dissipative effects in this thin layer are large enough to cause sufficiently high temperatures that satellites burn up on reentry.

8.6.2 Von Kármán Integral Equation

From the velocity profile in the boundary layer of Fig. 8.22 it is observed that the velocity goes from $u = 0.99U_\infty$ at $y = \delta$ to $u = 0$ at $y = 0$ over a very short distance (the thickness of the boundary layer). Hence it is not surprising that we can approximate the velocity profile, for both laminar and turbulent flow, with a good deal of accuracy. If the velocity profile can be assumed known, the continuity and momentum integral equations will enable us to predict the boundary layer thickness and the wall shear and thus the drag. So let us develop the integral equations for the boundary layer.

Consider an infinitesimal control volume, shown in Fig. 8.23a. The integral continuity equation allows us to find $\dot{m}_{\text{top}}$ (see Fig. 8.23b). It is,

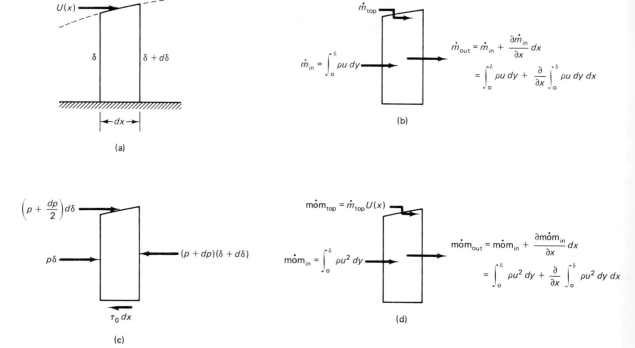

Figure 8.23 Control volume for a boundary layer: (a) control volume; (b) mass flux; (c) forces; (d) momentum flux.

assuming unit depth,

$$\dot{m}_{\text{top}} = \dot{m}_{\text{out}} - \dot{m}_{\text{in}}$$

$$= \frac{\partial}{\partial x} \int_0^\delta \rho u \, dy \, dx \qquad (8.6.1)$$

The integral momentum equation takes the form

$$\Sigma F_x = \dot{m}om_{\text{out}} - \dot{m}om_{\text{in}} - \dot{m}om_{\text{top}} \qquad (8.6.2)$$

where $\dot{m}om$ represents the momentum flux in the x-direction. Referring to Figs. 8.23c and d this becomes, neglecting high-order terms,

$$-\delta \, dp - \tau_0 dx = \frac{\partial}{\partial x} \int_0^\delta \rho u^2 \, dy \, dx - \left(\frac{\partial}{\partial x} \int_0^\delta \rho u \, dy \, dx \right) U(x) \quad (8.6.3)$$

where $\dot{m}_{\text{top}}$ is given in Eq. 8.6.1. Divide through by dx and obtain

$$\tau_0 + \delta \frac{dp}{dx} = U(x) \frac{d}{dx} \int_0^\delta \rho u \, dy - \frac{d}{dx} \int_0^\delta \rho u^2 \, dy \qquad (8.6.4)$$

where we have used ordinary derivatives since the integrals are only functions of x. This equation is often referred to as the **von Kámán integral equation.**

For flow over a flat plate with zero pressure gradient, so that $dp/dx = 0$ and $U(x) = U_\infty$, von Kármán's integral momentum equation takes the simplified form

$$\tau_0 = \frac{d}{dx} \int_0^\delta \rho u U_\infty \, dy - \frac{d}{dx} \int_0^\delta \rho u^2 \, dy$$

$$= \frac{d}{dx} \int_0^\delta \rho u (U_\infty - u) \, dy \qquad (8.6.5)$$

If the velocity profile can be assumed, this equation allows us to solve for both $\delta(x)$ and $\tau_0(x)$. This will be demonstrated in the following sections for both a laminar and a turbulent boundary layer.

Before we do this, however, there are two additional lengths often used in boundary layer theory. They are the **displacement thickness** δ_d and the **momentum thickness** θ, defined by

$$\delta_d = \frac{1}{U} \int_0^\delta (U - u) \, dy \qquad (8.6.6)$$

$$\theta = \frac{1}{U^2} \int_0^\delta u(U - u) \, dy \qquad (8.6.7)$$

The displacement thickness is the displacement of the streamlines in the free stream as a result of velocity deficits in the boundary layer, as can be shown by continuity considerations. The momentum layer thickness is the equivalent thickness of a fluid layer with velocity U with momentum equal to the momentum lost due to friction; the momentum thickness is often used as a characteristic length in turbulent boundary layer studies. The end-of-chapter problems will demonstrate the use of δ_d and θ. It should be noted, though, that von Kármán's integral equation (8.6.5) takes the form, assuming that $\rho = $ const.,

$$\tau_0 = \rho U_\infty^2 \frac{d\theta}{dx} \tag{8.6.8}$$

8.6.3 Approximate Solution to the Laminar Boundary Layer

It is possible to use the von Kármán integral equation and obtain a fairly accurate approximation to the laminar boundary layer on a flat plate with zero pressure gradient. We have four conditions that a proposed velocity profile should satisfy:

$$
\begin{aligned}
u &= 0 && \text{at} \quad y = 0 \\
u &= U_\infty && \text{at} \quad y = \delta \\
\frac{\partial u}{\partial y} &= 0 && \text{at} \quad y = \delta \\
\frac{\partial^2 u}{\partial y^2} &= 0 && \text{at} \quad y = 0
\end{aligned}
\tag{8.6.9}
$$

The first three of these conditions are obvious from a sketch of a velocity profile; the fourth condition comes from the x-component Navier–Stokes equation (4.3.14) since $u = v = 0$ at the wall, $\partial^2 u/\partial x^2 = 0$ on the wall, and $dp/dx = 0$ for the steady flow under consideration.

A cubic polynomial will satisfy the four conditions above; let us assume that

$$\frac{u}{U_\infty} = A + By + Cy^2 + Dy^3 \tag{8.6.10}$$

where A, B, C, and D can be functions of x. Using the four conditions, we see that

$$A = 0 \qquad B = \frac{3}{2\delta} \qquad C = 0 \qquad D = -\frac{1}{2\delta^3} \tag{8.6.11}$$

Hence a good approximation for the velocity profile in a laminar flow is

$$\frac{u}{U_\infty} = \frac{3}{2}\frac{y}{\delta} + \frac{1}{2}\left(\frac{y}{\delta}\right)^3 \tag{8.6.12}$$

Von Kármán's integral equation (8.6.5) gives

$$\tau_0 = \frac{d}{dx}\int_0^\delta \rho\left(\frac{3y}{2\delta} - \frac{y^3}{2\delta^3}\right)\left(1 - \frac{3y}{2\delta} - \frac{y^3}{2\delta^3}\right)U_\infty^2\,dy$$

$$= 0.139\rho U_\infty^2 \frac{d\delta}{dx} \tag{8.6.13}$$

At the wall we know that $\tau_0 = \mu\,\partial u/\partial y|_{y=0}$, or using the cubic profile (8.6.10),

$$\tau_0 = \mu\left(U_\infty\frac{3}{2\delta}\right) \tag{8.6.14}$$

Equating the foregoing expressions for $\tau_0(x)$, we find that

$$\delta\,d\delta = \frac{\frac{3}{2}\mu U_\infty}{0.139\rho U_\infty^2}\,dx = 10.8\frac{\nu}{U_\infty}\,dx \tag{8.6.15}$$

Using $\delta = 0$ at $x = 0$ (the leading edge), Eq. 8.6.15 is integrated to give

$$\delta = 4.65\sqrt{\frac{\nu x}{U_\infty}}$$

$$= 4.65\frac{x}{\sqrt{Re_x}} \tag{8.6.16}$$

This is substituted back into Eq. 8.6.14, giving the wall shear as

$$\tau_0 = 0.323\rho U_\infty^2\sqrt{\frac{\nu}{xU_\infty}}$$

$$= \frac{0.323\rho U_\infty^2}{\sqrt{Re_x}} \tag{8.6.17}$$

The shearing stress is made dimensionless by dividing by $\frac{1}{2}\rho U_\infty^2$. The **local skin friction coefficient** c_f results; it is

$$c_f = \frac{\tau_0}{\frac{1}{2}\rho U_\infty^2} = \frac{0.646}{\sqrt{U_\infty x/\nu}}$$

$$= \frac{0.646}{\sqrt{Re_x}} \tag{8.6.18}$$

If the wall shear is integrated over the length L, there results, per unit width,

$$\text{drag} = \int_0^L \tau_0 \, dx = 0.646\rho U_\infty \sqrt{U_\infty L \nu}$$

$$= \frac{0.646\rho U_\infty^2 L}{\sqrt{\text{Re}_L}}$$

(8.6.19)

or in terms of the **skin friction coefficient** C_f,

$$C_f = \frac{\text{drag}}{\frac{1}{2}\rho U_\infty^2 L}$$

$$= \frac{1.29}{\sqrt{U_\infty L/\nu}} = \frac{1.29}{\sqrt{\text{Re}_L}}$$

(8.6.20)

Note that the shearing stress τ_0 becomes unbounded as $x \to 0$. Hence we would not expect $\tau_0(x)$ to be a very good approximation to the wall shear near the leading edge. The expression for the drag is acceptable, however.

The results above are quite good when compared with results from a solution of the differential equations (refer to Section 8.6.6). The boundary layer thickness is 7% too low; the constant in Eq. 8.6.16 should be 5 if an exact solution were obtained. The wall shear is less than 3% too low; the constant in Eq. 8.6.17 should be 0.332.

EXAMPLE 8.13

Assume a parabolic velocity profile and calculate the boundary layer thickness and the wall shear. Compare with those calculated above.

Solution

The parabolic velocity profile is assumed to be

$$\frac{u}{U_\infty} = A + By + Cy^2$$

The fourth condition of (8.6.9) is omitted; this leaves

$$0 = A$$
$$1 = A + B\delta + C\delta^2$$
$$0 = B + 2C\delta$$

Thus

$$A = 0 \qquad B = \frac{2}{\delta} \qquad C = -\frac{1}{\delta^2}$$

The velocity profile is

$$\frac{u}{U_\infty} = 2\,\frac{y}{\delta} - \frac{y^2}{\delta^2}$$

This is substituted into von Kármán's integral equation (8.6.5) to obtain

$$\tau_0 = \frac{d}{dx} \int_0^\delta \rho U_\infty^2 \left(2\,\frac{y}{\delta} - \frac{y^2}{\delta^2}\right)\left(1 - \frac{2y}{\delta} + \frac{y^2}{\delta^2}\right) dy$$

$$= \frac{2}{15} \rho U_\infty^2 \frac{d\delta}{dx}$$

We also use $\tau_0 = \mu\, \partial u/\partial y|_{y=0}$; that is,

$$\tau_0 = \mu U_\infty \frac{2}{\delta}$$

Equating the two expressions above, we obtain

$$\delta\, d\delta = 15\,\frac{\nu}{U_\infty}\, dx$$

Using $\delta = 0$ at $x = 0$, this is integrated to

$$\delta = 5.48 \sqrt{\frac{\nu x}{U_\infty}}$$

This is 18% higher than the value using the cubic but only 10% higher than the more accurate result of $5\sqrt{\nu x/U_\infty}$.

The wall shear is found to be

$$\tau_0 = \frac{2\mu U_\infty}{\delta}$$

$$= 0.365 \rho U_\infty^2 \sqrt{\frac{\nu}{x U_\infty}}$$

This is 13% higher than the value using the cubic and 10% higher than the more accurate value of $0.332 \rho U_\infty^2 \sqrt{\nu/x U_\infty}$.

8.6.4 Turbulent Boundary Layer: Power-Law Form

For turbulent boundary layer flow we have two different methods for obtaining the information desired. Both methods utilize experimental data but the one that we present in this section is the simpler of the two. The second method, to be presented later, provides us with more information than we usually desire for most applications; however, it is more accurate.

In the method to be presented first we fit the data for the velocity profile with a power-law equation. The power-law form is

$$\frac{\bar{u}}{U_\infty} = \left(\frac{y}{\delta}\right)^{1/n} \qquad n = \begin{cases} 7 & \mathrm{Re}_x < 10^7 \\ 8 & 10^7 < \mathrm{Re}_x < 10^8 \\ 9 & 10^8 < \mathrm{Re}_x < 10^9 \end{cases} \qquad (8.6.21)$$

where

$$\mathrm{Re}_x = \frac{U_\infty x}{\nu} \qquad (8.6.22)$$

Von Kármán's integral equation can now be applied following the steps used for laminar flow, except when the shear stress at the wall is evaluated. The power-law form (8.6.21) yields $(\partial \bar{u}/\partial y)_{y=0} = \infty$; hence the profile gives poor results near the wall, especially for the shear stress. So rather than using $\tau_0 = (\mu\, \partial \bar{u}/\partial y)_{y=0}$, we use an empirical relation; the **Blasius formula,** which relates the local skin friction coefficient to the boundary layer thickness, is

$$c_f = 0.046 \left(\frac{\nu}{U_\infty \delta}\right)^{1/4} \qquad (8.6.23)$$

or, relating τ_0 to c_f using Eq. 8.6.18,

$$\tau_0 = 0.023 \rho U_\infty^2 \left(\frac{\nu}{U_\infty \delta}\right)^{1/4} \qquad (8.6.24)$$

Van Kármán's integral equation provides us with a second expression for τ_0; substitute the velocity profile of (8.6.21) with $\mathrm{Re}_x < 10^7$ into Eq. 8.6.5 and obtain

$$\begin{aligned} \tau_0 &= \frac{d}{dx} \int_0^\delta \rho U_\infty^2 \left(\frac{y}{\delta}\right)^{1/7} \left[1 - \left(\frac{y}{\delta}\right)^{1/7}\right] dy \\ &= \frac{7}{72} \rho U_\infty^2 \frac{d\delta}{dx} \end{aligned} \qquad (8.6.25)$$

Combining the two expressions above for τ_0, we find that

$$\delta^{1/4}\, d\delta = 0.237 \left(\frac{\nu}{U_\infty}\right)^{1/4} dx \qquad (8.6.26)$$

Assuming a turbulent flow from the leading edge (the laminar portion is often quite short), there results

$$\delta = 0.38x\left(\frac{\nu}{U_{\infty}x}\right)^{1/5}$$

$$= 0.38x\text{Re}_x^{-1/5} \qquad \text{Re}_x < 10^7 \qquad\qquad (8.6.27)$$

Substituting this expression for δ back into Eq. 8.6.23, we find that

$$c_f = 0.059\ \text{Re}_x^{-1/5} \qquad \text{Re}_x < 10^7 \qquad\qquad (8.6.28)$$

and

$$C_f = 0.074\ \text{Re}_L^{-1/5} \qquad \text{Re}_L < 10^7 \qquad\qquad (8.6.29)$$

where

$$\text{Re}_L = \frac{U_{\infty}L}{\nu} \qquad\qquad (8.6.30)$$

The relations above can be stretched to $\text{Re}_x \simeq 10^8$ without substantial error.

If there is a significant laminar portion on the leading part of a flat plate, the skin friction coefficient can be modified as

$$C_f = 0.074\text{Re}_L^{-1/5} - 1700\text{Re}_L^{-1} \qquad \text{Re}_L < 10^7 \qquad (8.6.31)$$

This relationship is based on transition occurring at $\text{Re}_{\text{crit}} = 5 \times 10^5$. If $\text{Re}_{\text{crit}} = 3 \times 10^5$, the constant of 1700 is replaced by 1060; if $\text{Re}_{\text{crit}} = 6 \times 10^5$, it becomes 2080.

EXAMPLE 8.14

Estimate the boundary layer thickness at the end of a 4-m-long flat surface if the free-stream velocity is $U_{\infty} = 5$ m/s. Use atmospheric air at 20°C. Also, predict the drag force if the surface is 5 m wide. (a) Neglect the laminar portion of the flow, and (b) account for the laminar portion using $\text{Re}_{\text{crit}} = 5 \times 10^5$.

Solution

(a) Let us first assume turbulent flow from the leading edge. The boundary layer thickness is given by Eq. 8.6.27. It is

$$\delta = 0.38x\text{Re}_x^{-1/5}$$

$$= 0.38 \times 4 \times \left(\frac{5 \times 4}{1.6 \times 10^{-5}}\right)^{-1/5} = 0.092 \text{ m}$$

The drag force is, using Eq. 8.6.29,

$$\text{drag} = C_f \times \tfrac{1}{2}\rho U_\infty^2 Lw$$

$$= 0.074 \left(\frac{5 \times 4}{1.6 \times 10^{-5}}\right)^{-1/5} \times \frac{1}{2} \times 1.20 \times 5^2 \times 4 \times 5 = 1.34 \text{ N}$$

The predictions above assume that $\text{Re}_L < 10^7$. The Reynolds number is

$$\text{Re}_L = \frac{5 \times 4}{1.6 \times 10^{-5}} = 1.25 \times 10^6$$

Hence the calculations are acceptable.

(b) Now let us account for the laminar portion of the boundary layer. Referring to the sketch below, the distance x_T is found from

$$\text{Re}_{\text{crit}} = 5 \times 10^5 = \frac{U_\infty x_T}{\nu}$$

$$\therefore \quad x_T = 5 \times 10^5 \times 1.6 \times \frac{10^{-5}}{5} = 1.6 \text{ m}$$

The boundary layer thickness at x_T is, replacing the constant of 4.65 in Eq. 8.6.16 with the more accurate value of 5.0,

$$\delta = 5\sqrt{\frac{x\nu}{U_\infty}}$$

$$= 5\sqrt{\frac{1.6 \times 1.6 \times 10^{-5}}{5}} = 0.0113 \text{ m}$$

The location of the fictitious origin of the turbulent flow (see the sketch below) is found using Eq. 8.6.27 to be

$$x'^{4/5} = \frac{\delta}{0.38}\left(\frac{U_\infty}{\nu}\right)^{1/5}$$

$$\therefore \quad x' = \left(\frac{0.0113}{0.38}\right)^{5/4}\left(\frac{5}{1.6 \times 10^{-5}}\right)^{1/4} = 0.292 \text{ m}$$

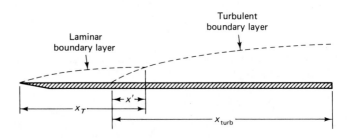

The distance x_{turb} is then $x_{\text{turb}} = 4 - 1.6 + 0.292 = 2.69$ m. Using Eq. 8.6.27, the thickness at the end of the surface is

$$\delta = 0.38x\left(\frac{\nu}{U_\infty x}\right)^{1/5}$$

$$= 0.38 \times 2.69 \times \left(\frac{1.6 \times 10^{-5}}{5 \times 2.69}\right)^{1/5} = 0.067 \text{ m}$$

The value of part (a) is 37% too high when compared with this more accurate value.

The more accurate drag force is found using Eq. 8.6.31 to be

$$\text{drag} = C_f \times \tfrac{1}{2}\rho U_\infty^2 Lw$$

$$= [0.074\text{Re}_L^{-1/5} - 1700\text{Re}_L^{-1}] \times \tfrac{1}{2}\rho U_\infty^2 Lw$$

$$= \left[0.074\left(\frac{5 \times 4}{1.6 \times 10^{-5}}\right)^{-1/5} - 1700\left(\frac{5 \times 4}{1.6 \times 10^{-5}}\right)^{-1}\right] \times \frac{1}{2} \times 1.20 \times 5^2 \times 4 \times 5$$

$$= 0.93 \text{ N}$$

The prediction of part (a) is 44% too high. For relatively short surfaces it is obvious that significant errors result if the laminar portion is neglected.

8.6.5 Turbulent Boundary Layer: Empirical Form

The second method of predicting turbulent flow quantities on a flat plate with zero pressure gradient is based entirely on data. It is more accurate than the power-law form but also more complicated. The time-average turbulent velocity profile can be divided into two regions, the **inner region** and the **outer region,** as shown in Fig. 8.24. The inner region is characterized by the **self-similar** (the dimensionless dependent variable depends on only one dimensionless independent variable) relation,

$$\frac{\bar{u}}{u_\tau} = f\left(\frac{u_\tau y}{\nu}\right) \tag{8.6.32}$$

in which u_τ is the **shear velocity,** given by[5]

$$u_\tau = \sqrt{\frac{\tau_0}{\rho}} \tag{8.6.33}$$

[5]The shear velocity is a fictitious velocity and is defined because the quantity $\sqrt{\tau_0/\rho}$ occurs often in empirical relations in turbulent boundary layer flows.

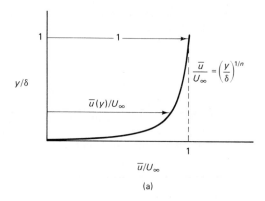

$$\frac{\overline{u}}{U_\infty} = \left(\frac{y}{\delta}\right)^{1/n}$$

(a)

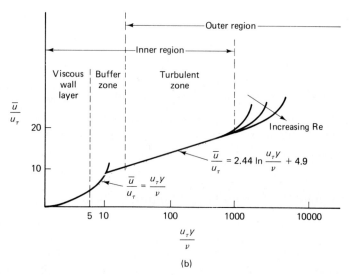

(b)

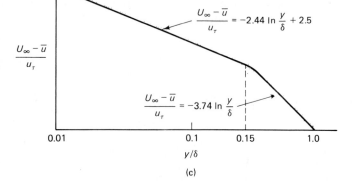

Figure 8.24 Velocity profile in a turbulent boundary layer: (a) common mode of presenting the mean velocity in the boundary layer; (b) inner region; (c) outer region.

(c)

The velocity profile in the outer region is given by the self-similar relation

$$\frac{U_\infty - \bar{u}}{u_\tau} = f\left(\frac{y}{\delta}\right) \tag{8.6.34}$$

where $(U_\infty - \bar{u})$ is called the **velocity defect.**

The inner region has three distinct zones: the viscous wall layer, the buffer zone, and the turbulent zone, as shown in Fig. 8.24b. The highly fluctuating **viscous wall layer** has a linear time-average profile given by

$$\frac{\bar{u}}{u_\tau} = \frac{u_\tau y}{\nu} \tag{8.6.35}$$

The quantity ν/u_τ is the characteristic length in the turbulent inner region; hence the dimensionless distance from the wall is denoted by

$$y^* = \frac{u_\tau y}{\nu} \tag{8.6.36}$$

The viscous wall layer is very thin, extending up to $y^* \simeq 5$. A logarithmic profile exists from $y^* \simeq 50$ to $y/\delta \simeq 0.15$. In this self-similar **turbulent zone,**

$$\frac{\bar{u}}{u_\tau} = 2.44 \ln \frac{u_\tau y}{\nu} + 4.9 \qquad 50 < \frac{u_\tau y}{\nu} \text{ and } \frac{y}{\delta} < 0.15 \tag{8.6.37}$$

The location of the outer edge of the turbulent zone is strongly dependent on the Reynolds number. The value of $u_\tau y/\nu$ that locates the outer edge increases as the Reynolds number increases, as shown. A **buffer zone,** with no specified velocity profile, connects the two self-similar zones.

The outer region relates the velocity defect to y/δ. In the turbulent zone the velocity-defect profile is sketched in Fig. 8.24c and is

$$\frac{U_\infty - \bar{u}}{u_\tau} = -2.44 \ln \frac{y}{\delta} + 2.5 \qquad 50 < \frac{u_\tau y}{\nu} \text{ and } \frac{y}{\delta} < 0.15 \tag{8.6.38}$$

Between $y/\delta = 0.15$ and $y/\delta = 1$ researchers fit the data with several relations; the one selected here is

$$\frac{U_\infty - \bar{u}}{u_\tau} = -3.74 \ln \frac{y}{\delta} \qquad \frac{y}{\delta} > 0.15 \tag{8.6.39}$$

The equations above involve the shear velocity u_τ, which depends on the wall shear τ_0. The equation that is valid at the wall is Eq. 8.6.35, which, using $\tau_0 = \mu\, \partial u/\partial y|_{y=0}$, simply provides us with an identity. It does not allow us to calculate τ_0. Hence a relationship is necessary to provide us with τ_0 (or equally c_f). There are several such relationships used; one that gives excellent results is

$$c_f = \frac{0.455}{(\ln 0.06 \mathrm{Re}_x)^2} \tag{8.6.40}$$

The local skin friction coefficient allows us to determine τ_0 and thus u_τ at any location of interest. Knowing u_τ, the velocity profiles can be used to calculate quantities of interest.

Assuming turbulent flow from the leading edge, the shear stress can be integrated to yield the drag. Then the skin friction coefficient becomes

$$C_f = \frac{0.523}{(\ln 0.06 \mathrm{Re}_L)^2} \tag{8.6.41}$$

This relation is very good and can be used up to $\mathrm{Re}_L = 10^9$ with an error of 2% or less. Even at $\mathrm{Re}_L = 10^{10}$ the error is about 4%. To account for a laminar portion, the same term included in Eq. 8.6.31 can be subtracted from Eq. 8.6.41.

In concluding this section, a very useful relationship can be obtained by combining the two logarithmic profiles for the common turbulent zone. Substitute Eq. 8.6.37 into Eq. 8.6.38 and obtain

$$\frac{U_\infty}{u_\tau} = 2.44 \ln \frac{u_\tau \delta}{\nu} + 7.4 \tag{8.6.42}$$

This equation allows an easy calculation of δ knowing u_τ.

EXAMPLE 8.15

Estimate the thickness δ_ν of the viscous wall layer and the boundary layer thickness at the end of a 2-m-long flat plate if $U_\infty = 30$ m/s in 20°C—atmospheric air. Also, calculate the drag force on one side if the plate is 3 m wide. Use the empirical data.

Solution

To find the viscous wall layer thickness, we must know the shear velocity and hence the wall shear. The wall shear, using Eq. 8.6.40, and the shear velocity are

$$\tau_0 = \tfrac{1}{2}\rho U_\infty^2 c_f$$

$$= \frac{1}{2}\rho U_\infty^2 \frac{0.455}{(\ln 0.06 \mathrm{Re}_x)^2}$$

$$= \frac{1}{2} \times 1.20 \times 30^2 \frac{0.455}{\left(\ln 0.06 \dfrac{30 \times 2}{1.6 \times 10^{-5}}\right)^2} = 1.62 \text{ Pa}$$

$$u_\tau = \sqrt{\frac{\tau_0}{\rho}}$$

$$= \sqrt{\frac{1.62}{1.20}} = 1.16 \text{ m/s}$$

The viscous wall layer thickness is determined using Eq. 8.6.36 with $y^* = 5$ as follows:

$$\frac{u_\tau \delta_\nu}{\nu} = 5$$

$$\therefore \quad \delta_\nu = \frac{5\nu}{u_\tau}$$

$$= \frac{5 \times 1.6 \times 10^{-5}}{1.16} = 6.9 \times 10^{-5} \text{ m}$$

The boundary layer thickness is found using Eq. 8.6.42. We have

$$\frac{U_\infty}{u_\tau} = 2.44 \ln \frac{u_\tau \delta}{\nu} + 7.4$$

$$\therefore \quad \frac{30}{1.16} = 2.44 \ln \frac{1.16\delta}{1.6 \times 10^{-5}} + 7.4 \qquad \therefore \quad \delta = 0.027 \text{ m}$$

The drag force is calculated using Eq. 8.6.41 to be

$$\text{drag} = C_f \times \tfrac{1}{2}\rho U_\infty^2 L w$$

$$= \frac{0.523}{(\ln 0.06 \text{Re}_L)^2} \times \frac{1}{2}\rho U_\infty^2 L w$$

$$= \frac{0.523}{\left(\ln 0.06 \dfrac{30 \times 2}{1.6 \times 10^{-5}}\right)^2} \times \frac{1}{2} \times 1.20 \times 30^2 \times 2 \times 3 = 11.2 \text{ N}$$

EXAMPLE 8.15 (English)

Estimate the thickness δ_ν of the viscous wall layer and the boundary layer thickness at the end of a 6-ft-long flat plate if $U_\infty = 100$ ft/sec in 60°F–atmospheric air. Also, calculate the drag force on one side if the plate is 10 ft wide. Use the empirical data.

Solution

To find the viscous wall layer thickness we must know the shear velocity and hence the wall shear. The wall shear, using Eq. 8.6.40, and the shear velocity are

$$\tau_0 = \tfrac{1}{2}\rho U_\infty^2 c_f$$

$$= \frac{1}{2}\rho U_\infty^2 \frac{0.455}{(\ln 0.06 \text{Re}_x)^2}$$

$$= \frac{1}{2} \times 0.0024 \times 100^2 \frac{0.455}{\left(\ln 0.06 \dfrac{100 \times 6}{1.6 \times 10^{-4}}\right)^2} = 0.036 \text{ psf}$$

$$u_\tau = \sqrt{\frac{\tau_0}{\rho}}$$

$$= \sqrt{\frac{0.036}{0.0024}} = 3.87 \text{ ft/sec}$$

The viscous wall layer thickness is determined using Eq. 8.6.36 with $y^* = 5$ as follows:

$$\frac{u_\tau \delta_\nu}{\nu} = 5$$

$$\therefore \quad \delta_\nu = \frac{5\nu}{u_\tau}$$

$$= \frac{5 \times 1.6 \times 10^{-4}}{3.87} = 2.07 \times 10^{-4} \text{ ft}$$

The boundary layer thickness is found using Eq. 8.6.42. We have

$$\frac{U_\infty}{u_\tau} = 2.44 \ln \frac{u_\tau \delta}{\nu} + 7.4$$

$$\therefore \quad \frac{100}{3.87} = 2.44 \ln \frac{3.87 \times \delta}{1.6 \times 10^{-4}} + 7.4 \qquad \therefore \quad \delta = 0.079 \text{ ft}$$

The drag force is calculated using Eq. 8.6.41 to be

$$\text{drag} = C_f \times \tfrac{1}{2}\rho U_\infty^2 Lw$$

$$= \frac{0.523}{(\ln 0.06 \text{Re}_L)^2} \times \frac{1}{2} \rho U_\infty^2 Lw$$

$$= \frac{0.523}{\left(\ln 0.06 \dfrac{100 \times 6}{1.6 \times 10^{-4}}\right)^2} \times \frac{1}{2} \times 0.0024 \times 100^2 \times 6 \times 10 = 2.48 \text{ lb}$$

EXAMPLE 8.16

Estimate the maximum boundary layer thickness and the drag on the side of a ship that measures 40 m long with a submerged depth of 8 m. The ship travels at 10 m/s. (a) Use the empirical methods, and (b) compare with the results using the one-seventh power law.

Solution

(a) The boundary layer thickness is found from Eq. 8.6.42. First we must find τ_0 from Eq. 8.6.40 and then u_τ as follows:

$$\tau_0 = \frac{1}{2} \rho U_\infty^2 \frac{0.455}{(\ln 0.06 \text{Re}_L)^2}$$

$$= \frac{1}{2} \times 1000 \times 10^2 \frac{0.455}{\left(\ln 0.06 \dfrac{10 \times 40}{10^{-6}}\right)^2} = 78.8 \text{ Pa}$$

$$\therefore \quad u_\tau = \sqrt{\frac{\tau_0}{\rho}}$$

$$= \sqrt{\frac{78.8}{1000}} = 0.28 \text{ m/s}$$

Using Eq. 8.6.42, the maximum boundary layer thickness is found from

$$\frac{U_\infty}{u_\tau} = 2.44 \ln \frac{u_\tau \delta}{\nu} + 7.4$$

$$\therefore \quad \frac{10}{0.28} = 2.44 \ln \frac{0.28\delta}{10^{-6}} + 7.4 \quad \therefore \quad \delta = 0.39 \text{ m}$$

The drag is

$$\text{drag} = C_f \times \tfrac{1}{2}\rho U_\infty^2 Lw$$

$$= \frac{0.523}{\left(\ln 0.06 \dfrac{10 \times 40}{10^{-6}}\right)^2} \times \frac{1}{2} \times 1000 \times 10^2 \times 40 \times 8 = 29\,000 \text{ N}$$

(b) Using Eq. 8.6.27, which has been obtained using the one-seventh power law of Eq. 8.6.21, we obtain

$$\delta = 0.38x\text{Re}_x^{-1/5}$$

$$= 0.38 \times 40 \times \left(\frac{10 \times 40}{10^{-6}}\right)^{-1/5} = 0.29 \text{ m}$$

This value is 26% too low. The drag force is found to be

$$\text{drag} = 0.074\text{Re}_L^{-1/5} \times \tfrac{1}{2}\rho U_\infty^2 Lw$$

$$= 0.074\left(\frac{10 \times 40}{10^{-6}}\right)^{-1/5} \times \frac{1}{2} \times 1000 \times 10^2 \times 40 \times 8 = 22\,500 \text{ N}$$

This value is 22% too low. Obviously, the power-law equations used with $\text{Re}_L = 4 \times 10^8$ are expected to be in substantial error.

8.6.6 Laminar Boundary Layer Equations

The solution presented in Section 8.6.3 for the laminar boundary layer was an approximate solution using a cubic polynomial to approximate the velocity profile. In this section we simplify the Navier–Stokes equations and present a more accurate solution for the laminar boundary layer on a flat plate with a zero pressure gradient.

The x-component Navier–Stokes equation for a steady, incompressible, plane flow is (see Eq. 5.3.14)

$$u \frac{\partial u}{\partial x} + v \frac{\partial u}{\partial y} = -\frac{1}{\rho}\frac{\partial p}{\partial x} + \nu\left(\frac{\partial^2 u}{\partial x^2} + \frac{\partial^2 u}{\partial y^2}\right) \qquad (8.6.43)$$

In boundary layer theory the boundary layer is assumed to be very thin (see Fig. 8.22), so that there is no pressure variation in the y-direction in the boundary layer, that is, $p = p(x)$. In addition (this is a very important point), the pressure $p(x)$ is given by the inviscid flow solution as the wall pressure; hence the pressure is not an unknown. This leaves only two unknowns, u and v. Equation 8.6.43 provides us with one equation and the continuity equation

$$\frac{\partial u}{\partial x} + \frac{\partial v}{\partial y} = 0 \tag{8.6.44}$$

provides us with the other. The y-component Navier–Stokes equation is not of use in boundary layer theory since all the terms are negligibly small.

In addition to the simplification that a known pressure gradient provides, it is obvious that $\partial^2 u/\partial x^2$ is much, much less than the large gradients that exist in the y-direction (refer to the sketch of Fig. 8.22); consequently, neglecting $\partial^2 u/\partial x^2$, the boundary layer equation that must be solved is

$$u\frac{\partial u}{\partial x} + v\frac{\partial u}{\partial y} = -\frac{1}{\rho}\frac{dp}{dx} + v\frac{\partial^2 u}{\partial y^2} \tag{8.6.45}$$

where the pressure gradient dp/dx is assumed to be known from the inviscid flow solution. This is often referred to as the **Prandtl boundary layer equation.** Neither term on the left can be neglected; the y-component v may be small, but the velocity gradient $\partial u/\partial y$ is obviously quite large; hence the product must be retained.

Let us focus on flow over a flat plate with zero pressure gradient. In addition, let us introduce the stream function:

$$u = \frac{\partial \psi}{\partial y} \qquad v = -\frac{\partial \psi}{\partial x} \tag{8.6.46}$$

The boundary layer equation becomes

$$\frac{\partial \psi}{\partial y}\frac{\partial^2 \psi}{\partial x \, \partial y} - \frac{\partial \psi}{\partial x}\frac{\partial^2 \psi}{\partial y^2} = v\frac{\partial^3 \psi}{\partial y^3} \tag{8.6.47}$$

In this form the x and y dependence cannot be separated. If we transform this equation (such transformations are selected by trial and error and experience) by letting

$$\phi = x \qquad \eta = y\,\sqrt{\frac{U_\infty}{vx}} \tag{8.6.48}$$

there results

$$-\frac{1}{2\phi}\left(\frac{\partial\psi}{\partial\eta}\right)^2 + \frac{\partial\psi}{\partial\eta}\frac{\partial^2\psi}{\partial\phi\,\partial\eta} - \frac{\partial\psi}{\partial\phi}\frac{\partial^2\psi}{\partial\eta^2} = \nu\frac{\partial^3\psi}{\partial\eta^3}\sqrt{\frac{U_\infty}{\nu\phi}} \qquad (8.6.49)$$

By observing the position of ϕ in this equation, we separate variables by letting

$$\psi(\phi, \eta) = \sqrt{U_\infty\nu\phi}\, F(\eta) \qquad (8.6.50)$$

An ordinary nonlinear, differential equation results; it is

$$F\frac{d^2F}{d\eta^2} + 2\frac{d^3F}{d\eta^3} = 0 \qquad (8.6.51)$$

The boundary conditions [$u(x, 0) = 0$, $v(x, 0) = 0$ and $u(x, y > \delta) = U_\infty$] take the form

$$F = F' = 0 \text{ at } \eta = 0 \qquad F' = 1 \text{ at large } \eta \qquad (8.6.52)$$

The velocity components can be shown to be

$$u = \frac{\partial\psi}{\partial y} = U_\infty F'(\eta)$$

$$v = -\frac{\partial\psi}{\partial x} = \frac{1}{2}\sqrt{\frac{\nu U_\infty}{x}}(\eta F' - F) \qquad (8.6.53)$$

The boundary value problem, consisting of the ordinary differential equation (8.6.51) and boundary conditions (8.6.52), can be solved numerically using a computer. The results are tabulated in Table 8.5.

TABLE 8.5   SOLUTION FOR THE LAMINAR BOUNDARY LAYER WITH $dp/dx = 0$

$\eta = y\sqrt{\dfrac{U_\infty}{\nu x}}$	F	$F' = u/U_\infty$	$\frac{1}{2}(\eta F' - F)$	F''
0	0	0	0	0.3321
1	0.1656	0.3298	0.0821	0.3230
2	0.6500	0.6298	0.3005	0.2668
3	1.397	0.8461	0.5708	0.1614
4	2.306	0.9555	0.7581	0.0642
5	3.283	0.9916	0.8379	0.0159
6	4.280	0.9990	0.8572	0.0024
7	5.279	0.9999	0.8604	0.0002
8	6.279	1.0000	0.8605	0.0000

Defining the boundary layer thickness to be the point where $u = 0.99U_\infty$, we see from Table 8.5 that this occurs where $\eta = 5$. Hence, letting $\eta = 5$ and $y = \delta$ in Eq. 8.6.48, we have

$$\delta = 5\sqrt{\frac{\nu x}{U_\infty}} \tag{8.6.54}$$

Using

$$\frac{\partial u}{\partial y} = \frac{\partial u}{\partial \eta}\frac{\partial \eta}{\partial y} = U_\infty F''\sqrt{\frac{U_\infty}{\nu x}} \tag{8.6.55}$$

the wall shear in a laminar boundary layer with $dp/dx = 0$ is

$$\tau_0 = \mu\left.\frac{\partial u}{\partial y}\right|_{y=0} = 0.332\rho U_\infty^2\sqrt{\frac{\nu}{xU_\infty}} \tag{8.6.56}$$

The local skin friction coefficient is

$$c_f = \frac{0.664}{\sqrt{\mathrm{Re}_x}} \tag{8.6.57}$$

and the skin friction coefficient is

$$C_f = \frac{1.33}{\sqrt{\mathrm{Re}_L}} \tag{8.6.58}$$

Numerically integrating Eqs. 8.6.6 and 8.6.7, the displacement and momentum thicknesses are found to be

$$\delta_d = 1.72\sqrt{\frac{\nu x}{U_\infty}} \qquad \theta = 0.644\sqrt{\frac{\nu x}{U_\infty}} \tag{8.6.59}$$

EXAMPLE 8.17

Atmospheric air at 20°C flows over a 8-m-long, 2-m-wide flat plate at 2 m/s. Assume that laminar flow exists in the boundary layer over the entire length. At $x = 8$ m, calculate (a) the maximum value of v, (b) the wall shear, and (c) the flow rate through the layer. (d) Also, calculate the drag force on the plate.

Solution

(a) The y-component of velocity has been assumed to be small in boundary layer theory. Its maximum value at $x = 8$ m is found using 8.6.53 to be

$$v = \sqrt{\frac{\nu U_\infty}{x}} \times \frac{1}{2}(\eta F' - F)$$

$$= \sqrt{\frac{1.6 \times 10^{-5} \times 2}{8}} \times 0.86 = 0.00172 \text{ m/s}$$

(b) The wall shear at $x = 8$ m is found using Eq. 8.6.56 to be

$$\tau_0 = 0.332 \rho U_\infty^2 \sqrt{\frac{\nu}{U_\infty x}}$$

$$= 0.332 \times 1.20 \times 2^2 \sqrt{\frac{1.6 \times 10^{-5}}{2 \times 8}} = 0.00159 \text{ Pa}$$

(c) The flow rate through the boundary layer at $x = 8$ m is given by

$$Q = \int_0^\delta u \times 2dy = 2 \sqrt{\frac{\nu x}{U_\infty}} \int_0^5 U_\infty F' \, d\eta$$

where we have substituted for u and y from Eqs. 8.6.53 and 8.6.48. Recognizing that $\int F' \, d\eta = F$, the flow rate is

$$Q = 2U_\infty \sqrt{\frac{\nu x}{U_\infty}} [F(5) - F(0)^0]$$

$$= 2 \times 2 \sqrt{\frac{1.6 \times 10^{-5} \times 8}{2}} \times 3.28 = 0.105 \text{ m}^3/\text{s}$$

(d) The drag force is determined as follows:

$$\text{drag} = \tfrac{1}{2}\rho U_\infty^2 L w C_f$$

$$= \frac{1}{2} \times 1.20 \times 2^2 \times 8 \times 2 \times \frac{1.33}{\sqrt{2 \times 8/1.70 \times 10^{-5}}} = 0.0511 \text{ N}$$

8.6.7 Pressure Gradient Effects

In the foregoing sections we have concentrated our study of boundary layers on the flat plate with a zero pressure gradient. This is the simplest boundary layer flow and allows us to model many flows of engineering interest. The inclusion of a pressure gradient, even though relatively low, markedly alters the boundary layer flow. In fact, a strong negative pressure gradient (such as flow in a contraction) may relaminarize a turbulent boundary layer; that is, the turbulence production in the viscous wall layer which sustains the turbulence ceases and a laminar boundary layer is reestablished. A positive pressure gradient quickly causes the boundary layer to thicken and eventually to separate. These two effects are shown in the photographs of Fig. 8.25.

The flow about any plane body with curvature, such as an airfoil, can be modeled as flow over a flat plate with a nonzero pressure gradient. The boundary layer thickness is so much smaller than the radius of curvature that the additional curvature terms drop out of the differential equations. The inviscid flow solution at the wall provides the pressure gradient dp/dx and the velocity $U(x)$ at the edge of the boundary layer. For axisymmetric flows, such as flow over the nose of an aircraft, boundary layer equations in cylindrical coordinates must be utilized.

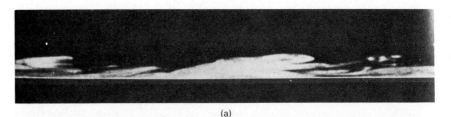

(a)

Figure 8.25 Influence of a strong pressure gradient on a turbulent flow: (a) a strong negative pressure gradient may relaminarize a flow; (b) a strong positive pressure gradient causes a strong boundary layer to thicken. (Photographs by R. E. Falco.)

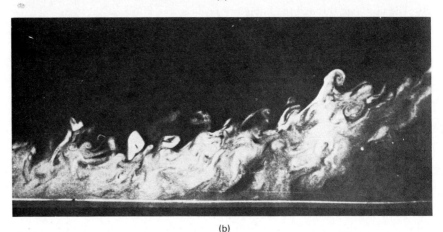

(b)

The pressure gradient determines the value of the second derivative $\partial^2 u/\partial y^2$ at the wall. From the boundary layer equation (8.6.45) at the wall, $u = v = 0$, so that

$$\frac{dp}{dx} = \mu \frac{\partial^2 u}{\partial y^2}\bigg|_{y=0} \qquad (8.6.60)$$

for either a laminar or a turbulent boundary layer flow. For a zero pressure gradient the second derivative is zero at the wall; then since the first derivative is greatest at the wall and decreases as y increases, the second derivative must be negative for positive y. The profiles are sketched in Fig. 8.26a.

For a negative (favorable) pressure gradient the slope of the velocity profile near the wall is relatively large with a negative second derivative at the wall and throughout the layer. The momentum near the wall is larger than that of the zero pressure gradient flow, as shown in Fig. 8.26b, and thus there is a reduced tendency for the flow to separate. Turbulence production is discouraged and relaminarization may occur for a sufficiently large negative pressure gradient.

If a positive (unfavorable) pressure gradient is imposed on the flow the second derivative at the wall will be positive and the flow will be as sketched in part (c) or (d). If the unfavorable pressure gradient acts over a sufficient distance, part (d) will probably represent the flow situation; the

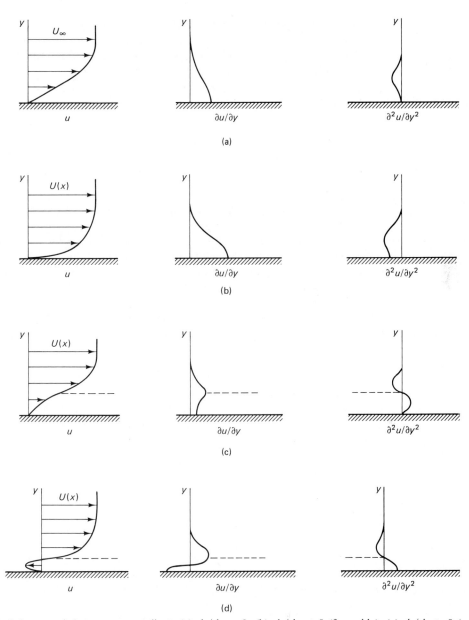

Figure 8.26   Influence of the pressure gradient: (a) $dp/dx = 0$; (b) $dp/dx < 0$ (favorable); (c) $dp/dx > 0$ (unfavorable); (d) $dp/dx > 0$ (separated flow).

flow will separate from the surface. Near the wall the higher pressure downstream will drive the low momentum flow near the wall in the up-stream direction, resulting in flow reversal, as shown. The point at which $\partial u/\partial y = 0$ at the wall locates the point of separation.

The problem of a laminar boundary layer with a pressure gradient can be solved using conventional numerical techniques with a computer. The procedure is relatively simple using the simplified boundary layer equation (8.6.45) with a known pressure gradient. For a turbulent flow the Reynolds stress term must be included; much work is being done to develop models of turbulent quantities that will result in acceptable numerical solutions. Experimental results are often necessary for turbulent flow problems, as was the situation for internal flows.

PROBLEMS

Separated Flows

8.1. Sketch the flow over an airfoil at a large angle of attack for both attached flow and separated flow. Also, sketch the expected pressure distributions on the top and bottom surfaces for both flows.

8.2. A spherical particle is moving in 20°C atmospheric air at a velocity of 20 m/s. What must its diameter be for Re = 5 and Re = 10^5? Sketch the expected flow field at these Reynolds numbers.

8.3. A 2-cm-diameter sphere is to move with Re = 5. At what velocity does it travel if it is submerged in:
(a) Water at 20°C?
(b) Water at 90°C?
(c) Standard air at 20°C?

8.3E. A 0.8-in.-diameter sphere is to move with Re = 5. At what velocity does it travel if it is submerged in:
(a) Water at 60°F?
(b) Water at 180°F?
(c) Standard air at 60°F?

8.4. Air at 20°C flows around a cylindrical body at a velocity of 20 m/s. Calculate the Reynolds number if the body is:
(a) A 6-m-diameter smokestack.
(b) A 6-cm-diameter flagpole.
(c) A 6-mm-diameter wire.
Use Re = VD/ν. Would a separated flow be expected?

8.5. If the drag coefficient for a 10-cm-diameter sphere is given by C_D = 1.0, calculate the drag if the sphere is falling in the atmosphere:
(a) At sea level.
(b) At 30 000 m.
(c) In 10°C water.

8.6. Sketch the flow that is expected over a semitruck (a tractor and trailer) where the trailer is substantially higher than the tractor with and without an air deflector attached to the roof of the tractor. Sketch a side view indicating any separated regions, boundary layers, and the wake.

8.7. Air is blowing by a long, rectangular building. Sketch the expected flow showing separated flow regions, the inviscid flow region, the boundary layers, and the wake. Sketch the top view with the wind blowing parallel to the long sides.

8.8. Calculate the drag on a smooth 50-cm-diameter sphere when subjected to a 20°C–atmospheric airflow of:
(a) 6 m/s.
(b) 15 m/s.

8.9. A 4.45-cm-diameter golf ball is roughened to reduce the drag during its flight. If the Reynolds number at which the sudden drop occurs is reduced from 3×10^5 to 6×10^4 by the roughening (dimples), would you expect this to lengthen the flight of a golf ball significantly? Justify your reasoning with appropriate calculations.

8.10. A 10-cm-diameter smooth sphere experiences a drag of 2.0 N when placed in atmospheric 20°C air.
(a) What is the velocity of the airstream?
(b) At what increased speed will the sphere experience the same drag?

8.10E. A 4-in.-diameter smooth sphere experiences a drag of 0.5 lb when placed in standard 60°F air.
(a) What is the velocity of the airstream?
(b) At what increased speed will the sphere experience the same drag?

8.11. A smooth, 20-cm-diameter sphere experiences a drag of 4.2 N when placed in a 20°C water channel. Calculate the drag coefficient and the Reynolds number.

8.12. A 2-m-diameter smokestack stands 60 m high. It is designed to resist a 40-m/s wind. At this speed, what total force would be expected, and what moment would the base be required to resist? Assume atmospheric 20°C air.

8.13. A flagpole is composed of three sections: a 5-cm-diameter top section that is 10 m long, a 7.5-cm-diameter middle section 15 m long, and a 10-cm-diameter bottom section 20 m long. Calculate the total force acting on the flagpole and the resisting moment provided by the base when subjected to a 25-m/s wind speed. Make the calculations for
(a) A winter day at −30°C.
(b) A summer day at 35°C.

8.14. A drag force of 42 N is desired at Re = 10^5 on a 2-m-long cylinder in a 20°C–atmospheric airflow. What velocity should be selected, and what should be the cylinder diameter?

8.14E. A drag force of 10 lb is desired at Re = 10^5 on a 6-ft-long cylinder in a 60°F–atmospheric airflow. What velocity should be selected, and what should be the cylinder diameter?

8.15. A 20-m-high structure is 2 m in diameter at the top and 8 m in diameter at the bottom. If the diameter varies linearly with height, estimate the total drag force due to a 30-m/s wind. Use atmospheric air at 20°C.

8.16. A steel sphere (S = 7.82) is dropped in water at 20°C. Calculate the terminal velocity if the diameter of the sphere is:
(a) 10 cm.
(b) 5 cm.
(c) 1 cm.
(d) 2 mm.

8.17. Estimate the terminal velocity of a 50-cm-diameter sphere as it falls in a 20°C atmosphere near the earth if it has a specific gravity of:
(a) 0.005.
(b) 0.02.
(c) 1.0.

8.17E. Estimate the terminal velocity of a 20-in.-diameter sphere as it falls in a 60°F atmosphere near the earth if it has a specific gravity of:
(a) 0.005.
(b) 0.02.
(c) 1.0.

8.18. Estimate the terminal velocity of a skydiver by making reasonable approximations of the arms, legs, head, and body. Assume air at 20°C.

8.19. Assuming that the drag on a modern automobile at high speeds is due primarily to form drag, estimate the power (horsepower) needed by an automobile with a 3.2-m² cross-sectional area to travel at:
(a) 80 km/h.
(b) 90 km/h.
(c) 100 km/h.

8.20. The 2 m × 3 m sign weighs 400 N. What wind speed is required to blow over the sign?

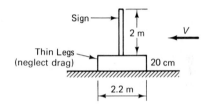

8.21. A rectangular car carrier has a 1.8 m × 0.6 m cross section. Estimate the minimum added power (horsepower) required to travel at 100 km/h because of the car carrier.

8.21E. A rectangular car carrier has a 6 ft × 2 ft cross section. Estimate the minimum added power (horsepower) required to travel at 60 mph because of the car carrier.

8.22. Assume that the velocity at the corners of an automobile where the rear-view mirror is located is 1.6 times the automobile's velocity. How much horsepower is required by the two 10-cm-diameter rearview mirrors for an automobile speed of 100 km/h?

8.23. An automobile with a cross-sectional area of 3 m² is powered by a 40-hp engine. Estimate the maximum possible speed if the drive train is 90% efficient. (The engine is rated by the power produced before the transmission.)

Vortex Shedding

8.24. Over what range of velocities would you expect vortex shedding from a telephone wire 3 mm in diameter? Could you hear any of the vorticies being shed?

(Human beings can hear frequencies between 20 and 20 000 Hz.)

8.25. A wire is being towed through 20°C water normal to its axis at a speed of 2 m/s. What diameter (both large and small) could the wire have so that vortex shedding would not occur?

8.25E. A wire is being towed through 60°F water normal to its axis at a speed of 6 ft/sec. What diameter (both large and small) could the wire have so that vortex shedding would not occur?

8.26. It is quite difficult to measure low velocities. To determine the velocity in a low-speed airflow, the vorticies being shed from a 10-cm-diameter cylinder are observed to occur at 0.2 Hz. Estimate the airspeed if the temperature is 20°C.

8.27. Motion pictures show that vorticies are shed from a 2-m-diameter cylinder at 0.001 Hz while it is moving in 20°C water. What is the cylinder velocity?

Streamlining

8.28. A 15-cm-diameter smokestack on a semitruck extends 2 m straight up. Estimate the horsepower required by the stack for a speed of 90 km/h. If the stack were streamlined, estimate the reduced horsepower.

8.28E. A 6-in.-diameter smokestack on a semitruck extends 6 ft straight up. Estimate the horsepower needed because of the stack for a speed of 60 mph. If the stack were streamlined, estimate the reduced horsepower.

8.29. A wind speed of 3 m/s blows normal to an 8-cm-diameter smooth cylinder which is 2 m long. Calculate the drag force. The cylinder is now streamlined. What is the percentage reduction in the drag? Assume that $T = 20°C$.

Cavitation

8.30. The critical cavitation number for a streamlined strut is 0.7. Find the maximum velocity of the body to which the strut is attached if cavitation is to be avoided. The body is traveling 5 m beneath a water surface.

8.31. A hydrofoil, designed to operate at a depth of 40 cm, has a 40 cm chord and is 10 m long. A lift force of 200 kN is desired at a speed of 12 m/s. Calculate the angle of attack and the drag force. Is cavitation present under these conditions?

8.31E. A hydrofoil, designed to operate at a depth of 16 in., has a 16-in. chord and is 30 ft long. A lift force of 50,000 lb is desired at a speed of 35 ft/sec. Calculate the angle of attack and the drag force. Is cavitation present under these conditions?

8.32. A body resembling a sphere has a diameter of approximately 0.8 m. It is towed at a speed of 20 m/s, 5 m below the water surface. Estimate the drag acting on the body.

Added Mass

8.33. A 40-cm-diameter sphere, which weighs 400 N, is released from rest while submerged in water. Calculate its initial acceleration:
(a) Neglecting the added mass.
(b) Including the added mass.

8.34. A submersible, whose length is twice its maximum diameter, resembles an ellipsoid. If its added mass is ignored, what is the percentage error in a calculation of its initial acceleration if its specific gravity is 1.2?

Lift and Drag on Airfoils

8.35. An aircraft with a mass of 1000 kg is designed to cruise at a speed of 80 m/s at 10 km. The wing area is approximately 15 m². Determine the lift coefficient and the angle of attack. What power is required for cruise?

8.36. A 1500-kg aircraft is designed to carry a payload of 3000 N when cruising at 80 m/s at 10 km. The wing area is 20 m². Calculate:
(a) The takeoff speed if an angle of attack of 10° is desired.
(b) The stall speed.
(c) The power required at cruise.

Vorticity

8.37. Take the curl of the Navier–Stokes equation and show that the vorticity equation (8.5.3) results. Use the vector identities from the Appendix.

8.38. Write the three component vorticity equations contained in Eq. 8.5.3 using rectangular coordinates. Use $\boldsymbol{\omega} = \omega_x \hat{i} + \omega_y \hat{j} + \omega_z \hat{k}$.

8.39. Simplify the vorticity equation (8.5.3) for a plane flow ($w = 0$ and $\partial/\partial z = 0$). Use $\boldsymbol{\omega} = (\omega_x, \omega_y, \omega_z)$. What conclusion can be made about the magnitude of ω_z in an inviscid, plane flow (such as flow through a short contraction) that contains vorticity?

Stream Function and Velocity Potential

8.40. An attempt is to be made to solve Laplace's equation for flow around a circular cylinder of radius r_c oriented in the center of a channel of height $2h$. The velocity profile far from the cylinder is uniform. State the necessary boundary conditions. Assume that $\psi = 0$ at $y = -h$. The origin of the coordinate system is located at the center of the cylinder.

8.41. State the stream function and velocity potential corresponding to a uniform velocity of $100\hat{i} + 50\hat{j}$ using rectangular coordinates.

8.42. A stream function is given by

$$\psi = 10y - 10y/(x^2 + y^2).$$

(a) Show that this satisfies $\nabla^2 \psi = 0$.
(b) Find the velocity potential $\phi(x, y)$.
(c) Assuming water to be flowing, find the pressure along the x-axis if $p = 50$ kPa at $x = -\infty$.
(d) Locate any stagnation points.

8.43. The velocity potential for a flow is

$$\phi = 10x + 5\ln(x^2 + y^2).$$

(a) Show that this function satisfies Laplace's equation.
(b) Find the stream function $\psi(x, y)$.
(c) Assume that water is flowing and find the pressure along the x-axis if $p = 100$ kPa at $x = -\infty$.
(d) Locate any stagnation points.
(e) Find the acceleration at $x = -2$ m, $y = 0$.

8.44. The velocity profile in a wide, 0.2-m-high channel is given by $u(y) = y - y^2/0.2$. Determine the stream function for this flow. Calculate the flow rate by integrating the velocity profile and by using $\Delta\psi$. Explain why a velocity potential does not exist by referring to Eq. 8.5.2.

Superposition of Simple Flows

8.45. Sketch the body formed by superimposing a source at the origin of strength $\pi/2$ m²/s and a uniform flow of 10 m/s.
(a) Locate any stagnation points.
(b) Find the y-intercept of the body.
(c) Find the thickness of the body at $x = \infty$.
(d) Find u at $x = -4$ cm, $y = 0$.

8.45E. Sketch the body formed by superimposing a source at the origin of strength 5π ft²/sec and a uniform flow of 30 ft/sec.
(a) Locate any stagnation points.
(b) Find the y-intercept of the body.
(c) Find the thickness of the body at $x = \infty$.
(d) Find u at $x = -12$ in., $y = 0$.

8.46. A source with strength π m²/s and a sink of equal strength are located at $(-1, 0)$ and $(1, 0)$, respectively. They are combined with a uniform flow $U_\infty =$ 10 m/s to form a **Rankin oval**. Calculate the length and maximum thickness of the oval. If $p = 10$ kPa at $x = -\infty$, find the minimum pressure if water is flowing.

8.47. An oval is formed from a source and sink of strengths 2π m²/s located at $(-1, 0)$ and $(1, 0)$, respectively, combined with a uniform flow of 2 m/s. Locate any stagnation points, and find the velocity at $(-4, 0)$ and $(0, 4)$.

8.48. Two sources of strength 2π m²/s are located at $(0, 1)$ and $(0, -1)$, respectively. Sketch the resulting flow and locate any stagnation points. Find the velocity at $(1, 1)$.

8.48E. Two sources of strength 20π ft²/sec are located at $(0, 6)$ and $(0, -6)$, respectively. Sketch the resulting flow and locate any stagnation points. Find the velocity at $(3, 3)$.

8.49. The two sources of Problem 8.48 are superimposed with a uniform flow. Sketch the flow, locate any stagnation points, and find the y-intercept of the body formed if:
(a) $U_\infty = 10$ m/s.
(b) $U_\infty = 1$ m/s.
(c) $U_\infty = 0.2$ m/s.

8.50. A cylinder is formed by combining a doublet of strength 40 m³/s with a uniform flow of 10 m/s.
(a) Sketch the velocity along the y-axis from the cylinder to $y = \infty$.
(b) Calculate the velocity at $(x = -4, y = 3)$.
(c) Calculate the drag coefficient for the cylinder assuming potential flow over the front half and constant pressure over the back half.

8.51. A 2-m-diameter cylinder is placed in a uniform water flow of 4 m/s.
(a) Sketch the velocity along the x-axis from the cylinder to $x = -\infty$.
(b) Find v_θ on the front half of the cylinder.
(c) Find $p(\theta)$ on the front half of the cylinder if $p = 50$ kPa at $x = -\infty$.
(d) Estimate the drag force on a 1-m-length of the cylinder if the pressure over the rear half is constant.

8.52. Superimpose a free stream $U_\infty = 10$ m/s, a doublet $\mu = 40$ m³/s, and a vortex $\Gamma = 200$ m²/s. Locate any stagnation points and predict the minimum and maximum pressure on the surface of the cylinder if $p = 0$ at $x = -\infty$ and atmospheric air is flowing.

8.52E. Superimpose a free stream $U_\infty = 30$ ft/sec, a doublet $\mu = 400$ ft²/sec, and a vortex $\Gamma = 2000$ ft²/sec. Locate any stagnation points and predict the minimum and maximum pressure on the surface of the cylinder if $p = 0$ at $x = -\infty$ and atmospheric air is flowing.

8.53. A 0.8-m-diameter cylinder is placed in a 20 m/s-atmospheric airflow. At what rotational speed should the cylinder rotate so that only one stagnation point

exists on its surface? Calculate the minimum pressure acting on the cylinder if $p = 0$ at $x = -\infty$.

8.54. A 1.2-m-diameter cylinder rotates at 120 rpm in a 3-m/s-atmospheric airstream. Locate any stagnation points and calculate the minimum and maximum pressures on the cylinder if $p = 0$ at $x = -\infty$.

8.55. The circulation around a 20-m-long airfoil (measured tip to tip) is calculated to have a value of 1500 m²/s. Estimate the lift generated by the airfoil if the aircraft is flying at 10 000 m with a speed of 100 m/s.

8.55E. The circulation around a 60-ft-long airfoil (measured tip to tip) is calculated to have a value of 15 000 ft²/sec. Estimate the lift generated by the airfoil if the aircraft is flying at 30 000 ft with a speed of 350 ft/sec. Assume the flow to be incompressible.

8.56. The velocity field due to a source of strength 2π m²/s per meter, located at $(2, 2)$ in a 90° corner, is desired. Use the **method of images,** that is, add one or more sources at the appropriate locations, and determine the velocity field by finding $u(x, y)$ and $v(x, y)$.

8.57. Flow through a porous media is modeled with Laplace's equation and an associated velocity potential function. Natural gas can be stored in certain underground rock structures for use at a later time. A well is placed next to an impervious rock formation, as shown. If the well is to extract 0.2 m³/s per meter, predict the velocity to be expected at the point $(4, 3)$. See Problem 8.56 for the method of images.

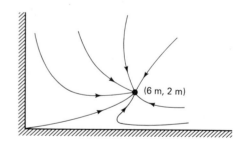

Boundary Layers

8.58. How far from the leading edge can turbulence be expected on an airfoil traveling at 100 m/s if the elevation is:
(a) 0 m?
(b) 4000 m?
(c) 10 000 m?
Use $\text{Re}_{crit} = 6 \times 10^5$.

8.58E. How far from the leading edge can turbulence be expected on an airfoil traveling at 300 ft/sec if the elevation is:
(a) 0 ft?
(b) 12 000 ft?
(c) 30 000 ft?
Use $\text{Re}_{crit} = 6 \times 10^5$.

8.59. A laminar region is desired to be at least 2 m long on a smooth flat plate. A wind tunnel and a water channel are available. What maximum speed can be selected for each? Assume low free-stream fluctuation intensity.

8.60. Assuming inviscid, uniform flow of air through the contraction shown, estimate $U(x)$ and dp/dx, which are necessary to solve for the boundary layer growth on the flat plate. Assume a one-dimensional flow with $\rho = 1.0 \text{ kg/m}^3$.

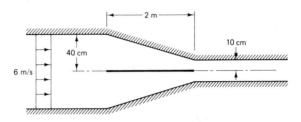

Von Kármán Integral Equation

8.61. Show that the von Kármán integral equation (8.6.4) can be put in the form

$$\tau_0 = -\delta \frac{dp}{dx} + \rho \frac{d}{dx} \int_0^\delta u(U - u) \, dy - \rho \frac{dU}{dx} \int_0^\delta u \, dy$$

Note that the quantity $\int_0^\delta u \, dy$ is only a function of x.

8.62. Show that the von Kármán integral equation of Problem 8.61 can be written as

$$\tau_0 = \rho \frac{d}{dx} (\theta U^2) + \rho \delta_d U \frac{dU}{dx}$$

Note: Bernoulli's equation $p + \rho U^2/2 = \text{const.}$ can be differentiated to yield

$$\frac{dp}{dx} = -\rho U \frac{dU}{dx} = -\frac{1}{\delta} \rho \frac{dU}{dx} \int_0^\delta U \, dy$$

8.63. Assume that $u = U_\infty \sin(\pi y/2\delta)$ in a zero pressure gradient boundary layer. Calculate:
(a) $\delta(x)$.
(b) $\tau_0(x)$.
(c) v at $y = \delta$ and $x = 3$ m.

8.64. Assume a linear velocity profile and find $\delta(x)$ and $\tau_o(x)$. Compute the percentage error when compared with the exact expressions for a laminar flow. Use $dp/dx = 0$.

8.65. If the walls in a wind-tunnel test section are parallel, the velocity in the center portion of the tunnel will accelerate as shown. To maintain a constant tunnel velocity so that $dp/dx = 0$, show that the walls should be displaced outward a distance $\delta_d(x)$. If a wind tunnel were square, how far should one wall be displaced outward for $dp/dx = 0$?

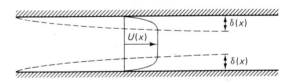

8.66. Find δ_d and θ for a laminar boundary layer assuming:
(a) A cubic profile.
(b) A parabolic profile.
(c) That $u = U_\infty \sin(\pi y/2\delta)$.
Compute percentage errors when compared with the exact values of $\delta_d = 1.72\sqrt{\nu x/U_\infty}$ and $\theta = 0.644\sqrt{\nu x/U_\infty}$.

8.67. A laminar flow is maintained in a boundary layer on a 6-m-long, 5-m-wide flat plate with 20°C atmospheric air flowing at 4 m/s. Assuming a cubic profile, calculate:
(a) δ at $x = 6$ m.
(b) τ_0 at $x = 6$m.
(c) The drag force on one side.
(d) v at $y = \delta$ and $x = 3$ m.

8.67E. A laminar flow is maintained in a boundary layer on a 20-ft-long, 15-ft-wide flat plate with 60°F atmospheric air flowing at 12 ft/sec. Assuming a cubic profile, calculate:
(a) δ at $x = 20$ ft.
(b) τ_0 at $x = 20$ ft.
(c) The drag force on one side.
(d) v at $y = \delta$ and $x = 10$ ft.

8.68. Work Problem 8.67, but assume a parabolic profile.

Laminar and Turbulent Boundary Layers

8.69. Atmospheric air at 20°C flows at 10 m/s over a 2-m-long, 4-m-wide flat plate. Calculate the maximum boundary layer thickness and the drag force on one side assuming:
(a) Laminar flow over the entire length.
(b) Turbulent flow over the entire length.

8.70. Fluid flows over a flat plate at 20 m/s. Determine δ and τ_0 at $x = 6$ m if the fluid is:
(a) Atmospheric air at 20°C.
(b) Water at 20°C.
Neglect the laminar portion.

8.71. Assume a turbulent velocity profile $\bar{u} = U_\infty (y/\delta)^{1/7}$. Does this profile satisfy the conditions at $y = \delta$? Can it give the shear stress at the wall? Plot both a cubic laminar profile and the one-seventh power law profile on the same graph assuming the same boundary layer thickness.

8.72. Estimate the drag on one side of a 4-m-long, 5-m-wide flat plate if atmospheric air at 20°C is flowing with a velocity of 6 m/s. Assume that:
(a) $\text{Re}_{\text{crit}} = 3 \times 10^5$.
(b) $\text{Re}_{\text{crit}} = 5 \times 10^5$.
(c) $\text{Re}_{\text{crit}} = 6 \times 10^5$.

8.72E. Estimate the drag on one side of a 12-ft-long, 15-ft-wide flat plate if atmospheric air at 60°F is flowing with a velocity of 20 ft/sec. Assume that:
(a) $\text{Re}_{\text{crit}} = 3 \times 10^5$.
(b) $\text{Re}_{\text{crit}} = 5 \times 10^5$.
(c) $\text{Re}_{\text{crit}} = 6 \times 10^5$.

8.73. A 1-m-long flat plate that is 2 m wide is towed parallel to itself in 20°C water at 1.2 m/s. Estimate the total drag if:
(a) $\text{Re}_{\text{crit}} = 3 \times 10^5$.
(b) $\text{Re}_{\text{crit}} = 5 \times 10^5$.
(c) $\text{Re}_{\text{crit}} = 6 \times 10^6$.

8.74. Air moving at 60 km/h is considered to have a zero boundary layer thickness at a distance of 100 km from shore. At the beach estimate the boundary layer thickness and the wall shear using:
(a) A one-seventh power law.
(b) Empirical data.
Use $T = 20$°C.

8.75. For the conditions of Problem 8.74, calculate:
(a) The thickness of the viscous wall layer.
(b) The displacement thickness at the beach.

8.76. Atmospheric air at 20°C flows over a flat plate at 100 m/s. At $x = 6$ m estimate:
(a) The local skin friction coefficient.
(b) The wall shear.
(c) The viscous wall layer thickness.
(d) The boundary layer thickness.

8.76E. Atmospheric air at 60°F flows over a flat plate at 300 ft/sec. At $x = 20$ ft, estimate:
(a) The local skin friction coefficient.
(b) The wall shear.
(c) The viscous wall layer thickness.
(d) The boundary layer thickness.

8.77. Water at 20°C flows over a flat plate at 10 m/s. At $x = 3$ m, estimate:
(a) The viscous wall layer thickness.
(b) The velocity at the edge of the viscous wall layer.
(c) The value of y at the outer edge of the turbulent zone.
(d) The boundary layer thickness.

8.78. Estimate the total shear drag on a ship traveling at 10 m/s if the sides are 10 m × 100 m. What is the maximum boundary layer thickness?

8.79. A dirigible (blimp) is shaped like a long cigar, 30 m in diameter and 300 m long. If the drag is due primarily to wall shear, estimate the power needed to propel the dirigible at 20 m/s at an elevation of 1000 m. Neglect the drag over the nose.

8.79E. A dirigible (blimp) is shaped like a long cigar, 100 ft in diameter and 1000 ft long. If the drag is due primarily to wall shear, estimate the power needed to propel the dirigible at 60 ft/sec at an elevation of 3000 ft. Neglect the drag over the nose.

Laminar Boundary Layer Equations

8.80. Assuming that $dp/dx = 0$, show that Eq. 8.6.47 follows from Eq. 8.6.45.

8.81. Show that Eq. 8.6.53 follows from the preceding equations.

8.82. A laminar boundary layer exists on a flat plate with atmospheric air at 20°C moving at 5 m/s. At $x = 2$ m, find:
(a) The wall shear.
(b) The boundary layer thickness.
(c) The maximum value of v.
(d) The flow rate through the boundary layer.

8.82E. A laminar boundary layer exists on a flat plate with atmospheric air at 60°F moving at 15 ft/sec. At $x = 6$ ft, find:
(a) The wall shear.
(b) The boundary layer thickness.
(c) The maximum value of v.
(d) The flow rate through the boundary layer.

8.83. Sketch the boundary layer velocity profile near the end of the flat plate of Problem 8.67 and show a Blasius profile of the same thickness on the same sketch.

Pressure Gradient Effects

8.84. Sketch the velocity profiles near and normal to the cylinder's surface at each of the points indicated. The flow separates at C.

8.85. Sketch the expected boundary layer profiles at each of the points indicated, showing relative thicknesses. The flow undergoes transition to turbulence just after point A. It separates at D. Show all profiles on the same plot. Indicate the sign of the pressure gradient at each point.

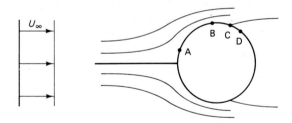

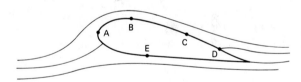

NINE

Compressible Flow

9.1 INTRODUCTION

In this chapter we consider flows in which the density of fluid changes significantly between points on a streamline; such flows are called **compressible flows.** It will be necessary to use the integral equations: the continuity equation, the energy equation, and for some problems, the momentum equation.

Not all gas flows are compressible flows, neither are all compressible flows gas flows. At low speeds, less than a Mach number (M = $V/\sqrt{kRT}$) of about 0.3, gas flows may be treated as incompressible flows. This is justified because the density variations caused by the flow are negligible (less than about 3%). Incompressible gas flows occur in a large number of situations of engineering interest; many of these have been considered in earlier chapters. There are many flows, however, in which the density variations must be accounted for. Included among these are airflows around commerical and military aircraft, airflow though jet engines, and the fluid flow in compressors and turbines. There are examples of compressible effects important in liquid flows; water hammer and compression waves due to underwater blasts are examples of compressible liquid flows. Compressibility of rock accounts for the propagation through the earth's surface of transverse and longitudinal waves due to an earthquake. In this chapter we are concerned primarily with compressibility effects in gas flows.

We will introduce the effects of compressibility into only the simplest of flow situations. The velocity at a given streamwise location in a conduit is assumed to be uniform and hence does not vary normal to the flow direction. This is referred to as **uniform flow.** For this simple flow we recall that the continuity equation takes the form (see Eq. 4.3.5)

$$\dot{m} = \rho_1 A_1 V_1 = \rho_2 A_2 V_2 \tag{9.1.1}$$

The momentum equation for the one-dimensional, compressible flow takes the form (see Eq. 4.5.6)

$$\Sigma \mathbf{F} = \dot{m}(\mathbf{V}_2 - \mathbf{V}_1) \tag{9.1.2}$$

The energy equation, neglecting potential energy changes, is written (see Eq. 4.4.17)

$$\frac{\dot{Q} - \dot{W}_S}{\dot{m}} = \frac{V_2^2 - V_1^2}{2} + h_2 - h_1 \tag{9.1.3}$$

where we have used $h = \bar{u} + p/\rho$. Assuming an ideal gas with constant specific heats, the energy equation takes the form

$$\frac{\dot{Q} - \dot{W}_S}{\dot{m}} = \frac{V_2^2 - V_1^2}{2} + c_p(T_2 - T_1) \tag{9.1.4}$$

or

$$\frac{\dot{Q} - \dot{W}_S}{\dot{m}} = \frac{V_2^2 - V_1^2}{2} + \frac{k}{k-1}\left(\frac{p_2}{\rho_2} - \frac{p_1}{\rho_1}\right) \qquad (9.1.5)$$

where we have used the thermodynamic relations

$$h_2 - h_1 = c_p(T_2 - T_1) \qquad c_p = R + c_v \qquad k = \frac{c_p}{c_v} \qquad (9.1.6)$$

and the ideal-gas law

$$p = \rho R T \qquad (9.1.7)$$

Should we have an interest in calculating the entropy change between two sections, we will use the definition of entropy as

$$\Delta S = \int \frac{\delta Q}{T}\bigg|_{\text{reversible}} \qquad (9.1.8)$$

where δQ represents the differential heat transfer for an ideal gas with constant specific heat. Using the first law, this becomes

$$\Delta s = c_p \ln \frac{T_2}{T_1} - R \ln \frac{p_2}{p_1} \qquad (9.1.9)$$

If a process is isentropic ($\Delta s = 0$), obviously Eq. 9.1.8 shows that the heat transfer is zero. If the flow is isentropic, the relationship above can be used, along with the ideal-gas law, to show that

$$\frac{T_2}{T_1} = \left(\frac{p_2}{p_1}\right)^{(k-1)/k} \qquad \frac{p_2}{p_1} = \left(\frac{\rho_2}{\rho_1}\right)^{k} \qquad (9.1.10)$$

Let us now introduce a parameter, the Mach number, that will be of special interest throughout our study of compressible flow.

9.2 SPEED OF SOUND AND THE MACH NUMBER

The speed of sound is the speed at which a pressure disturbance of small amplitude travels through a fluid. It is analogous to the small ripple, a gravity wave, that travels radially outward when a pebble is dropped into a pond. To determine the speed of sound consider a small pressure disturbance, called a **sound wave,** to be passing through a pipe, as shown in Fig. 9.1. It travels with a velocity c relative to a stationary observer as shown

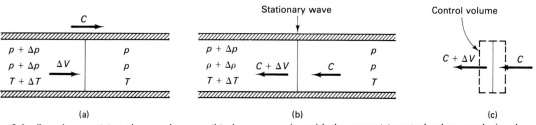

Figure 9.1   Sound wave: (a) stationary observer; (b) observer moving with the wave; (c) control volume enclosing the wave.

in part (a); the pressure, density, and temperature will change by the small amounts Δp, $\Delta \rho$, and ΔT, respectively. There will also be an induced velocity ΔV in the fluid immediately behind the sound wave. To simplify the problem we will create a steady flow by having the observer travel at the speed of the wave so that the sound wave appears to be stationary. The flow will then approach the wave from the right, with the speed of sound c, as shown in Fig. 9.1b. The flow properties will all change across the wave and the velocity in the downstream flow will be expressed as $c + \Delta V$, where ΔV is the small change in velocity. If the flow speed is reduced, the wave would move to the right, and if the flow speed is increased it would move to the left. For the flow speed equal to the speed of sound, $V = c$, the sound wave would be stationary, as shown.

Let us apply the continuity equation and the energy equation to a small control volume enclosing the sound wave, shown in Fig. 9.1c. The continuity equation (9.1.1) takes the form

$$\rho A c = (\rho + \Delta \rho) A (c + \Delta V) \tag{9.2.1}$$

This can be rewritten as

$$\rho \, \Delta V = -c \, \Delta \rho \tag{9.2.2}$$

where we have neglected the higher-order term $\Delta \rho \, \Delta V$; that is, $\Delta \rho$ represents a small percentage change in ρ so that $\Delta \rho \ll \rho$.

The streamwise-component momentum equation yields

$$pA - (p + \Delta p)A = \rho A c (c + \Delta V - c) \tag{9.2.3}$$

which simplifies to, neglecting higher-order terms,

$$-\Delta p = \rho c \, \Delta V \tag{9.2.4}$$

Combining this with Eq. 9.2.2, results in

$$c = \sqrt{\frac{\Delta p}{\Delta \rho}} \tag{9.2.5}$$

Since the changes Δp and $\Delta \rho$ are quite small, we can write

$$\frac{\Delta p}{\Delta \rho} \simeq \frac{dp}{d\rho} \tag{9.2.6}$$

Small-amplitude, moderate-frequency waves (up to about 18 000 Hz) travel with no change in entropy (isentropically), so that

$$\frac{p}{\rho^k} = \text{const.} \tag{9.2.7}$$

This can be differentiated to give

$$\frac{dp}{d\rho} = k \frac{p}{\rho} \tag{9.2.8}$$

Using this in Eq. 9.2.5, the speed of sound c is given by

$$c = \sqrt{\frac{kp}{\rho}} \tag{9.2.9}$$

or, using the ideal-gas law,

$$c = \sqrt{kRT} \tag{9.2.10}$$

At high frequency, sound waves generate friction and the process ceases to be isentropic. It is better approximated by an isothermal process. For an ideal gas an isothermal approximation would lead to

$$c = \sqrt{RT} \tag{9.2.11}$$

For small waves traveling through liquids or solids the **bulk modulus** is used; it has dimensions of pressure and is equal to $\rho \, dp/d\rho$. For water it has a nominal value of 2110 MPa. It does vary slightly with temperature and pressure. Using Eq. 9.2.5, this leads to a speed of propagation of 1450 m/s for a small-amplitude pressure wave in water.

An important quantity used in the study of compressible flow is the dimensionless velocity called the **Mach number,** introduced in Eq. 3.3.3 as

$$\text{M} = \frac{V}{c} \tag{9.2.12}$$

If $\text{M} < 1$, the flow is a **subsonic** flow, and if $\text{M} > 1$, it is a **supersonic** flow.

If a source of sound waves is at a fixed location, the waves travel radially away from the source with the speed of sound c. Figure 9.2a shows the position of the sound waves after a time increment Δt and after multiples of Δt. Part (b) shows a source moving with a velocity V which is

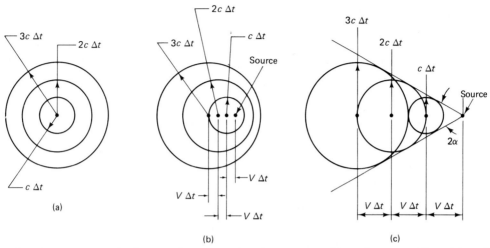

Figure 9.2   Sound waves propagating from a noise source: (a) stationary source; (b) moving source: $V < c$; (c) moving source, $V > c$.

less than the speed of sound. Note that the sound waves always propagate ahead of the source so that an airplane traveling at a speed less than the speed of sound will always "announce" its approach. This is not true, however, for an object traveling at a speed greater than the speed of sound, as shown in Fig. 9.2c. The zone outside the cone is a **zone of silence,** so that an approaching object moving at a supersonic speed could not be heard until it passed overhead and the **Mach cone,** the cone shown, intercepted the observer. From the figure the angle α of the Mach cone is given by

$$\alpha = \sin^{-1}\frac{c}{V} = \sin^{-1}\frac{1}{M} \qquad (9.2.13)$$

The discussion above is limited to small-amplitude sound waves, often called **Mach waves.** They are formed on the needle nose of an aircraft, or on the leading edge of an airfoil if it has a sufficiently sharp leading edge. If the nose is blunt or if the leading edge is not sufficiently sharp, a supersonic aircraft will produce a large-amplitude wave called a **shock wave.** The shock wave will also form a zone of silence, but the initial angle at the source created by the shock wave will be larger than that of a Mach wave. Shock waves will be considered in subsequent sections.

EXAMPLE 9.1

A needle-nose projectile traveling at a speed with $M = 3$ passes 200 m above an observer. Calculate the projectile's velocity and determine how far beyond the observer the projectile will first be heard.

Solution

At a Mach number of 3 the velocity is

$$V = Mc = M\sqrt{kRT}$$

$$= 3\sqrt{1.4 \times 287 \times 288} = 1021 \text{ m/s}$$

where a standard temperature of 15°C has been assumed. Using h as the height and L as the distance beyond the observer, we have

$$\sin \alpha = \frac{h}{\sqrt{L^2 + h^2}} = \frac{1}{M}$$

With the information given,

$$\frac{200}{\sqrt{L^2 + 200^2}} = \frac{1}{3}$$

giving

$$L = 566 \text{ m}$$

Note: The units on kRT are $\dfrac{N \cdot m}{kg \cdot K} \times K = \dfrac{N \cdot m}{N \cdot s^2/m} = \dfrac{m^2}{s^2}$. The quantity k is dimensionless.

EXAMPLE 9.1 (English)

A needle-nose projectile traveling at a speed with $M = 3$ passes 600 ft above an observer. Calculate the projectile's velocity and determine how far beyond the observer the projectile will first be heard.

Solution

At a Mach number of 3 the velocity is

$$V = Mc = M\sqrt{kRT}$$

$$= 3\sqrt{1.4 \times 1716 \times 519} = 3350 \text{ ft/sec}$$

where a standard temperature of 59°F has been assumed. Using h as the height and L as the distance beyond the observer, we have

$$\sin \alpha = \frac{h}{\sqrt{L^2 + h^2}} = \frac{1}{M}$$

With the information given,

$$\frac{600}{\sqrt{L^2 + 600^2}} = \frac{1}{3}$$

giving

$$L = 1697 \text{ ft}$$

Note: The units on kRT are $\dfrac{\text{ft-lb}}{\text{slug} - °\text{R}} \times °\text{R} = \dfrac{\text{ft-lb}}{\text{lb} - \text{sec}^2/\text{ft}} = \dfrac{\text{ft}^2}{\text{sec}^2}$. The quantity k is dimensionless.

9.3 ISENTROPIC NOZZLE FLOW

There are many applications where gas flows through a section of tube or conduit that has a changing area in which a steady, uniform, isentropic flow is a good approximation to the actual flow situation. The diffuser near the front of a jet engine, exhaust gases passing through the blades of a turbine, the nozzles on a rocket engine, a broken natural gas line, and gas flow measuring devices are all examples of situations that can be modeled with a steady, uniform, isentropic flow. Consider the flow through the infinitesimal control volume shown in Fig. 9.3. With a changing area the continuity equation

$$\rho A V = \text{const.} \qquad (9.3.1)$$

applied between two sections a distance dx apart takes the form

$$\rho A V = (\rho + d\rho)(A + dA)(V + dV) \qquad (9.3.2)$$

or, keeping only first-order terms in the differential quantities,

$$\frac{dV}{V} + \frac{dA}{A} + \frac{d\rho}{\rho} = 0 \qquad (9.3.3)$$

The energy equation can be written (see Eq. 9.1.5)

$$\frac{V^2}{2} + \frac{k}{k-1}\frac{p}{\rho} = \text{const.} \qquad (9.3.4)$$

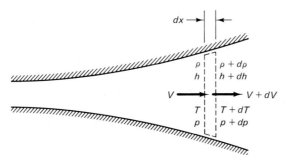

Figure 9.3 Uniform, isentropic flow.

For the present application we have

$$\frac{V^2}{2} + \frac{k}{k-1}\frac{p}{\rho} = \frac{(V+dV)^2}{2} + \frac{k}{k-1}\frac{p+dp}{\rho+d\rho} \tag{9.3.5}$$

or, to first order,

$$V\,dV + \frac{k}{k-1}\frac{\rho\,dp - p\,d\rho}{\rho^2} = 0 \tag{9.3.6}$$

For an isentropic process we use Eq. 9.2.8 and there results

$$V\,dV + k\frac{p}{\rho^2}\,d\rho = 0 \tag{9.3.7}$$

Substituting for $d\rho/\rho$ from Eq. 9.3.3, the above becomes

$$\frac{dV}{V}\left(\frac{\rho V^2}{kp} - 1\right) = \frac{dA}{A} \tag{9.3.8}$$

In terms of the speed of sound this is written as

$$\frac{dV}{V}\left(\frac{V^2}{c^2} - 1\right) = \frac{dA}{A} \tag{9.3.9}$$

Introduce the Mach number and we have the very important relationship

$$\frac{dV}{V}(M^2 - 1) = \frac{dA}{A} \tag{9.3.10}$$

From this equation we can make the following observations:

1. If the area is increasing, $dA > 0$, and $M < 1$, we see that dV must be negative, that is, $dV < 0$. The flow is decelerating for this subsonic flow.
2. If the area is increasing and $M > 1$, we see that $dV > 0$; hence the flow is accelerating in the diverging section for this supersonic flow.
3. If the area is decreasing and $M < 1$, then $dV > 0$, resulting in an accelerating flow.
4. If the area is decreasing and $M > 1$, then $dV < 0$, indicating a decelerating flow.
5. At a throat where $dA = 0$, either $dV = 0$ or $M = 1$, or possibly both.

If we define a **nozzle** as a device that accelerates the flow, we see that observations 2 and 3 describe a nozzle and observations 1 and 4 describe a **diffuser,** a device that decelerates the flow. The supersonic flow leads to rather surprising results: an accelerating flow in an enlarging area and a decelerating flow in a decreasing area. This is, in fact, the situation encountered in a traffic flow on a freeway; hence a supersonic flow could be used to model a traffic flow.

Note that the observations above prohibit a supersonic flow in a converging section attached to a reservoir. If a supersonic flow is to be generated by releasing a gas from a reservoir, there must be a converging section in which a subsonic flow accelerates to the throat where M = 1 followed by a diverging section in which the flow continues to accelerate with M > 1. This is shown in Fig. 9.4.

The isentropic flow in the nozzle will now be considered in more detail. The energy equation between the reservoir and any section can be written in the form

$$c_p T_0 = \frac{V^2}{2} + c_p T \tag{9.3.11}$$

Quantities with a zero subscript are often called **stagnation quantities** since they occur at a location where $V = 0$. Recognizing that $V = \mathrm{M}c$, $c_p/c_v = k$, $c_p = c_v + R$, and $c = \sqrt{kRT}$, we can write the energy equation as

$$\frac{T_0}{T} = 1 + \frac{k-1}{2}\,\mathrm{M}^2 \tag{9.3.12}$$

For our isentropic flow the pressure and density ratios are expressed as

$$\frac{p_0}{p} = \left(1 + \frac{k-1}{2}\,\mathrm{M}^2\right)^{k/(k-1)}$$

$$\frac{\rho_0}{\rho} = \left(1 + \frac{k-1}{2}\,\mathrm{M}^2\right)^{1/(k-1)}$$

$$\tag{9.3.13}$$

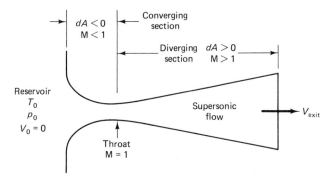

Figure 9.4 A supersonic nozzle.

If a supersonic flow occurs downstream of the throat, then M = 1 at the throat and denoting this **critical area** with an asterisk (*) superscript, we have the following critical ratios:

$$\frac{T^*}{T_0} = \frac{2}{k + 1}$$

$$\frac{p^*}{p_0} = \left(\frac{2}{k + 1}\right)^{k/(k-1)}$$

$$\frac{\rho^*}{\rho_0} = \left(\frac{2}{k + 1}\right)^{1/(k-1)}$$

(9.3.14)

We often make reference to the critical area even though an actual throat does not occur; we can imagine a throat occurring and call it the critical area. In fact, the isentropic flow table, Table D.1 in Appendix D, includes just such an area. For $k = 1.4$ the critical values are

$$p^* = 0.5283p_0 \qquad T^* = 0.8333T_0 \qquad \rho^* = 0.6340\rho_0 \qquad (9.3.15)$$

We can determine an expression for the mass flux through the nozzle from the equation

$$\dot{m} = \rho A V$$

$$= \frac{p}{RT} AM\sqrt{kRT} = \frac{p}{\sqrt{T}}\sqrt{\frac{k}{R}}\,AM$$

(9.3.16)

Using Eqs. 9.3.12 and 9.3.13, this can be expressed as

$$\dot{m} = \frac{p_0\left(1 + \dfrac{k - 1}{2}M^2\right)^{k/(1-k)}}{\sqrt{T_0}\left(1 + \dfrac{k - 1}{2}M^2\right)^{-1/2}}\sqrt{\frac{k}{R}}\,AM$$

$$= p_0\sqrt{\frac{k}{RT_0}}\,MA\left(1 + \frac{k - 1}{2}M^2\right)^{(k+1)/2(1-k)}$$

(9.3.17)

If we choose the critical area where M* = 1, we see that

$$\dot{m} = p_0 A^*\sqrt{\frac{k}{RT_0}}\left(\frac{k + 1}{2}\right)^{(k+1)/2(1-k)}$$

(9.3.18)

This shows that the mass flux in the nozzle is only dependent on the reservoir conditions and the critical area A^*. By combining Eqs. 9.3.17

and 9.3.18, the area ratio A/A^* can be written in terms of the Mach number as

$$\frac{A}{A^*} = \frac{1}{M} \left[\frac{2 + (k - 1)M^2}{k + 1} \right]^{(k+1)/2(k-1)} \tag{9.3.19}$$

This ratio is included in Table D.1 for air flow.

As a final consideration in our study of isentropic nozzle flow, we will present the influence of reservoir pressure and receiver pressure on the mass flux. First, the converging nozzle will be presented; then the converging–diverging nozzle will follow. The converging nozzle is assumed to be attached to a reservoir, as shown in Fig. 9.5a, with fixed conditions; the pressure in the receiver can be lowered to provide an increasing mass flux through the nozzle, as shown by the left curve in Fig. 9.5b. When the receiver pressure p_r reaches the critical pressure (for air $p_r = 0.5283p_0$) the Mach number M_e at the throat (the exit) is unity. As p_r is reduced below this critical value the mass flux will not increase, and the condition of **choked flow** occurs as shown by the left curve in Fig. 9.5b. If the throat is a critical area, that is, $M = 1$ at the throat, then $\dot{m}$ is only dependent on the throat area A^* and the reservoir conditions, as indicated by Eq. 9.3.18. Hence, reducing p_r below the critical pressure has no effect on the upstream flow. This is reasonable since disturbances travel at the speed of sound; if p_r is reduced the disturbances that would travel upstream thereby changing conditions cannot do so since the velocity of the stream at the exit is equal to the speed of sound thereby preventing any disturbances from propagating upstream.

If, in the converging nozzle, we keep p_r constant and increase the reservoir pressure (keep T_0 constant also), a choked flow again occurs

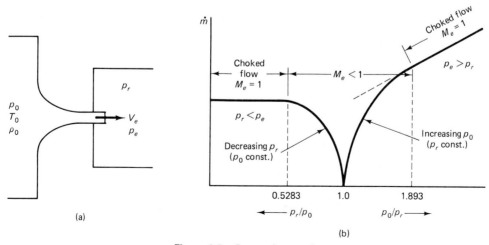

Figure 9.5 Converging nozzle.

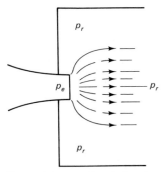

Figure 9.6   Nozzle exit flow for
$p_e > p_r$.

when $M_e = 1$; however, when p_0 is increased still further, we see from Eq.
9.3.18 that the mass flux will increase, as shown by the right curve of Fig.
9.5b. The exit pressure p_e for the choked flow condition will be greater
than the receiver pressure p_r.

A note may be in order explaining how it is possible for the flow exit
pressure p_e to exceed the receiver pressure p_r. If $p_e > p_r$, the flow exiting
the nozzle is able to turn rather sharply, causing a flow pattern sketched in
Fig. 9.6. This possible flow situation will be studied in Section 9.8.

Now, consider the converging–diverging nozzle, shown in Fig. 9.7,
with reservoir and receiver as indicated. For this nozzle we sketch the
pressure ratio p/p_0 as a function of location in the nozzle for various
receiver pressure ratios p_r/p_0. If $p_r/p_0 = 1$, no flow occurs, corresponding
to curve A. If p_r is reduced a small amount, curve B results and a subsonic
flow exists throughout the nozzle with a minimum pressure occurring at
the throat. As the pressure is reduced still further, a pressure is reached
that will result in the Mach number at the throat just being unity, as
sketched by curve C; the flow remains subsonic throughout, however.
Another particular receiver pressure exists, considerably below that of
curve C, that will also produce an isentropic flow throughout; it results in
curve D. Any receiver pressure in between these two particular pressures
will result in a nonisentropic flow in the nozzle; a shock wave, to be
studied later, occurs which renders our assumption of isentropic flow
invalid. If the receiver pressure is below that associated with curve D, we
again find that the nozzle exit pressure p_e is greater than the receiver
pressure p_r. The mass flux in the nozzle increases from curve A to curve
C; but as the receiver pressure is reduced below that of curve C, no
increase in mass flux can occur. The mass flux is given by Eq. 9.3.17.

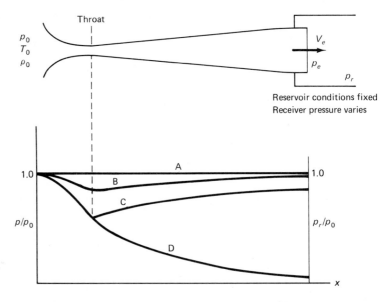

Figure 9.7   Converging–diverging nozzle.

Final notes regard nozzle and diffuser effectiveness. The purpose of a nozzle is to convert enthalpy (which can be thought of as stored thermal energy) into kinetic energy. The efficiency η_n of a nozzle is defined as

$$\eta_n = \frac{(\Delta KE)_{\text{actual}}}{(\Delta KE)_{\text{isentropic}}}$$
$$= \frac{h_0 - h_e}{h_0 - h_{es}} \tag{9.3.20}$$

where h_e is the actual exit enthalpy and h_{es} is the isentropic exit enthalpy. Efficiencies are between 90 and 99% with larger nozzles having the higher percentages because the viscous wall effects that account for most of the losses are relatively small with the larger nozzles.

The purpose of a diffuser is to slow down the fluid and recover the pressure. For a diffuser we define the **pressure recovery factor** C_p to be

$$C_p = \frac{\Delta p_{\text{actual}}}{\Delta p_{\text{isentropic}}} \tag{9.3.21}$$

Such factors vary between 40% when the flow actually separates from the wall (the included angle should be less than about 10° for the subsonic section to avoid this separation) to 85%. Viscous effects are significantly greater in the diffuser than in the nozzle because of the thicker viscous wall layers. The reader may think that the flow in the supersonic diverging nozzle may also tend to separate from the wall; this is not the case. Expansion fans, to be studied in a subsequent section, allow the supersonic flow to turn rather sharp angles so that supersonic nozzles are constructed with large included angles.

EXAMPLE 9.2

Air exits from a reservoir maintained at 20°C and 500 kPa absolute into a receiver maintained at (a) 300 kPa absolute and (b) 200 kPa absolute. Estimate the mass flux if the exit area is 10 cm². Use the equations first and then the isentropic flow table, Table D.1.

Solution

To estimate the mass flux we will assume isentropic flow. For air the receiver pressure that would result in $M_e = 1$ is

$$p_r = 0.5283\, p_0 = 0.5283 \times 500 = 264.2 \text{ kPa}$$

For part (a) $M_e < 1$, and for part (b) choked flow occurs and $M_e = 1$.

(a) To find the exit Mach number Eq. 9.3.13 gives

$$1 + \frac{k-1}{2} M^2 = \left(\frac{p_0}{p}\right)^{(k-1)/k}$$

or

$$M_e^2 = \frac{2}{k-1} \left(\frac{p_0}{p}\right)^{(k-1)/k} - \frac{2}{k-1}$$

$$= \frac{2}{0.4} \left(\frac{500}{300}\right)^{0.2857} - \frac{2}{0.4} = 0.7857$$

This gives

$$M_e = 0.8864$$

The mass flux is given by Eq. 9.3.17 and is found to be

$$\dot{m} = p_0 \sqrt{\frac{k}{RT_0}} MA \left(1 + \frac{k-1}{2} M^2\right)^{(k+1)/2(1-k)}$$

$$= 500\,000 \sqrt{\frac{1.4}{287 \times 293}} \times 0.8864 \times 0.001 \left(1 + \frac{0.4}{2} \times 0.8864^2\right)^{-2.4/0.8}$$

$$= 1.167 \text{ kg/s}$$

(b) Choked flow occurs and thus $M_e = 1$ and Eq. 9.3.18 yields

$$\dot{m} = p_0 A^* \sqrt{\frac{k}{RT_0}} \left(\frac{k+1}{2}\right)^{(k+1)/2(1-k)}$$

$$= 500\,000 \times 0.001 \sqrt{\frac{1.4}{287 \times 293}} \left(\frac{2.4}{2}\right)^{-2.4/0.8} = 1.181 \text{ kg/s}$$

Now, let us use the isentropic flow table (Table D.1) and solve parts (a) and (b).
(a) For a pressure ratio of $p/p_0 = 300/500 = 0.6$, we interpolate to find

$$M_e = \frac{0.6041 - 0.6}{0.6041 - 0.5913} \times 0.02 + 0.88 = 0.886$$

$$\frac{T_e}{T_0} = \frac{0.6041 - 0.6}{0.6041 - 0.5913} (0.8606 - 0.8659) + 0.8659 = 0.864$$

$$T_e = 0.864 \times 293 = 253 \text{ K}$$

The velocity and density are, respectively,

$$V = Mc = 0.886 \sqrt{1.4 \times 287 \times 253} = 282 \text{ m/s}$$

$$\rho = \frac{p}{RT} = \frac{300}{0.287 \times 253} = 4.13 \text{ kg/m}^3$$

The mass flux is then

$$\dot{m} = \rho A V$$

$$= 4.13 \times 0.001 \times 282 = 1.165 \text{ kg/s}$$

(b) For choked flow we know that $p_e/p_0 = 0.5283$. The table gives

$$\frac{T_e}{T_0} = 0.8333 \quad \text{and} \quad M_e = 1$$

Thus the temperature, velocity, and density are, respectively,

$$T = 0.8333 \times 293 = 244.2 \text{ K}$$

$$V = Mc = 1 \times \sqrt{1.4 \times 287 \times 244.2} = 313.2 \text{ m/s}$$

$$\rho = \frac{p}{RT} = \frac{0.5283 \times 500}{0.287 \times 244.2} = 3.769 \text{ kg/m}^3$$

The mass flux is calculated to be

$$\dot{m} = \rho A V$$

$$= 3.769 \times 0.001 \times 313.2 = 1.180 \text{ kg/s}$$

EXAMPLE 9.3

A converging–diverging nozzle, with an exit area of 40 cm^2 and a throat area of 10 cm^2, is attached to a reservoir with $T = 20°C$ and $p = 500$ kPa absolute. Determine the two exit pressures that result in $M = 1$ at the throat for an isentropic flow. Also, determine the associated exit temperatures and velocities.

Solution

The exit pressures we seek are associated with curves C and D of Fig. 9.7. The area ratio is

$$\frac{A}{A^*} = \frac{40}{10} = 4$$

We could solve Eq. 9.3.19 for M using a trial-and-error technique; however, it is simpler to use the isentropic flow table, Table D.1. There are two entries for $A/A^* = 4$. Interpolation gives

$$\left(\frac{p}{p_0}\right)_C = \frac{4.182 - 4.0}{4.182 - 3.673}(0.9823 - 0.9864) + 0.9864 = 0.9849$$

$$\left(\frac{p}{p_0}\right)_D = \frac{4.0 - 3.999}{4.076 - 3.999}(0.02891 - 0.02980) + 0.02980 = 0.02979$$

Hence the two exit pressures that will result in isentropic flow are

$$p = 492.4 \text{ kPa} \quad \text{and} \quad 14.9 \text{ kPa}$$

Note the very small pressure difference (7.6 kPa) between receiver and reservoir necessary to create the flow condition of curve C.

The temperature ratios and Mach numbers are interpolated to be

$$\left(\frac{T}{T_0}\right)_C = 0.3576(0.9949 - 0.9961) + 0.9961 = 0.9957$$

$$\left(\frac{T}{T_0}\right)_D = 0.01299(0.3633 - 0.3665) + 0.3665 = 0.3665$$

$$(M)_C = 0.3576 \times 0.02 + 0.14 = 0.147$$

$$(M)_D = 0.01299 \times 0.02 + 2.94 = 2.94$$

The exit temperatures associated with curves C and D are thus

$$(T)_C = 0.9957 \times 293 = 291.7 \text{ K}$$

$$(T)_D = 0.3665 \times 293 = 107.4 \text{ K}$$

The velocities are found from $V = Mc$ to be

$$(V)_C = 0.147\sqrt{1.4 \times 287 \times 291.7} = 50.3 \text{ m/s}$$

$$(V)_D = 2.94\sqrt{1.4 \times 287 \times 107.4} = 611 \text{ m/s}$$

EXAMPLE 9.4

Gas flows are assumed to be incompressible flows at Mach numbers less than about 0.3. Determine the error involved in calculating the stagnation pressure for an airflow with M = 0.3.

Solution

For an incompressible airflow the energy equation (4.4.20) with no losses would give

$$p_0 = p_1 + \rho\frac{V_1^2}{2}$$

where $V_0 = 0$ at the stagnation point.

The isentropic flow equation (9.3.13), with $k = 1.4$, gives

$$p_0 = p_1(1 + 0.2M_1^2)^{3.5}$$

Use the binomial theorem $(1 + x)^n = 1 + nx + n(n - 1)x^2/2! + \cdots$ and express this as, letting $x = 0.2M_1^2$,

$$p_0 = p_1(1 + 0.7M_1^2 + 0.175M_1^4 + \cdots)$$

This can be written as (see Eqs. 9.2.9 and 9.2.12)

$$p_0 - p_1 = p_1 M_1^2 (0.7 + 0.175 M_1^2 + \cdots)$$

$$= \frac{p_1 V_1^2}{1.4}(0.7 + 0.175 M_1^2 + \cdots)$$

$$= \rho_1 \frac{V_1^2}{2}(1 + 0.25 M_1^2 + \cdots)$$

Substituting $M_1 = 0.3$, we see that

$$p_0 - p_1 = \rho_1 \frac{V_1^2}{2}(1 + 0.0225 + \cdots)$$

Comparing this with the incompressible flow equation, we see that the error is only slightly greater than 2%. Hence it is reasonable to approximate a gas flow below $M = 0.3$ with an incompressible flow.

9.4 NORMAL SHOCK WAVE

Small-amplitude disturbances travel at the speed of sound, as was established in a preceding section. In this section a large-amplitude disturbance will be studied. Its speed of propagation and its effect on other flow properties, such as pressure and temperature, will be considered. Large-amplitude disturbances occur in a number of situations. Examples include the flow in a gun barrel ahead of the projectile, the exit flow from a rocket or jet engine nozzle, the airflow around a supersonic aircraft, and the expanding front due to an explosion. Such large disturbances propagating through a gas are called **shock waves.** They can be oriented normal to the flow or at oblique angles. In this section we consider only the normal shock wave that occurs in a tube or directly in front of blunt object; the photograph in Fig. 9.8 shows the shock wave in front of a sphere.

    The property changes that occur across a shock wave take place over an extremely short distance. For usual conditions the distance is only several mean free paths of the molecules, on the order of 10^{-4} mm. Phenomena such as viscous dissipation and heat conduction that occur within the shock wave will not be studied here. We will treat the shock wave as a zero-thickness discontinuity in the flow and allow our integral control volume equations to relate the quantities of interest.

    Consider a normal shock wave moving with velocity V_1. We can make it stationary by moving the flow in a tube at velocity V_1, as shown in Fig. 9.9. The continuity equation, recognizing that $A_1 = A_2$, is

$$\rho_1 V_1 = \rho_2 V_2 \qquad (9.4.1)$$

The energy equation (9.1.5), with $\dot{Q} = \dot{W}_s = 0$, is

$$\frac{V_2^2 - V_1^2}{2} + \frac{k}{k-1}\left(\frac{p_2}{\rho_2} - \frac{p_1}{\rho_1}\right) = 0 \qquad (9.4.2)$$

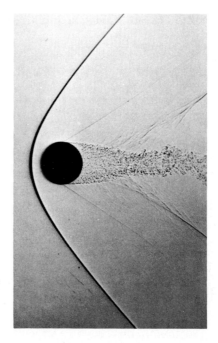

Figure 9.8 A shock wave is observed in front of a sphere at M = 1.53. (Photograph by A. C. Charters.)

The momentum equation (9.1.2), with only pressure forces, becomes

$$p_1 - p_2 = \rho_1 V_1 (V_2 - V_1) \tag{9.4.3}$$

where the areas have divided out. These three equations allow us to determine three unknown quantities; if ρ_1, V_1, and p_1 are known, we can find ρ_2, V_2, p_2, and subsequently T_2 and M_2, using the appropriate equations.

It is convenient, however, to express the equations in terms of the Mach numbers M_1 and M_2. This results in a set of equations that are simpler to solve than solving the three simultaneous equations listed above. To do this we write Eq. 9.4.3, using $\rho_1 V_1 = \rho_2 V_2$, in the form

$$p_1 \left(1 + \frac{\rho_1 V_1^2}{p_1} \right) = p_2 \left(1 + \frac{\rho_2 V_2^2}{p_2} \right) \tag{9.4.4}$$

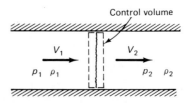

Figure 9.9 Stationary shock wave in a tube.

Introducing $M^2 = V^2 \rho / pk$, the momentum equation becomes

$$\frac{p_2}{p_1} = \frac{1 + kM_1^2}{1 + kM_2^2} \tag{9.4.5}$$

Similarly, the energy equation (9.4.2), with $p = \rho RT$, is written as

$$T_1\left(1 + \frac{k-1}{kRT_1}\frac{V_1^2}{2}\right) = T_2\left(1 + \frac{k-1}{kRT_2}\frac{V_2^2}{2}\right) \tag{9.4.6}$$

or, substituting $M^2 = V^2/kRT$,

$$\frac{T_2}{T_1} = \frac{1 + \dfrac{k-1}{2}M_1^2}{1 + \dfrac{k-1}{2}M_2^2} \tag{9.4.7}$$

If we substitute $\rho = p/RT$ into the continuity equation (9.4.1), we have

$$\frac{p_1 V_1}{RT_1} = \frac{p_2 V_2}{RT_2} \tag{9.4.8}$$

which becomes, using $V = M\sqrt{kRT}$,

$$\frac{p_2}{p_1}\frac{M_2}{M_1}\sqrt{\frac{T_2}{T_1}} = 1 \tag{9.4.9}$$

Substituting for the pressure and temperature ratios from Eqs. 9.4.5 and 9.4.7, the continuity equation takes the form

$$\frac{M_1\left(1 + \dfrac{k-1}{2}M_1^2\right)^{1/2}}{1 + kM_1^2} = \frac{M_2\left(1 + \dfrac{k-1}{2}M_2^2\right)^{1/2}}{1 + kM_2^2} \tag{9.4.10}$$

Hence the downstream Mach number is related to the upstream Mach number by

$$M_2^2 = \frac{M_1^2 + \dfrac{2}{k-1}}{\dfrac{2k}{k-1}M_1^2 - 1} \tag{9.4.11}$$

This allows us to express the pressure and temperature ratios in terms of M_1 only. The momentum equation (9.4.5) takes the form

$$\frac{p_2}{p_1} = \frac{2k}{k+1}M_1^2 - \frac{k-1}{k+1} \tag{9.4.12}$$

and the energy equation becomes

$$\frac{T_2}{T_1} = \frac{\left(1 + \frac{k-1}{2} M_1^2\right)\left(\frac{2k}{k-1} M_1^2 - 1\right)}{\frac{(k+1)^2}{2(k-1)} M_1^2} \tag{9.4.13}$$

For air, with $k = 1.4$, the preceding 3 equations reduce to

$$M_2^2 = \frac{M_1^2 + 5}{7M_1^2 - 1}$$

$$\frac{p_2}{p_1} = \frac{7M_1^2 - 1}{6} \tag{9.4.14}$$

$$\frac{T_2}{T_1} = \frac{(M_1^2 + 5)(7M_1^2 - 1)}{36M_1^2}$$

From the first of these three equations, we observe:

- If $M_1 = 1$, then $M_2 = 1$ and no shock wave exists.
- If $M_1 > 1$, then $M_2 < 1$ and the normal shock wave converts a supersonic flow into a subsonic flow.
- If $M_1 < 1$, then $M_2 > 1$ and a subsonic flow appears to be converted into a supersonic flow by the presence of a normal shock wave. This possibility is eliminated by the second law since it would demand a decrease in entropy by a process in an isolated system, an impossibility.

The impossibility stated above is observed by considering the entropy increase, given by

$$s_2 - s_1 = c_p \ln \frac{T_2}{T_1} - R \ln \frac{p_2}{p_1}$$

$$= c_p \ln \frac{2 + (k-1)M_2^2}{2 + (k-1)M_1^2} - R \ln \frac{1 + kM_1^2}{1 + kM_2^2} \tag{9.4.15}$$

For air, with $k = 1.4$, this is plotted in Fig. 9.10, relating M_2 to M_1, with Eq. 9.4.11. Note the impossible negative entropy change whenever $M_1 < 1$.

The relationship among the thermodynamic properties, listed in the equations above, can be demonstrated with reference to the T–s diagram in Fig. 9.11. The conditions upstream of the normal shock are designated by state 1, and downstream by state 2. Note the dashed line from state 1 to state 2 for the irreversible process that occurs inside the shock wave.

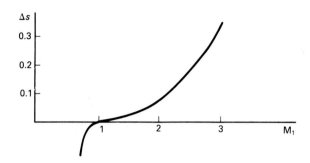

Figure 9.10 Entropy change for a normal shock in air.

The energy equation, with $\dot{Q} = \dot{W}_s = 0$, can be written as

$$\frac{V_1^2}{2c_p} + T_1 = \frac{V_2^2}{2c_p} + T_2 \qquad (9.4.16)$$

The stagnation temperature is defined as the temperature that would exist if the flow is brought to rest isentropically. Thus the energy equation gives

$$T_{01} = T_{02} \qquad (9.4.17)$$

as shown on the figure. The substantial decrease in stagnation pressure, $p_{02} < p_{01}$, is also observed in Fig. 9.11. If the entropy increases from state 1 to state 2, as it must, the stagnation pressure p_{02} must decrease, as shown, if we are to maintain $T_{01} = T_{02}$.

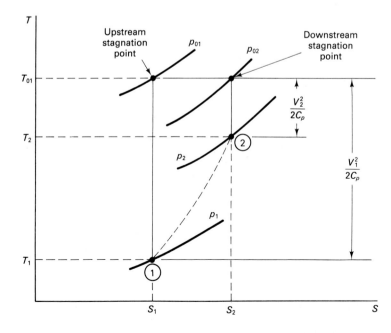

Figure 9.11 T–s diagram for a normal shock wave.

Gas tables are available that give the pressure ratio, the temperature ratio, the downstream Mach number, and the stagnation pressure ratio as a function of the upstream Mach number. Table D.2 is such a table for $k = 1.4$ and includes the ratios given by Eq. 9.4.14. Note that M_2 is always less than unity, p_2 is always greater than p_1, T_2 is always greater than T_1, and p_{02} is always less than p_{01}.

EXAMPLE 9.5

A normal shock wave passes through stagnant air at 20°C and atmospheric pressure of 80 kPa with a speed of 500 m/s. Calculate the pressure and temperature downstream of the shock wave. Use (a) the equations and (b) the gas tables.

Solution

(a) To use the simplified equations (9.4.14), we must know the upstream Mach number. It is

$$M_1 = \frac{V_1}{c_1} = \frac{V_1}{\sqrt{kRT_1}}$$

$$= \frac{500}{\sqrt{1.4 \times 287 \times 293}} = 1.457$$

The pressure and temperature are then found to be

$$p_2 = \frac{p_1(7M_1^2 - 1)}{6}$$

$$= \frac{80(7 \times 1.457^2 - 1)}{6} = 184.8 \text{ kPa}$$

$$T_2 = \frac{T_1(M_1^2 + 5)(7M_1^2 - 1)}{36M_1^2}$$

$$= \frac{293(1.457^2 + 5)(7 \times 1.457^2 - 1)}{36 \times 1.457^2} = 378.5 \text{ K}$$

(b) From part (a), we use $M_1 = 1.457$. Interpolation in Table D.2 yields

$$\frac{p_2}{p_1} = \frac{1.457 - 1.44}{1.46 - 1.44}(2.320 - 2.253) + 2.253 = 2.310$$

$$\frac{T_2}{T_1} = \frac{1.457 - 1.44}{1.46 - 1.44}(1.294 - 1.281) + 1.281 = 1.292$$

Using the given information, we have

$$p_2 = 80 \times 2.310 = 184.8 \text{ kPa}$$

$$T_2 = 293 \times 1.292 = 378.6 \text{ K}$$

These results are essentially the same as those found in part (a).

EXAMPLE 9.5 (English)

A normal shock wave passes through stagnant air at 60°F and atmospheric pressure of 12 psi with a speed of 1500 ft/sec. Calculate the pressure and temperature downstream of the shock wave. Use (a) the equations and (b) the gas tables.

Solution

(a) To use the simplified equations (9.4.14) we must know the upstream Mach number. It is

$$\text{M}_1 = \frac{V_1}{c_1} = \frac{V_1}{\sqrt{kRT_1}}$$

$$= \frac{1500}{\sqrt{1.4 \times 1716 \times 520}} = 1.342$$

The pressure and temperature are then found to be

$$p_2 = \frac{p_1(7\text{M}_1^2 - 1)}{6}$$

$$= \frac{12(7 \times 1.342^2 - 1)}{6} = 23.21 \text{ psia}$$

$$T_2 = \frac{T_1(\text{M}_1^2 + 5)(7\text{M}_1^2 - 1)}{36\text{M}_1^2}$$

$$= \frac{520(1.342^2 + 5)(7 \times 1.342^2 - 1)}{36 \times 1.342^2} = 633.1°R$$

(b) From part (a), we use $\text{M}_1 = 1.342$. Interpolation in Table D.2 yields

$$\frac{p_2}{p_1} = \frac{1.342 - 1.34}{1.36 - 1.34}(1.991 - 1.928) + 1.928 = 1.934$$

$$\frac{T_2}{T_1} = \frac{1.342 - 1.34}{1.36 - 1.34}(1.229 - 1.216) + 1.216 = 1.217$$

Using the information given, we have

$$p_2 = 12 \times 1.934 = 23.21 \text{ psia}$$
$$T_2 = 520 \times 1.217 = 632.8°R$$

EXAMPLE 9.6

A normal shock wave propagates through otherwise stagnant air at standard conditions at a speed of 700 m/s. Determine the speed induced in the air immediately behind the shock wave.

Solution

For standard conditions the temperature is 15° C. The upstream Mach number is thus

$$M_1 = \frac{V_1}{c_1} = \frac{V_1}{\sqrt{kRT_1}}$$

$$= \frac{700}{\sqrt{1.4 \times 287 \times 288}} = 2.06$$

Using Eq. 9.4.14, we find that

$$M_2 = \left(\frac{M_1^2 + 5}{7M_1^2 - 1}\right)^{1/2}$$

$$= \left(\frac{2.06^2 + 5}{7 \times 2.06^2 - 1}\right)^{1/2} = 0.567$$

$$T_2 = \frac{T_1(M_1^2 + 5)(7M_1^2 - 1)}{36M_1^2}$$

$$= \frac{288(2.06^2 + 5)(7 \times 2.06^2 - 1)}{36 \times 2.06^2} = 500.2 \text{ K}$$

This allows us to calculate

$$V_2 = M_2 c_2$$

$$= 0.567 \sqrt{1.4 \times 287 \times 500.2} = 254.4 \text{ m/s}$$

This velocity assumes a flow with the shock wave stationary and the air approaching the shock wave at 700 m/s. If we superimpose a velocity of 700 m/s moving opposite to V_1, we find the induced velocity to be

$$V_{\text{induced}} = V_2 - V_1$$

$$= 254.4 - 700 = -446 \text{ m/s}$$

where the negative sign means that the induced velocity would be moving to the left if V_1 is to the right. The induced velocity would be in the same direction as the propagation of the shock wave.

9.5 SHOCK WAVES IN CONVERGING–DIVERGING NOZZLES

The converging–diverging nozzle has already been presented for an isentropic flow; for receiver-to-reservoir pressure ratios between those of curves C and D of Figs. 9.7 and 9.12, shock waves exist in the flow either inside or outside the nozzle. If $p_r/p_0 = a$, then a normal shock wave would

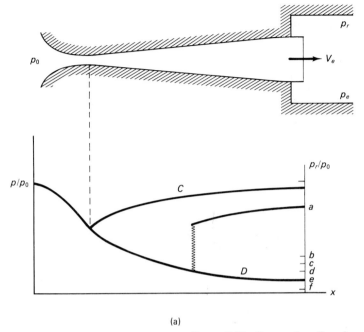

Figure 9.12   Converging–diverging nozzle.

exist at an internal location in the diverging portion of the nozzle. Usually, the location of the shock wave is prescribed in student problems since to locate the shock a trial-and-error solution is necessary. When $p_r/p_0 = b$ the normal shock wave is located in the exit plane of the nozzle. For a pressure ratio less than b but greater than e, two types of oblique shock wave patterns are observed, one with a central normal shock wave, as sketched for $p_r/p_0 = c$, and one with only oblique waves, as sketched for $p_r/p_0 = d$. These pressure ratios that result in oblique shock waves will not be considered here. As we move from d to e, the oblique shock waves become weaker and weaker until the isentropic flow is again realized at $p_r/p_0 = e$ with all shock waves absent. For pressure ratios below e a very complicated flow exists. The flow turns the corner at the nozzle exit rather abruptly due to expansion waves (isentropic waves to be considered in a subsequent section), then turns back due to the same expansion waves, resulting in a billowing out of the exhaust flow, as is visible from high-altitude satellite rocket engines. Let us work some examples now for the converging–diverging nozzle; no new equations are necessary.

EXAMPLE 9.7

A converging–diverging nozzle has a throat diameter of 5 cm and an exit diameter of 10 cm. The reservoir is the laboratory, maintained at atmospheric conditions of 20°C and 90 kPa absolute. Air is constantly pumped from a receiver so that a normal shock wave stands across the exit plane of the nozzle. Determine the receiver pressure and the mass flux.

Solution

Isentropic flow occurs from the reservoir, to the throat, to the exit plane in front of the normal shock wave at state 1. Supersonic flow occurs downstream of the throat making the throat the critical area. Hence

$$\frac{A_1}{A^*} = \frac{10^2}{5^2} = 4$$

Interpolation in the isentropic flow table (Table D.1) gives

$$M_1 = 2.94 \qquad \frac{p_1}{p_0} = 0.0298$$

Hence the pressure in front of the normal shock is

$$p_1 = p_0 \times 0.0298$$
$$= 90 \times 0.0298 = 2.68 \text{ kPa}$$

From the normal shock table (Table D.2), using $M_1 = 2.94$, we find that

$$\frac{p_2}{p_1} = 9.918$$
$$\therefore \quad p_2 = 9.918 \times 2.68 = 26.6 \text{ kPa}$$

This is the receiver pressure needed to orient the shock across the exit plane as shown for $p_r/p_0 = b$ in Fig. 9.12.

To find the mass flux through the nozzle, we need only consider the throat. Recognizing that $M_t = 1$, so that $V_t = c_t$, we can write

$$\dot{m} = \rho_t A_t V_t = \frac{p_t}{RT_t} A_t \sqrt{kRT_t} = p_t A_t \sqrt{\frac{k}{RT_t}}$$

The isentropic flow table yields

$$\frac{p_t}{p_0} = 0.5283 \qquad \frac{T_t}{T_0} = 0.8333$$

Thus the mass flux becomes

$$\dot{m} = (0.5283 \times 90\,000) \times \frac{\pi \times 0.05^2}{4} \sqrt{\frac{1.4}{287 \times (0.8333 \times 293)}}$$

$$= 0.417 \text{ kg/s}$$

Remember, the pressure must be measured in pascals in the equation above.

EXAMPLE 9.8

Air flows from a reservoir at 20°C and 200 kPa absolute through a 5-cm-diameter throat and exits from a 10 cm-diameter nozzle. Calculate the pressure needed to locate a normal shock wave at a position where the diameter is 7.5 cm.

Solution

We will use the gas tables. The throat is a critical area since for a supersonic flow $M_t = 1$. The area ratio is

$$\frac{A_1}{A^*} = \frac{7.5^2}{5^2} = 2.25$$

For this area ratio we find, from Table D.1, that

$$M_1 = 2.33$$

Then from Table D.2, for this Mach number, we obtain

$$M_2 = 0.531 \qquad \frac{p_{02}}{p_{01}} = 0.570$$

The reservoir pressure $p_0 = p_{01}$. Thus

$$p_{02} = 0.570 \times 200 = 114 \text{ kPa}$$

Isentropic flow occurs from state 2 immediately after the normal shock wave to the exit. Hence, for $M_2 = 0.531$, we find from Table D.1 that

$$\frac{A_2}{A^*} = 1.285$$

so that, if A_e is the exit area,

$$\frac{A_e}{A^*} = \frac{A_2}{A^*} \times \frac{A_e}{A_2}$$

$$= 1.285 \times \frac{10^2}{7.5^2} = 2.284$$

The Mach number and pressure ratio corresponding to this area ratio are

$$M_e = 0.265 \qquad \frac{p_e}{p_{0e}} = 0.952$$

For our isentropic flow between the shock and the exit, we know that $p_{02} = p_{0e}$; thus

$$p_e = p_{02} \times 0.952$$

$$= 114 \times 0.952 = 109 \text{ kPa}$$

Note the usefulness of the critical area ratio in obtaining the desired results.

EXAMPLE 9.8 (English)

Air flows from a reservoir at 60°F and 30 psia through a 2-in.-diameter throat and exits from a 4-in.-diameter nozzle. Calculate the pressure needed to locate a normal shock wave at a position where the diameter is 3 in.

Solution

We will use the gas tables. The throat is a critical area since for a supersonic flow $M_t = 1$. The area ratio is

$$\frac{A_1}{A^*} = \frac{3^2}{2^2} = 2.25$$

For this area ratio we find, from Table D.1,

$$M_1 = 2.33$$

Then for this Mach number, Table D.2 gives

$$M_2 = 0.531 \qquad \frac{p_{02}}{p_{01}} = 0.570$$

The reservoir pressure $p_0 = p_{01}$. Thus

$$p_{02} = 0.570 \times 30 = 17.1 \text{ psia}$$

Isentropic flow occurs from state 2 immediately after the normal shock wave to the exit. Hence for $M_2 = 0.531$, we find from Table D.1

$$\frac{A_2}{A^*} = 1.285$$

so that, if A_e is the exit area,

$$\frac{A_e}{A^*} = \frac{A_2}{A^*} \times \frac{A_e}{A_2}$$

$$= 1.285 \times \frac{4^2}{3^2} = 2.284$$

The Mach number and pressure ratio corresponding to this area ratio are

$$M_e = 0.265 \qquad \frac{p_e}{p_{0e}} = 0.952$$

For our isentropic flow between the shock and the exit, we know that $p_{02} = p_{0e}$; thus

$$p_e = p_{02} \times 0.952$$

$$= 17.1 \times 0.952 = 16.28 \text{ psia}$$

Note the usefulness of the critical area ratio in obtaining the desired results.

EXAMPLE 9.9

A pitot probe, the device used to measure the stagnation pressure in a flow, is inserted into an airstream and measures 300 kPa absolute, as shown. The pressure in the flow is measured to be 75 kPa absolute. If the temperature at the stagnation point of the probe is measured as 150°C, determine the free-stream velocity V.

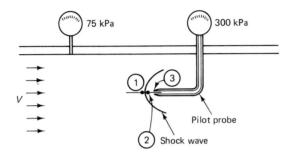

Solution

When a blunt object is placed in a supersonic flow a shock wave forms around the object, as it does around the front of the pitot probe shown. The flow that meets the front of the pitot probe at the stagnation point passes through a normal shock wave from state 1 to state 2; the subsonic flow at state 2 then decelerates isentropically to state 3, the stagnation point.

For the isentropic flow from 2 to 3 we can use Eq. 9.3.13,

$$\frac{p_3}{p_2} = \left(1 + \frac{k-1}{2} M_3^2\right)^{k/(k-1)}$$

Across the normal shock we know that (see Eq. 9.4.12)

$$\frac{p_2}{p_1} = \frac{2k}{k+1} M_1^2 - \frac{k-1}{k+1}$$

Also, the Mach numbers are related by Eq. 9.4.11,

$$M_3^2 = \frac{M_1^2 + 2/(k-1)}{2k M_1^2/(k-1) - 1}$$

The three equations above can be combined, with some algebraic manipulation, to yield the **Rayleigh-pitot-tube formula** for supersonic flows, namely,

$$\frac{p_3}{p_1} = \frac{\left(\dfrac{k+1}{2} M_1^2\right)^{k/(k-1)}}{\left(\dfrac{2k M_1^2}{k+1} - \dfrac{k-1}{k+1}\right)^{1/(k-1)}}$$

Substituting $k = 1.4$, $p_3 = 300$ kPa, and $p_1 = 75$ kPa, we have

$$\frac{300}{75} = \frac{(1.2 M_1^2)^{3.5}}{(1.167 M_1^2 - 0.1667)^{2.5}}$$

This can be solved by trial and error to give

$$M_1 = 1.65$$

The Mach number after the normal shock is interpolated from the shock table to be

$$M_2 = 0.654$$

Using the isentropic flow table with this Mach number, we interpolate the temperature at state 2 to be

$$T_2 = T_3 \times 0.921$$

$$= 423 \times 0.921 = 389.6 \text{ K}$$

The temperature in front of the normal shock is found by using the normal shock table as follows:

$$\frac{T_2}{T_1} = 1.423$$

$$\therefore \quad T_1 = \frac{T_2}{1.423} = 274 \text{ K}$$

Finally, the velocity before the normal shock is given by

$$V_1 = M_1 c_1 = M_1 \sqrt{kRT_1}$$

$$= 1.65 \sqrt{1.4 \times 287 \times 274} = 547 \text{ m/s}$$

9.6 VAPOR FLOW THROUGH A NOZZLE

The flow of vapor through a nozzle forms a very important engineering problem. High-pressure steam flows through the nozzles of turbines in electrical generating plants; in this section we present the technique for analyzing such a problem. We recall, however, that vapor that is not substantially superheated does not behave as an ideal gas; the vapor tables must be consulted since c_p and c_v cannot be assumed to be constant. Consider the problem of a superheated vapor entering the nozzle of Fig. 9.12. The flow would be isentropic unless a shock wave were encountered, and the expansion would be as sketched on the T–s diagram of Fig. 9.13. Assume that the flow initiates in a reservoir with stagnation conditions, flows through a throat indicated by state t, and exits out a diverging section to the exit at state e.

Note that the exit state could very likely be in the quality region with the possibility of condensation of liquid droplets; however, for sufficiently high flow velocities there may not be sufficient time for the formation of the droplets and the associated heat transfer process. This produces a situation called **supersaturation** and a condition of **metastable equilibrium**

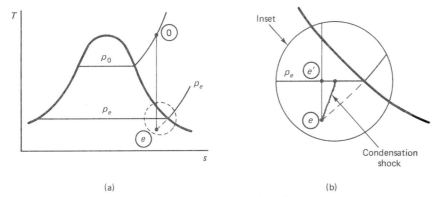

Figure 9.13   Isentropic expansion of a vapor.

exists; that is, the nonequilibrium state e is reached rather than the equilibrium state e'. To estimate the temperature of the metastable state e, we assume that $Tp^{k/(k-1)} = $ const. If the exit state is sufficiently far into the quality region, a **condensation shock** will be experienced, a phenomenon not considered in this book.

Because of supersaturation it is possible to model the isentropic flow of a vapor through a nozzle with acceptable accuracy by considering the ratio of specific heats to be constant. For steam $k = 1.3$ gives acceptable results over a considerable range of temperatures. The critical pressure ratio given by Eq. 9.3.14 for steam becomes

$$\frac{p^*}{p_0} = \left(\frac{2}{k+1}\right)^{k/(k-1)} = 0.546 \qquad (9.6.1)$$

If saturated steam enters the nozzle, some liquid droplets will be formed and entrained by the vapor; provided that a condensation shock does not exist, a good approximation to the critical pressure ratio is 0.577 corresponding to $k = 1.14$.

EXAMPLE 9.10

Steam is to be expanded isentropically from reservoir conditions of 300°C and 800 kPa absolute to an exit condition of 100 kPa absolute. If supersonic flow is desired, calculate the necessary throat and exit diameters if a mass flux of 2 kg/s is demanded.

Solution

From the steam tables (found in any thermodynamics textbook) we find that

$$s_0 = s_e = 7.2336 \text{ kJ/kg} \cdot \text{K}$$
$$h_0 = 3056.4 \text{ kJ/kg}$$

To estimate the temperature of the metastable exit state, we use

$$T_e = T_0 \left(\frac{p_e}{p_0} \right)^{(k-1)/k}$$

$$= 593 \left(\frac{100}{800} \right)^{0.3/1.3} = 367 \text{ K} \quad \text{or} \quad 94°C$$

Using the steam tables at this temperature, we interpolate, using the exit quality x_e, to find that

$$7.2336 = 1.239 + 6.90x_e$$

$$\therefore \quad x_e = 0.968$$

Thus we have the enthalpy and specific volume at the exit:

$$h_e = 394 + 0.968 \times 2273 = 2594 \text{ kJ/kg}$$

$$v_e = 0.001 + 0.968 \times (2.06 - 0.001) = 1.99 \text{ m}^3/\text{kg}$$

Using the energy equation, the exit velocity is estimated as follows:

$$\frac{\cancel{V_0^2}^{\,0}}{2} + h_0 = \frac{V_e^2}{2} + h_e$$

$$\therefore \quad V_e = \sqrt{2(h_0 - h_e)}$$

$$= \sqrt{2(3056 - 2594) \times 1000} = 961 \text{ m/s}$$

where the 1000 converts kJ to J. From the definition of mass flux we have

$$m_e = \rho_e A_e V_e$$

$$2 = \frac{1}{1.99} \times \frac{\pi d_e^2}{4} \times 961$$

$$\therefore \quad d_e = 0.0726 \text{ m} \quad \text{or} \quad 7.26 \text{ cm}$$

To determine the diameter of the throat, we recognize the throat to be the critical area; thus Eq. 9.6.1 gives

$$p^* = 0.546 \, p_0 = 437 \text{ kPa}$$

Using this pressure and $s^* = s_0 = 7.2336 \text{ kJ/kg} \cdot \text{K}$ we could use the steam tables to find h^* and v^*; the energy equation would then allow us to find V^* and thus d_t. However, a simpler, approximate technique, assuming constant specific heats, is to use Eq. 9.3.18 with $k = 1.3$ and obtain the following:

$$\dot{m} = p_0 A^* \sqrt{\frac{1.3}{RT_0}} \left(\frac{2.3}{2} \right)^{2.3/-0.6}$$

$$2 = 800\,000 \, \frac{\pi d_t^2}{4} \sqrt{\frac{1.3}{287 \times 573}} \times 0.585$$

$$\therefore \quad d_t = 0.044 \text{ m} \quad \text{or} \quad 4.4 \text{ cm}$$

This is reasonable since we have already assumed constant specific heats in predicting T_e. Obviously, the above is approximate; using $k = 1.3$ does, though, give reasonable predictions.

9.7 OBLIQUE SHOCK WAVE

In this section we investigate the oblique shock wave, a finite-amplitude wave that is not normal to the incoming flow. The flow approaching the oblique shock wave will be assumed to be in the x-direction. After the oblique shock the velocity vector will have a component normal to the flow direction. We will continue to assume that the flow before and after the oblique shock is uniform and steady.

Oblique shock waves form on the leading edge of a supersonic airfoil or in an abrupt corner, as sketched in Fig. 9.14. Oblique shock waves may also be found on axisymmetric bodies such as a nose cone or a bullet traveling at supersonic speeds. In this book, we consider only plane flows.

The function of the oblique shock wave is to turn the flow so that the velocity vector $\mathbf{V}_2$ is parallel to the plane wall. The angle between the two velocity vectors introduces another variable into our analysis. The problem remains solvable, however, with the additional tangential momentum equation.

To analyze the oblique shock wave, consider a control volume enclosing a portion of the shock as shown in Fig. 9.15. The velocity vector upstream is assumed to be in the x-direction only; the oblique shock wave makes an angle β with the upstream velocity vector and turns the flow through the **deflection angle** or **wedge angle** θ. The components of the velocity vectors are shown normal and tangential to the oblique shock wave. The tangential components do not cause fluid to flow through the shock; hence the continuity equation, with $A_1 = A_2$, gives

$$\rho_1 V_{1n} = \rho_2 V_{2n} \tag{9.7.1}$$

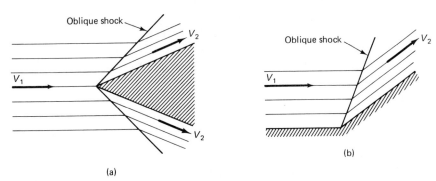

(a)

(b)

Figure 9.14 Oblique shock waves in a supersonic flow: (a) flow over a wedge; (b) flow in a corner.

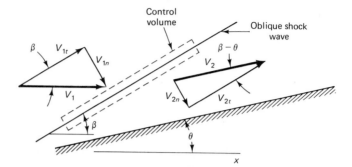

Figure 9.15 Control volume enclosing a small portion of an oblique shock wave.

The pressure forces act normal to the oblique shock and produce no tangential components. Thus the momentum equation expressed in the tangential direction requires that the tangential momentum into the control volume equal the tangential momentum leaving the control volume, that is,

$$\dot{m}_1 V_{1t} = \dot{m}_2 V_{2t} \tag{9.7.2}$$

or, using $\dot{m}_1 = \dot{m}_2$, we demand that

$$V_{1t} = V_{2t} \tag{9.7.3}$$

The normal momentum equation takes the form

$$p_1 - p_2 = \rho_2 V_{2n}^2 - \rho_1 V_{1n}^2 \tag{9.7.4}$$

The energy equation, using $V^2 = V_n^2 + V_t^2$, may be written as

$$\frac{V_{1n}^2}{2} + \frac{k}{k-1}\frac{p_1}{\rho_1} = \frac{V_{2n}^2}{2} + \frac{k}{k-1}\frac{p_2}{\rho_2} \tag{9.7.5}$$

where the tangential component terms have been canceled from both sides. Note that the tangential components of the two velocity vectors do not enter the continuity, normal momentum, or energy equations, the three equations used in the solution of the normal shock wave. Hence we may substitute V_{1n} and V_{2n} for V_1 and V_2, respectively, of the normal shock wave equations and obtain a solution. Either the normal shock wave equations or the normal shock wave table (Table D.2) may be used. Of course, we also replace M_1 and M_2 with M_{1n} and M_{2n}, respectively.

It is useful to relate the oblique shock angle β to the deflection angle θ. Using the continuity equation (9.7.1), with reference to Fig. 9.15, there results

$$\frac{\rho_2}{\rho_1} = \frac{V_{1n}}{V_{2n}} = \frac{V_{1t}\tan\beta}{V_{2t}\tan(\beta-\theta)} = \frac{\tan\beta}{\tan(\beta-\theta)} \tag{9.7.6}$$

From the normal shock wave equations (9.4.12) and (9.4.13) we can find the density ratio to be

$$\frac{\rho_2}{\rho_1} = \frac{p_2 T_1}{p_1 T_2} = \frac{(k+1)M_{1n}^2}{(k-1)M_{1n}^2 + 2} \tag{9.7.7}$$

Substituting this into Eq. 9.7.6 gives

$$\tan(\beta - \theta) = \frac{\tan \beta}{k+1}\left[k - 1 + \frac{2}{M_1^2 \sin^2 \beta}\right] \tag{9.7.8}$$

For a flow with a given M_1, this equation relates the oblique shock angle β to the wedge or corner angle θ. The three variables β, θ, and M_1 in the equation above are often plotted as in Fig. 9.16. We can observe several phenomena by studying the figure.

- For a specified upstream Mach number M_1 and a given wedge angle θ there are two possible oblique shock angles β, the large one corresponding to a "strong" shock and the smaller corresponding to a "weak" shock.
- For a given wedge angle θ there is a minimum Mach number for which there is only one oblique shock angle β.
- For a given wedge angle θ, if M_1 is less than the minimum for that particular curve, no oblique shock wave exists and the shock wave

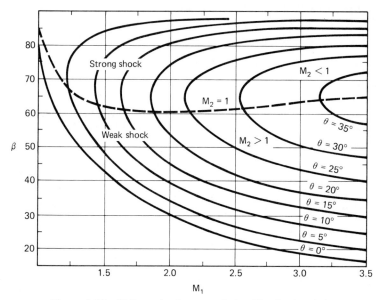

Figure 9.16 Oblique shock wave relationships for $k = 1.4$.

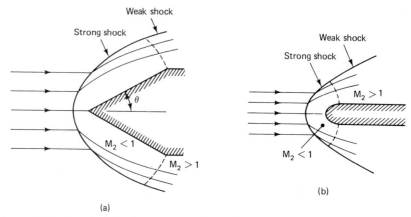

Figure 9.17 Detached shock waves: (a) flow around a wedge; (b) flow around a blunt object.

becomes detached, as shown in Fig. 9.17. Also, for a given M_1 there is a sufficiently large θ that will result in a detached shock wave.

The pressure rise across the oblique shock wave determines whether a weak shock or a strong shock occurs. For a relatively small pressure rise a weak shock will occur with $M_2 > 1$. If the pressure rise is relatively large a strong shock occurs with $M_2 < 1$. Note that for the detached shocks around bodies a normal shock exists for the stagnation streamline; this is followed by the strong oblique shock, then the weak oblique shock, and eventually a Mach wave. For blunt bodies moving at supersonic speeds the shock wave is always detached.

EXAMPLE 9.11

Air flows over a wedge with $M_1 = 3$ as shown. A weak shock reflects from the wall. Determine the values of M_3 and β_3 for the reflected wave.

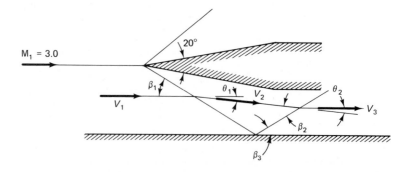

Solution

From Fig. 9.16 with $\theta_1 = 10°$ and $M_1 = 3.0$, we find for the weak shock that $\beta_1 = 27.5°$. This yields

$$M_{1n} = 3 \sin 27.5° = 1.39$$

From the shock table we interpolate to find

$$M_{2n} = 0.744 = M_2 \sin(27.5° - 10°)$$

$$\therefore \quad M_2 = 2.48$$

The reflected shock must again turn the flow through an angle of 10°, that is, $\theta_2 = 10°$. For this wedge angle and $M_2 = 2.48$ from Fig. 9.16 for a weak shock, we see that $\beta_2 = 33°$. This results in

$$M_{2n} = 2.48 \sin 33° = 1.35$$

From the shock table

$$M_{3n} = 0.762 = M_3 \sin 23°$$

$$\therefore \quad M_3 = 1.95$$

The desired angle is calculated to be

$$\beta_3 = \beta_2 - 10° = 23°$$

Note that Fig. 9.16 does not allow precise calculations. Equation 9.7.8 could be used, by trial and error, to improve the accuracy of the β's and hence the quantities that follow.

9.8 ISENTROPIC TURNING-EXPANSION WAVES

In this section we consider the supersonic flow around a convex corner, as shown in Fig. 9.18. Let us first attempt to accomplish such a flow with the finite amplitude wave of part (a). The flow must turn the angle θ so that V_2 is parallel to the wall. The tangential component must be conserved because of momentum conservation. This would result in $V_2 > V_1$, as is obvious from the sketch. This would be the situation if a subsonic flow, $M_{1n} < 1$, could experience a finite increase to a supersonic flow, $M_{2n} > 1$. This, of course, is impossible because of the second law, as was noted in the discussion associated with Fig. 9.10. Consequently, we consider the turning of a finite wave an impossibility.

Consider a second possible mechanism that would allow the flow to turn the corner, a fan composed of an infinite number of Mach waves, emanating from the corner, as shown in Fig. 9.18b. The second law would

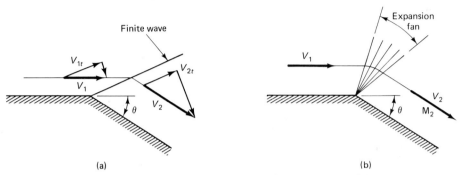

Figure 9.18   Supersonic flow around a convex corner: (a) single finite wave; (b) infinite number of Mach waves.

not be violated with such a mechanism since each Mach wave is an isentropic wave. We will determine the effect of a single Mach wave on the flow and then integrate to obtain the total effect. Figure 9.19 shows the infinitesimal velocity change due to a single Mach wave. For the control volume enclosing the Mach wave, we know that the tangential momentum is conserved; thus the tangential velocity component remains unchanged as shown, as in the oblique shock wave. From the triangles in the figure we can write

$$V_t = V \cos \mu = (V + dV) \cos(\mu + d\theta) \tag{9.8.1}$$

Since $d\theta$ is small, this becomes,[1] using $\cos(\mu + d\theta) = \cos \mu - d\theta \sin \mu$,

$$V \sin \mu \, d\theta = \cos \mu \, dV \tag{9.8.2}$$

Substituting $\sin \mu = 1/M$ (see Eq. 9.2.13) and $\cos \mu = \sqrt{(M^2 - 1)}/M$, we have

$$d\theta = \sqrt{M^2 - 1} \, \frac{dV}{V} \tag{9.8.3}$$

We know that $V = M \sqrt{kRT}$. This can be differentiated to give

$$\frac{dV}{V} = \frac{dM}{M} + \frac{1}{2} \frac{dT}{T} \tag{9.8.4}$$

The energy equation, in the form $V^2/2 + kRT/(k - 1) = \text{const.}$, may also be differentiated to yield

$$\frac{dV}{V} + \frac{1}{(k - 1)M^2} \frac{dT}{T} = 0 \tag{9.8.5}$$

---

[1]Recall the trigonometric identity, $\cos(\alpha + \beta) = \cos \alpha \cos \beta - \sin \alpha \sin \beta$. Then using $\cos d\theta = 1$ and $\sin d\theta = d\theta$, we have $\cos(\mu + d\theta) = \cos \mu - d\theta \sin \mu$.

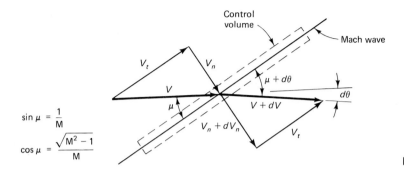

$$\sin \mu = \frac{1}{M}$$

$$\cos \mu = \frac{\sqrt{M^2 - 1}}{M}$$

Figure 9.19   Mach wave.

Eliminating dT/T by combining the two preceding equations, there results

$$\frac{dV}{V} = \frac{2}{2 + (k - 1)M^2} \frac{dM}{M} \tag{9.8.6}$$

This can be substituted into Eq. 9.8.3, allowing a relationship between θ and M to be obtained. We have

$$d\theta = \frac{2\sqrt{M^2 - 1}}{2 + (k - 1)M^2} \frac{dM}{M} \tag{9.8.7}$$

This can be integrated, using $\theta = 0$ at M $= 1$, to provide a relationship between the resulting Mach number (M$_2$ in Fig. 9.18) and the angle, provided that the incoming Mach number is unity; the relationship is

$$\theta = \left(\frac{k + 1}{k - 1}\right)^{1/2} \tan^{-1}\left[\frac{k - 1}{k + 1}(M^2 - 1)\right]^{1/2} - \tan^{-1}(M^2 - 1)^{1/2} \tag{9.8.8}$$

The angle θ, which is a function of M, is the **Prandtl–Meyer function.** It is tabulated for $k = 1.4$ in Table D.3 so that trial-and-error solutions to Eq. 9.8.8 are not necessary. Other changes that may be desired, such as the pressure or temperature changes, can be found from the isentropic-flow equations.

We will see, in working the examples and problems, that both the Mach number and the velocity increase as supersonic flow turns the convex corner. The flow remains attached to the wall as it turns the corner, even for large angles, a phenomenon not observed in a subsonic flow; a subsonic flow would separate from the abrupt corner, even for small angles. If we substitute M $= \infty$ in Eq. 9.8.8, we find the angle of maximum turning to be $\theta = 130.5°$. This would mean that both the temperature and pressure would be absolute zero; obviously, the gas would become a liquid before this would be possible. The angle of 130.5° is, however, an upper limit. The point is, rather large turning angles are possible in supersonic flows, angles that may exceed 90°. This introduces a design constraint on the exhaust nozzle of rocket engines that exhaust into the vacuum of space; the exhaust gases may turn through an angle so that impingement on the spacecraft body is possible.

EXAMPLE 9.12

Air at a Mach number of 2.0 and a temperature and pressure of 500°C and 200 kPa absolute, respectively, flows around a corner with a convex angle of 20°. Find M_2, p_2, T_2, and V_2.

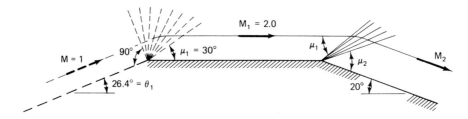

Solution

Table D.3 uses $M = 1$ as a reference condition; thus we visualize the flow as originating from a flow with $M = 1$ and turning through the angle θ_1 to $M_1 = 2$, as shown in the sketch. From the table we find, adding an additional 20° to the deflection angle, that $\theta_2 = 46.4°$. This would be equivalent to the flow at $M = 1$ turning a convex corner with $\theta = 46.4°$. Since the flow is isentropic, we can simply superimpose in this manner. Now, for an angle of $\theta = 46.4°$ from the table, we find that

$$M_2 = 2.83$$

From the isentropic flow table (Table D.1) we find that

$$p_2 = p_1 \frac{p_0}{p_1} \frac{p_2}{p_0}$$

$$= 200 \times \frac{1}{0.1278} \times 0.0352 = 55.1 \text{ kPa}$$

$$T_2 = T_1 \frac{T_0}{T_1} \frac{T_2}{T_0}$$

$$= 773 \times \frac{1}{0.5556} \times 0.3844 = 534.8 \text{ K} \quad \text{or} \quad 261.8°\text{C}$$

The velocity V_2 is found to be

$$V_2 = M_2 \sqrt{kRT_2}$$

$$= 2.83 \sqrt{1.4 \times 287 \times 534.8} = 1312 \text{ m/s}$$

PROBLEMS

9.1. Show that $c_p = Rk/(k - 1)$.

Speed of Sound

9.2. Show that for a small adiabatic disturbance in a steady flow, the energy equation takes the form $\Delta h = -c\,\Delta V$.

9.3. Two rocks are slammed together on one side of a lake. An observer on the other side with his head under water "hears" the disturbance 0.6 sec later. How far is it across the lake?

9.4. Calculate the Mach number for an aircraft traveling at 200 m/s if it is flying at:
(a) Sea level.
(b) 5000 m.
(c) 10 000 m.
(d) 20 000 m.
(e) 35 000 m.

9.4E. Calculate the Mach number for an aircraft traveling at 600 ft/sec if it is flying at:
(a) Sea level.
(b) 15,000 ft.
(c) 30,000 ft.
(d) 60,000 ft.
(e) 100,000 ft.

9.5. A wood chopper is chopping wood some distance away. You carefully observe, using your numerical wrist timer, that it takes 1.21 sec for the sound of the axe to reach your ears. Calculate the distance you are from the chopper if the temperature is $-10°C$.

9.6. A needle-nosed projectile passes over you on a military testing field at a speed of 1000 m/s. You know that it has an elevation of 1000 m where $T = -10°C$. How long after it passes overhead will you hear its sound? How far away will it be? Calculate its Mach number.

9.7. A small-amplitude wave passes through the standard atmosphere at sea level with a pressure rise measured to be 15 Pa. Estimate the associated induced velocity and temperature rise.

9.7E. A small-amplitude wave passes through the standard atmosphere at sea level with a pressure rise of 0.3 psf. Estimate the associated induced velocity and temperature rise.

Isentropic Flow

9.8. A pitot probe, an instrument that measures stagnation pressure, is used to determine the velocity of an airplane. Such a probe, attached to an aircraft, measures 10 kPa. Determine the speed of the aircraft if it is flying at an altitude of:
(a) 3000 m.
(b) 10 000 m.
Assume an isentropic process from the free stream to the stagnation point.

9.9. A pitot probe, used to measure the stagnation pressure, indicates a pressure of 4 kPa at the nose of a surface vehicle traveling in 15°C-atmospheric air. Calculate its speed assuming:
(a) Isentropic flow.
(b) Incompressible flow.
Compute the percent error for part (b).

9.10. A venturi tube is used to measure the mass flux of air in a pipe by reducing the diameter from 10 cm to 5 cm and then back to 10 cm. The pressure at the inlet is 300 kPa and at the minimum diameter section is

240 kPa. If the upstream temperature is 20°C, determine the mass flux.

9.10E. A venturi tube is used to measure the mass flux of air in a pipe by reducing the diameter from 4 in. to 2 in. and then back to 4 in. The pressure at the inlet is 45 psi and at the minimum diameter section it is 36 psi. If the upstream temperature is 60°F, determine the mass flux.

9.11. Air flows from a reservoir maintained at 30°C and 200 kPa absolute through a converging–diverging nozzle that has a 10-cm-diameter throat. Determine the diameter where M = 3. Use equations only.

9.12. Air flows in a converging–diverging nozzle from a reservoir maintained at 20°C and 500 kPa absolute. The throat and exit diameters are 5 cm and 15 cm, respectively. What two receiver pressures will result in M = 1 at the throat if isentropic flow occurs throughout? Use equations only.

9.13. Rework Problem 9.12 using the isentropic flow table.

9.14. A converging nozzle with an exit diameter of 2 cm is attached to a reservoir maintained at 25°C and 200 kPa absolute. Using equations only, determine the mass flux of air if the receiver pressure is:
(a) 100 kPa absolute.
(b) 130 kPa absolute.

9.14E. A converging nozzle with an exit diameter of 1 in. is attached to a reservoir maintained at 70°F and 30 psia. Using equations only, determine the mass flux of air if the receiver pressure is
(a) 15 psia.
(b) 20 psia.

9.15. Rework Problem 9.14 using the isentropic flow table.

9.15E. Rework Problem 9.14E using the isentropic flow table.

9.16. Air flows from a reservoir ($T_0 = 30$°C, $p_0 = 400$ kPa absolute) out a converging nozzle with a 10-cm-diameter exit. What exit pressure would just result in $M_e = 1$? Determine the mass flux for this condition. Use the isentropic flow table.

9.17. Air flows from a converging nozzle attached to a reservoir with $T_0 = 10$°C. What reservoir pressure is necessary to just cause $M_e = 1$ if the 6-cm-diameter nozzle exits to atmospheric pressure? Calculate the mass flux for this condition. Now double the reservoir pressure and determine the increased mass flux.

9.17E. Air flows from a converging nozzle attached to a reservoir with $T_0 = 40$°F. What reservoir pressure is necessary to just cause $M_e = 1$ if the 2.5-in.-diameter nozzle exits to atmospheric pressure? Calculate the mass flux for this condition. Now double the reservoir pressure and determine the increased mass flux.

9.18. A 25-cm-diameter air line is pressurized to 500 kPa absolute and suddenly bursts. The exit area is later measured to be 30 cm². If 6 minutes elapsed before the 10°C air was turned off, estimate the cubic meters of air that were lost.

9.19. A converging nozzle is attached to a reservoir containing helium with $T_0 = 27$°C and $p_0 = 200$ kPa absolute. Determine the receiver pressure that will just give $M_e = 1$. Now attach a diverging section with a 15-cm-diameter exit to the 6-cm-diameter throat. What is the highest receiver pressure that will give $M_t = 1$?

9.20. Air flows from a nozzle with a mass flux of 6 kg/s. If $T_0 = 27$°C, $p_0 = 800$ kPa absolute, and $p_e = 100$ kPa absolute, calculate the throat and exit diameters for an isentropic flow. Also, determine the exit velocity.

9.20E. Air flows from a nozzle with a mass flux of 1.0 slug/sec. If $T_0 = 60$°F, $p_0 = 120$ psia, and $p_e = 15$ psia, calculate the throat and exit diameters for an isentropic flow. Also, determine the exit velocity.

9.21. Air flows from a reservoir, maintained at 20°C and 2 MPa absolute, and exits from a nozzle with $M_e = 4$. The receiver pressure is then raised until the flow is just subsonic throughout the entire nozzle. Calculate this receiver pressure.

9.22. Air at 30°C flows in a 10-cm-diameter pipe at a velocity of 150 m/s. A venturi tube is used to measure the flow rate. What should be the minimum diameter of the tube so that choked flow does not occur?

9.23. For a nozzle efficiency of 96%, rework Problem 9.16.

9.24. Nitrogen enters a diffuser at 100 kPa absolute and 100°C with a Mach number of 3.0. The mass flux is 10 kg/s and the exit velocity is small. Sketch the diffuser, then determine the throat area and the exit pressure and temperature, assuming isentropic flow.

9.24E. Nitrogen enters a diffuser at 15 psia and 200°F with a Mach number of 3.0. The mass flux is 0.2 slug/sec and the exit velocity is small. Sketch the diffuser, then determine the throat area and the exit pressure and temperature, assuming isentropic flow.

9.25. A rocket has a mass of 80 000 kg and is to be lifted vertically from a platform with six nozzles exiting exhaust gases with $T_e = 1000$°C. What should the exit velocity be from each 50-cm-diameter nozzle if the exhaust gases are assumed to be carbon dioxide?

9.26. A 100-kg man straps a small air-breathing jet engine on his back and lifts off the ground vertically. The engine has an exit area of 200 cm². With what velocity must the 600°C exhaust gases exit the engine?

9.27. A converging–diverging nozzle is bolted into a reservoir at a diameter of 40 cm. The throat and exit diameters are 5 cm and 10 cm, respectively. If $T_0 = 27$°C and $p_e = 100$ kPa absolute and isentropic flow of air occurs throughout the supersonic nozzle, calculate the force necessary to hold the nozzle unto the reservoir.

Normal Shock

9.28. The pressure, temperature, and velocity before a normal shock wave are 80 kPa absolute, 10°C, and 1000 m/s, respectively. Calculate M_1, M_2, p_2, T_2, and ρ_2 for air.
(a) Use basic equations.
(b) Use the normal shock table.

9.28E. The pressure, temperature, and velocity before a normal shock wave are 12 psia, 40°F, and 3000 ft/sec, respectively. Calculate M_1, M_2, p_2, T_2, and ρ_2 for air.
(a) Use basic equations.
(b) Use the normal shock table.

9.29. Derive the **Rankine–Hugoniot relationship**

$$\frac{\rho_2}{\rho_1} = \frac{(k+1)p_2/p_1 + k - 1}{(k-1)p_2/p_1 + k + 1}$$

which relates the density ratio to the pressure ratio across a normal shock wave. Find the limiting density ratio for air across a strong shock for which $p_2/p_1 \gg 1$.

9.30. An explosion occurs just above the surface of the earth, producing a shock wave that travels radially outward. At a given location it has a Mach number of 2.0. Determine the pressure just behind the shock and the induced velocity.

9.31. Air at 200 kPa absolute and 20°C passes through a normal shock wave with strength so that $M_2 = 0.5$. Calculate V_1, p_2, and ρ_2.

9.31E. Air at 30 psia and 60°F passes through a normal shock wave with strength so that $M_2 = 0.5$. Calculate V_1, p_2, and ρ_2.

9.32. A blunt object travels at 1000 m/s at an elevation of 10 000 m. The flow approaching the stagnation point passes through a normal shock wave and then decelerates isentropically to the stagnation point. Calculate p_0 and T_0 at the stagnation point.

9.33. A pitot probe is inserted in an airflow in a pipe in which $p = 800$ kPa absolute, $T = 40$°C, and M = 3.0. What pressure does it measure?

9.34. Air flows from a 25°C reservoir to the atmosphere through a nozzle with a 5-cm-diameter throat and a 10-cm-diameter exit. What reservoir pressure will just result in M = 1 at the throat? Also, calculate the mass flux. Maintaining this reservoir pressure, reduce the throat diameter to 4 cm and determine the resulting mass flux. Sketch the pressure distribution as in Fig. 9.12.

9.35. Air flows from a 20°C reservoir to the atmosphere through a nozzle with a 5-cm-diameter throat and a 10-cm-diameter exit. What reservoir pressure is necessary to locate a normal shock wave at the exit? Also, calculate the velocity and pressure at the throat, before the shock, and after the shock.

9.35E. Air flows from a 60°F reservoir to the atmosphere through a nozzle with a 2-in.-diameter throat and a 4-in.-diameter exit. What reservoir pressure is necessary to locate a normal shock wave at the exit? Also, calculate the velocity and pressure at the throat, before the shock, and after the shock.

9.36. Air flows from a receiver maintained at 25°C and 500 kPa absolute. It flows through a nozzle with throat and exit diameters of 5 cm and 10 cm, respectively. What receiver pressure is necessary to locate a normal shock wave at a location where the diameter is 8 cm? Also, determine the velocity before the shock and at the exit.

Vapor Flow

9.37. Steam flows at a rate of 4 kg/s from reservoir conditions of 400°C and 1.2 MPa absolute through a converging–diverging nozzle to the atmosphere. Determine the throat and exit diameters if supersonic, isentropic flow exists throughout the diverging section.

9.38. Steam flows from reservoir conditions of 350°C and 1000 kPa absolute to the atmosphere at the rate of 15 kg/s. Estimate the converging nozzle exit diameter.

9.38E. Steam flows from reservoir conditions of 700°F and 150 psia to the atmosphere at the rate of 0.25 slug/sec. Estimate the converging nozzle exit diameter.

9.39. A header supplies steam at 400°C and 1.2 MPa absolute to a set of nozzles with throat diameters of 1.5 cm. The nozzles exhaust to a pressure of 120 kPa absolute. If the flow is approximately isentropic, calculate the mass flux and the exit temperature.

Oblique Shock Wave

9.40. An airflow with velocity, temperature, and pressure of 800 m/s, 30°C, and 40 kPa absolute, respectively, is turned with an oblique shock wave emanating from the wall, which makes an abrupt 20° corner.
(a) Find the downstream Mach number, pressure, and velocity for a weak shock.
(b) Find the downstream Mach number, pressure, and velocity for a strong shock.
(c) If the concave corner angle were 35°, sketch the corner flow situation.

9.41. Two oblique shocks intersect as shown. Determine the angle of the reflected shocks if the airflow must leave parallel to its original direction. Also find M_3.

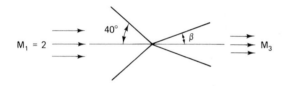

9.42. An oblique shock wave at a 35° angle is reflected from a plane wall. The upstream Mach number M_1 is

3.5 and $T_1 = 0°C$. Find V_3 after the reflected oblique shock wave for the airflow.

9.42E. An oblique shock wave at a 35° angle is reflected from a plane wall. The upstream Mach number M_1 is 3.5 and $T_1 = 30°F$. Find V_3 after the reflected oblique shock wave for the airflow.

9.43. A supersonic inlet can be designed to have a normal shock wave oriented at the inlet, or a wedge can be used to provide a weak oblique shock wave, as shown. Compare the pressure p_3 from the flow shown to the pressure that would exist behind a normal shock wave with no oblique shock.

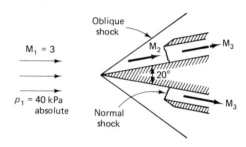

Expansion Waves

9.44. A supersonic airflow with $M_1 - 3$, $T_1 = -20°C$, and $p_1 = 20$ kPa absolute turns a convex corner of 25°. Calculate M_2, p_2, T_2, and V_2 after the expansion fan. Also calculate the included angle of the fan.

9.45. A supersonic airflow with $M_1 = 2$, $T_1 = 0°C$, and $p_1 = 20$ kPa absolute turns a convex corner. If $M_2 = 4$, what angle θ should the corner have? Also calculate T_2 and V_2.

9.45E. A supersonic airflow with $M_1 = 2$, $T_1 = 30°F$, and $p_1 = 5$ psia turns a convex corner. If $M_2 = 4$, what angle θ should the corner have? Also calculate T_2 and V_2.

9.46. The flat plate shown is used as an airfoil at an angle of attack of 5°. Oblique shock waves and expansion fans allow the air to remain attached to the plate with the flow behind the airfoil parallel to the original direction. Calculate:
(a) The pressures on the upper and lower sides of the plate.
(b) The downstream Mach numbers M_{2u} and M_{2l}.
(c) The lift coefficient defined by $C_L = \text{lift}/(\frac{1}{2}\rho_1 V_1^2 A)$, where A is the area of one side of the flat plate.
(d) The drag coefficient where $C_D = \text{drag}/(\frac{1}{2}\rho_1 V_1^2 A)$. Note that $\rho_1 V_1^2 = kM_1^2 p_1$.

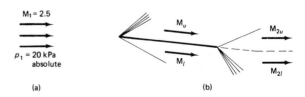

9.47. The supersonic airfoil shown is to fly at zero angle of attack. Calculate the drag coefficient $C_D = $ drag$/(\frac{1}{2}\rho_1 V_1^2 A)$. Note that $\rho_1 V_1^2 = kM_1^2 p_1$.

9.48. The air foil in Problem 9.47 flies at an angle of attack of 5°. Determine the lift and drag coefficients, C_L and C_D. See Problem 9.46 for definitions of C_L and C_D.

$M_1 = 4$

$p_1 = 20\ \text{kPa}$
absolute

5°   5°

TEN

Flow in Open Channels

10.1 INTRODUCTION

Free-surface flow is probably the most commonly occurring flow phenomenon that we encounter on the surface of the earth. Ocean waves, river currents, and overland flow of rainfall are examples that occur in nature. Human-induced situations include flows in canals and culverts, drainage over impervious materials, such as roofs and parking lots, and wave motion in harbors.

In all of these situations, the flow is characterized by an interface between the air and the upper layer of water, which is termed the **free surface.** At the free surface, the pressure is constant, and for many situations, it is atmospheric. In such a case, the hydraulic grade line and the liquid free surface coincide. In engineering practice, most open channels convey water as the fluid. However, the principles developed and implemented in this chapter are also applicable for other liquids that flow with a free surface.

Generally, the elevation of the free surface does not remain constant; it can vary along with the fluid velocities. Another complexity is that the flow is often three-dimensional. Fortunately, there are many instances in which two-dimensional and even one-dimensional simplifications can be made. The flow patterns in estuaries[1] under certain circumstances can be treated as two-dimensional in the horizontal plane, averaged vertically over the depth. River and channel flows are usually treated as one-dimensional with respect to the position coordinate along the streambed. In this chapter we restrict our consideration to one-dimensional flows.

Figure 10.1a shows a representative centerline velocity distribution in a channel. The velocity profile is three-dimensional at a given cross section, Fig. 10.1b. The boundary shear stress is nonuniform; at the free surface the shear stress is negligible, yet a shear stress varies around the wetted perimeter. In some circumstances the presence of secondary currents will force the maximum velocity to occur slightly below the free surface. By convention, y is defined as the depth from the deepest location to the free surface; note that y is not a coordinate. The mean velocity is given by the relation

$$V = \frac{1}{A} \int_A v \, dA \qquad (10.1.1)$$

In the one-dimensional model, we assume the velocity to be equal to V everywhere at a given cross section. This model provides excellent results and is used widely. Flows in open channels are most likely to be turbulent, and the velocity profile can be assumed to be approximately

[1]An estuary is the lower course of a river that is influenced by the ocean tides.

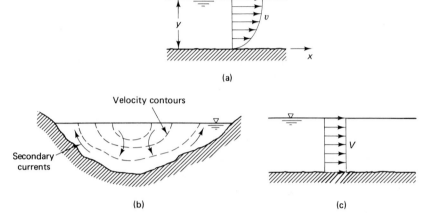

Figure 10.1 Free-surface flow: (a) centerline velocity distribution; (b) cross section; (c) one-dimensional model.

constant, as in Fig. 10.1c, without significant error. Hence the one-dimensional model is employed.

10.2 OPEN CHANNEL FLOWS

10.2.1 Classification of Free-Surface Flows

Flow in a channel is characterized by the mean velocity, even though a velocity profile exists at a given section, as shown in Fig. 10.1. The flow is classified as a combination of steady or unsteady, and uniform or nonuniform. Steady flow signifies that the mean velocity V, as well as the depth y, is independent of time, whereas unsteady flow necessitates that time be considered as an independent variable. Uniform flow implies that V and y are independent of the position coordinate in the direction of flow; nonuniform flow signifies that V and y vary in magnitude along that coordinate. The possible combinations are shown in Table 10.1; the position coordinate is designated as x.

Uniform flow is the situation where terminal velocity has been reached in a channel of constant cross section; not only is the mean velocity constant, but the depth is invariant as well. Steady, nonuniform

TABLE 10.1 COMBINATIONS OF ONE-DIMENSIONAL
FREE-SURFACE FLOWS

Type of flow	Average velocity	Depth
Steady, uniform	$V = \text{const.}$	$y = \text{const.}$
Steady, nonuniform	$V = V(x)$	$y = y(x)$
Unsteady, uniform	$V = V(t)$	$y = y(t)$
Unsteady, nonuniform	$V = V(x, t)$	$y = y(x, t)$

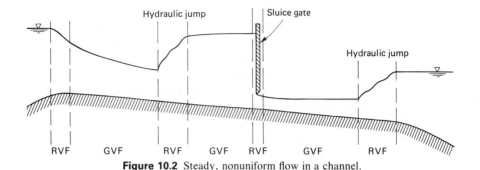

Figure 10.2 Steady, nonuniform flow in a channel.

flow is a common occurrence in rivers and in man-made channels. In those situations, it will be found to occur in two ways. In relatively short reaches called **transitions** there is a rapid change in depth and velocity; such flow is termed **rapidly varied flow.** Examples are the hydraulic jump, flow entering a steep channel from a lake or reservoir, flow close to a free outfall from a channel, and flow in the vicinity of an obstruction such as a bridge pier or a sluice gate.

Along more extensive reaches of channel the velocity and depth may not vary rapidly, but rather, change in a slow manner. Here the water surface can be considered continuous and the regime is called **gradually varied flow.** Examples of gradually varied, steady flow are the backwater created by a dam placed in a river, and the drawdown of a water surface as flow approaches a falls. Figure 10.2 illustrates how both rapidly varied flow (RVF) and gradually varied flow (GVF) can occur simultaneously in a reach of channel. Note that the vertical scale is larger than the horizontal scale; such scale distortion is common when representing open-channel flow situations.

Unsteady uniform flow rarely occurs, but unsteady nonuniform flow is common. Flood waves in rivers, hydraulic bores, and regulated flow in canals are all examples of the latter category. In many situations, these flows may be considered to behave sufficiently like steady uniform flow or steady nonuniform flow to justify treating them as such. Unsteady flows are beyond the scope of a fundamental treatment and will not be presented in this book; the one exception is the **hydraulic bore**—a moving hydraulic jump—which can be analyzed in a quasi-steady fashion.

10.2.2 Significance of Froude Number

The primary mechanism for sustaining flow in an open channel is gravitational force. For example, the elevation difference between two reservoirs will cause water to flow through a canal that connects them. The parameter that represents this gravitational effect is the Froude number,

$$\text{Fr} = \frac{V}{\sqrt{gL}} \qquad (10.2.1)$$

which has been observed in Chapter 6 to be the ratio of inertial force to gravity force. In the context of open-channel flow, V is the mean cross-sectional velocity, and L is a representative length parameter. For a channel of rectangular cross section, L becomes the depth y of the flow.

The Froude number plays the dominant role in open-channel flow analysis. It appears in a number of relations that will be developed later in this chapter. Furthermore, by knowing its magnitude, one can ascertain significant characteristics regarding the flow regime. For example, if $Fr > 1$, the flow possesses a relatively high velocity and shallow depth; on the other hand, when $Fr < 1$, the velocity is relatively low and the depth is relatively deep. Except in the vicinity of rapids, cascades, and waterfalls, most rivers experience a Froude number less than unity. Constructed channels may be designed for Froude numbers to be greater or less than unity, or to vary from greater than unity to less than unity along the length of the channel.

10.2.3 Hydrostatic Pressure Distribution

Consider a channel in which the flow is nearly horizontal, as shown in Fig. 10.3. In this case, there is little or no vertical acceleration of fluid within the reach, and the streamlines remain nearly parallel. Such a condition is common in many open-channel flows, and indeed, if slight variations exist, the streamlines are assumed to behave as if they are parallel. Since the vertical accelerations are nearly zero, one can conclude that in the vertical direction the pressure distribution is hydrostatic. As a result, the sum $(p + \gamma z)$ remains a constant at any depth, and the hydraulic grade line coincides with the water surface. It is customary in open-channel flow to designate z as the elevation of the channel bottom, and y as the depth of flow. Since at the channel bottom $p/\gamma = y$, the hydraulic grade line is given by the sum $(y + z)$. The concepts in this chapter are developed assuming a hydrostatic pressure distribution.

10.3 UNIFORM FLOW

Before we study nonuniform flow, let us focus our attention on the simpler condition of uniform flow. Such a flow is rare, but if it were to occur

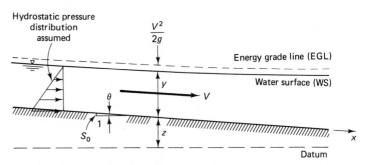

Figure 10.3 Reach of open-channel flow.

in a channel, the depth and velocity would not vary along its length, or in other words, terminal conditions would have been reached. In addition to uniform flow, this section covers the cross-sectional geometry of open channels; the formulations will apply to both uniform and nonuniform flow. Some of that material is covered in Section 7.7; it is reviewed here for completeness.

10.3.1 Channel Geometry

Channel cross sections can be considered to be either regular or irregular. A **regular section** is one whose shape does not vary along the length of the channel, whereas an **irregular section** will have changes in its geometry. In this chapter we consider mostly regular channel shapes; three common geometries are shown in Fig. 10.4.

The simplest channel shape is a rectangular section. The cross-sectional area is given by

$$A = by \qquad (10.3.1)$$

in which b is the width of the channel bottom (see Fig. 10.4a). Additional parameters of importance for open-channel flow are the wetted perimeter, the hydraulic radius, and the width of the free surface. The **wetted perimeter** P is the length of the line of contact between the liquid and the channel; for a rectangular channel it is

$$P = b + 2y \qquad (10.3.2)$$

The **hydraulic radius** R is the area divided by the wetted perimeter, that is,

$$R = \frac{A}{P} = \frac{by}{b + 2y} \qquad (10.3.3)$$

The **free-surface width** B is equal to the bottom width b for a rectangular section.

A trapezoidal section (Fig. 10.4b) has the added feature of sloped side walls. If m_1 is the ratio of the horizontal to vertical change of the wall on one side, and m_2 is the corresponding quantity on the other wall, the

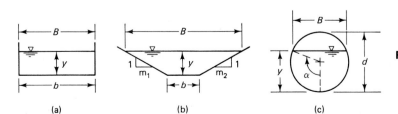

Figure 10.4 Representative regular cross sections: (a) rectangular; (b) trapezoidal; (c) circular.

area, wetted perimeter, and free-surface width are given as

$$A = by + \tfrac{1}{2}y^2(m_1 + m_2) \tag{10.3.4}$$

$$P = b + y(\sqrt{1 + m_1^2} + \sqrt{1 + m_2^2}) \tag{10.3.5}$$

$$B = b + y(m_1 + m_2) \tag{10.3.6}$$

Note that the rectangular section is embodied in the trapezoidal definition, since for vertical side walls m_1 and m_2 are zero, and the relations for
A, P, and B become identical. In addition, if b is set equal to zero, Eqs.
10.3.4 to 10.3.6 describe the geometry of a triangular-shaped channel.

The circular cross section is an important one to consider, since
many free-surface flows in drainage and sewer systems are conveyed in
circular conduits. If d is the conduit diameter, the area, wetted perimeter,
and free-surface width are given by

$$A = \frac{d^2}{4}(\alpha - \sin \alpha \cos \alpha) \tag{10.3.7}$$

$$P = \alpha d \tag{10.3.8}$$

$$B = d \sin \alpha \tag{10.3.9}$$

where

$$\alpha = \cos^{-1}\left(1 - 2\frac{y}{d}\right) \tag{10.3.10}$$

The angle α is defined in Fig. 10.4c.

A generalized cross-sectional geometry, such as that shown in Fig.
10.5a, can be expressed in functional form as $A(y)$, $P(y)$, $R(y)$, and $B(y)$.
The functional representations encompass all of the analytical forms
given above, and they also may be used to describe an irregular channel.
For example, at a river section, one can describe the area and wetted
perimeter in tabular form and utilize techniques such as curve fitting or

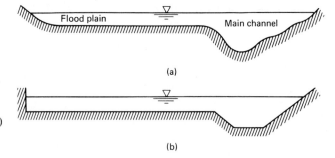

Figure 10.5 Generalized section representa-
tion: (a) actual
cross section; (b)
composite cross
section.

interpolation to extract the numerical information as functions of the depth. Such procedures are useful for computer-based analyses.

A composite section is one made up of several subsections; usually these subsections are of analytic form. The example shown in Fig. 10.5b consists of a main channel and a floodplain. The main channel is approximated by a trapezoid and the floodplain by a rectangle. One could derive analytical expressions for such a composite section; however, it may be more useful to consider the functional forms for the geometric parameters. Note that the functions will be discontinuous at depths where the two sections are matched.

Most of the theoretical developments in this chapter focus on cross sections that are rectangular. Such an assumption allows one to simplify the mathematics associated with open-channel flow analysis. Even though the equations will be simplified relative to more complicated geometries, the physical understanding of the phenomena and conclusions reached will apply to most generalized prismatic cross sections. A clear distinction will be made between rectangular and other types of geometry when various developments and concepts are presented.

10.3.2 Equation for Uniform Flow

Uniform flow occurs in a channel when the depth and velocity do not vary along its length, that is, when terminal conditions have been reached in the channel. Under such conditions, the energy grade line, water surface, and channel bottom are all parallel. It was seen in Section 7.7 that uniform flow can be predicted by an equation of the form

$$V = C\sqrt{RS_0} \qquad (10.3.11)$$

in which S_0 is the slope of the channel bottom and C is the Chezy coefficient, independent of the Reynolds number since the flow is considered completely turbulent. It has become common engineering practice to relate C to the channel roughness and the hydraulic radius by use of the Manning relation

$$C = \frac{c_1}{n} R^{1/6} \qquad (10.3.12)$$

where $c_1 = 1$ for SI units and $c_1 = 1.49$ for English units. Combining Eqs. 10.3.11 and 10.3.12 with the definition of discharge results in the Chezy–Manning equation

$$Q = \frac{c_1}{n} AR^{2/3} \sqrt{S_0} \qquad (10.3.13)$$

Values of the Manning coefficient n are given in Table 7.3.

The depth associated with uniform flow is designated y_0; it is called either **uniform depth** or **normal depth.** Uniform flow rarely occurs in rivers because of the irregularity of the geometry. In manmade channels it is also uncommon, since the presence of controls such as sluice gates, weirs, or outfalls will cause the flow to become gradually varied. It is, however, necessary to determine y_0 when analyzing gradually varied flow conditions, since it provides in part a basis for evaluating the type of water surface that may exist in the channel. The design of gravity flow sewer networks is often based on assuming uniform flow and the use of Eq. 10.3.13, even though much of the time the flow in such systems may be nonuniform.

An examination of Eq. 10.3.13 reveals that it can be solved explicitly for Q, n, or S_0. Examples 7.17 and 7.18 provide illustrations. A trial-and-error solution or graphical procedure is necessary when it is required to find y_0 and the remaining parameters are given. A computer algorithm, based on the interval halving method to find the normal depth, is given in Section 10.8 as a subroutine.

EXAMPLE 10.1

Water is flowing at a rate of 4.5 m^3/s in a trapezoidal channel whose bottom width is 2.4 m and side slopes are 1 vertical to 2 horizontal. Compute y_0 if $n = 0.012$ and $S_0 = 0.0001$.

Solution

Given geometrical data are $b = 2.4$ m and $m_1 = m_2 = 2$. Rearrange Eq. 10.3.13, noting that $R = A/P$ and $c_1 = 1$:

$$\frac{A^{5/3}}{P^{2/3}} = \frac{Qn}{\sqrt{S_0}}$$

Substituting in the known data and trapezoidal geometry, one has

$$\frac{[2.4y_0 + \frac{1}{2}y_0^2(2 + 2)]^{5/3}}{[2.4 + y_0(2\sqrt{1 + 2^2})]^{2/3}} = \frac{4.5 \times 0.012}{\sqrt{0.0001}}$$

Solving by trial and error gives $y_0 = 1.28$ m.

10.3.3 Most Efficient Section

Design of a channel to convey uniform flow typically consists of selecting or specifying the appropriate geometrical cross section provided that Q, n, and S_0 are known. Once chosen, an optimum sizing of the cross section can be based on the criteria of minimum resistance to flow. The flow resistance per unit length equals surface shear stress times wetted perimeter. Using a control volume for uniform flow, and assuming a small slope S_0 so that $\sin \theta \simeq S_0$ (see Fig. 10.3), one can show that the resistance per unit length is

$$\tau_0 P = \gamma A S_0 \tag{10.3.14}$$

Hence a least resistance criterion is equivalent to requiring a minimum cross-sectional area with respect to the parameters defining the area. Furthermore, we have to satisfy the Chezy-Manning equation. Since Q, n, and S_0 are given and $R = A/P$, Eq. 10.3.13 can be written as

$$P = cA^{5/2} \tag{10.3.15}$$

in which c is a constant. As an example, consider a rectangular channel with width b and depth y. The best hydraulic cross section is obtained by rewriting Eq. 10.3.15 such that A becomes a function of b only. There-fore, we express P in terms of A and b, using $A = by$ and $P = b + 2y$:

$$P = b + \frac{2A}{b} \tag{10.3.16}$$

Substitution of this equation into 10.3.15 yields

$$b + \frac{2A}{b} = cA^{5/2} \tag{10.3.17}$$

Now, differentiate Eq. 10.3.17 with respect to b:

$$1 + \frac{2}{b}\frac{dA}{db} - \frac{2A}{b^2} = \frac{5}{2} cA^{3/2} \frac{dA}{db} \tag{10.3.18}$$

and set $dA/db = 0$, since the objective is to find the value of b that minimizes A. The result is

$$\frac{2A}{b^2} = 1 \tag{10.3.19}$$

or, using $A = by$,

$$b = 2y \tag{10.3.20}$$

Thus, if the width of a rectangular channel is twice the depth of the flowing water, the water will flow most efficiently.

For a trapezoidal cross section, it is simpler to begin with Eq. 10.3.15, eliminate b, and express P as a function of A, y, and m, where $m = m_1 = m_2$. Then by considering A as a function of m and y, the minimum of A is found by setting the gradient vector of A to zero. The result is $m = \sqrt{3}/3$, or a side slope angle of 60° with the horizontal. The resulting hexagonal shape is a trapezoid that best approximates a semicircle. The evaluation is left as an exercise for the reader.

The optimum criterion based on flow resistance cannot always be used, and in fact may be less significant than other design factors. Additional aspects to be considered are the type of excavation, and if the channel is unlined, the slope stability of the side walls and the possibility of erosion of the bed.

10.4 ENERGY CONCEPTS IN OPEN-CHANNEL FLOW

The energy at any position along the channel is the sum of **flow energy head**[2] y, potential energy head z and kinetic energy head $V^2/2g$. That sum defines the energy grade line and is termed the **total energy** H:

$$H = z + y + \frac{V^2}{2g} \tag{10.4.1}$$

The kinetic-energy-correction factor associated with the $V^2/2g$ term is assumed to be unity (see Section 4.4.4); this is common practice for most prismatic channels of simple geometry since the velocity profiles are nearly uniform for the turbulent flows involved. The fundamental energy equation, developed in Section 4.4, states that losses will occur for a real fluid between any two sections of the channel, and hence the total energy will not remain constant. The energy balance is given simply by the relation

$$H_1 = H_2 + h_L \tag{10.4.2}$$

in which h_L is the head loss. The only manner in which energy can be added to an open-channel flow system is for mechanical pumping or lifting of the liquid to take place. Equation 10.4.2 is applicable for rapidly varied as well as gradually varied flow situations; it will be used in conjunction with the momentum and continuity equations in a number of applications.

10.4.1 Specific Energy E

It is convenient in open-channel flow to measure the energy relative to the bottom of the channel; it provides a useful means to analyze complex flow situations. Such a measure is termed **specific energy** and is designated

---

[2]This terminology is adopted since $y = p/\gamma$ and the pressure term ($\dot{m}\,\Delta p/\rho$) in the energy equation is often called the "flow work."

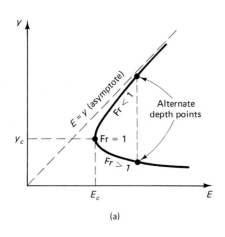
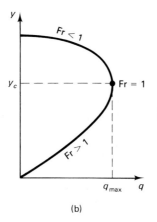

Figure 10.6 Variation of specific energy and specific discharge with depth: (a) E versus y for constant q; (b) q versus y for constant E.

as E:

$$E = y + \frac{V^2}{2g}$$
(10.4.3)

Specific energy then is the sum of flow energy head y and kinetic energy head $V^2/2g$.

Rectangular Sections

For a rectangular section, the specific energy can be expressed as a function of the depth y. The **specific discharge** q is defined as the total discharge divided by the channel width, that is,

$$q = \frac{Q}{b} = Vy$$
(10.4.4)

The specific energy for a rectangular channel can thus be put in the form

$$E = y + \frac{q^2}{2gy^2}$$
(10.4.5)

This E–y relation is shown in Fig. 10.6a. We may observe that a specific discharge requires at least a minimum energy; this minimum energy is referred to as **critical energy,** E_c. The corresponding depth y_c is called the **critical depth.** If the specific energy is greater than E_c, two depths are possible; those depths are referred to as **alternate depths.**[3] For constant q, Eq. 10.4.5 is a cubic equation in y for a given value of E which is greater

---

[3]The E–y curve falls between the two asmyptotes $E = y$ and $y = 0$. Another curve defined by the relation exists for negative y; it is not considered since it has no physical meaning for open-channel flow.

than E_c. The two positive solutions of y are the alternate depths. Another way to express Eq. 10.4.5 is to consider E constant and vary q. Equation 10.4.5 can be solved for q as

$$q = \sqrt{2gy^2(E - y)} \qquad (10.4.6)$$

This relation is shown in Fig. 10.6b. This form of the energy relation is useful for analyzing flows in which the specific energy remains nearly constant throughout the transition region; examples are a change in the width of a channel and the variation of depth with discharge at the entrance to a channel.

The critical depth y_c can be evaluated from Eq. 10.4.5 by setting the derivative of E with respect to y equal to zero:

$$\frac{dE}{dy} = 1 - \frac{q^2}{gy^3} = 0 \qquad (10.4.7)$$

Since $q = Vy$, the condition of minimum E is

$$1 - \frac{V^2}{gy} = 1 - \mathrm{Fr}^2 = 0 \qquad (10.4.8)$$

where the Froude number in a rectangular channel is

$$\mathrm{Fr} = \frac{V}{\sqrt{gy}} = \frac{q}{\sqrt{gy^3}} \qquad (10.4.9)$$

Thus from Eq. 10.4.8 the Froude number is equal to unity for minimum energy. Solving for the depth in Eq. 10.4.7 in terms of q, we have

$$y = \left(\frac{q^2}{g}\right)^{1/3} = y_c \qquad (10.4.10)$$

This relation gives the critical depth in terms of the specific discharge.

On the E–y curve, for any depth greater than y_c, the flow is relatively slow or tranquil, and $\mathrm{Fr} < 1$; such a state is termed **subcritical flow.** Conversely, for a depth less than critical, the flow is relatively rapid or shooting, $\mathrm{Fr} > 1$, and the regime is one of **supercritical flow.**

At critical flow conditions, E_c can conveniently be expressed by combining Eqs. 10.4.5 and 10.4.10 to eliminate q, resulting in

$$E_c = \tfrac{3}{2}y_c \qquad (10.4.11)$$

The E–y diagram is a representation of the change in specific energy as the depth is varied, given a constant specific discharge. It is possible for q to vary, as when the width of a rectangular section changes in a transition region. As q increases, the $E-y$ curve shifts to the right in Fig. 10.6a. It is left as an exercise for the reader to show that for a given specific energy, maximization of q of Eq. 10.4.6 will produce critical conditions at maximum discharge (see Fig. 10.6b).

Generalized Cross Section

For a generalized section the specific energy can be written in terms of the total discharge Q and the cross-sectional area A:

$$E = y + \frac{Q^2}{2gA^2} \qquad (10.4.12)$$

The minimum-energy condition is obtained by differentiating Eq. 10.4.12 with respect to y to obtain

$$\frac{dE}{dy} = 1 - \frac{Q^2}{gA^3}\frac{dA}{dy} \qquad (10.4.13)$$

For incremental changes in depth, the corresponding change in area is $dA = B\,dy$. Thus the minimum-energy condition becomes

$$1 - \frac{Q^2B}{gA^3} = 0 \qquad (10.4.14)$$

By analogy to Eq. 10.4.8, the second term in Eq. 10.4.14 is the Froude number, given by

$$Fr = \sqrt{\frac{Q^2B}{gA^3}} = \frac{Q/A}{\sqrt{gA/B}} \qquad (10.4.15)$$

The ratio A/B is termed the **hydraulic depth,** and its use allows the definition of the Froude number to be generalized. The reader can easily verify that A/B is equal to y for a rectangular channel.

The specific energy diagram of Fig. 10.6 provides a useful means of visualizing the solution to a transition problem. Even though one probably would solve the problem numerically, an assessment of the graphical solution may provide useful physical insight as well as prevent the incorrect root from being selected.

EXAMPLE 10.2

Water is flowing in a triangular channel with $m_1 = m_2 = 1.0$ at a discharge of $Q = 3$ m³/s. If the water depth is 2.5 m, determine the specific energy, Froude number, hydraulic depth, and alternate depth.

Solution

The flow area and top width are computed from Eqs. 10.3.4 and 10.3.6 as follows:

$$A = \tfrac{1}{2}y^2(m_1 + m_2)$$
$$= \tfrac{1}{2} \times 2.5^2 \times (1 + 1) = 6.25 \text{ m}^2$$
$$B = (m_1 + m_2)y$$
$$= (1 + 1) \times 2.5 = 5.0 \text{ m}$$

Using Eqs. 10.4.12 and 10.4.15, E and Fr are found to be

$$E = y + \frac{Q^2}{2gA^2}$$
$$= 2.5 + \frac{3^2}{2 \times 9.81 \times 6.25^2} = 2.51 \text{ m}$$

$$\text{Fr} = \sqrt{\frac{Q^2B}{gA^3}}$$
$$= \sqrt{\frac{3^2 \times 5}{9.81 \times 6.25^3}} = 0.137$$

The hydraulic depth is

$$\frac{A}{B} = \frac{6.25}{5.0} = 1.25 \text{ m}$$

The alternate depth is calculated using the energy equation. Recognizing that $A = y^2$, we have

$$2.51 = y + \frac{3^2}{2 \times 9.81 \times (y^2)^2}$$
$$= y + \frac{0.459}{y^4}$$

By trial and error, $y = 0.71$ m.

EXAMPLE 10.2 (English)

Water is flowing in a triangular channel with $m_1 = m_2 = 1.0$ at a discharge of $Q = 100$ ft³/sec. If the water depth is 8 ft, determine the specific energy, Froude number, hydraulic depth, and alternate depth.

Solution

The flow area and top width are computed from Eqs. 10.3.4 and 10.3.6 as follows:

$$A = \tfrac{1}{2}y^2(m_1 + m_2)$$
$$= \tfrac{1}{2} \times 8^2 \times (1 + 1) = 64 \text{ ft}^2$$

$$B = (m_1 + m_2)y$$

$$= (1 + 1) \times 8 = 16 \text{ ft}$$

Using Eqs. 10.4.12 and 10.4.15, E and Fr are found to be

$$E = y + \frac{Q^2}{2gA^2}$$

$$= 8 + \frac{100^2}{2 \times 32.2 \times 64^2} = 8.04 \text{ ft}$$

$$\text{Fr} = \sqrt{\frac{Q^2 B}{gA^3}}$$

$$= \sqrt{\frac{100^2 \times 16}{32.2 \times 64^3}} = 0.138$$

The hydraulic depth is

$$\frac{A}{B} = \frac{64}{16} = 4.0 \text{ ft}$$

The alternate depth is calculated using the energy equation. Recognizing that $A = y^2$, we have

$$8.04 = y + \frac{100^2}{2 \times 32.2 \times (y^2)^2}$$

$$= y + \frac{155.3}{y^4}$$

By trial and error $y = 2.28$ ft.

10.4.2 Use of the Energy Equation in Transitions

A **transition** is a relatively short reach of channel where the depth and velocity change, creating a nonuniform, rapidly varied flow situation. The mechanism for such changes in the flow is usually an alteration of one or more geometric parameters of the channel.[4] Within such regions, the energy equation can be used effectively to analyze the flow in a transition or to aid in the design of a transition. Two applications are given in this section.

Channel Constriction

Consider a rectangular channel whose bottom is raised by a distance h over a short region (Fig. 10.7a). The change in depth in the transition can be analyzed by use of the energy equation. As a first approximation, one can neglect losses in the transition. Assume that the specific energy upstream of the transition is known. Recognizing that $H = E + z$, Eq.

[4]One notable exception to this is the hydraulic jump, which requires use of the momentum principle and will be considered in Section 10.5.2.

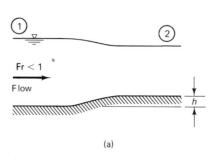
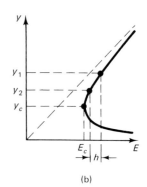

Figure 10.7 Channel constriction: (a) raised channel bottom; (b) specific energy diagram.

10.4.2 is applied from location 1 to the end of the transition region, location 2:

$$E_1 = E_2 + h \qquad (10.4.16)$$

The depth y_2 at the end of the transition can be visualized by inspection of the E–y diagram (Fig. 10.7b). If the flow at location 1 is subcritical, y_1 is located on the upper leg as shown in the specific energy diagram. The magnitude of h is selected to be relatively small such that $y_2 > y_c$, hence the flow at location 2 is similarly subcritical; however, as h is increased still further, a state of minimum energy is ultimately reached in the transition. Any further increase in h would result in a change in the flow rate. The condition of minimum energy is sometimes referred to as a "choking condition" or as "choked flow." Once choked flow occurs, as h increases, the variations in depth and velocity are no longer localized in the vicinity of the transition. Influences may be observed for significant distances both upstream and downstream of the transition.

  The reader should verify that in a transition region, a narrowing of the channel width will create a situation similar to an elevation of the channel bottom. The most general transition region is one that possesses both a change in width and in bottom elevation.

EXAMPLE 10.3

A rectangular channel 3 m wide is conveying water at a depth $y_1 = 1.55$ m, and velocity $V_1 = 1.83$ m/s. The flow enters a transition region as shown in Fig. 10.7, in which the bottom elevation is raised by $h = 0.20$ m. Determine the depth and velocity in the transition, and the value of h for choking to occur.

Solution

Use Eq. 10.4.4 to find the specific discharge to be

$$q = V_1 y_1$$

$$= 1.83 \times 1.55 = 2.84 \text{ m}^2/\text{s}$$

The Froude number at location 1 is

$$Fr = \frac{V_1}{\sqrt{gy_1}}$$

$$= \frac{1.83}{\sqrt{9.81 \times 1.55}} = 0.47$$

which is less than unity. Hence the flow at location 1 is subcritical. The specific energy at location 1 is found, using Eq. 10.4.3, to be

$$E_1 = y_1 + \frac{V_1^2}{2g}$$

$$= 1.55 + \frac{1.83^2}{2 \times 9.81} = 1.72 \text{ m}$$

The specific energy at location 2 is found, using Eq. 10.4.16, to be

$$E_2 = E_1 - h$$

$$= 1.72 - 0.20 = 1.52 \text{ m}$$

If $E_2 > E_c$, it is possible to find the depth y_2. Therefore, E_c is calculated first. From Eqs. 10.4.10 and 10.4.11 the critical conditions are

$$y_c = \left(\frac{q^2}{g}\right)^{1/3}$$

$$= \left(\frac{2.84^2}{9.81}\right)^{1/3} = 0.94 \text{ m}$$

$$E_c = \frac{3y_c}{2}$$

$$= 3 \times \frac{0.94}{2} = 1.41 \text{ m}$$

Hence, since $E_2 > E_c$, we can proceed with calculating y_2. The depth y_2 can be evaluated by substituting known values into Eq. 10.4.5:

$$1.52 = y_2 + \frac{2.84^2}{2 \times 9.81 \times y_2^2}$$

Solving by trial and error gives the solution

$$y_2 = 1.26 \text{ m}$$

$$\therefore \ V_2 = \frac{q}{y_2}$$

$$= \frac{2.84}{1.26} = 2.25 \text{ m/s}$$

The flow at location 2 is subcritical since there is no way in which the flow can become supercritical in the transition with the given geometry.

The value of h for critical flow to appear at location 2 is determined by setting $E_2 = E_c$ in Eq. 10.4.16:

$$h = E_1 - E_c$$

$$= 1.72 - 1.40 = 0.31 \text{ m}$$

Channel Entrance with Critical Flow

Consider flow entering a channel from a lake or reservoir over a short rounded crest (see Fig. 10.8). If the channel slope is steep, the flow will discharge freely into the channel and supercritical flow will occur downstream of the entrance region. Upstream of the crest the flow may be considered as subcritical. Since the flow regime at the crest changes from subcritical to supercritical, the flow at crest must be critical. To determine the discharge, we assume that rapidly varied flow takes place over the crest in conjunction with the critical flow condition at the crest. An example illustrates the procedure.

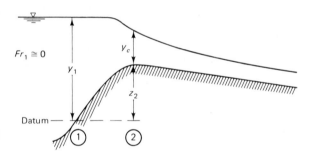

Figure 10.8 Outflow from a reservoir with critical flow at the channel entrance.

EXAMPLE 10.4

Water flows freely from a reservoir into a trapezoidal channel with bottom width $b = 5.0$ m and side slope parameters $m_1 = m_2 = 2.0$. The elevation of the water surface in the reservoir is 2.3 m above the entrance crest. Assuming negligible losses in the transition and a negligible velocity in the reservoir upstream of the entrance, find the critical depth at the transition and the discharge into the channel.

Solution

The total energy at location 1 in Fig. 10.8 is y_1 since the kinetic energy in the reservoir is negligible. Equating the total energies at locations 1 and 2 gives

$$y_1 = E_2 + z_2$$

Since critical conditions occur at location 2, Eqs. 10.4.12 and 10.4.14 can be combined to eliminate the discharge, with the result

$$E_2 = y_c + \frac{A}{2B}$$

Elimination of E_2 in the two equations yields the expression

$$y_1 - z_2 = y_c + \frac{A}{2B}$$

$$= y_c + \frac{by_c + \frac{1}{2}(m_1 + m_2)y_c^2}{2[b + (m_1 + m_2)y_c]}$$

or, with the given data, the expression becomes

$$2.3 = y_c + \frac{5y_c + \frac{1}{2}(2 + 2)y_c^2}{2[5 + (2 + 2)y_c]}$$

The relation above is a quadratic in y_c. The positive root is chosen, which gives

$$y_c = 1.70 \text{ m}$$

Subsequently, one can find that $A = 14.28 \text{ m}^2$ and $B = 11.80 \text{ m}$. Use Eq. 10.4.14 to find the discharge to be

$$Q = \sqrt{\frac{gA^3}{B}}$$

$$= \sqrt{\frac{9.8 \times 14.3^3}{11.8}} = 49.3 \text{ m}^3/\text{s}$$

For more complex sections such a problem may require a trial-and-error solution.

Energy Losses

Energy losses in expansions and contractions are known to be relatively small when the flow is subcritical; however, it may be necessary under certain circumstances that the losses be considered. Equation 10.4.2 includes a loss term that can account for transitional losses. The following experimentally derived formulas can be employed. For a channel expansion use

$$h_L = K_e\left(\frac{V_1^2}{2g} - \frac{V_2^2}{2g}\right) \qquad (10.4.17)$$

and for a channel contraction use

$$h_L = K_c\left(\frac{V_2^2}{2g} - \frac{V_1^2}{2g}\right) \qquad (10.4.18)$$

In Eq. 10.4.17, K_e is an expansion coefficient; it has been suggested (King and Brater, 1963) to use $K_e = 1.0$ for sudden or abrupt expansions, and $K_e = 0.2$ for well-designed, or rounded expansions. For the contraction coefficient K_c in Eq. 10.4.18, use $K_c = 0.5$ for sudden contractions, and $K_c = 0.10$ for well-designed contractions. When the flow is supercritical, significant standing-wave patterns can be generated throughout and downstream of the transition region; for such flows, proper design requires that wave mechanics be considered (Chow, 1959; Henderson, 1966).

10.4.3 Flow Measurement

The most common means of metering discharge in an open channel is to use a weir. Basically, a **weir** is an obstruction placed in the channel that forces the flow through an opening or aperture designed to measure the discharge. Specialized weirs have been designed for specific needs; in this section two fundamental types—broad-crested and sharp-crested—will be presented.

   A properly designed weir will exhibit subcritical flow upstream of the structure, and the flow will converge and accelerate to a critical condition near the top or crest of the weir. As a result, a correlation between discharge and a depth upstream of the weir can be made. The downstream overflow is termed the **nappe,** which usually discharges freely into the atmosphere. There are a number of factors that affect the performance of a weir; significant among them are the three-dimensional flow pattern, the effects of turbulence, frictional resistance, surface tension, and the amount of ventilation beneath the nappe. The simplified derivations presented here are based on the Bernoulli equation; the other effects can be accounted for by modifying the ideal discharge with a **discharge coefficient,** C_d; the actual discharge is the ideal discharge multiplied by the discharge coefficient. When possible, it is advantageous to calibrate a particular weir in place in order to obtain the desired accuracy.

Broad-crested Weir

   A broad-crested weir is shown in Fig. 10.9. It has sufficient elevation above the channel bottom to choke the flow, and it is long enough so that

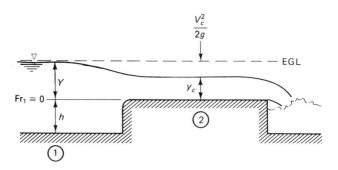

Figure 10.9 Broad-crested weir.

the overflow streamlines become parallel, resulting in a hydrostatic pressure distribution. Then, at some position, say location 2, a critical flow condition exists.

Assume a rectangular horizontal channel and let the height of the weir be h. Location 1 is a point upstream of the weir where the flow is relatively undisturbed; Y is the vertical distance from the top of the weir to the free surface at that point. Applying Bernoulli's equation from location 1 to location 2 at the free surface and neglecting the kinetic energy head at location 1, there results

$$h + Y = h + y_c + \frac{V_c^2}{2g} \tag{10.4.19}$$

Solving for V_c,

$$V_c = \sqrt{2g(Y - y_c)} \tag{10.4.20}$$

For a weir whose breadth normal to the flow is b, the ideal discharge is

$$Q = by_cV_c = by_c\sqrt{2g(Y - y_c)} \tag{10.4.21}$$

Recognizing that $Y = E_c$, Eq. 10.4.11 is used to relate y_c to Y, resulting in

$$Q = \tfrac{2}{3}\sqrt{\tfrac{2}{3}g}\, bY^{3/2} \tag{10.4.22}$$

For a properly rounded upstream edge on the weir, Eq. 10.4.22 is accurate to within several percent of the actual flow; hence a discharge coefficient is not applied.

Sharp-crested Weir

A sharp-crested weir is a vertical plate placed normal to the flow containing a sharp-edge crest so that the nappe will behave as a free jet. Figure 10.10 shows a rectangular weir with a horizontal crest extending across the entire width of the channel. Because of the presence of the side walls, lateral contractions are not present.

Let us define an idealized flow situation: The flow in the vertical

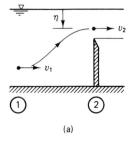

(a)

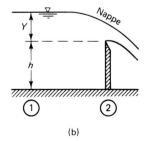

(b)

Figure 10.10 Rectangular sharp-crested weir: (a) ideal flow; (b) actual flow.

plane does not contract as it passes over the crest so that the streamlines are parallel, atmospheric pressure is present in the nappe, and uniform flow exists at location 1 with negligible kinetic energy. The Bernoulli equation is applied along a representative streamline (Fig. 10.10a) and solved for v_2, the local velocity in the nappe:

$$v_2 = \sqrt{2g\eta} \qquad (10.4.23)$$

If b is the width of the crest normal to the flow, the ideal discharge is given as

$$Q = b \int_0^Y v_2 \, d\eta = b \int_0^Y \sqrt{2g\eta} \, d\eta = b\tfrac{2}{3}\sqrt{2g} \; Y^{3/2} \qquad (10.4.24)$$

Experiments have shown that the magnitude of the exponent is nearly correct but that a discharge coefficient C_d must be applied to accurately predict the real flow, shown in Fig. 10.10b:

$$Q = C_d\tfrac{2}{3}\sqrt{2g} \; bY^{3/2} \qquad (10.4.25)$$

Generally, the discharge coefficient accounts for the effect of contraction, velocity of approach, viscosity, and surface tension. An experimentally (Chow, 1959) obtained formula for C_d has been given in the form

$$C_d = 0.61 + 0.08 \frac{Y}{h} \qquad (10.4.26)$$

Normally, for small Y/h ratio, $C_d = 0.61$. If the weir crest does not extend to the side walls but allows for end contractions to appear, as in Fig. 10.11a, the effective width of the weir can be approximated by $(b - 0.2Y)$.

The V-notch weir (Fig. 10.11b) is more accurate than the rectangular weir for measurement of low discharge. In a manner similar to the development of the rectangular weir relation, the idealized discharge is found by integrating the local velocity throughout the nappe above the crest. A

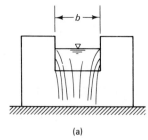
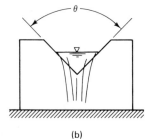

Figure 10.11 Contracted rectangular and V-notch weirs: (a) rectangular; (b) V-notch.

(a)   (b)

discharge coefficient is applied to yield

$$Q = C_d \frac{8}{15} \sqrt{2g} \left(\tan \frac{\theta}{2} \right) Y^{5/2} \qquad (10.4.27)$$

For use with water, and for θ varying from 22.5° to 120°, experiments (King and Brater, 1963) have shown that a value of $C_d = 0.58$ is acceptable for engineering calculations. Derivation of Eq. 10.4.27 is left as an exercise.

EXAMPLE 10.5

Determine the discharge of water over a rectangular sharp-crested weir, $b = 1.25$ m, $Y = 0.35$ m, $h = 1.47$ m, with side walls and with end contractions. If a 90° V-notch weir were to replace the rectangular weir, what would be the required Y for a similar discharge?

Solution

For the rectangular weir, using Eq. 10.4.26, the discharge coefficient is

$$C_d = 0.61 + 0.08 \frac{Y}{h}$$

$$= 0.61 + 0.08 \frac{0.35}{1.47}$$

$$= 0.63$$

Substitute into Eq. 10.4.25 and calculate

$$Q = C_d \frac{2}{3} \sqrt{2g} \, bY^{3/2}$$

$$= 0.63 \times \tfrac{2}{3} \times \sqrt{2 \times 9.81} \times 1.25 \times 0.35^{3/2}$$

$$= 0.48 \text{ m}^3/\text{s}$$

With end contractions the effective width of the weir is reduced by $0.2Y$, resulting in

$$Q = C_d \frac{2}{3} \sqrt{2g} \, (b - 0.2Y)Y^{3/2}$$

$$= 0.63 \times \tfrac{2}{3} \times \sqrt{2 \times 9.81} \times (1.25 - 0.2 \times 0.35) \times 0.35^{3/2}$$

$$= 0.45 \text{ m}^3/\text{s}$$

With a discharge of $Q = 0.48$ m³/s, use Eq. 10.4.27 to find Y for the 90° V-notch weir:

$$Y = \left[\frac{Q}{C_d \times \frac{8}{15} \times \sqrt{2g} \, \tan(\theta/2)} \right]^{2/5}$$

$$= \left[\frac{0.482}{0.58 \times \frac{8}{15} \times \sqrt{2 \times 9.81} \times \tan 45°} \right]^{2/5} = 0.66 \text{ m}$$

EXAMPLE 10.5 (English)

Determine the discharge of water over a rectangular sharp-crested weir, $b = 4.0$ ft, $Y = 1.15$ ft, $h = 4.8$ ft, with side walls and with end contractions. If a 90° V-notch weir were to replace the rectangular weir, what would be the required Y for a similar discharge?

Solution

For the rectangular weir, using Eq. 10.4.26, the discharge coefficient is

$$C_d = 0.61 + 0.08 \frac{Y}{h}$$

$$= 0.61 + 0.08 \frac{1.15}{4.8}$$

$$= 0.63$$

Substitute into Eq. 10.4.25 and calculate

$$Q = C_d \tfrac{2}{3} \sqrt{2g} \; bY^{3/2}$$

$$= 0.63 \times \tfrac{2}{3} \times \sqrt{2 \times 32.2} \times 4 \times 1.15^{3/2}$$

$$= 16.6 \; \text{ft}^3/\text{sec}$$

Without side walls, the effective width of the weir is reduced by $0.2Y$, resulting in

$$Q = C_d \tfrac{2}{3} \sqrt{2g} \; (b - 0.2Y)Y^{3/2}$$

$$= 0.63 \times \tfrac{2}{3} \times \sqrt{2 \times 32.2} \times (4 - 0.2 \times 1.15) \times 1.15^{3/2}$$

$$= 15.7 \; \text{ft}^3/\text{sec}$$

With a discharge of $Q = 16.6$ ft³/sec, use Eq. 10.4.27 to find Y for the 90° V-notch weir:

$$Y = \left[\frac{Q}{C_d \times \tfrac{8}{15} \times \sqrt{2g} \; \tan(\theta/2)} \right]^{2/5}$$

$$= \left[\frac{16.6}{0.58 \times \tfrac{8}{15} \times \sqrt{2 \times 32.2} \times \tan 45°} \right]^{2/5} = 2.14 \; \text{ft}$$

Additional Methods of Flow Measurement

Other types of weirs include those whose faces are inclined in the upstream and downstream direction (triangular, trapezoidal, irregular). In addition, the spillway section of a dam can be considered as a weir with a rounded crest. King and Brater (1963) give details on the selection and use of these types. The **Parshall flume** is a special type of "venturi flume" based on a standarized section; extensive calibrations have established reliable empirical formulas to predict the discharge.

In a natural section of a river it may be impractical to place a weir; in that case stream gaging can be used to measure the discharge. A control location is established upstream of the gaging site, and for a given depth, or **stage,** of the river, the two-dimensional velocity profile is measured using current meters. Subsequently, the profile is numerically integrated to yield the discharge. A series of such measurements will produce a stage-discharge curve, which can then be employed to estimate the discharge by measuring the river stage.

10.5 MOMENTUM CONCEPTS IN OPEN-CHANNEL FLOW

In the preceding section we have seen how the energy equation is applied to rapidly varied flow situations, and in particular, how it is utilized to analyze flows in transition regions. The momentum equation is also employed to study certain phenomena in those situations. When used in conjunction with the energy and continuity relations, the momentum equation gives the user a concise means of analyzing nearly all significant transition problems, including problems that involve a hydraulic jump.

10.5.1 Momentum Equation

Consider the open-channel reach with supercritical flow upstream of a submerged obstacle, as depicted in Fig. 10.12a. It represents a rapidly varied flow with an abrupt change in depth but no change in width. In general, such a change may result from an obstacle in the flow or from a hydraulic jump. The upstream flow is supercritical and the downstream flow is subcritical. Other flow regimes may also be considered; for exam-

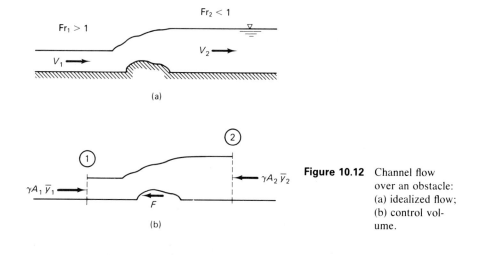

Figure 10.12 Channel flow over an obstacle: (a) idealized flow; (b) control volume.

ple, the conditions could be subcritical throughout the entire control volume. Each situation should be approached as a unique formulation.

The generalized flow situation may be used to develop the equation of motion for transition regions. The control volume corresponding to Fig. 10.12a is shown in Fig. 10.12b. The pressure distribution is assumed hydrostatic, and the resultant hydrostatic forces are given by $\gamma A \bar{y}$, where the distance $\bar{y}$ to the centroid of the cross-sectional area is measured from the free surface. The submerged obstacle imparts a force F on the control volume with a direction opposite to the direction of flow. The linear momentum equation, presented in Section 4.5, is applied to the control volume in the x-direction to give

$$\gamma A_1 \bar{y}_1 - \gamma A_2 \bar{y}_2 - F = \rho Q(V_2 - V_1) \tag{10.5.1}$$

Note that frictional forces have not been included in Eq. 10.5.1; they are usually quite small relative to the other terms, so they can be ignored. Likewise, gravitational forces in the flow direction are negligible for the small channel slopes considered. Equation 10.5.1 can be rearranged in the form

$$M_1 - M_2 = \frac{F}{\gamma} \tag{10.5.2}$$

in which M_1 and M_2 are terms that contain the hydrostatic force and the momentum flux at locations 1 and 2, respectively. The quantity M is called the **momentum function,** and for a general prismatic section it is given by

$$M = A\bar{y} + \frac{Q^2}{gA} \tag{10.5.3}$$

For a rectangular section, $A\bar{y} = by^2/2$, and the momentum function is

$$M = \frac{by^2}{2} + \frac{bq^2}{gy}$$
$$= b\left(\frac{y^2}{2} + \frac{q^2}{gy}\right) \tag{10.5.4}$$

Equation 10.5.4 is sketched in Fig. 10.13. Two positive roots y exist for a given M and q; they are termed either **conjugate** or **sequent** depths. The upper leg of the curve ($y > y_c$) applies to subcritical flow and the lower leg ($y < y_c$) to supercritical flow. The values of M_1 and M_2 for the example of Fig. 10.12a are shown, indicating supercritical flow upstream and subcritical flow downstream of the submerged obstacle. The horizontal distance between M_1 and M_2 is equal to F/γ. The downstream flow would be

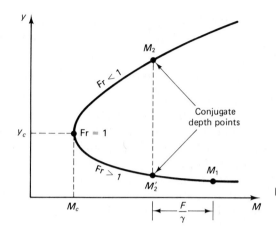

Figure 10.13 Variation of the momentum function with depth.

supercritical if no hydraulic jump occurred; the corresponding value of the momentum function is indicated by M_2'.

The depth associated with a minimum M is found by differentiating M with respect to y in Eq. 10.5.3:

$$\frac{dM}{dy} = A - \frac{BQ^2}{gA^2} = 0 \tag{10.5.5}$$

Note[5] that $d(A\bar{y})/dy = A$. The condition for minimum M is thus

$$Q^2 B = gA^3 \tag{10.5.6}$$

[5]This can be observed from the definition $\bar{y} = (1/A) \int (y - \eta)\, dA$. Differentiation of $(\bar{y}A)$ using Leibnitz's rule from calculus yields

$$\frac{d}{dy}(\bar{y}A) = \frac{d}{dy} \int_0^y (y - \eta) w(\eta)\, d\eta$$

$$= (y - \eta)w(\eta)\Big|_{\eta=y} + \int_0^y w(\eta)\, d\eta$$

$$= 0 + A = A$$

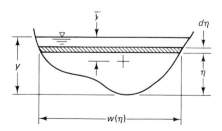

This result is identical to Eq. 10.4.14. Hence the condition of minimum M is equivalent to that of minimum energy: critical flow with a Froude number equal to unity.

Momentum Equation Applied to a Transition Region

Quite often the momentum equation is applied in situations where it is desired to determine the resultant force acting at a specific location, or to find the change in depth or velocity when there is a significant undefined loss across the transition region. It is important to remember that the energy and continuity equations are also at our disposal, and we must determine which relations are necessary. In some instances, along with the continuity equation, both energy and momentum equations must be applied. An example is given below to illustrate the technique.

EXAMPLE 10.6

In a rectangular 5-m-wide channel, water is discharging at 14.0 m³/s. Find the force exerted on the sluice gate[6] when $y_1 = 2$ m and $y_2 = 0.5$ m.

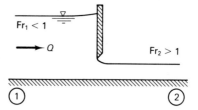

Solution

Using Eq. 10.5.3, the momentum functions at 1 and 2 are

$$M_1 = A_1 \bar{y}_1 + \frac{Q^2}{gA_1}$$

$$= 5 \times 2 \times 1 + \frac{(14)^2}{9.81 \times 5 \times 2} = 12.0 \ \text{m}^3$$

$$M_2 = A_2 \bar{y}_2 + \frac{Q^2}{gA_2}$$

$$= 5 \times 0.5 \times 0.25 + \frac{(14)^2}{9.81 \times 5 \times 0.5} = 8.62 \ \text{m}^3$$

The resultant force acting on the fluid control volume is determined, using Eq. 10.5.2, to be

$$F = \gamma(M_1 - M_2)$$

$$= 9800 \times (12.0 - 8.62) = 33 \ 100 \ \text{N}$$

[6]A sluice is an artificial channel having a gate at its upstream end which can be raised or lowered to regulate the flow.

Hence the force on the gate acts in the downstream sense with a magnitude of 33.1 kN.

10.5.2 Hydraulic Jump

A hydraulic jump is a phenomenon in which fluid flowing at a supercritical state will undergo a transition to a subcritical state. Boundary conditions upstream and downstream of the jump will dictate its strength as well as its location. An idealized hydraulic jump is shown in Fig. 10.14. The strength of the jump varies widely, as shown in Fig. 10.15, with relatively mild disturbances occurring at one extreme, and significant separation and eddy formation taking place at the other. As a result, the energy loss associated with the jump is considered to be unknown, so the energy equation is not used in the initial analysis. Assuming no friction along the bottom and no submerged obstacle; that is, setting F equal to zero, Eq. 10.5.2 shows that $M_1 = M_2$. For a rectangular section, Eq. 10.5.4 can be substituted into the relation, allowing one to obtain

$$\frac{q^2}{g}\left(\frac{1}{y_1} - \frac{1}{y_2}\right) = \frac{1}{2}\left(y_2^2 - y_1^2\right) \tag{10.5.7}$$

Rearranging, factoring, and noting that $\mathrm{Fr}_1^2 = q^2/(gy_1^3)$ there results

$$\mathrm{Fr}_1^2 = \frac{1}{2}\frac{y_2}{y_1}\left(\frac{y_2}{y_1} + 1\right) \tag{10.5.8}$$

This equation is dimensionless and it relates the Froude number upstream of the jump with the ratio of the downstream to upstream depths. It can be seen that Eq. 10.5.8 is quadratic with respect to y_2/y_1 provided that Fr_1 is known. Solving for y_2/y_1, one obtains

$$\frac{y_2}{y_1} = \frac{1}{2}\left(\sqrt{1 + 8\mathrm{Fr}_1^2} - 1\right) \tag{10.5.9}$$

The positive sign in front of the radical has been chosen to yield a physically meaningful solution. It is worth noting that Eq. 10.5.9 is also

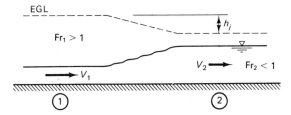

EGL

$\mathrm{Fr}_1 > 1$

h_j

V_1

V_2 $\mathrm{Fr}_2 < 1$

①   ②

Figure 10.14 Idealized hydraulic jump.

Upstream Fr	Type	Description	
1.0–1.7	Undular	Ruffled or undular water surface; surface rollers form near Fr = 1.7	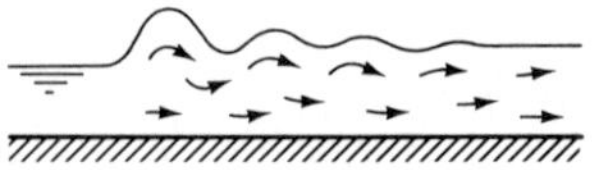
1.7–2.5	Weak	Prevailing smooth flow; low energy loss	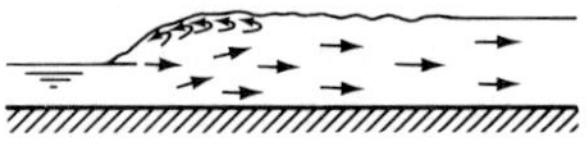
2.5–4.5	Oscillating	Intermittent jets from bottom to surface, causing persistent downstream waves	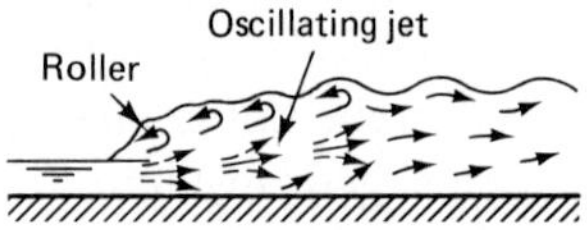
4.5–9.0	Steady	Stable and well-balanced; energy dissipation contained in main body of jump	
> 9.0	Strong	Effective, but with rough, wavy surface downstream	

Figure 10.15 Hydraulic jumps in horizontal rectangular channels. (Adapted with permission from Chow, 1959.)

valid if the subscripts on the depths and Froude number are reversed:

$$\frac{y_1}{y_2} = \frac{1}{2} \left(\sqrt{1 + 8Fr_2^2} - 1 \right) \qquad (10.5.10)$$

The theoretical energy loss associated with a hydraulic jump in a rectangular channel can be determined once the depths and flows at locations 1 and 2 are known. The energy equation is applied from 1 to 2, including the head loss h_j across the jump. Combining the relation with Eq. 10.5.1 and the continuity equation, one can show after some algebra that

$$h_j = \frac{(y_2 - y_1)^3}{4y_1 y_2} \qquad (10.5.11)$$

Equations 10.5.8 through 10.5.11 are useful forms for solving most hydraulic jump problems in a rectangular channel.

Figure 10.15 shows the various forms that a jump may assume relative to the upstream Froude number. A steady, well-established jump, with $4.5 < \text{Fr} < 9.0$, is often used as an energy dissipator downstream of a dam or spillway. It is characterized by the existence of breaking waves and rollers accompanied by a submerged jet with significant turbulence and dissipation of energy in the main body of the jump; downstream, the water surface is relatively smooth. For Froude numbers outside the range 4.5 to 9.0, less desirable jumps exist and may create undesirable downstream surface waves. The length of a jump is the distance from the front face to just downstream where smooth water exists; a steady jump has an approximate length of six times the upstream depth.

EXAMPLE 10.7

A hydraulic jump is situated in a 4-m-wide rectangular channel. The discharge in the channel is 7.5 m³/s, and the depth upstream of the jump is 0.20 m. Determine the depth downstream of the jump, the upstream and downstream Froude numbers, and the rate of energy dissipated by the jump.

Solution

Find the unit discharge and upstream Froude number:

$$q = \frac{Q}{b}$$

$$= \frac{7.5}{4} = 1.88 \text{ m}^2/\text{s}$$

$$\text{Fr}_1 = \frac{q}{\sqrt{gy_1^3}}$$

$$= \frac{1.88}{\sqrt{9.81 \times 0.20^3}} = 6.71$$

The downstream depth is computed, using Eq. 10.5.9, to be

$$y_2 = \frac{y_1}{2}\left(\sqrt{1 + 8\text{Fr}_1^2} - 1\right)$$

$$= \frac{0.20}{2}\left(\sqrt{1 + 8 \times 6.71^2} - 1\right) = 1.80 \text{ m}$$

The downstream Froude number is

$$\text{Fr}_2 = \frac{q}{\sqrt{gy_2^3}}$$

$$= \frac{1.88}{\sqrt{9.81 \times 1.80^3}} = 0.25$$

The head loss in the jump is given by Eq. 10.5.11:

$$h_j = \frac{(y_2 - y_1)^3}{4y_1y_2}$$

$$= \frac{(1.80 - 0.20)^3}{4 \times 0.20 \times 1.80} = 2.84 \text{ m}$$

Hence the rate of energy dissipation in the jump is

$$\gamma Q h_j = 9800 \times 7.5 \times 2.84 = 2.09 \times 10^5 \text{ W} \quad \text{or} \quad 209 \text{ kW}$$

Translating Hydraulic Jump

A translating hydraulic jump, alternately termed a **surge** or a **hydraulic bore**, is shown in Fig. 10.16a; it is termed a positive surge wave in the sense that it maintains a stable front as it propagates at speed w into an undisturbed region. Such a wave can be generated by abruptly lowering a downstream gate or by rapidly releasing water at an upstream location into a channel. This unsteady rapidly varied flow situation can be conveniently analyzed as a steady-state problem by superposing the bore speed w in the opposite sense on the control volume (Fig. 10.16b). The front appears stationary and the relative velocities at locations 1 and 2 are equal to $V_1 + w$ and $V_2 + w$, respectively.

Assume a bore translating in a rectangular, frictionless, horizontal channel. Equation 10.5.9 can be applied by substituting $V_1 + w$ for V_1 in the definition of Fr_1 to obtain

$$\frac{y_2}{y_1} = \frac{1}{2}\left[\sqrt{1 + 8\frac{(V_1 + w)^2}{gy_1}} - 1\right] \tag{10.5.12}$$

The continuity relation applied to the control volume of Fig. 10.16b is

$$y_1(V_1 + w) = y_2(V_2 + w) \tag{10.5.13}$$

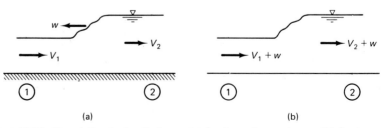

(a)   (b)

Figure 10.16 Translating hydraulic jump: (a) front moving upstream; (b) front appears stationary by superposition.

Equations 10.5.12 and 10.5.13 contain five parameters: y_1, y_2, V_1, V_2, and w. Three of them must be known to solve for the remaining two. Depending on which variables are unknown, the solution of Eqs. 10.5.12 and 10.5.13 will be either explicit or based on a trial-and-error procedure.

EXAMPLE 10.8

Water flows in a rectangular channel with a velocity of 2.5 m/s and a depth of 1.5 m. A gate is suddenly completely closed, forming a surge that travels upstream. Find the speed of the surge and the depth behind the surge.

Solution

Since the gate is closed, the downstream velocity is $V_2 = 0$. Equations 10.5.12 and 10.5.13 contain the two unknowns w and y_2. Combining them to eliminate y_2 and substituting $V_1 = 2.5$ and $y_1 = 1.5$ results in the relation

$$2.5 + w = \frac{w}{2}\left[\sqrt{1 + 8\frac{(2.5 + w)^2}{9.81 \times 1.5}} - 1\right]$$

Solving by trial and error yields $w = 3.41$ m/s. Use Eq. 10.5.13 to compute y_2:

$$y_2 = y_1\frac{V_1 + w}{V_2 + w}$$

$$= 1.5 \times \frac{2.5 + 3.41}{3.41} = 2.60 \text{ m}$$

Drag on Submerged Objects

If an object is submerged in the flow it is possible to describe the drag force in the manner

$$F = C_D A\rho\,\frac{V^2}{2} \tag{10.5.14}$$

in which C_D is the drag coefficient and A is the projected area normal to the flow. A discussion of drag coefficients for various immersed objects is given in Section 8.3.1. In open channel flow, since a free surface is present, the drag coefficient must be modified to account for wave drag as well as drag due to friction and separation. Examples of submerged objects in open channel flow include pipelines and bridge piers; in these situations, a hydraulic jump may not be present.

A design application is illustrated in Fig. 10.17. Baffle blocks are devices placed in a reach of channel known as a **stilling basin** to stabilize

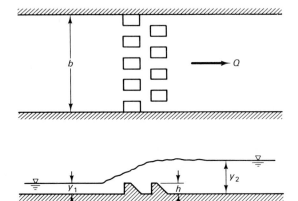

Figure 10.17 Channel with baffle blocks.

the location of a hydraulic jump and to aid in the dissipation of flow energy. Typically they are used for Fr_1 greater than 4.5. Example 10.9 demonstrates how baffle blocks reduce the magnitude of the momentum function downstream of a hydraulic jump, resulting in reduced downstream depth and increased energy dissipation. In practice, Eq. 10.5.14 is seldom employed to analyze or design a stilling basin, since other factors such as additional appurtenances, approach velocity, scour, and cavitation must also be considered. Instead, design standards have been established based on observations of existing basins and laboratory model studies (U.S. Dept. of Interior, 1974; Roberson, et al., 1988).

EXAMPLE 10.9

In the flow situation presented in Example 10.7, a series of baffle blocks is placed in the channel as shown in Fig. 10.17. Laboratory experimentation has shown that the arrangement has an effective drag coefficient of 0.25, provided that the blocks are submerged in the flow. If the blocks are 0.15 m high, and if the discharge and upstream depth remain the same, determine the depth downstream of the jump and the rate of energy dissipated by the jump.

Solution

It is necessary to use Eq. 10.5.2 since obstacles (i.e., the baffle blocks) are placed within the control volume. The upstream velocity is

$$V_1 = \frac{Q}{A_1}$$

$$= \frac{7.5}{4 \times 0.2} = 9.38 \text{ m/s}$$

The force F due to the presence of the baffle blocks is computed using Eq. 10.5.14:

$$F = C_D A \rho \frac{V_1^2}{2}$$

$$= 0.25 \times (4 \times 0.15) \times 1000 \times \frac{9.38^2}{2} = 6600 \text{ N}$$

Note that the frontal area is the width of the channel multiplied by the height of the blocks. Substituting known conditions into Eq. 10.5.2, making use of Eq. 10.5.4 which defines M for a rectangular channel, and noting that $q = 7.5/4 = 1.88 \text{ m}^2/\text{s}$, we find

$$b\left(\frac{y_1^2}{2} + \frac{q^2}{gy_1}\right) - b\left(\frac{y_2^2}{2} + \frac{q^2}{gy_2}\right) = \frac{F}{\gamma},$$

$$4\left(\frac{0.2^2}{2} + \frac{1.88^2}{9.81 \times 0.2}\right) - 4\left(\frac{y_2^2}{2} + \frac{1.88^2}{9.81 y_2}\right) = \frac{6600}{9800}$$

The relation reduces to

$$y_2^2 + \frac{0.721}{y_2} = 3.31$$

By trial and error, y_2 is found to be approximately 1.70 m. The change in specific energy between locations 1 and 2 is

$$E_1 - E_2 = y_1 + \frac{q^2}{2gy_1^2} - \left(y_2 + \frac{q^2}{2gy_2^2}\right)$$

$$= 0.2 + \frac{1.88^2}{2 \times 9.81 \times 0.2^2} - \left(1.70 + \frac{1.88^2}{2 \times 9.81 \times 1.70^2}\right)$$

$$= 2.94 \text{ m}$$

The rate of energy dissipation, therefore, is

$$\gamma Q(E_1 - E_2) = 9800 \times 7.5 \times 2.94$$

$$= 2.16 \times 10^5 \text{ W} \quad \text{or} \quad 216 \text{ } kW$$

EXAMPLE 10.9 (English)

In the flow situation presented in Example 10.7, a series of baffle blocks is placed in the channel as shown in Fig. 10.17. Laboratory experimentation has shown that the arrangement has an effective drag coefficient of 0.25, provided that the blocks are submerged in the flow. If the blocks are 0.5 ft high, and if the discharge and upstream depth remain the same, determine the depth downstream of the jump and the rate of energy dissipated by the jump.

Solution

Referring to Example 10.7, in English units, the channel is 13.1 ft wide, the discharge is 265 ft³/sec, and the depth upstream of the jump is 0.66 ft.

It is necessary to use Eq. 10.5.2 since obstacles (i.e., the baffle blocks) are placed within the control volume. The upstream velocity is

$$V_1 = \frac{Q}{A_1}$$

$$= \frac{265}{13.1 \times 0.66} = 30.65 \text{ ft/sec}$$

The force F due to the presence of the baffle blocks is computed using Eq. 10.5.14:

$$F = C_D A \rho \frac{V_1^2}{2}$$

$$= 0.25 \times (13.1 \times 0.5) \times 1.94 \times \frac{30.65^2}{2} = 1492 \text{ lb}$$

Note that the frontal area is the width of the channel multiplied by the height of the blocks. Substituting known conditions into Eq. 10.5.2, making use of Eq. 10.5.4 which defines M for a rectangular channel, and noting that $q = 265/13.1 = 20.2$ ft²/sec, we find

$$b\left(\frac{y_1^2}{2} + \frac{q^2}{gy_1}\right) - b\left(\frac{y_2^2}{2} + \frac{q^2}{gy_2}\right) = \frac{F}{\gamma},$$

$$13.1\left(\frac{0.66^2}{2} + \frac{20.2^2}{32.2 \times 0.66}\right) - 13.1\left(\frac{y_2^2}{2} + \frac{20.2^2}{32.2y_2}\right) = \frac{1492}{62.4}$$

The relation reduces to

$$y_2^2 + \frac{25.3}{y_2} = 35.2$$

By trial and error, y_2 is found to be approximately 5.54 ft. The change in specific energy between locations 1 and 2 is

$$E_1 - E_2 = y_1 + \frac{q^2}{2gy_1^2} - \left(y_2 + \frac{q^2}{2gy_2^2}\right)$$

$$= 0.66 + \frac{20.2^2}{2 \times 32.2 \times 0.66^2} - \left(5.54 + \frac{20.2^2}{2 \times 32.2 \times 5.54^2}\right)$$

$$= 9.46 \text{ ft}$$

The rate of energy dissipation, therefore, is

$$\gamma Q(E_1 - E_2) = 62.4 \times 265 \times 9.46$$

$$= 1.56 \times 10^5 \text{ ft-lb/sec or 284 horsepower}$$

10.5.3 Numerical Solution of the Momentum Equation

For nonrectangular channels the momentum relation can be used directly to analyze the hydraulic jump or other problems requiring the momentum equation; the technique is demonstrated as follows. Consider a trapezoidal channel with conditions known at location 1 upstream of the jump. Consequently, M_1 and F are evaluated as constants and Eq. 10.5.2 can be written in the form

$$M_2 - M_1 + \frac{F}{\gamma} = 0 \qquad (10.5.15)$$

in which M_2 is a function of y_2. Introducing the trapezoidal geometry at location 2, with $m_1 = m_2 = m$, the relation above can be written as

$$\frac{y_2^2}{6}(2my_2 + 3b) + \frac{Q^2}{g(by_2 + my_2^2)} - M_1 + \frac{F}{\gamma} = 0 \qquad (10.5.16)$$

This can be solved for y_2 by a suitable numerical technique such as interval halving, false position, or Newton's method, which are discussed in any book on numerical methods (see, e.g., Chapra and Canale, 1988). Note that by setting $F = 0$, it becomes the relation for finding the conditions downstream of a hydraulic jump, and that by additionally letting $m = 0$, a rectangular hydraulic jump problem can be solved.

EXAMPLE 10.10

A hydraulic jump occurs in a triangular channel with $m_1 = m_2 = 2.5$. The discharge is 20 m³/s and $y_c = 1.67$ m. Upstream of the jump the following parameters are given: $y_1 = 0.75$ m, $Fr_1 = 7.42$, and $M_1 = 29.35$ m³. Determine the conjugate depth y_2 downstream of the jump.

Solution

Use Eq. 10.5.16 with $F = 0$:

$$\frac{y_2^2}{6} \times 2 \times 2.5y_2 + \frac{20^2}{9.81 \times 2.5y_2^2} - 29.35 = 0$$

The relation reduces to

$$f(y_2) = y_2^3 + \frac{19.58}{y_2^2} - 35.23 = 0$$

The false position method is chosen to find y_2. The first step is to set the upper and lower limits, termed y_u and y_l. Since $Fr_1 > 1$, $y_2 > y_c$, and an appropriate lower limit is $y_l = y_c = 1.67$ m. The upper limit is assumed to be 5 m.

Iteration	y_l	y_u	$f(y_l)$	$f(y_u)$	y_r	$f(y_r)$	Sign of $f(y_l) \times f(y_r)$	ε
1	5	1.67	90.55	−23.55	2.357	−18.61	—	
2	5	2.357	90.55	−18.61	2.808	−10.61	—	
3	5	2.808	90.55	−10.61	3.038	−5.076	—	
4	5	3.038	90.55	−5.067	3.142	−2.231	—	
5	5	3.142	90.55	−2.231	3.187	−0.944	—	
6	5	3.187	90.55	−0.944	3.205	−0.393	—	
7	5	3.205	90.55	−0.393	3.213	−0.162	—	2.5×10^{-3}
8	5	3.213	90.55	−0.162	3.216	−0.0672	—	9.4×10^{-4}
9	5	3.216	90.55	−0.0672	3.218	−0.0277	—	4.9×10^{-4}

The solution is tabulated above. In each iteration a new estimate y_r of the root is made:

$$y_r = \frac{y_u f(y_l) - y_l f(y_u)}{f(y_l) - f(y_u)}$$

The product $f(y_l) \times f(y_r)$ is formed to determine on which subinterval the root will be found. If $f(y_l) \times f(y_r) < 0$, then $y_u = y_r$; otherwise, $y_l = y_r$. It is required that initially $f(y_u)$ and $f(y_l)$ be of opposite sign. Iterations continue until a relative error ε, defined by

$$\varepsilon = \left| \frac{y_r^{new} - y_r^{old}}{y_r^{old}} \right|$$

is less than a specified value, which in the example is 0.0005. The result after nine iterations is $y_2 = 3.22$ m, rounded off to three significant figures.

10.6 NONUNIFORM, GRADUALLY VARIED FLOW

The evaluation of many open-channel flow situations must include accurate analyses of relatively long reaches where the depth and velocity may vary but do not exhibit rapid or sudden changes. In the preceding two sections we emphasized nonuniform, rapidly varied flow phenomena that occur over relatively short reaches, or transitions, in open channels. Attention is now focused on nonuniform, gradually varied flow, where the water surface is continuously smooth. One significant difference between the two is that for rapidly varied flow, losses may often be neglected without severe consequences, whereas for gradually varied flow, it is necessary to include losses due to shear stress distributed along the channel length. The shear stress is the primary mechanism that resists the flow.

Gradually varied flow is a type of steady, nonuniform flow in which

y and V do not exhibit sudden or rapid changes, but instead vary so gradually that the water surface can be considered continuous. As a result, it is possible to develop a differential equation that can describe the incremental variation of y with respect to x, the distance along the channel. An analysis of this relation will enable one to predict the various trends that the water surface profile may assume based on the channel geometry, magnitude of the discharge, and the known boundary conditions. Numerical evaluation of the same equation will provide engineering design criteria.

10.6.1 Differential Equation for Gradually Varied Flow

A representative nonuniform, gradually varied flow is shown in Fig. 10.18. Over the incremental distance Δx, the depth and velocity are known to change slowly. The slope of the energy grade line is designated as S. In contrast to uniform flow, the slopes of the energy grade line, water surface, and channel bottom are no longer parallel. Since the changes in y and V are gradual, the energy loss over the incremental length Δx can be represented by the Chezy–Manning equation. This means that Eq. 10.3.13, which is valid for uniform flow, can also be used to evaluate S for a gradually varied flow situation, and that the roughness coefficients presented in Table 7.3 are applicable. Additional assumptions include a regular cross section, small channel slope, hydrostatic pressure distribution, and one-dimensional flow.

The energy equation is applied from location 1 to location 2, with the loss term h_L given by $S \, \Delta x$. If the total energy at location 2 is expressed as the energy at location 1 plus the incremental change in energy over the distance Δx, Eq. 10.4.2 becomes

$$H_1 = H_2 + S \, \Delta x = H_1 + \frac{dH}{dx} \Delta x + S \, \Delta x \qquad (10.6.1)$$

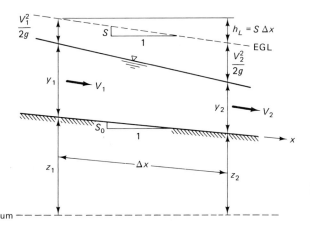

Figure 10.18 Nonuniform, gradually varied flow.

Substitute $H = y + z + V^2/2g$ and $dz/dx = -S_0$ into this relation and rearrange to find

$$S - S_0 = -\frac{d}{dx}\left(y + \frac{V^2}{2g}\right) \qquad (10.6.2)$$

The right-hand term is $-dE/dx$; with the use of Eqs. 10.4.13 and 10.4.15, it is transformed to

$$\frac{dE}{dx} = \frac{dE}{dy}\frac{dy}{dx} = (1 - \text{Fr}^2)\frac{dy}{dx} \qquad (10.6.3)$$

Finally, upon substitution into the energy relation and solving for the slope of the water surface, dy/dx, one finds that

$$\frac{dy}{dx} = \frac{S_0 - S}{1 - \text{Fr}^2} \qquad (10.6.4)$$

This is the differential equation for gradually varied flow and is valid for any regular channel shape.

EXAMPLE 10.11

Using an appropriate control volume for gradually varied flow, show that the slope S of the energy grade line is equivalent to $\tau_0/\gamma R$.

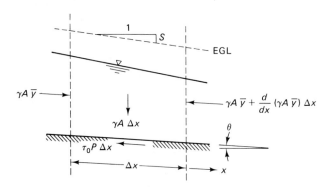

Solution

The control volume is shown in the above figure. The resultant force acting on the control volume is due to the incremental change in hydrostatic pressure $[\gamma d(A\bar{y})/dx]\,\Delta x$, the component of weight in the x-direction $\gamma A \sin\theta \,\Delta x$, and the resistance term $\tau_0 P\,\Delta x$. Using the momentum equation

$$\Sigma F_x = \dot{m}(V_{2x} - V_{1x})$$

with $V_{2x} - V_{1x} = (dV/dx) \Delta x$ results in

$$- \gamma \frac{d}{dx} (A\bar{y}) \Delta x + \gamma A \sin \theta \Delta x - \tau_0 P \Delta x = \rho V A \frac{dV}{dx} \Delta x$$

This relation can be simplified by noting that

$$\frac{d(A\bar{y})}{dx} = \frac{d(A\bar{y})}{dy} \frac{dy}{dx} = A \frac{dy}{dx}$$

and $P = A/R$. Substitute and divide the equation by $\gamma A \Delta x$, the weight of the control volume, and find that

$$- \frac{dy}{dx} + \sin \theta - \frac{\tau_0}{\gamma R} = \frac{V}{g} \frac{dV}{dx}$$

Since $\sin \theta \simeq S_0$ for small θ, the equation above can be rearranged in the form

$$\frac{\tau_0}{\gamma R} - S_0 = - \frac{dy}{dx} - \frac{V}{g} \frac{dV}{dx}$$

$$= - \frac{d}{dx} \left(y + \frac{V^2}{2g} \right)$$

Upon comparison with Eq. 10.6.2, it is seen that the right-hand side is equivalent to $S - S_0$, and consequently,

$$\frac{\tau_0}{\gamma R} - S_0 = S - S_0$$

or

$$S = \frac{\tau_0}{\gamma R}$$

10.6.2 Water Surface Profiles

It is possible to identify a series of water surface profiles based on an evaluation (Bakhmeteff, 1932) of Eq. 10.6.4. Essential to the development is the determination of normal and critical depths. Note that y_0 and y_c are uniquely determined once the channel properties and discharge are established. Table 10.2 shows the classification of the water surface profiles. Associated with y_c is a critical slope S_c, which is found by substituting y_c into the Chezy–Manning equation and solving for the slope. The channel slope can be designated as mild, steep, or critical, depending on whether S_0 is less than, greater than, or equal to S_c, respectively. A horizontal slope exists when $S_0 = 0$ and an adverse slope occurs when $S_0 < 0$.

TABLE 10.2   Classification of Surface Profiles

Channel slope	Profile type	Depth range	Fr	$\dfrac{dy}{dx}$	$\dfrac{dE}{dx}$	
Mild $S_o < S_c$ $y_o > y_c$	M_1	$y > y_o > y_c$	<1	>0	>0	
	M_2	$y_o > y > y_c$	<1	<0	<0	
	M_3	$y_o > y_c > y$	>1	>0	<0	
Steep $S_o > S_c$ $y_o < y_c$	S_1	$y > y_c > y_o$	<1	>0	>0	
	S_2	$y_c > y > y_o$	>1	<0	>0	
	S_3	$y_c > y_o > y$	>1	>0	<0	
Critical $S_o = S_c$ $y_o = y_c$	C_1	$y > y_c$ or y_o	<1	>0	>0	
	C_3	y_c or $y_o > y$	>1	>0	<0	
Horizontal $S_o = 0$ $y_o \to \infty$	H_2	$y > y_c$	<1	<0	<0	
	H_3	$y_c > y$	>1	>0	<0	
Adverse $S_o < 0$ y_o undefined	A_2	$y > y_c$	<1	<0	<0	
	A_3	$y_c > y$	>1	>0	<0	

Inspection of Table 10.2 shows that there are 12 possible profiles. Each profile is classified by a letter/number combination. The letter refers to the channel slope: M for mild, S for steep, C for critical, H for horizontal, and A for adverse. The numerical subscript designates the range of y relative to y_0 and y_c. Flow can occur at depths above or below y_c and at depths above or below y_0.

The variation of y with respect to x for each profile in Table 10.2 can now be found. For a given Q, n, S_0, and channel geometry, the analysis reduces to the determination of how S and Fr vary with y. An inspection of the Chezy–Manning equation will reveal that S decreases with increasing y; similarly, the Froude number decreases as y increases. The numerator in Eq. 10.6.4 will assume the following inequalities: $(S_0 - S) > 0$ for $y > y_0$, and $(S_0 - S) < 0$ for $y < y_0$. In addition, the denominator varies in the manner $(1 - Fr^2) > 0$ for $y > y_c$ and $(1 - Fr^2) < 0$ for $y < y_c$. With these criteria, the sign of dy/dx can be evaluated. In addition, with the use of Eq. 10.6.2, the sign of dE/dx is revealed.

The boundaries of the profiles can be established as follows[7]:

1. As y tends to y_0, S tends to S_0. Hence dy/dx approaches zero, or in other words, the water surface approaches y_0 asymptotically. This applies to curves M_1, M_2, S_2, and S_3.

2. As y becomes large, the velocity becomes small and S and Fr tend to zero, so that dy/dx approaches S_0. Hence the surface approaches a horizontal asymptote; curves M_1, S_1, and C_1 are of this type.

3. As y approaches y_c, dy/dx becomes infinite, a limit that is never reached. For supercritical flow, as the M_3, H_3, and A_3 curves approach y_c, a hydraulic jump will form and create a discontinuity in the water surface; at the beginning of the S_2 curve, rapid acceleration occurs with nonparallel streamlines. When the flow is subcritical (M_2, H_2, A_2), rapid drawdown takes place close to y_c, and the streamlines are no longer parallel. For all of these situations, Eq. 10.6.4 is invalid, since the flow is no longer one-dimensional.

For the profiles shown in Table 10.2, there is no physical significance to the theoretical limit of y approaching zero, since a finite depth is necessary for the existence of flow. It is left as an exercise to show that for the critical slope condition, $dy/dx > S_c$ as y approaches y_c from either a C_1 or C_2 profile, and that dy/dx approaches a horizontal asymptote as y becomes very large.

---

[7]Note that x is measured parallel to the channel bottom, and that y is measured vertically from the channel bottom; hence dy/dx and dE/dx are evaluated relative to the channel bottom and not to a horizontal datum.

EXAMPLE 10.12

By assuming a wide rectangular channel, develop the right-hand side of Eq. 10.6.4 to show how dy/dx varies with y.

Solution

For a wide rectangular channel, assume that $b >> y$, so that the wetted perimeter is approximated by $P \approx b$. The hydraulic radius then becomes $R = A/P \approx$

$(by)/b = y$. Noting that $Q = qb$, the Chezy–Manning equation, used to evaluate S, simplifies to

$$S = \frac{(qbn)^2}{(by^{5/3})^2} = \frac{(qn)^2}{y^{10/3}}$$

It is assumed that in the Chezy–Manning equation $c_1 = 1$. For a rectangular section the square of the Froude number is

$$Fr^2 = \frac{q^2}{gy^3}$$

Substituting into Eq. 10.6.4 gives

$$\frac{dy}{dx} = \frac{S_0 - (qn)^2/y^{10/3}}{1 - q^2/(gy)^3}$$

Since $(qn)^2 y_0^{-10/3} = S_0$ and $Fr_c^2 = q^2/(gy_c^3) = 1$, the relation can be written as

$$\frac{dy}{dx} = S_0 \frac{1 - (y_0/y)^{10/3}}{1 - (y_c/y)^3}$$

This equation can be used as an alternative to Eq. 10.6.4 to evaluate the water surface profiles shown in Table 10.2.

10.6.3 Controls and Critical Flow

The existence of the various profiles shown in Table 10.2 depend on the boundary conditions that are specified at given locations in the channel. Quite often, a control will define the boundary condition. A **control** exists when a depth–discharge relationship can be established at a section. The manner in which the control section affects the water surface away from its specific location can be studied by examining flow near the critical state in a rectangular channel.

Equation 10.5.12 is the expression for a surge of finite magnitude translating in the upstream direction in a rectangular channel. As y_2 approaches y_1, the magnitude of the surge becomes infinitesimal; these waves may be generated by the presence of control structures and other transition sections that tend to "disturb" the flow. In Eq. 10.5.12, one can replace y_1 and y_2 by y, and V_1 by V. In addition, let the wave travel in the downstream direction, so that the equation becomes

$$1 = \frac{1}{2}\left[\sqrt{1 + 8\frac{(V - w)^2}{gy}} - 1 \right] \tag{10.6.5}$$

Solving for w there results

$$w = V \pm \sqrt{gy} = V \pm c \tag{10.6.6}$$

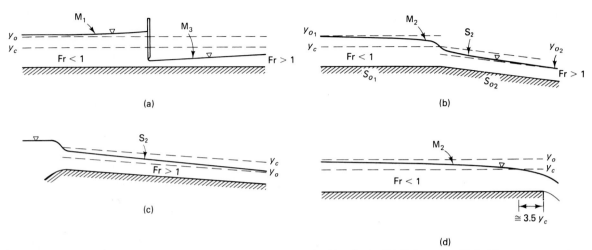

Figure 10.19   Representative controls: (a) sluice gate; (b) change in slope from mild (S_{01}) to steep (S_{02}); (c) entrance to a steep channel; (d) free outfall.

in which $c = \sqrt{gy}$, termed the **celerity**.[8] Thus if a disturbance is created at a midstream location in a channel, two infinitesimal waves are generated. One wave front will tend to propagate upstream at the speed $V - c$ and the second will move downstream at the speed $V + c$. These observations are made relative to an observer in a fixed position, that is, to one who is standing on the shore observing the wave motion.

In a rectangular channel with critical flow conditions, Fr $=$ $V/\sqrt{gy} = 1$, or $V = \sqrt{gy} = c$. Hence at critical flow, the first wave front generated by the disturbance would not move upstream but would appear stationary and become a so-called "standing wave." The opposite front would be swept downstream at the speed $2c$. If Fr < 1, the first wave would travel upstream, and the opposite wave would travel downstream at a speed less than $2c$. For Fr > 1 in the channel, since $V > c$, both waves are swept downstream. Thus, since it contains a mechanism that disturbs the flow, a control will influence upstream conditions only when the flow is subcritical (Fr < 1). Similarly, when the flow is supercritical (Fr > 1), the control can influence only the downstream flow conditions. This is illustrated in Figure 10.19a, where a sluice gate is positioned in a channel with subcritical flow upstream and supercritical flow downstream. An M_1 profile is generated above the gate and an M_3 profile exists below it. Any movement of the gate would influence the nature of the two profiles; lowering the gate would lengthen their range, and raising it would produce the opposite effect.

[8]The celerity is the speed at which an infinitesimal wave will travel into an undisturbed region, that is, a region with zero velocity. For a nonrectangular channel, $c = \sqrt{gA/B}$.

The profiles shown in Table 10.2 are all influenced by the presence of controls. In Fig. 10.19, several controls are shown which create a variety of profiles. Critical depth often is associated with an effective control. Examples include sluice gates, weirs, dams, and flumes, all of which force critical flow to occur somewhere in the transition region. In addition, a control can be located at a break in channel slope from mild upstream to steep downstream (Fig. 10.19b). The entrance to a steep channel (Fig. 10.19c) is an example of an upstream control, with critical flow occurring at the crest. Critical flow will occur a short distance upstream from a free outfall on a mild slope (Fig. 10.19d). Other controls, not shown in Fig. 10.19, are a channel constriction acting as a choke, and for a mild slope, the existence of uniform flow at some location.

10.6.4 Profile Synthesis

Identification of controls and their interaction with the possible profiles is a requirement for successful understanding and correct design and analysis of open-channel flow. Since controls are essentially transition sections, the rapidly varied flow principles presented in Sections 10.4 and 10.5 can be used to determine the necessary depth–discharge relations. Once the controls have been identified, the profiles can be selected and the range of influence of the controls established.

As an example, consider the situation shown in Fig. 10.20. Flow enters the steep channel from a reservoir, so that critical flow exists at the channel entrance. Both the magnitude of the discharge and the S_2 profile are influenced by the depth at the entrance; therefore, the depth acts as a control. At the downstream end of the channel, the lower reservoir acts as a control to establish an S_1 profile that projects upstream. At some interior location, a hydraulic jump occurs to allow the flow to pass from a super-critical to a subcritical state. The location of the jump can be found by plotting a curve of the depth conjugate to the S_2 profile and finding its point of intersection with the S_1 curve. A lowering of the lower reservoir elevation would cause the jump to move downstream and ultimately be swept out of the channel. Increasing the lower reservoir elevation would move the jump upstream; if it were to move into the entrance region, the upstream control would no longer exist. Additional examples of profile synthesis are given in Section 10.7.

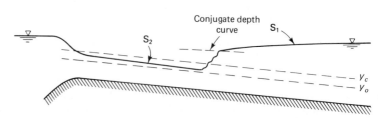

Figure 10.20 Example of profile synthesis.

EXAMPLE 10.13

In a rectangular channel, $b = 3$ m, $n = 0.015$, $S_0 = 0.0005$, and $Q = 5$ m³/s. At the entrance to the channel, flow issues from a sluice gate at a depth of 0.15 m. The channel is sufficiently long that uniform flow conditions are established away from the entrance region. Find the nature of the water surface profile in the vicinity of the entrance.

Solution

First find y_0 and y_c to determine the type of channel. To find y_0, follow the method shown in Example 10.1. Substitute known data into the Chezy–Manning equation:

$$\frac{(3y_0)^{5/3}}{(3 + 2y_0)^{2/3}} = \frac{0.015 \times 5}{\sqrt{0.0005}} = 3.354$$

Solving by trial and error gives $y_0 = 1.39$ m. Next, the critical depth is computed to be

$$y_c = \left(\frac{q^2}{g}\right)^{1/3}$$

$$= \left[\frac{(5/3)^2}{9.81}\right]^{1/3} = 0.66 \text{ m}$$

Since $y_0 > y_c$, a mild slope condition exists. The gate is a control and there will be an M_3 profile beginning at the entrance, terminated by a hydraulic jump. Downstream of the jump, the condition of uniform flow acts as a control, so at that location the depth is y_0, and the Froude number is

$$Fr_0 = \frac{q}{\sqrt{gy_0^3}}$$

$$= \frac{5/3}{\sqrt{9.81 \times 1.39^3}} = 0.325$$

Using Eq. 10.5.10, the depth before the jump is

$$y_1 = \frac{y_0}{2}\left(\sqrt{1 + 8Fr_0^2} - 1\right)$$

$$= \frac{1.39}{2}\left(\sqrt{1 + 8 \times 0.325^2} - 1\right) = 0.25 \text{ m}$$

The depths y_c and y_1 have been calculated to two significant figures, since the Manning coefficient is known to only two significant figures.

EXAMPLE 10.13 (English)

In a rectangular channel, $b = 10$ ft, $n = 0.015$, $S_0 = 0.0005$, and $Q = 175$ ft³/sec. At the entrance to the channel, flow issues from a sluice gate at a depth of 0.50 ft. The channel is sufficiently long that uniform flow conditions are established away from the entrance region. Find the nature of the water surface profile in the vicinity of the entrance.

Solution

First find y_0 and y_c to determine the type of channel. To find y_0, follow the method shown in Example 10.1. Substitute known data into the Chezy–Manning equation:

$$\frac{(10y_0)^{5/3}}{(10 + 2y_0)^{2/3}} = \frac{0.015 \times 175}{1.49 \times \sqrt{0.0005}} = 78.79$$

Solving by trial and error gives $y_0 = 4.45$ ft. Next, the critical depth is computed to be

$$y_c = \left(\frac{q^2}{g}\right)^{1/3}$$

$$= \left[\frac{(175/10)^2}{32.2}\right]^{1/3} = 2.1 \text{ ft}$$

Since $y_0 > y_c$, a mild slope condition exists. The gate is a control and there will be an M_3 profile beginning at the entrance, terminated by a hydraulic jump. Downstream of the jump, the condition of uniform flow acts as a control, so at that location the depth is y_0, and the Froude number is

$$\text{Fr}_0 = \frac{q}{\sqrt{gy_0^3}}$$

$$= \frac{175/10}{\sqrt{32.2 \times 4.45^3}} = 0.328$$

Using Eq. 10.5.10, the depth before the jump is

$$y_1 = \frac{y_0}{2}\left(\sqrt{1 + 8\text{Fr}_0^2} - 1\right)$$

$$= \frac{4.45}{2}\left(\sqrt{1 + 8 \times 0.328^2} - 1\right) = 0.81 \text{ ft}$$

The depths y_c and y_1 have been calculated to two significant figures, since the Manning coefficient is known to only two significant figures.

10.7 NUMERICAL ANALYSIS OF WATER SURFACE PROFILES

Section 10.6 dealt with the understanding and interpretation of various aspects of gradually varied flow. An important procedure was outlined: A water surface profile can be synthesized, or predicted, by making use of relevant information about the channel geometry, roughness, and flow, and by determining or assuming the appropriate controls. Once a satisfactory profile synthesis has been conducted, we are in a position to numerically calculate the desired water surface profiles and accompanying energy gradelines.

Regardless of the type of method chosen to numerically evaluate gradually varied flow, complete analysis of a water surface profile on a reach of channel with constant slope usually follows these steps:

1. The channel geometry, channel slope S_0, roughness coefficient n, and discharge Q are given or assumed.
2. Determine normal depth y_0 and critical depth y_c.
3. Establish the controls (i.e., the depth of flow) at the upstream and downstream ends of the channel reach.
4. Integrate Eq. 10.6.4 to find y and subsequently E as functions of x, allowing for the possibility of a hydraulic jump to occur within the reach.

For a prismatic channel, y_0 can be evaluated by applying the Chezy–Manning equation and finding the root of the function

$$\frac{Qn}{c_1 A R^{2/3} \sqrt{S_0}} - 1 = 0 \qquad (10.7.1)$$

Similarly, y_c is found from application of the Froude number in the form

$$\frac{Q^2 B}{g A^3} - 1 = 0 \qquad (10.7.2)$$

Numerical methods, such as the false position or interval halving (Chapra and Canale, 1988), can be used to solve Eqs. 10.7.1 and 10.7.2.

The Chezy–Manning equation, given here in the form of Eq. 10.7.1, is used to represent the depth–discharge variation for nonuniform as well as for uniform flow. Thus one can replace S_0 by the slope S of the energy grade line in Eq. 10.7.1 and solve for S in the manner

$$S(y) = \frac{Q^2 n^2}{c_1^2 A^2 R^{4/3}} \qquad (10.7.3)$$

Since Q and n are either given or assumed, the right-hand side of Eq. 10.7.3 is a function of y alone.

Until the advent of electronic data processing, the numerical evaluation of water surface profiles was a tedious and time-consuming task. Today, one has a choice of equipment available to rapidly execute the desired numerical computations, ranging from hand-held calculators and microcomputers to the mainframe computer. Two numerical methods are presented in this section to solve Eq. 10.6.4 or its counterpart, Eq. 10.6.2. The first, termed the **standard step method**, can be adapted for use with a calculator or with any type of programmable computer; it is probably the most widely used technique found today in commercially available software. The second method is one that is computer based and uses an established numerical integration scheme.

10.7.1 Standard Step Method

To develop a numerical procedure for solving gradually varied flow problems, we use Eq. 10.6.2; it is applied over the reach of channel shown in Fig. 10.21, resulting in

$$\frac{dE}{dx} = \frac{d}{dx}\left(y + \frac{V^2}{2g}\right)$$
$$= S_0 - S(y) \tag{10.7.4}$$

For small changes in y and V between x_i and x_{i+1}, Eq. 10.7.4 can be approximated by

$$E_{i+1} - E_i = \int_{x_i}^{x_{i+1}} [S_0 - S(y)]\, dx$$
$$\simeq (x_{i+1} - x_i)[S_0 - S(y_m)] \tag{10.7.5}$$

in which $y_m = (y_{i+1} + y_i)/2$. Equation 10.7.5 is solved for x_{i+1} to yield

$$x_{i+1} = x_i + \frac{E_{i+1} - E_i}{S_0 - S(y_m)} \tag{10.7.6}$$

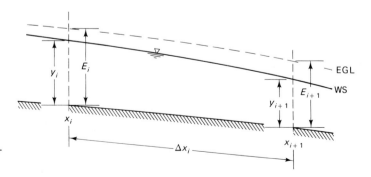

Figure 10.21 Notation for computing gradually varied flow.

The calculations take place stepwise, beginning at a control point or other location where the depth is known. Assume that it is desired to generate a water surface profile along the channel, and compute values of x_i, y_i, and E_i, for $i = 1, \ldots, k$, where k is the location at the opposite end of the reach. Other than x_k, it is not necessary that location x_i be fixed, so it can take on any value. Beginning at location i, the evaluation of conditions at location $i + 1$ proceeds as follows:

1. Choose y_{i+1}.
2. Compute E_{i+1} from Eq. 10.4.12, y_m knowing y_i and y_{i+1}, and $S(y_m)$ from Eq. 10.7.3.
3. Compute x_{i+1} from Eq. 10.7.6.
4. At location k, trial values of y_k are assumed until Eq. 10.7.6 is satisfied for the known value of x_k.

The standard step method can be carried out using a programmable calculator or by use of a microcomputer "spreadsheet" algorithm.

EXAMPLE 10.14

Water is flowing at $Q = 22$ m³/s in a long trapezoidal channel, $b = 7.5$ m, $m_1 = m_2 = 2.5$. A free overfall is located at the downstream end of the channel, where $x = 2000$ m. For $n = 0.015$, $S_0 = 0.0006$, find the water surface profile and energy grade line for a distance of approximately 800 m upstream from the free outfall.

Solution

Equations 10.7.1 and 10.7.2 are used to evaluate y_0 and y_c by substituting in known data:

$$\frac{22 \times 0.015 \times [7.5 + 2y_0 \sqrt{1 + (2.5)^2}]^{2/3}}{[7.5y_0 + \frac{1}{2}y_0^2(2.5 + 2.5)]^{5/3} \sqrt{0.0006}} - 1 = 0$$

$$\frac{(22)^3 \times [7.5 + y_c(2.5 + 2.5)]}{9.8 \times [7.5y_c + \frac{1}{2}y_c^2(2.5 + 2.5)]^3} - 1 = 0$$

By trial and error, one finds $y_0 = 1.29$ m and $y_c = 0.86$ m. Hence the channel is a mild type, and control will be close to the free overfall at the downstream end of the channel. Without any serious loss of accuracy, one can assume that critical conditions will exist at the free overfall. Referring to Table 10.2, the profile upstream of the overfall will be of type M_2. Calculations are shown in the following table. The calculations proceed from location 1 to location 5 in a straightforward manner, with arbitrary values of y_1 to y_5 selected and placed in the first column. Then x_1 to x_5 are evaluated without any trial and error, since they involve interior locations. The last set of calculations requires trial and error, since estimates of y_6 must be made until x_6 is nearly equal to 1200 m. Three trials, the first

i	y_i (m)	A_i (m²)	V_i (m/s)	E_i (m)	y_m (m)	$S(y_m)$	Δx (m)	x_i (m)
1	0.865	8.358	2.632	1.218				2000
					0.9075	2.165×10^{-3}	-8	
2	0.950	9.381	2.345	1.230				1992
					1.000	1.527	-41	
3	1.050	10.63	2.069	1.268				1951
					1.100	1.081	-114	
4	1.150	11.93	1.844	1.323				1837
					1.200	0.7865	-359	
5	1.250	13.28	1.656	1.390				1478
					(1.265	0.6478	-460)	
(6	1.280	13.70	1.606	1.412)				(1018)
					(1.263	0.6526	-342)	
(6	1.275	13.63	1.614	1.408)				(1136)
					1.261	0.6554×10^{-3}	-272	
6	1.271	13.57	1.621	1.405				1206

two shown in parentheses, are required to find the depth $y_6 = 1.27$ m at $x_6 = 1206$ m, which is sufficiently close to the desired value of 1200 m.

10.7.2 Irregular Channels

The method demonstrated in Example 10.14 works well with regular channels. When calculating the profile for a reach of river channel in which the cross-sectional data are irregular and usually given at fixed locations, a trial solution is necessary at each location. Henderson (1966) outlines a method of computation that includes corrections for composite sections and eddy losses that occur at bends and expansions. For a composite section, the slope of the energy grade line is given by

$$S = \frac{Q^2}{(\Sigma \ K_i)^2} \tag{10.7.7}$$

in which K_i is termed the **conveyance** of subsection i:

$$K_i = \left(\frac{c_1 A R^{2/3}}{n}\right)_i \tag{10.7.8}$$

A kinetic-energy coefficient must be applied to the kinetic-energy term; for a composite section use

$$\alpha = \frac{(\Sigma \ A_i)^2}{(\Sigma \ K_i)^3} \Sigma \left(\frac{K_i^3}{A_i^2}\right) \tag{10.7.9}$$

With such complexity, it is most convenient to use the step method procedure in computer-coded form. The United States Army Corps of Engineers has developed the algorithm "HEC-2 Water Surface Profiles" for use in natural channels. The solution is based on the standard step method and is available in versions for use on either a large mainframe computer or a microcomputer.

10.7.3 Computer Method for Regular Channels

There are a number of standard methods available to solve Eq. 10.6.4 for channels with regular cross sections, any of which would provide a solution with sufficient accuracy for design and analysis purposes. Worthy of mention are integration techniques using trapezoidal or Simpson's rules, and solution of the differential form using a Runge–Kutta procedure. The underlying theory of these methods and algorithms for their implementation can be found in a number of texts on numerical methods. McBean and Perkins (1975) discuss convergence problems associated with use of the trapezoidal rule applied to gradually varied flow.

A useful integration scheme is the two-point Gauss–Legendre quadrature (Chapra and Canale, 1988). It is particularly well suited for computer use and is usually more accurate than either the trapezoidal or Simpson's rule methods. Equation 10.6.4 is written in the integral form

$$x_{i+1} = x_i + \int_{y_i}^{y_{i+1}} \frac{1 - Fr^2}{S_0 - S} \, dy$$

$$= x_i + \int_{y_i}^{y_{i+1}} G(y) \, dy \tag{10.7.10}$$

The last integral is approximated by the Gauss–Legendre formula, giving the result

$$\int_{y_i}^{y_{i+1}} G(y) \, dy = \frac{y_{i+1} - y_i}{2} \left[G\left(\frac{y_{i+1} + y_i - \sqrt{3}/3(y_{i+1} - y_i)}{2}\right) \right.$$

$$\left. + G\left(\frac{y_{i+1} + y_i + \sqrt{3}/3(y_{i+1} - y_i)}{2}\right) \right] \tag{10.7.11}$$

Documentation is provided in Section 10.8 for a program written in BASIC, suitable for microcomputer use, which evaluates the water surface and energy grade line for a single reach of a prismatic channel. The following example illustrates use of the program.

EXAMPLE 10.15

A trapezoidal channel 300 m in length conveys water flowing at $Q = 25$ m³/s. The inverse side slopes of the cross section are $m_1 = m_2 = 2.5$, the bottom width is 5 m, the bottom slope is $S_0 = 0.005$, and the Manning coefficient is $n = 0.013$. Compute the water surface profile and energy grade line if the upstream depth $y_1 = 0.55$ m, and the downstream depth $y_2 = 2.15$ m.

Solution

The computer output for the program listed in Section 10.8 is shown below. Note that since $y_c > y_0$, the slope is steep. One control point is at the upstream end, and an S_3 profile exists in the upper portion of the channel. The supercritical flow is

```
Q =   25              N = 0.013           S0 = 0.005
L = 300               B = 5               M1 = 2.5
M2 =    2.5          Y1 = 0.55            Y2 = 2.15
H =   10           UNITS =  SI
```

```
        YO =    0.864              EO =   1.697

        YC =    1.124              EC =   1.537
```

X	Y	E	YCJ
0.000	0.550	3.141	1.954
16.281	0.581	2.844	1.888
33.458	0.613	2.601	1.825
51.758	0.644	2.401	1.766
71.516	0.676	2.236	1.710
93.259	0.707	2.099	1.657
117.875	0.738	1.985	1.606
147.032	0.770	1.891	1.558
184.483	0.801	1.813	1.511
242.165	0.832	1.749	1.467
462.742	0.864	1.697	1.424
300.000	2.150	2.214	0.468
280.918	2.047	2.122	0.509
262.144	1.945	2.031	0.554
243.767	1.842	1.944	0.604
225.910	1.740	1.860	0.659
208.749	1.637	1.781	0.719
192.541	1.534	1.708	0.785
177.683	1.432	1.643	0.858
164.827	1.329	1.590	0.938
155.130	1.227	1.552	1.027
150.946	1.124	1.537	1.124

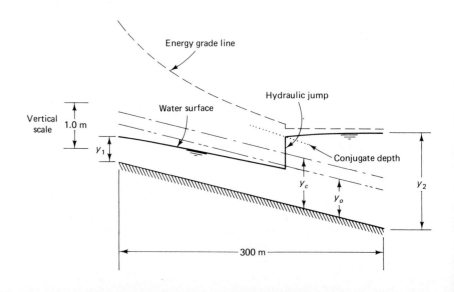

terminated by a hydraulic jump approximately 190 m downstream from the channel entrance. Downstream of the jump the flow is subcritical in the form of an S_1 curve. The computed profile is plotted in the accompanying figure; the hydraulic jump is located by plotting the depth conjugate to the S_3 profile and finding its intersection with the S_1 profile as sketched. The tables include the input data and the computed output.

10.7.4 Direct Integration Methods

Equation 10.6.4 can be integrated in a straightforward manner if one assumes a wide rectangular, horizontal channel. A more general integration procedure is available for nonhorizontal prismatic channels of various shapes (Chow, 1959). Assume that in the Chezy–Manning equation the product $A^2 R^{4/3}$ is proportional to y^N, and in the Froude number the ratio A^3/B is proportional to y^M, where M and N are constants. Then Eq. 10.6.4 can be written

$$dx = \frac{1}{S_0} \left[\frac{1 - (y_c/y)^M}{1 - (y_0/y)^N} \right] dy \qquad (10.7.12)$$

The solution of Eq. 10.7.12 is

$$x = \frac{y_0}{S_0} \left[u - F(u, N) + \left(\frac{y_c}{y_0}\right)^M \frac{J}{N} F(v, J) \right] \qquad (10.7.13)$$

in which $u = y/y_0$, $v = u^{N/J}$, $J = N/(N - M + 1)$, and F is the varied flow function

$$F(u, N) = \int_0^u \frac{d\eta}{1 - \eta^N} \qquad (10.7.14)$$

Equation 10.7.14 can be evaluated numerically, and tables are given by Chow (1959) or Henderson (1966). The method is mentioned here for completeness and will not be employed further in the book. It is of use when a computer is not available or when it is desired to find a single location of x for a given y, since stepwise integration is unnecessary.

10.8 COMPUTER PROGRAM FOR GRADUALLY VARIED FLOW

Listed in Figure 10.22 is a BASIC program that calculates water-surface data for gradually varied nonuniform flow. Input data include the geometrical and roughness parameters, flow properties, and depths of flow at the upstream and downstream ends of the reach. A trapezoidal section, which includes a triangular or rectangular section as a limiting situation, can be analyzed. Consideration is given to either a mild slope or a steep slope

```
1000 CLS
1010 REM ***** NONUNIFORM FLOW IN A TRAPEZOIDAL CHANNEL *****
1020 OPEN "DATA.OUT" FOR OUTPUT AS #1
1030 REM INITIALIZE FUNCTIONS
1040 DEF FNA(Y)=Y*(B+.5*Y*(M1+M2))
1050 DEF FNFM(Y)=Q*Q/(G*FNA(Y))+((M1+M2)*Y/6!+B/2!)*Y*Y
1060 DEF FNNUM(Y)=1-Q*Q*(B+Y*(M1+M2))/(G*FNA(Y)^3)
1070 DEF FNDEN(Y)=S0-(Q*N/(C1*FNA(Y)*(FNA(Y)/(B+Y*(X1+X2)))^.6667))^2
1080 DEF FNENE(Y)=Y+(Q/FNA(Y))^2/(2!*G)
1090 READ Q,N,S0,L,B,M1,M2,Y1,Y2,H,UN$
1100 PRINT#1,"     Q =",Q,"     N =",N,"     S0 =",S0
1110 PRINT#1,"     L =",L,"     B =",B,"     M1 =",M1
1120 PRINT#1,"    M2 =",M2,"    Y1 =",Y1,"    Y2 =",Y2
1130 PRINT#1,"     H =",H,"UNITS =",UN$
1140 IF UN$="EN" THEN G=32.2:C1=1.49 ELSE G=9.810001:C1=1!
1150 X1=SQR(1!+M1*M1)
1160 X2=SQR(1!+M2*M2)
1170 REM ***** COMPUTE Y0 AND YC *****
1180 DEF FNF(Y)=FNDEN(Y)
1190 YL=.001
1200 YR=100!
1210 GOSUB 1890
1220 Y0=YT
1230 E=FNENE(Y0)
1240 PRINT#1,
1250 PRINT#1,
1260 PRINT#1,"    Y0 =";:PRINT#1, USING "####.###";Y0;
1262 PRINT#1,"             E0 =";:PRINT#1, USING "####.###";E
1270 DEF FNF(Y)=FNNUM(Y)
1280 YL=.001
1290 YR=100!
1300 GOSUB 1890
1310 YC=YT
1320 E=FNENE(YC)
1330 PRINT#1,
1340 PRINT#1,"    YC =";:PRINT#1, USING "####.###";YC;
1342 PRINT#1,"             EC =";:PRINT#1, USING "####.###";E
1350 REM ***** COMPUTE WATER SURFACE PROFILE *****
1360 PRINT#1,
1370 PRINT#1,
1380 PRINT#1,"          X          Y          E          YCJ"
1390 IF YC<=Y0 THEN 1430
1400     IF YC<Y1 THEN Y1=YC
1410     DY=(Y0-Y1)/H
1420     GOTO 1450
1430 IF YC<=Y1 THEN 1630
1440 DY=(YC-Y1)/H
1450 Y=Y1
1460 X=0!
```

Figure 10.22 Computer program for gradually varied nonuniform flow.

condition; it is not necessary to know the channel classification before entering the data. The output provides the normal and critical depths and their associated specific energies, and a table of the depth, specific energy, and conjugate depth at computed intervals along the channel.

The critical, normal, and conjugate depths are computed by the

```
1470 E=FNENE(Y)
1480 GOSUB 2020
1490 PRINT#1,USING "######.###";X;Y;E;YCJ
1500 IF X>=L THEN 1630
1510 Y =Y+DY
1520 IF YO>YC AND Y>YC THEN 1630
1530 IF YO<YC AND Y1>=YO AND Y<=YO THEN 1560
1540 IF YO<YC AND Y1<=YO AND Y>=YO THEN 1560
1550 GOTO 1600
1560 DX=L/H
1570 X=X+DX
1580 Y=YO
1590 GOTO 1470
1600 GOSUB 2110
1610 X=X+DX
1620 GOTO 1470
1630 PRINT#1,
1640 IF YC<YO THEN 1680
1650 IF YC>Y2 THEN 1880
1660 DY=(YC-Y2)/H
1670 GOTO 1700
1680 IF YC>Y2 THEN Y2=YC
1690 DY=(YO-Y2)/H
1700 Y=Y2
1710 X=L
1720 E=FNENE(Y)
1730 GOSUB 2020
1740 PRINT#1,USING "######.###";X;Y;E;YCJ
1750 IF X<=0! THEN 1880
1760 Y=Y+DY
1770 IF Y<=YC THEN 1880
1780 IF Y>YC AND Y2>=YO AND Y<=YO THEN 1810
1790 IF YO>YC AND Y2<=YO AND Y>=YO THEN 1810
1800 GOTO 1850
1810 DX=L/H
1820 X=X-DX
1830 Y=YO
1840 GOTO 1720
1850 GOSUB 2110
1860 X=X+DX
1870 GOTO 1720
1880 END
1890 REM ***** INTERVAL-HALVING SUBROUTINE TO FIND ROOT OF FUNCTION *****
1900 FYL=FNF(YL)
1910 FYR=FNF(YR)
1920 IF FYL*FYR>0! THEN PRINT#1,"*** NO ROOT ON STARTING INTERVAL":STOP
1930 IF FYL*FYR<0! THEN GOTO 1950
1940 IF FYL<>0! THEN YT=YR:FT=0!:RETURN ELSE YT=YL:FT=0!:RETURN
1950 FOR I=1 TO 40
1960      YT=.5*(YL+YR)
```

Figure 10.22 (*continued*)

interval-halving method (lines 1890–2010); the appropriate limits for the interval-halving method to compute the conjugate depth are defined in lines 2020–2100. The water surface profile is calculated using a two-point Gauss–Legendre quadrature (lines 2110–2190). The program logic, found in lines 1390–1880, is described in Section 10.8.3.

```
1970        FT=FNF(YT)
1980        IF ABS(FT)<.000001 THEN RETURN
1990        IF FT*FYL<0! THEN YR=YT:FYR=FT:GOTO 2000 ELSE YL=YT:FYL=FT
2000 NEXT I
2010 RETURN
2020 REM ***** SUBROUTINE TO COMPUTE CONJUGATE DEPTH *****
2030 F1=FNFM(Y)
2040 IF Y>YC THEN YL=.001:YR=YC:GOTO 2070
2050 IF Y=YC THEN YCJ=YC:RETURN
2060 IF Y<YC THEN YL=YC:YR=100!
2070 DEF FNF(Y)=FNFM(Y)-F1
2080 GOSUB 1890
2090 YCJ=YT
2100 RETURN
2110 REM ***** TWO-POINT GAUSS-LEGENDRE QUADRATURE SUBROUTINE *****
2120 ARG=Y-.2113248*DY
2130 G1=FNNUM(ARG)/FNDEN(ARG)
2140 ARG=Y-.7886751*DY
2150 G2=FNNUM(ARG)/FNDEN(ARG)
2160 DX=.5*DY*(G1+G2)
2170 CLS
2180 PRINT "Computing..."
2190 RETURN
2200 REM .. Q .. N ... S0 ... L ... B .. M1 .. M2 .. Y1 .. Y2 .. H .. UN$
2210 DATA    25, .013, .005,   300,  5,  2.5,  2.5,  0.55, 2.15, 10,  SI
```

Figure 10.22 (*continued*)

10.8.1 Description of Input Data

Figure 10.23 defines input parameters. The data are entered in line 2210 of the program and summarized below (dimensions are shown in brackets):

$$Q = \text{discharge } [L^3/T]$$
$$N = \text{Manning coefficient}$$
$$S0 = \text{channel slope}$$
$$L = \text{length of channel } [L]$$
$$B = \text{bottom width } [L]$$
$$M1, M2 = \text{ratio of horizontal to vertical change in side walls}$$
$$Y1 = \text{depth at upstream end of channel } [L]$$
$$Y2 = \text{depth at downstream end of channel } [L]$$

Figure 10.23 Input parameters for computation of gradually varied flow: (a) cross section; (b) profile.

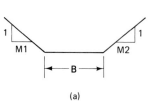

(a)

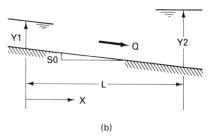

(b)

H = parameter used to set the increment DY in the integration procedure; normally, use $10 < H < 30$

UN\$ = "EN" for English or "SI" for metric units

10.8.2 Description of Output

The numerical output will be placed on a file named DATA.OUT and consists of the following:

1. Echo of input data.
2. Computed normal depth Y0, critical depth YC, and their associated specific energies, E0 and EC.
3. Table of computed water-surface profiles:
   X = distance along channel $[L]$
   Y = depth $[L]$
   E = specific energy $[L]$
   YCJ = conjugate depth $[L]$

Depending on the inputted parameters, an M_3, S_2, or S_3 profile is generated from the upstream end and terminates when either Y approaches YC or X reaches L. In addition, a possible M_1, M_2, or S_1 profile is produced beginning at the downstream end, terminating when either Y approaches YC or X reaches the upstream end of the channel.

10.8.3 Program Logic

For the case where Y0 > YC:

1. If Y1 < YC, an M_3 profile is computed downstream from location 1 (the upstream end of the channel), or if Y1 > YC, continue to step 2.
2. If Y2 > Y0, an M_1 profile is computed upstream from location 2 (the downstream end of the channel), or if $Y0 > Y2 \geqq YC$, an M_2 profile is computed upstream from location 2.

For the case where Y0 < YC:

3. If Y1 > YC, set Y1 = YC, and an S_2 profile is computed downstream from location 1, or if $YC > Y1 > Y0$, an S_2 profile is computed downstream from location 1, or if Y1 < Y0, an S_3 profile is computed downstream from location 1.
4. If Y2 > YC, an S_1 profile is computed upstream from location 2, or if Y2 < YC, the program is terminated.

PROBLEMS

10.1. Give an example of each of the following types of flow for both closed-conduit and free-surface conditions. Include a sketch with each representation:
(a) Steady, uniform.
(b) Unsteady, nonuniform.
(c) Steady, nonuniform.
(d) Unsteady, uniform.

10.2. Water is flowing with a free surface at 4 m³/s in a circular conduit. Find the diameter d such that the critical depth $y_c = 0.3d$. Under critical flow conditions, $Q^2B/(gA^3) = 1$.

10.3. Classify the following flows as steady or unsteady, and uniform or nonuniform:
(a) Water flowing through a logjam in a river with the observer standing on the shore.
(b) Water flowing through a river rapids with the observer on a raft traveling with the current.
(c) A flood wave traveling in a river which is generated by a gentle rainfall/runoff event, with the observer standing on the bank.
(d) A moving hydraulic jump in a prismatic channel, with the observer running along side the bank at the same speed as the jump.

Uniform Flow

10.4. For each regular channel section, construct the curves R versus y and $AR^{2/3}$ versus y:
(a) Circular, $d = 2$ m.
(b) Trapezoidal, $b = 3$ m, $m_1 = m_2 = 2.5$.
(c) Composite (see the figure).

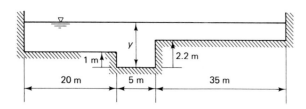

10.5. Determine the most efficient section based on flow resistance for a trapezoidal channel. Assume equal side slopes, that is, $m_1 = m_2$.

10.6. Determine the uniform depth y_0, area A, and wetted perimeter P for the following conditions:
(a) Trapezoidal channel, $m_1 = 2.5$, $m_2 = 3.5$, $b = 4.5$ m, $Q = 35$ m³/s, $n = 0.015$, $S_0 = 0.00035$.
(b) Circular channel, $d = 2.1$m, $Q = 4.25$ m³/s, $n = 0.012$, $S_0 = 0.001$.
(c) Circular channel, $d = 6$ ft, $Q = 120$ ft³/sec, $n = 0.012$, $S_0 = 0.001$.
(d) Trapezoidal channel, $Q = 15$ m³/s, $n = 0.011$, $S_0 = 0.0013$, "most efficient" cross section.
(e) Rectangular channel, $b = 25$ ft, $Q = 1200$ ft³/sec, $n = 0.020$, $S_0 = 0.0004$.

10.7. Uniform flow occurs in a trapezoidal channel with $m_1 = m_2 = 1.75$. The channel is to convey a discharge of 4.0 m³/s, with an average velocity of 1.4 m/s. If $n = 0.015$ and $S_0 = 0.001$, find the bottom width and the uniform depth.

Energy Concepts

10.8. For a given specific energy in a rectangular channel, show that critical flow exists when the specific discharge is at its maximum value q_{max}.

10.9. The specific energy relation for a rectangular section, Eq. 10.4.5, can be made dimensionless by normalizing it with respect to critical depth y_c. Prepare such a plot with y/y_c as the ordinate and E/y_c as the abscissa. Locate the minimum point mathematically.

10.10. Water is flowing in a rectangular channel at a velocity of 3 m/s and a depth of 2.5 m. Determine the changes in water surface elevation for the following alterations in the channel bottom:
(a) An increase (upward step) of 20 cm, neglecting losses.
(b) A "well-designed" increase of 15 cm.
(c) The maximum increase allowable for the specified upstream flow conditions to remain unchanged, neglecting losses.
(d) A "well-designed" decrease (downward step) of 20 cm.

10.10E. Water is flowing in a rectangular channel at a velocity of 10 ft/sec and a depth of 8 ft. Determine the changes in water surface elevation for the following alterations in the channel bottom:

(a) An increase (upward step) of 8 in., neglecting losses.

(b) A "well-designed" increase of 6 in.

(c) The maximum increase allowable for the specified upstream flow conditions to remain unchanged, neglecting losses.

(d) A "well-designed" decrease (downward step) of 8 in.

10.11. Water is flowing in a rectangular channel whose width is 5 m. The depth of flow is 2 m and the discharge is 25 m³/s. Determine the changes in depth for the following alterations in the channel width:

(a) An increase of 50 cm, neglecting losses.

(b) A decrease of 25 cm, assuming a "well-designed" transition.

10.12. Flow in a 3-m-wide rectangular channel occurs at a depth of 3 m and a velocity of 3 m/s. There is an upward step of 70 cm in the channel. Neglecting losses, determine the change in width that must take place simultaneously for critical flow conditions to occur at the downstream location.

10.12E. Flow in a 10-ft-wide rectangular channel occurs at a depth of 10 ft and a velocity of 10 ft/sec. There is an upward step of 2.3 ft in the channel. Neglecting losses, determine the change in width that must take place simultaneously for critical flow conditions to occur at the downstream location.

10.13. Water is flowing in a rectangular channel 2.0 m wide. At a transition section the channel bottom is lowered by $h = 0.1$ m for a short distance, and then is raised back to the original elevation. If $y = 1.22$ m and $Q = 4.8$ m³/s, then, with losses neglected:

(a) Find the change in channel width necessary to maintain a horizontal water surface through the transition.

(b) What change in width would cause critical flow to occur in the transition?

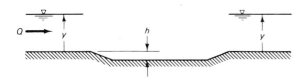

10.13E. Work Problem 10.13 for a channel width of 6.5 ft, $h = 4$ in., $y = 4$ ft, and $Q = 170$ ft³/sec.

10.14. A lake discharges into a steep channel. At the channel entrance the lake level is 2.5 m above the channel bottom. Neglecting losses, find the discharge for the following geometries:

(a) Rectangular section, $b = 4$ m.

(b) Trapezoidal section, $b = 3$ m, $m_1 = m_2 = 2.5$.

(c) Circular section, $d = 3.5$ m.

10.15. Steep flow conditions exist at the outlet from a reservoir into an open canal. The canal width is not yet fixed, but the required depth at the entrance is 1 m and the required discharge is 18 m³/s. Determine the necessary widths for the following geometries:

(a) Rectangular section.

(b) Trapezoidal section, side slopes $m_1 = m_2 = 3$.

10.15E. Steep flow conditions exist at the outlet from a reservoir into an open canal. The canal width is not yet fixed, but the required depth at the entrance is 3 ft and the required discharge is 635 ft³/sec. Determine the necessary widths for the following geometries:

(a) Rectangular section.

(b) Trapezoidal section, side slopes $m_1 = m_2 = 3$.

10.16. Rapidly varied flow occurs over a rectangular sill, with free outfall conditions at location C. Throughout the reach losses can be neglected.

(a) Compute the discharge.

(b) Evaluate the depths at locations A, B, and C.

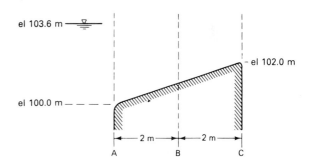

10.17. Prepare a computer algorithm that evaluates the critical depth of a regular channel using the false position method of solution. Verify the program with the following data:

(a) Circular section, $Q = 5$ m³/s, $d = 2.5$ m.

(b) Rectangular section, $Q = 125$ ft³/sec, $b = 12$ ft.

(c) Trapezoidal section, $Q = 120$ m³/s, $b = 10$ m, $m_1 = m_2 = 5$.

(d) Triangular section, $Q = 250$ ft³/sec, $m_1 = 2.5$, $m_2 = 3.0$.

10.18. For the composite channel section shown below, find the critical depth y_c if:
(a) $Q = 55$ m³/s.
(b) $Q = 3.5$ m³/s.

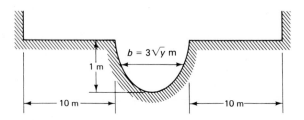

10.19. A parabolic channel is described by the function $\eta = 0.1x^2$. If $Q = 25$ m³/s and $y = 2.00$ m, find the alternate depth.

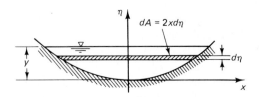

10.20. Prepare a computer algorithm that determines the alternate depth, given the geometry of a prismatic channel, the discharge, and the depth of flow. Use the interval-halving technique of solution.

10.21. Derive the relation for a V-notch weir that expresses Q as a function of Y, Eq. 10.4.27.

10.22. Develop the expression for the theoretical discharge for the weir shown.

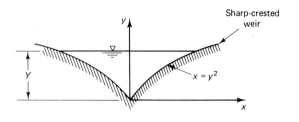

10.23. Design a broad-crested weir to convey a river discharge that varies between 0.15 and 30 m³/s. The maximum water depth upstream of the weir is not to exceed 1.75 m and the minimum depth is not to be less than 1.05 m.

10.23E. Design a broad-crested weir to convey a river discharge that varies between 5 and 1000 ft³/sec. The maximum water depth upstream of the weir is not to exceed 5.75 ft and the minimum depth is not to be less than 3.45 ft.

10.24. Design a canal to deliver water between the two reservoirs shown below. Elevation A = 501.8 m and elevation B = 500.2 m. The horizontal distance between the reservoirs is $L = 1500$ m. It is required that uniform flow exist throughout the reach. The water surface in reservoir A is to be no more than $h = 2$ m above the bottom of the canal at the entrance. The canal is rectangular in cross section, 2.5 m wide, and made of concrete. Neglect entrance and exit losses, and assume uniform, subcritical flow conditions at the channel entrance.

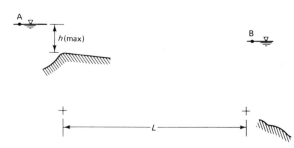

10.24E. Design a canal to deliver water between the two reservoirs shown in Problem 24. Elevation A = 1646 ft and elevation B = 1641 ft. The horizontal distance between the reservoirs is $L = 4920$ ft. It is required that uniform flow exist throughout the reach. The water surface in reservoir A is to be no more than $h = 6.5$ ft above the bottom of the canal at the entrance. The canal is rectangular in cross section, 8 ft wide, and made of concrete. Neglect entrance and exit losses, and assume uniform, subcritical flow conditions at the channel entrance.

Momentum Concepts

10.25. The momentum function for a rectangular section, Eq. 10.5.4, can be made dimensionless by normalizing it with respect to $b(y_c)^2$. Prepare such a plot using y/y_c as ordinate and $M/(by_c)^2$ as abscissa.

10.26. Rework Problem 10.20 to evaluate the conjugate depth in place of the alternate depth.

10.27. A flow of 1.5 m³/s per meter width occurs in a wide rectangular channel at a depth of 1.80 m. A construction project over this channel requires that cofferdams spaced at 6 m on centerlines be placed in the channel (see the figure below). Assuming that the cofferdams create no energy losses between locations 1 and 2, determine:

(a) The maximum permissible diameter of the cofferdams without creating upstream backwater effects (i.e., without increasing the upstream depth).

(b) The resultant drag force on each cofferdam (assume that $C_D = 0.15$).

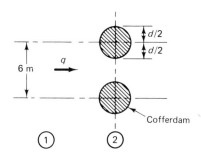

10.27E. A flow of 16 ft³/sec per foot width occurs in a wide rectangular channel at a depth of 6 ft. A construction project over this channel requires that cofferdams spaced at 20 ft on centerlines be placed in the channel (see the figure in Problem 10.27). Assuming that the cofferdams create no energy losses between locations 1 and 2, determine:

(a) The maximum permissible diameter of the cofferdams without creating upstream backwater effects (i.e., without increasing the upstream depth).

(b) The resultant drag force on each cofferdam (assume that $C_D = 0.15$).

10.28. A temporary pipeline (diameter = 200 mm $C_D = 0.30$) is to be placed on a river bed normal to the direction of flow. The river is approximately rectangular in section, and is 100 m wide. The depth of flow downstream of the pipe is 2.5 m and the mean velocity is 3 m/s. Neglecting bed slope and resistance, determine the following:

(a) Flow conditions (i.e., depth and velocity) upstream of the pipe once it is in place.

(b) Resultant drag force on the pipe.

10.29. A hydraulic jump occurs over a sill located in a triangular channel with water flowing as shown. The inverse side slopes are $m_1 = m_2 = 3$, the drag coefficient $C_D = 0.40$, and the height of the sill $h = 0.3$ m. Determine the discharge Q if $y_1 = 0.50$ m and $y_2 = 1.8$ m.

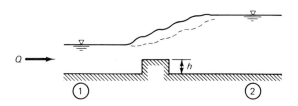

10.30. Water is flowing in a rectangular channel at a depth of 1.6 m and a velocity of 0.85 m/s. At a downstream location the discharge is suddenly reduced to zero, causing a surge to propagate upstream. Find the depth and velocity behind the surge, and the speed of the surge wave.

10.31. In a rectangular channel, water is flowing at a depth of 1.5 m and a velocity of 1 m/s. At a downstream location, the discharge is suddenly reduced by 60% (i.e., to 40% of the original value), causing a surge to propagate upstream. Determine the depth and velocity behind the surge, and the speed of the surge wave.

10.31E. In a rectangular channel, water is flowing at a depth of 5 ft and a velocity of 3 ft/sec. At a downstream location, the discharge is suddenly reduced by 60% (i.e., to 40% of the original value), causing a surge to propagate upstream. Determine the depth and velocity behind the surge, and the speed of the surge wave.

10.32. Water enters a reach of rectangular channel where $y_1 = 0.5$ m, $b = 7.5$ m, and $Q = 20$ m³/s. It is desired that a hydraulic jump occur upstream (location 2) of the sill and on the sill critical conditions exist (location 3). Other than across the jump, losses can be neglected. Determine the following:

(a) Depths at locations 2 and 3.

(b) Required height of the sill, h.

(c) Sketch the water surface and energy grade line.

(d) Describe the nature and character of the jump.

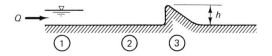

10.33. A transition section is located at the entrance to a rectangular channel as shown. At location 1, the depth is sufficiently large so that the velocity is negligible. If $b = 3$ m, $Fr_3 = 0.75$, and $Q = 5.55$ m³/s, determine the following:
(a) The water surface elevation relative to the datum at locations 1, 2, and 3. Sketch the water surface and energy grade lines between locations 1 and 3.
(b) The resultant horizontal force acting on the channel bottom between locations 2 and 3.

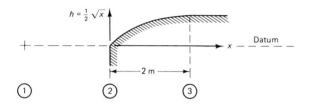

10.34. Water is flowing in a trapezoidal channel, $b = 5$ m, $m_1 = m_2 = 3$. A stationary hydraulic jump occurs, with the upstream depth $y_1 = 1.1$ m and the discharge $Q = 60$ m³/s. Find the downstream depth y_2 and the power dissipated by the jump.

10.35. A rectangular channel has a sudden upward step of 0.17 m. A hydraulic jump occurs above the step. At location 2, downstream of the jump, the depth is $y_2 = 1.5$ m, and the Froude number there is $Fr_2 = 0.40$. If the width of the channel is 5 m and the drag coefficient on the step is $C_D = 0.35$, find the discharge Q and the depth y_1 upstream of the jump.

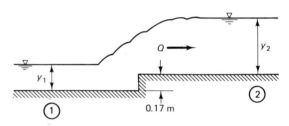

Nonuniform, Gradually Varied Flow

10.36. The channel whose bottom profile is shown below has a rectangular cross section, with $b = 8$ m, $n = 0.014$, and $S_0 = 0.004$. Determine the following (*Hint:* Assume critical flow conditions at the channel entrance):
(a) The discharge in the channel.
(b) Sketch the water surface and energy grade line.
(c) The possible existence of a hydraulic jump.

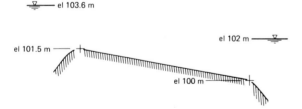

10.37. Uniform flow occurs in a long rectangular channel with the following properties: $Q = 0.35$ m³/s, $S_0 = 0.0163$, $n = 0.012$, and $b = 1.80$ m. The channel is altered by raising the channel bottom by 100 mm for a short distance. The alteration is intended to act as a "smooth transition region" so that losses can be considered negligible. Determine what changes take place

due to the presence of the transition. In your analysis include all relevant rapidly varied flow calculations, identify any gradually varied flow profiles, and sketch the water surface and energy grade line.

10.37E. Uniform flow occurs in a long rectangular channel with the following properties: $Q = 12.5$ ft³/sec, $S_0 = 0.0163$, $n = 0.012$, $b = 6$ ft. The channel is altered by raising the channel bottom by 4 in. for a short distance. The alteration is intended to act as a "smooth transition region" so that losses can be considered negligible. Determine what changes take place due to the presence of the transition. In your analysis include all relevant rapidly varied flow calculations, identify any gradually varied flow profiles, and sketch the water surface and energy grade line.

10.38. Rework Problem 10.37 or Problem 10.37E if $S_0 = 0.0013$.

10.39. Flow of water takes place in a rectangular channel with $b = 4$ m, $n = 0.012$, $L = 500$ m, $S_0 = 0.00087$. The discharge is 33 m³/s. At the entrance to the channel, the depth is 0.68 m, and at the downstream end a free-outfall condition exists. Perform a profile synthesis to determine the nature of the water surface and energy grade line.

10.40. A flow of 8.5 m³/s occurs in a long rectangular channel 3 m wide with $y_0 = 1.54$ m. There is a smooth constriction in the channel to 1.8 m width.
(a) Determine the depths to be expected in and just upstream of the constriction, neglecting losses. Show the solution on an E–y diagram.
(b) Classify the gradually varied flow profile upstream of the constriction.

10.41. Water is discharging at 20 m³/s in a triangular channel with $m_1 = 3.5$, $m_2 = 2.5$, $S_0 = 0.001$, $n = 0.014$. At the entrance to the channel the depth is 0.50 m, and at a distance 300 m downstream the depth is 2.5 m. Determine the nature of the water surface over the 300-m reach.

10.41E. Water is discharging at 700 ft³/sec in a triangular channel with $m_1 = 3.5$, $m_2 = 2.5$, $S_0 = 0.001$, $n = 0.014$. At the entrance to the channel the depth is 1.65 ft, and at a distance 1000 ft downstream the depth is 8 ft. Determine the nature of the water surface over the 1000-ft reach.

10.42. A long channel has a change in bottom slope 500 m from its downstream end; upstream of the transition the slope is $S_{01} = 0.0003$, and downstream of it the slope is $S_{02} = 0.005$. The downstream reach of channel terminates at a reservoir whose elevation is 3 m above the channel bottom at that location. The upper reach is quite long, and very far upstream from the transition normal flow conditions exist. The channel section is trapezoidal, with a bottom width of 3 m, $m_1 = m_2 = 1.8$, $n = 0.012$, and a discharge of 17.5 m³/s. Determine the nature of the water surface and energy grade line from a location far upstream of the transition to the end of the channel.

10.43. A rectangular channel with $b = 4$ m has a change in slope such that the normal depths are $y_{01} = 0.93$ m and $y_{02} = 1.42$ m. At locations far upstream and far downstream of the transition, flow occurs at the respective uniform depths that serve as controls. The discharge is 15 m³/s. Find the variation of the water surface and energy grade line throughout the region.

10.44. Water is flowing at a discharge of $Q = 2.5$ m³/s in a 500-m-long circular conduit with $d = 2.5$ m, $S_0 = 0.001$, and $n = 0.015$. At the inlet the water depth is 0.4 m and at the outlet a free overfall exists. Determine the nature of the water surface in the conduit.

10.44E. Water is flowing at a discharge of $Q = 88$ ft³/sec in a 1600-ft-long circular conduit with $d = 8$ ft, $S_0 = 0.001$, and $n = 0.015$. At the inlet the water depth

is 1.3 ft and at the outlet a free overfall exists. Determine the nature of the water surface in the conduit.

10.45. With the results from Problem 10.42, compute the water surface profile and energy grade line using a numerical method of your choice.

10.46. Gradually varied flow occurs over a reach of rectangular channel with $n = 0.013$, $S_0 = 0.005$, and $b = 2.5$ m. At location 1 the depth is 1.05 m and at location 2 the depth is 1.2 m. If the two locations are 50 m apart, find the discharge and identify the profile type.

10.47. Over a portion of a very long rectangular channel, uniform flow occurs with $Q = 5.5$ m³/s, $y_0 = 0.5$ m, and $b = 3$ m. At the end of the channel the flow is terminated by a free outfall. In addition, at the outlet the channel width is reduced to 1.5 m for a very short distance. Determine and describe as completely as possible the changes that take place in the water surface between locations A and B. Use an E–y diagram for illustration.

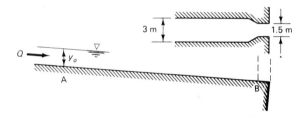

10.47E. Over a portion of a very long rectangular channel, uniform flow occurs with $Q = 200$ ft³/sec, $y_0 = 1.65$ ft, and $b = 10$ ft. At the end of the channel the flow is terminated by a free outfall. In addition, at the outlet the channel width is reduced to 5 ft for a very short distance. Determine and describe as completely as possible the changes that take place in the water surface between locations A and B. Use an E–y diagram for illustration.

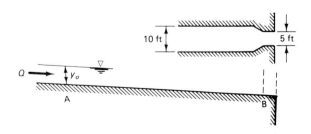

10.48. A sluice gate is placed in a long rectangular channel, $b = 4$ m, $n = 0.014$, and $S_0 = 0.0008$. The depth upstream of the gate is $y_1 = 1.85$ m and downstream of the gate the depth is $y_2 = 0.35$ m. Neglecting losses, identify and compute the water surface profiles on either side of the gate.

10.49. With the following data, identify the profiles and plot the water surface and energy grade line to scale. SI units prevail.

Q =	35	N = 0.011		SO = 0.001	
L =	200	B = 5		M1 = 2.5	
M2 =	2.5	Y1 = 0.8		Y2 = 2	
H =	10	UNITS = SI			

YO =	1.442	EO =	1.848
YC =	1.356	EC =	1.838

X	Y	E	YCJ
0.000	0.800	2.791	2.087
36.173	0.856	2.529	1.993
72.348	0.911	2.331	1.904
108.334	0.967	2.181	1.821
143.885	1.022	2.069	1.744
178.673	1.078	1.986	1.670
212.227	1.133	1.925	1.601
200.000	2.000	2.156	0.851
141.180	1.944	2.114	0.885
81.636	1.888	2.074	0.921
21.133	1.833	2.035	0.959
-40.675	1.777	1.999	0.998

10.50. With the following data, identify the profiles and plot the water surface and energy grade line to scale. SI units prevail.

Q =	3.5	N = 0.013		SO = 0.005	
L =	100	B = 2.5		M1 = 0	
M2 =	0	Y1 = 0.3		Y2 = 0.95	
H =	5	UNITS = SI			

YO =	0.508	EO =	0.895
YC =	0.585	EC =	0.877

X	Y	E	YCJ
0.000	0.300	1.410	1.014
14.119	0.342	1.198	0.924
29.404	0.383	1.064	0.848
46.797	0.425	0.979	0.781
69.358	0.466	0.926	0.722
89.358	0.508	0.895	0.669
109.358	0.508	0.895	0.669
100.000	0.950	1.061	0.329
86.757	0.877	1.007	0.366
74.102	0.804	0.958	0.410
62.436	0.731	0.918	0.459
52.665	0.658	0.889	0.517

10.50E. With the following data, identify the profiles and plot the water surface and energy grade line to scale. English units prevail.

Q = 125		N = 0.013		SO = 0.005	
L = 300		B = 8		M1 = 0	
M2 =	0	Y1 = 1		Y2 = 3	
H =	5	UNITS = EN			

YO =	1.710	EO =	3.007
YC =	1.965	EC =	2.947

X	Y	E	YCJ
0.000	1.000	4.791	3.426
48.059	1.142	4.049	3.118
99.969	1.284	3.584	2.854
158.860	1.426	3.291	2.625
234.897	1.568	3.110	2.424
450.359	1.710	3.007	2.244
300.000	3.000	3.421	1.203
263.407	2.793	3.279	1.320
228.701	2.586	3.153	1.452
197.072	2.379	3.049	1.602
171.096	2.172	2.975	1.771

10.51. A rectangular channel has a change in slope as shown. The channel is 3.66 m wide, with $n = 0.017$, $S_{02} = 0.00228$, and $Q = 15.38$ m³/s.

(a) Determine the depth that must exist in the downstream channel for a hydraulic jump to terminate at uniform flow conditions.

(b) If $y_{01} = 0.6$ m, calculate the length to the jump, L_j, using several increments of depth in a step calculation.

(c) Sketch the water surface and energy grade line, and identify all gradually varied flow profiles.

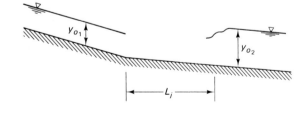

10.52. In the three channels shown, normal depths and critical depths are indicated by dashed (---) and dotted (. . .) lines, respectively. Sketch one possible composite profile for each system, and identify all gradually varied surfaces.

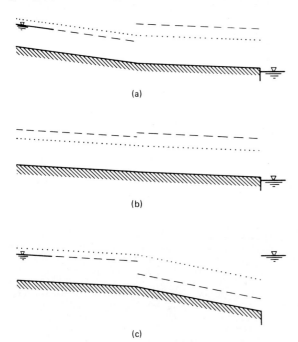

(a)

(b)

(c)

ELEVEN

Flows in Piping Systems

11.1   INTRODUCTION

Internal flows in pipelines and ducts are commonly encountered in all parts of our industrialized society. From delivering potable water to transporting chemicals and other industrial liquids, engineers have designed and constructed untold kilometers of relatively large-scale piping systems. Smaller piping units are also in abundance: in hydraulic controls, in heating and air conditioning systems, and in cardiovascular and pulmonary flow systems, to name only a few. These flows can be either steady or unsteady, uniform or nonuniform. The fluid can be either incompressible or compressible, and the piping material can be elastic, inelastic, or perhaps, viscoelastic. In this chapter we concentrate primarily on incompressible, steady flows in rigid piping. The piping system may be relatively simple, such that the variables can be solved rather easily using a calculator, or it may be sufficiently complicated so it is more convenient to use a computer.

Piping systems are considered to be composed of elements and components. Basically, pipe **elements** are reaches of constant-diameter piping and the **components** consist of valves, tees, bends, reducers, or any other device that may create a loss to the system. In addition to components and elements, pumps add energy to the system and turbines extract energy. The elements and components are linked at junctions. Figure 11.1 illustrates several types of piping systems.

Following a discussion of piping losses, several pipe systems, including series, branch, and parallel configurations, are analyzed. Attention is then directed to comprehensive network systems, where several methods of solution are presented. Most of the piping problems analyzed are those where the discharge is the unknown variable; this type of problem is classified as category 2 in Section 7.6.4.

11.2   LOSSES IN PIPING SYSTEMS

Losses can be divided into two categories: (a) those due to wall shear in pipe elements, and (b) those due to piping components. The former are distributed along the length of pipe elements. The latter are treated as discrete discontinuities in the hydraulic grade line and the energy grade line and are commonly referred to as minor losses; they are due primarily to separated or secondary flows.

The fundamental mechanics of wall shear and development of the empirical relations relating to pipe losses are treated in Chapter 7. Minor losses are covered in detail in Section 7.6.5 and are not discussed further here. The following material focuses on the treatment of losses in the analysis of piping systems.

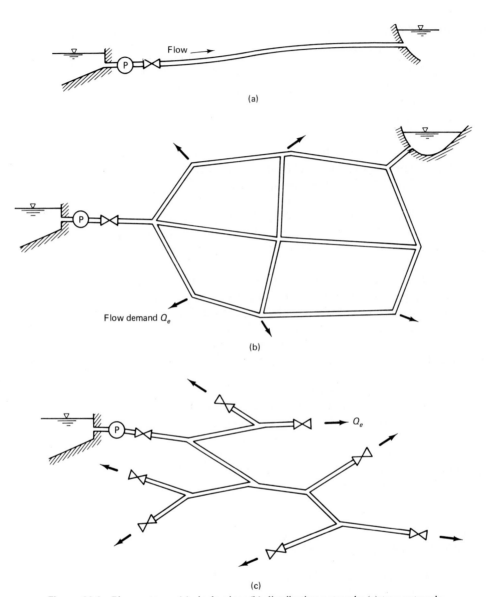

Figure 11.1 Pipe systems: (a) single pipe; (b) distribution network; (c) tree network.

11.2.1 Frictional Losses in Pipe Elements

It is convenient to express the pipe element frictional loss in the exponential form

$$h_L = RQ^x \qquad (11.2.1)$$

in which h_L is the head loss over length L of pipe, R is the **resistance coefficient**, Q is the discharge in the pipe, and x is an exponent. Depending on the formulation chosen, the resistance coefficient may be a function of pipe roughness, Reynolds number, or length and diameter of the pipe element. In particular, the Darcy–Weisbach relation, Eq. 7.6.23, can be substituted into Eq. 11.2.1. Then $x = 2$, and the resulting expression for R is

$$R = \frac{fL}{2gDA^2}$$

$$= \frac{8fL}{g\pi^2 D^5}$$

(11.2.2)

where f is the friction factor. The characteristics of f for commercial pipe flows are developed in Section 7.6.4. Specifically, the Moody diagram, Fig. 7.13, presents a comprehensive picture of just how the friction factor varies over a wide range of Reynolds numbers and for a range of relative roughnesses. For pipe network analysis, it is convenient to express the behavior of f using approximate, equivalent empirical formulas in which the friction factor can be obtained directly in terms of the Reynolds number and relative roughness. A number of relations have been developed and shown to be reasonably accurate for engineering calculations (Benedict, 1980). In particular, the formulas of Swamee and Jain are presented in Section 7.6.4 and are shown to accurately represent the Colebrook relation, Eq. 7.6.28. The friction factor formula developed by Swamee and Jain is

$$f = 1.325 \left\{ \ln \left[0.27 \left(\frac{e}{D} \right) + 5.74 \left(\frac{1}{\text{Re}} \right)^{0.9} \right] \right\}^{-2}$$

(11.2.3)

Combining Eqs. 11.2.2 and 11.2.3, one finds that

$$R = 1.07 \left(\frac{L}{gD^5} \right) \left\{ \ln \left[0.27 \left(\frac{e}{D} \right) + 5.74 \left(\frac{1}{\text{Re}} \right)^{0.9} \right] \right\}^{-2}$$

(11.2.4)

Equations 11.2.3 and 11.2.4 are valid over the ranges $0.01 > e/D > 10^{-8}$, and $10^8 > \text{Re} > 5000$. The fully rough regime, where Re has a negligible effect on f, begins at a Reynolds number given by

$$\text{Re} = \frac{200D}{e\sqrt{f}}$$

(11.2.5)

For values of Re greater than this, the friction factor is a function only of e/D, and is given by

$$f = 1.325 \left\{ \ln \left[0.27 \left(\frac{e}{D} \right) \right] \right\}^{-2}$$

(11.2.6)

TABLE 11.1 NOMINAL VALUES OF THE
HAZEN–WILLIAMS COEFFICIENT C

Type of pipe	C
Extremely smooth; asbestos-cement	140
New or smooth cast iron; concrete	130
Wood stave; newly welded steel	120
Average cast iron; newly riveted steel; vitrified clay	110
Cast iron or riveted steel after some years of use	95–100
Deteriorated old pipes	60–80

Two additional expressions for pipe frictional losses, which have found wide use, are the Hazen–Williams and Chezy–Manning formulas. For water flow, the value of R in Eq. 11.2.1 for the Hazen–Williams relation is

$$R = \frac{K_1 L}{C^x D^m} \tag{11.2.7}$$

in which the exponents are $x = 1.85$, $m = 4.87$, and C is the Hazen–Williams coefficient dependent only on the roughness. The constant K_1 depends on the system of units; for SI units K_1 has a magnitude of 10.59, and for English units the magnitude of K_1 is 4.72. Values of the Hazen–Williams roughness coefficient C are given in Table 11.1.

The Chezy–Manning equation is more commonly associated with open-channel flow. However, in sewage and drainage systems in particular, it has been applied to conduits flowing under **surcharge,** that is, under pressurized conditions. The Chezy–Manning equation was introduced in Section 7.7. For a circular pipe flowing full, Eq. 7.7.6 can be substituted into Eq. 11.2.1 and solved for R:

$$R = \frac{10.29 n^2 L}{K_2 D^{5.33}} \tag{11.2.8}$$

in which n is the Manning roughness coefficient and $K_2 = 1$ for SI units or $K_2 = 2.22$ for English units. In Eq. 11.2.1, the exponent $x = 2$.

An advantage to using Eq. 11.2.7 or Eq. 11.2.8 as opposed to Eq. 11.2.4 is that in the first two, C and n are dependent on roughness only, while in the last, f depends on the Reynolds number as well as the relative roughness. However, Eq. 11.2.4 is preferred since it provides a more precise representation of pipe frictional losses. Note that the Hazen–Williams and Manning relations are dimensionally inhomogeneous,[1]

[1]In a dimensionally inhomogeneous equation the constants in the equation have units assigned to them. In a dimensionally homogeneous equation the constants are dimensionless.

whereas the Swamee–Jain equation is dimensionally homogeneous and contains the two parameters e/D and Re, which appropriately influence the losses.

The limitations of the Hazen–Williams and Chezy–Manning formulas are demonstrated as follows. Beginning with Eq. 11.2.1, the head loss h_L based on the Darcy–Weisbach resistance coefficient, Eq. 11.2.2, with the exponent $x = 2$ can be equated with h_L based on the Hazen–Williams coefficient, Eq. 11.2.7, with $x = 1.85$. Introducing the Reynolds number to eliminate Q and solving for the friction factor f results in

$$f = \frac{1.28gK_1}{C^{1.85}D^{0.02}(\text{Re } \nu)^{0.15}}$$ (11.2.9)

With SI units and for water at 20°C, Eq. 11.2.9 reduces to

$$f = \frac{1056}{C^{1.85}D^{0.02}\text{Re}^{0.15}}$$ (11.2.10)

Note that f is weakly dependent upon D in this equation. In a similar fashion, Eq. 11.2.8 can be substituted into Eq. 11.2.1 with $x = 2$ and equated to h_L based on the Darcy–Weisbach formulation to yield

$$f = \frac{124.5n^2}{D^{0.33}}$$ (11.2.11)

Comparisons between the original Colebrook formula (Eq. 7.6.28), the formula of Swamee and Jain (Eq. 11.2.3), and the equivalent expressions for the Hazen–Williams and Moody coefficients (Eqs. 11.2.10 and 11.2.11) are shown in Fig. 11.2 for a 1-m-diameter concrete pipe flowing

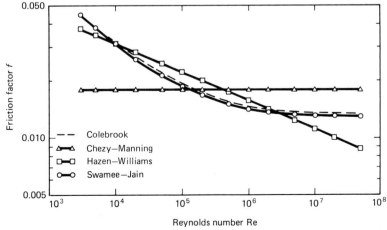

Figure 11.2 Comparison of several approximate formulas with the original Colebrook formula.

full with water ($C = 130$, $n = 0.012$, $e/D = 0.00015$). It is evident that the Hazen–Williams and Manning relations are valid over a limited range of Re, and that Eq. 11.2.3 provides a more versatile and accurate estimate of pipe losses.

EXAMPLE 11.1

A pipeline is conveying 0.05 m³/s of water, at 30°C. The length of the line is 300 m, and the diameter is 0.25 m. Estimate the head loss due to friction using the three formulas (a) Darcy–Weisbach ($e = 0.5$ mm), (b) Hazen–Williams ($C = 110$), and (c) Manning ($n = 0.012$).

Solution

The kinematic viscosity of water at 30°C is $\nu = 0.804 \times 10^{-6}$ m²/s. The relative roughness, velocity, and Reynolds number are computed to be

$$\frac{e}{D} = 0.002 \qquad V = 1.02 \text{ m/s} \qquad \text{Re} = 3.17 \times 10^5$$

(a) Substitute values into Eq. 11.2.4:

$$R = 1.07 \frac{300}{9.81 \times (0.25)^5}$$

$$\times \{\ln[0.27 \times 0.002 + 5.74(3.17 \times 10^5)^{-0.9}]\}^{-2} = 610$$

Then, with Eq. 11.2.1, the friction loss based on the Darcy–Weisbach formulation is

$$h_L = RQ^2$$
$$= 610 \, (0.05)^2 = 1.52 \text{ m}$$

(b) For the Hazen–Williams formulation, use Eq. 11.2.7:

$$R = \frac{10.59 \times 300}{(110)^{1.85}(0.25)^{4.87}} = 454$$

$$h_L = 454 \, (0.05)^{1.85} = 1.78 \text{ m}$$

(c) Use Eq. 11.2.8 for the Chezy–Manning formulation:

$$R = \frac{10.29(0.012)^2(300)}{1 \times (0.25)^{5.33}} = 719$$

$$h_L = 719 \, (0.05)^2 = 1.80 \text{ m}$$

Part (a) with $h_L = 1.52$ m is the most accurate. The Hazen–Williams result is 17% too high, and the Manning result is 18% too high.

11.3 SIMPLE PIPE SYSTEMS

The analysis of a single pipeline is presented in Section 7.6.4; it would be appropriate here for the reader to review the three categories of pipe problems in that section. For more complex piping systems, the methodology is similar. With relatively simple networks, such as series, parallel and branching systems, ad hoc solutions can be developed that are suitable for use with calculators or spreadsheet algorithms. Such approaches are relevant since they make use of the solver's ingenuity and require an understanding of the nature of the flow and piezometric head distributions for the particular piping arrangement. For systems of greater complexity, an alternative means of solving such problems is the Hardy Cross method, presented in Section 11.4.

The fundamental principle in the ad hoc approach is to identify all of the unknowns and write an equivalent number of independent equations to be solved. Subsequently, the system is simplified by eliminating as many unknowns as possible and reducing the problem to a series of single pipe problems—either category 1 or category 2, Section 7.6.4. It is likely that the resulting expression or series of expressions will be nonlinear, so an iterative solution will be required. In some instances, numerical methods such as interval halving, false position, Newton's, or successive substitution can be used to find the solution. In other cases, trial-and-error solutions will be satisfactory. It may be advantageous to employ a graphical procedure to aid in visualizing convergence of the solution.

11.3.1 Series Piping

Consider the series system shown in Fig. 11.3. It consists of N pipe elements and minor-loss components ΣK associated with each ith pipe element. A single minor loss is described in Section 7.6.5 to be equal to $h_L = KV^2/2g$. It is convenient here to express the minor loss in terms of the discharge rather than the velocity, so that $h_L = KQ^2/2gA^2$. For many flow situations, it is common practice to neglect the kinetic-energy terms at the inlet and outlet; they would be significant only if the velocities were relatively high. Assuming that the exponent in Eq. 11.2.1 is $x = 2$, the energy equation applied from location A to location B of Fig. 11.3 is

$$\left(\frac{p}{\gamma} + z\right)_A - \left(\frac{p}{\gamma} + z\right)_B = \left(R_1 + \frac{\Sigma K}{2gA_1^2}\right)Q_1^2 + \left(R_2 + \frac{\Sigma K}{2gA_2^2}\right)Q_2^2$$

$$+ \cdots + \left(R_N + \frac{\Sigma K}{2gA_N^2}\right)Q_N^2 \qquad (11.3.1)$$

$$= \sum_{i=1}^{N}\left(R_i + \frac{\Sigma K}{2gA_i^2}\right)Q_i^2$$

in which R_i is the resistance coefficient for pipe i.

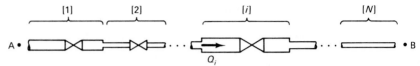

Figure 11.3 Series piping system.

The statement of continuity for the series system is that the discharge in every element is identical, or

$$Q_1 = Q_2 = \cdots = Q_i = \cdots = Q_N = Q \qquad (11.3.2)$$

Replacing Q_i with Q, Eq. 11.3.1 becomes

$$\left(\frac{p}{\gamma} + z\right)_A - \left(\frac{p}{\gamma} + z\right)_B = \left[\sum_{i=1}^{N}\left(R_i + \frac{\Sigma K}{2gA_i^2}\right)\right]Q^2 \qquad (11.3.3)$$

Equation 11.2.4 can be substituted for R_i, or alternatively, one may use the Moody diagram, Fig. 7.13, to obtain values of the friction factor. The reader should recognize that in a series system, the discharge remains constant from one pipe element to another, and the losses are accumulative, that is, they are the sum of the minor loss components and the pipe frictional losses.

For a category 1 problem, the right-hand side of Eq. 11.3.3 is known and the solution is straightforward. For a category 2 problem, in which Q is unknown, a trial-and-error solution is required, since the Reynolds number, in terms of the unknown discharge (Re = $4Q/\pi\nu D$), is present in the friction factor relation. Note that if flow in the fully rough zone is assumed, f is independent of Q and Eq. 11.3.3 reduces to a quadratic in Q. A category 3 problem is not encountered in this type of analysis. A series piping system with minor losses and constant-diameter piping is applied to a category 1 problem in Example 7.13. The solution for a somewhat more complex series system is illustrated in Example 11.2.

EXAMPLE 11.2

For the system shown, find the required power to pump 100 L/s of liquid (S = 0.85, $\nu = 10^{-5}$ m^2/s). The pump is operating at an efficiency $\eta = 0.75$. Pertinent data are given in the figure.

Line 1: L = 10 m, D = 0.20 m, e = 0.05 mm, K_1 = 0.5, K_v = 2
Line 2: L = 500 m, D = 0.25 m, e = 0.05 mm, K_e = 0.25, K_2 = 1

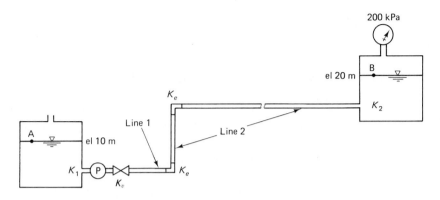

Solution

This is a category 1 problem. The energy relation, Eq. 11.3.3, for the system is

$$\left(\frac{p}{\gamma} + z\right)_A + H_P = \left(\frac{p}{\gamma} + z\right)_B + \left[R_1 + \frac{K_1 + K_v}{2gA_1^2} + R_2 + \frac{2K_e + K_2}{2gA_2^2}\right]Q^2$$

The resistance coefficients, R_1 and R_2, are calculated with Eq. 11.2.4 after first evaluating Re and e/D:

$$Re_1 = \frac{4Q}{\pi D_1 \nu} = \frac{4 \times 0.10}{\pi \times 0.20 \times 10^{-5}} = 6.37 \times 10^4 \qquad \left(\frac{e}{D}\right)_1 = \frac{0.05}{200} = 0.00025$$

$$Re_2 = \frac{4Q}{\pi D_2 \nu} = \frac{4 \times 0.10}{\pi \times 0.25 \times 10^{-5}} = 5.09 \times 10^4 \qquad \left(\frac{e}{D}\right)_2 = \frac{0.05}{250} = 0.0002$$

$$R_1 = 1.07\left(\frac{10}{9.81 \times (0.20)^5}\right)$$

$$\times \ \{\ln[0.27 \times 0.00025 + 5.74 \times (6.37 \times 10^4)^{-0.9}]\}^{-2} = 53.4$$

$$R_2 = 1.07\left(\frac{500}{9.81 \times (0.25)^5}\right)$$

$$\times \ \{\ln[0.27 \times 0.0002 + 5.74 \times (5.09 \times 10^4)^{-0.9}]\}^{-2} = 904$$

The minor loss coefficient terms are calculated to be

$$\frac{K_1 + K_v}{2gA_1^2} = \frac{0.5 + 2}{2 \times 9.81 \times [\pi/4 \times (0.20)^2]^2} = 129.1$$

$$\frac{2K_e + K_2}{2gA_2^2} = \frac{2 \times 0.25 + 1}{2 \times 9.81 \times [\pi/4 \times (0.25)^2]^2} = 31.7$$

Substitute these values into the energy equation and obtain

$$0 + 10 + H_P = \frac{200 \times 10^3}{0.85 \times 9800} + 20 + (53.4 + 129.1 + 904 + 31.7)(0.1)^2$$

This relation reduces to

$$10 + H_P = 24 + 20 + 11.2$$

Solving for the head across the pump gives $H_P = 45.2$ m. The required input power is

$$\dot{W}_P = \frac{\gamma Q H_P}{\eta}$$

$$= \frac{(9800 \times 0.85) \times 0.10 \times 45.2}{0.75}$$

$$= 5.0 \times 10^4 \text{ W} \quad \text{or} \quad 50 \text{ kW}$$

11.3.2 Parallel Piping

A parallel piping arrangement is shown in Fig. 11.4; it is essentially an arrangement of N pipe elements in series with ΣK minor loss components associated with each pipe element i. The continuity equation applied at either location A or B is given by

$$Q = \sum_{i=1}^{N} Q_i \tag{11.3.4}$$

The algebraic sum of the energy grade line around any defined loop must be zero. As in the case of series piping, it is customary to assume that $V^2/2g \ll (p/\gamma + z)$. Hence, for any pipe element i, the energy equation from location A to B is

$$\left(\frac{p}{\gamma} + z\right)_A - \left(\frac{p}{\gamma} + z\right)_B = \left(R_i + \frac{\Sigma K}{2gA_i^2}\right) Q_i^2 \quad i = 1, \ldots, N \tag{11.3.5}$$

The unknowns in Eqs. 11.3.4 and 11.3.5 are the discharges Q_i and the difference in piezometric head between A and B; the discharge Q into the system is known. It is possible to convert the minor loss terms using an

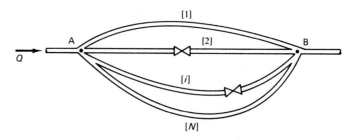

Figure 11.4 Parallel piping
system.

equivalent length as defined in Section 7.6.5. For each pipe element i the equivalent length L_e for ΣK minor loss components is

$$(L_e)_i = \frac{D_i}{f_i} \Sigma K \tag{11.3.6}$$

Thus Eq. 11.3.5 simplifies to the form

$$\left(\frac{P}{\gamma} + z\right)_A - \left(\frac{P}{\gamma} + z\right)_B = \bar{R}_i Q_i^2 \tag{11.3.7}$$

in which the modified pipe resistance coefficient $\bar{R}_i$ is given by

$$\bar{R}_i = \frac{8 f_i [L_i + (L_e)_i]}{g \pi^2 D_i^5} \tag{11.3.8}$$

A solution employing the method of successive substitution is developed in the following manner. Define the variable W to be the change in hydraulic grade line between A and B; that is, $W = (p/\gamma + z)_A - (p/\gamma + z)_B$. Then Eq. 11.3.7 can be solved for Q_i in terms of W as

$$Q_i = \left(\frac{W}{\bar{R}_i}\right)^{1/2} \tag{11.3.9}$$

Equations 11.3.4 and 11.3.9 are combined to eliminate the unknown discharges Q_i, resulting in

$$Q = \sum_{i=1}^{N} \left(\frac{W}{\bar{R}_i}\right)^{1/2} = \sqrt{W} \sum_{i=1}^{N} (\bar{R}_i)^{-1/2} \tag{11.3.10}$$

The remaining unknown W is taken out of the summation sign since it is the same in all pipes. Solving for W in Eq. 11.3.10 results in

$$W = \left(\frac{Q}{\sum\limits_{i=1}^{N} (\bar{R}_i)^{-1/2}}\right)^2 \tag{11.3.11}$$

An iterative procedure can be formulated to solve for W and the discharges Q_i as follows:

Step 1:   Assume flows in each line to be in the completely rough zone, and compute an initial estimate of the friction factors in each line using Eq. 11.2.6.

Step 2:   Compute $\bar{R}_i$ for each pipe and evaluate W with Eq. 11.3.11.

Step 3:   Compute Q_i in each pipe with Eq. 11.3.9.

Step 4:   Update the estimates of the friction factors in each line using the current values of Q_i and Eq. 11.2.3.

Step 5:   Repeat steps 2 to 4 until the unknowns W and Q_i do not vary according to a desired tolerance.

Note that if friction factors are in the completely rough zone so that they are independent of the discharge, steps 4 and 5 are unnecessary and a solution results on the first iteration. The technique is illustrated in Example 11.3.

EXAMPLE 11.3

Find the distribution of flow and the drop in hydraulic grade line for the three-parallel-pipe arrangement. Use variable friction factors with $\nu = 10^{-6}$ m²/s. The total water discharge is $Q = 0.020$ m³/s.

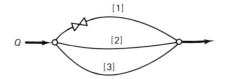

Pipe	L (m)	D (m)	e (mm)	ΣK
1	100	0.05	0.1	10
2	150	0.075	0.2	3
3	200	0.085	0.1	2

Solution

The initial estimates of f are based on Eq. 11.2.6. Preliminary calculations provide the following:

Pipe	e/D	f (Eq. 11.2.6)	L_e (Eq. 11.3.6)	$\bar{R}$ (Eq. 11.3.8)
1	0.002	0.023	21.7	7.40×10^5
2	0.0027	0.025	9.0	1.38×10^5
3	0.0012	0.021	8.1	8.14×10^4

Apply Eq. 11.3.11, and the first estimate of W is

$$W = \left[\frac{0.020}{(7.40 \times 10^5)^{-1/2} + (1.38 \times 10^5)^{-1/2} + (8.14 \times 10^4)^{-1/2}} \right]^2$$

$$= 7.39 \text{ m}$$

Then with Eq. 11.3.9, estimates of Q_i are made:

$$Q_1 = \left(\frac{7.39}{7.40 \times 10^5}\right)^{1/2} = 0.00316 \text{ m}^3/\text{s}$$

$$Q_2 = \left(\frac{7.39}{1.38 \times 10^5}\right)^{1/2} = 0.00732 \text{ m}^3/\text{s}$$

$$Q_3 = \left(\frac{7.39}{8.14 \times 10^4}\right)^{1/2} = 0.00953 \text{ m}^3/\text{s}$$

A continuity check is made using Eq. 11.3.4:

$$\sum_{i=1}^{3} Q_i = 0.00316 + 0.00732 + 0.00953 = 0.0200 \text{ m}^3/\text{s}$$

Even though the sum satisfies the solution, another iteration will be carried out to study the convergence of the solution technique. First, the $\bar{R}$-values are updated using Eq. 11.2.3 to evaluate the friction factors.

Pipe	$\text{Re} = \dfrac{4Q}{\pi D \nu}$	f (Eq. 11.2.3)	$\bar{R}$ (Eq. 11.3.8)
1	8.05×10^4	0.026	8.20×10^5
2	1.24×10^5	0.027	1.49×10^5
3	1.43×10^5	0.022	8.51×10^4

Evaluating W using Eq. 11.3.11 yields $W = 7.88$ m, and application of Eq. 11.3.9 gives the new estimates of Q_i:

$$Q_1 = 0.00310 \text{ m}^3/\text{s} \qquad Q_2 = 0.00727 \text{ m}^3/\text{s} \qquad Q_3 = 0.00962 \text{ m}^3/\text{s}$$

A continuity check shows that

$$\sum_{i=1}^{3} Q_i = 0.0200 \text{ m}^3/\text{s}$$

which is the same as in the first iteration.

11.3.3 Branch Piping

The branching network, illustrated in Fig. 11.5a, is made up of three elements connected to a single junction. In contrast to the parallel system shown in Fig. 11.4, no closed loops exist. In the analysis, one assumes the direction of flow in each element; then the energy equation for each

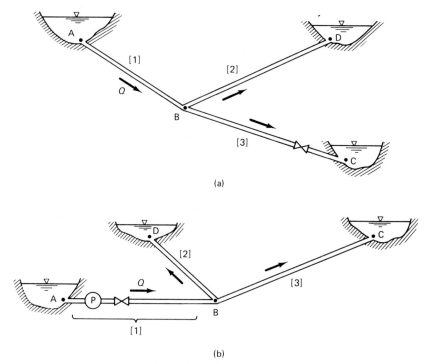

Figure 11.5 Branch piping systems: (a) gravity flow; (b) pump-driven flow.

element is written using an equivalent length to account for minor losses:

$$\left(\frac{p}{\gamma} + z\right)_A - \left(\frac{p}{\gamma} + z\right)_B = \bar{R}_1 Q_1^2 \qquad (11.3.12)$$

$$\left(\frac{p}{\gamma} + z\right)_B - \left(\frac{p}{\gamma} + z\right)_C = \bar{R}_2 Q_2^2 \qquad (11.3.13)$$

$$\left(\frac{p}{\gamma} + z\right)_B - \left(\frac{p}{\gamma} + z\right)_D = \bar{R}_3 Q_3^2 \qquad (11.3.14)$$

The piezometric heads at locations A, C, and D are considered known.
The unknowns are the piezometric head at B and the discharges Q_1, Q_2,
and Q_3. The additional relation is the continuity balance at location B,
which is

$$Q_1 - Q_2 - Q_3 = 0 \qquad (11.3.15)$$

Thus there are four equations with four unknowns. One convenient ad
hoc method of solution is outlined below and illustrated in Example 11.4:

Step 1:   Assume a discharge Q_1 in element 1 (with or without a pump). Establish the piezometric head H at the junction by solving Eq. 11.3.12 (a category 1 problem).

Step 2:   Compute the discharge Q_i in the remaining branches using Eqs. 11.3.13 and 11.3.14 (a category 2 problem).

Step 3:   Substitute the Q_i into Eq. 11.3.15 to check for continuity balance. Generally, the flow imbalance at the junction ΔQ will be nonzero. In Eq. 11.3.15, $\Delta Q = Q_1 - Q_2 - Q_3$.

Step 4:   Adjust the flow Q_1 in element 1 and repeat steps 2 and 3 until ΔQ is within desired limits.

If a pump exists in pipe 1 (see Fig. 11.5b). Eq. 11.3.12 is altered in the manner

$$\left(\frac{p}{\gamma} + z\right)_A - \left(\frac{p}{\gamma} + z\right)_B + H_P = \bar{R}_1 Q_1^2 \qquad (11.3.16)$$

An additional unknown, namely, the pump head H_P, is introduced. The additional necessary relationship is the head-discharge curve for the pump; see, for example, Fig. 7.18. The solution can proceed in a manner similar to that described in Example 11.5. It may be convenient to follow the solution graphically by plotting the assumed discharge in line 1 versus either the piezometric head at B or the imbalance of flow at B. Such a step is helpful to determine in what manner the discharge in line 1 should be altered for the next iteration.

An alternative method of solution for a single branching system is to eliminate all of the variables except the piezometric head H $(= p/\gamma + z)$ at the junction, location B in Fig. 11.5. Then a standard numerical iterative technique can be employed. An additional requirement is to assume the direction of flow in each pipe. It may become necessary to correct the sign in one or more of the equations if during the solution, H moves from below one of the reservoir elevations to above it, or vice versa. Example 11.4 illustrates the procedure.

Examples 11.4 and 11.5 represent a level of complexity that one may not want to exceed employing a calculator-based solution. To include additional pumps, reservoirs or pipes would make either ad hoc or simplified numerical approaches too cumbersome. For such systems, the more generalized network analysis described in Section 11.4 is recommended.

EXAMPLE 11.4

A three-branch piping system with given data is shown. Determine the flow rates Q_i and the piezometric head H at the junction. Assume constant friction factors.

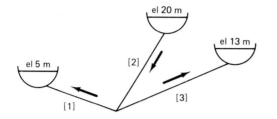

Pipe	L (m)	D (m)	f	ΣK
1	500	0.10	0.025	3
2	750	0.15	0.020	2
3	1000	0.13	0.018	7

Solution

The equivalent lengths and resistance coefficients are

$$(L_e)_1 = \frac{0.10}{0.025} \times 3 = 12 \text{ m} \qquad \bar{R}_1 = \frac{8 \times 0.025 \times 512}{9.81 \times \pi^2 \times (0.10)^5} = 1.06 \times 10^5 \text{ s}^2/\text{m}^5$$

$$(L_e)_2 = \frac{0.15}{0.020} \times 2 = 15 \text{ m} \qquad \bar{R}_2 = \frac{8 \times 0.020 \times 765}{9.81 \times \pi^2 \times (0.15)^5} = 1.66 \times 10^4 \text{ s}^2/\text{m}^5$$

$$(L_e)_3 = \frac{0.13}{0.018} \times 7 = 51 \text{ m} \qquad \bar{R}_3 = \frac{8 \times 0.018 \times 1051}{9.81 \times \pi^2 \times (0.13)^5} = 4.21 \times 10^4 \text{ s}^2/\text{m}^5$$

With the flow directions assumed as shown, the energy equation is written for each pipe and solved for the unknown discharge:

$$Q_1 = \left(\frac{H - 5}{\bar{R}_1}\right)^{1/2} \qquad Q_2 = \left(\frac{20 - H}{\bar{R}_2}\right)^{1/2} \qquad Q_3 = \left(\frac{H - 13}{\bar{R}_3}\right)^{1/2}$$

The continuity equation is $-Q_1 + Q_2 - Q_3 = 0$. Eliminating Q_1, Q_2, and Q_3 with the energy relations results in an algebraic equation in terms of H:

$$w(H) = -\left(\frac{H - 5}{1.06 \times 10^5}\right)^{1/2} + \left(\frac{20 - H}{1.66 \times 10^4}\right)^{1/2} - \left(\frac{H - 13}{4.21 \times 10^4}\right)^{1/2} = 0$$

Even though this can be written as a quadratic, the method of false position is chosen to compute H, which would be required if the friction factors varied. The procedure is presented in Example 10.10. In the present example the recurrence formula is

$$H_r = \frac{H_l w(H_u) - H_u w(H_l)}{w(H_u) - w(H_l)}$$

The solution is shown in the table below. Note that with the initial guesses of H_l and H_u, the sign conventions in w require that $20 > H > 13$. Iteration continues until the convergence criterion shown in the last column becomes less than the arbitrary value 0.005.

Iteration	H_l	H_u	$w(H_l)$	$w(H_u)$	H_r	$w(H_r)$	Sign of $w(H_l) \times w(H_r)$	$\varepsilon = \left\|\dfrac{H_r^{\text{new}} - H_r^{\text{old}}}{H_r^{\text{old}}}\right\|$
1	18	13	−0.01100	0.01185	15.59	−0.00154	+	
2	15.59	13	−0.00154	0.01185	15.29	−0.000384	+	0.019
3	15.29	13	−0.000384	0.01185	15.22	−0.000112	+	0.0046

Hence $H \simeq 15.2$ m. The discharges are now computed:

$$Q_1 = \left(\frac{15.2 - 5}{1.06 \times 10^5}\right)^{1/2} = 0.0098 \text{ m}^3/\text{s}$$

$$Q_2 = \left(\frac{20 - 15.2}{1.66 \times 10^4}\right)^{1/2} = 0.0170 \text{ m}^3/\text{s}$$

$$Q_3 = \left(\frac{15.2 - 13}{4.21 \times 10^4}\right)^{1/2} = 0.0072 \text{ m}^3/\text{s}$$

Note that continuity is satisfied.

EXAMPLE 11.5

Determine the flow distribution Q_i of water and the piezometric head H at the junction. The fluid power input by the pump is constant, equal to $\gamma Q H_P = 20$ kW. Assume constant friction factors.

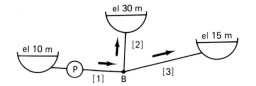

Pipe	L (m)	D (m)	f	Σ K
1	50	0.15	0.02	2
2	100	0.10	0.015	1
3	300	0.10	0.025	1

Solution

The equivalent lengths and resistance coefficients are computed from Eqs. 11.3.6 and 11.3.8 to be

$$(L_e)_1 = \frac{0.15}{0.02} \times 2 = 15 \text{ m} \qquad \bar{R}_1 = \frac{8 \times 0.02 \times 65}{9.81 \times \pi^2 \times (0.15)^5} = 1.42 \times 10^3 \text{ s}^2/\text{m}^5$$

$$(L_e)_2 = \frac{0.10}{0.015} \times 1 = 6.7 \text{ m} \qquad \bar{R}_2 = \frac{8 \times 0.015 \times 106.7}{9.81 \times \pi^2 \times (0.10)^5} = 1.32 \times 10^4 \text{ s}^2/\text{m}^5$$

$$(L_e)_3 = \frac{0.10}{0.025} \times 1 = 4 \text{ m} \qquad \bar{R}_3 = \frac{8 \times 0.025 \times 304}{9.81 \times \pi^2 \times (0.10)^5} = 6.28 \times 10^4 \text{ s}^2/\text{m}^5$$

Assume flow directions as shown. The energy equation for pipe 1 from the reservoir to the junction B is

$$z_1 + H_P = H + \bar{R}_1 Q_1^2$$

in which H is the piezometric head at B. Substituting in known parameters and solving for H results in

$$H = 10 + \frac{20 \times 10^3}{9800 Q_1} - 1.42 \times 10^3 Q_1^2$$

$$= 10 + \frac{2.04}{Q_1} - 1420 Q_1^2$$

An iterative solution is shown in the accompanying table. For each iteration, a value of Q_1 is estimated. Then H is calculated and Q_2 and Q_3 are evaluated from the relations

$$Q_2 = \left(\frac{H - z_2}{\bar{R}_2}\right)^{1/2} = \left(\frac{H - 30}{1.32 \times 10^4}\right)^{1/2}$$

$$Q_3 = \left(\frac{H - z_3}{\bar{R}_3}\right)^{1/2} = \left(\frac{H - 15}{6.28 \times 10^4}\right)^{1/2}$$

In the last column of the table, a continuity balance is employed to check the accuracy of the estimate of Q_1. The third estimate of Q_1 is based on a linear interpolation by setting $\Sigma Q = 0$ and using values of Q_1 and ΔQ from the first two iterations.

Iteration	Q_1	H	Q_2	Q_3	$\Delta Q = Q_1 - Q_2 - Q_3$
1	0.050	47.25	0.0362	0.0227	−0.0089
2	0.055	42.80	0.0311	0.0210	+0.0029
3	0.054	43.64	0.0322	0.0214	+0.0004

The approximate solution is $H = 43.6$ m, $Q_1 = 54$ L/s, $Q_2 = 32$ L/s, and $Q_3 = 21$ L/s. If greater precision is desired, a solution should be employed similar to that shown in Example 11.4.

11.4 ANALYSIS OF PIPE NETWORKS

Piping systems more complicated than those considered in Section 11.3 are best analyzed by formulating the solution for a network. Before considering a generalized set of network equations, it is worthwhile to examine a specific example of a piping network to observe the degree of complexity involved.

Figure 11.6a shows a relatively simple network consisting of seven pipes, two reservoirs, and one pump. The hydraulic grade lines at A and F are assumed known; these locations are termed **fixed-grade nodes.** Outflow demands are present at nodes C and D. Nodes C and D, along with nodes B and E are called **interior nodes** or **junctions.** Flow directions, even though not initially known, are assumed to take place as shown. The system equations are given as follows:

1. Energy balance for each pipe (seven equations):

$$H_A - H_B + H_P(Q_1) = \bar{R}_1 Q_1^2$$

$$H_B - H_D = \bar{R}_2 Q_2^2$$

$$H_C - H_D = \bar{R}_3 Q_3^2$$

$$H_B - H_C = \bar{R}_4 Q_4^2$$

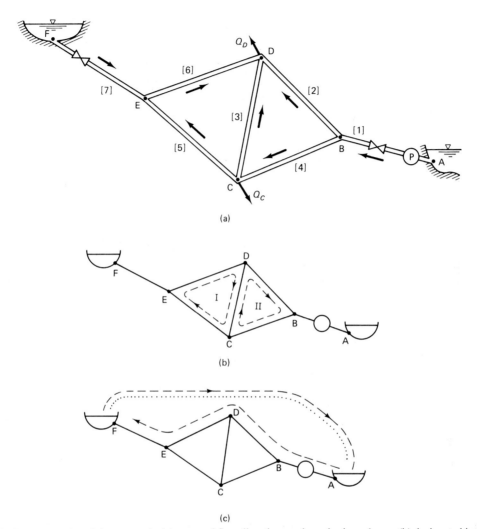

Figure 11.6 Representative piping network: (a) assumed flow directions and numbering scheme; (b) designated interior loops; (c) path between two fixed-grade nodes.

$$H_C - H_E = \bar{R}_5 Q_5^2$$
$$H_E - H_D = \bar{R}_6 Q_6^2 \qquad (11.4.1)$$
$$H_F - H_E = \bar{R}_7 Q_7^2$$

2. Continuity balance for each interior node (four equations):

$$Q_1 - Q_2 - Q_4 = 0$$
$$Q_2 + Q_3 + Q_6 = Q_D$$
$$Q_4 - Q_3 - Q_5 = Q_C \qquad (11.4.2)$$
$$Q_5 - Q_6 + Q_7 = 0$$

3. Approximation of pump curve (one equation):

$$H_P(Q_1) = a_0 + a_1 Q_1 + a_2 Q_1^2 \qquad (11.4.3)$$

in which a_0, a_1, and a_2 are known constants.

The unknowns are $Q_1, \ldots, Q_7$, H_B, H_C, H_D, H_E, and H_P. Thus there are 12 unknowns and 12 equations to solve simultaneously. Since the energy equations and the pump equation are nonlinear, it is necessary to resort to some type of successive iteration solution. The 12 equations can be reduced in number by combining the energy equations along special paths. Let the drop in hydraulic grade line for any pipe element i be designated as W_i. Then

$$W_i = \bar{R}_i Q_i^2 \qquad (11.4.4)$$

For the system under consideration, two closed paths, or **interior loops,** can be identified (see Fig. 11.6b). Flow is considered positive in a clockwise sense around each loop. Energy balances, written around loops I and II, are

$$W_6 - W_3 + W_5 = 0,$$
$$W_3 - W_2 + W_4 = 0 \qquad (11.4.5)$$

To account for the flow in pipes 1 and 7, a path can be defined along nodes A, B, D, E, and F in Fig. 11.6c. Then with the addition of the pump head, the energy balance from A to F is

$$H_A + H_P - W_1 - W_2 + W_6 + W_7 = H_F \qquad (11.4.6)$$

Note that the path energy equation connects two fixed-grade nodes. Such a path is sometimes termed a **pseudoloop,** since an imaginary pipe with infinite resistance, or no flow, can be considered to connect the two reservoirs. The imaginary pipe is shown by a dotted line in Fig. 11.6c. Substituting the pump equation and the friction equation into the energy relations above results in the following reduced set of equations:

$$-\bar{R}_3 Q_3^2 + \bar{R}_5 Q_5^2 + \bar{R}_6 Q_6^2 = 0$$
$$-\bar{R}_2 Q_2^2 + \bar{R}_3 Q_3^2 + \bar{R}_4 Q_4^2 = 0$$
$$-\bar{R}_1 Q_1^2 + (a_0 + a_1 Q_1 + a_2 Q_1^2) - \bar{R}_2 Q_2^2 + \bar{R}_6 Q_6^2 + \bar{R}_7 Q_7^2 + H_A - H_F = 0$$
$$Q_1 - Q_2 - Q_4 = 0$$
$$Q_2 + Q_3 + Q_6 = Q_D$$
$$Q_4 - Q_3 - Q_5 = Q_C$$
$$Q_5 - Q_6 + Q_7 = 0$$

$$(11.4.7)$$

There are now seven unknowns $(Q_1, \ldots, Q_7)$ and seven equations to solve. The energy relations are nonlinear since the loss terms and the pump head are represented as polynomials with respect to the discharges.

11.4.1 Generalized Network Equations

Networks of piping such as those shown in Figs. 11.6 and 11.7 can be represented by the following equations.

1. Continuity at the jth interior node:

$$\Sigma(\pm)_j Q_j - Q_e = 0 \qquad (11.4.8)$$

   in which the subscript j refers to the pipes connected to a node, and Q_e is the external demand. The algebraic plus or minus sign convention pertains to the assumed flow direction: Use the positive sign for flow into the junction, and the negative sign for flow out of the junction.

2. Energy balance around an interior loop:

$$\Sigma(\pm)_i W_i = 0 \qquad (11.4.9)$$

   in which the subscript i pertains to the pipes that make up the loop. There will be a relation for each of the loops. Here it is assumed that there are no pumps located in the interior of the network. The plus sign is used if the flow in the element is positive in the clockwise sense; otherwise, the minus sign is employed.

3. Energy balance along a unique path or pseudoloop connecting two fixed-grade nodes:

$$\Sigma(\pm)_i [W_i - (H_P)_i] + \Delta H = 0 \qquad (11.4.10)$$

   where ΔH is the difference in magnitude of the two fixed-grade nodes in the path ordered in a clockwise fashion across the imaginary pipe in the pseudoloop. If F is the number of fixed-grade nodes, there will be $(F - 1)$ unique path equations. The plus and minus signs in Eq. 11.4.10 follow the same argument given for Eq. 11.4.9.

   Let P be the number of pipe elements in the network, J the number of interior nodes, and L the number of interior loops. Then the following relation will hold if the network is properly defined:

$$P = J + L + F - 1 \qquad (11.4.11)$$

In Fig. 11.6 of the introductory example, $J = 4$, $L = 2$, $F = 2$, so that $P = 4 + 2 + 2 - 1 = 7$.

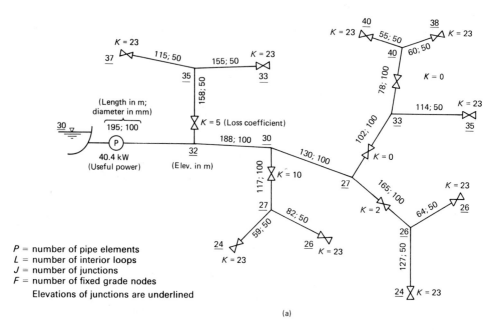

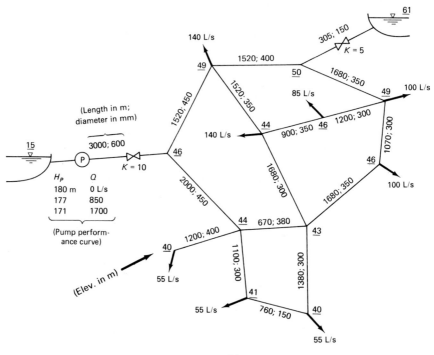

(b)

Figure 11.7   Two piping networks: (a) tree watering system: $P = 17$, $L = 0$, $J = 8$, $F = 10$; (b) water distribution system: $P = 17$, $L = 4$, $J = 12$, $F = 2$ (Wood, 1981).

An additional necessary formulation is the relation between discharge and loss in each pipe; it is

$$W = RQ^x + BQ^2 \qquad (11.4.12)$$

in which the term $B = \Sigma\, K/(2gA^2)$ accounts for the minor losses in each pipe. If the minor losses can be defined in terms of an equivalent length, Eq. 11.4.12 can be replaced by[2]

$$W = \bar{R}Q^x \qquad (11.4.13)$$

An approximate pump head-discharge representation is given by the polynomial

$$H_P(Q) = a_0 + a_1 Q + a_2 Q^2 \qquad (11.4.14)$$

The coefficients a_0, a_1, and a_2 are assumed known; typically, they can be found by substituting three known data points from a specified pump curve and solving the three resulting equations simultaneously. In place of defining the pump head-discharge curve, an alternative means of including a pump in a line is to specify the useful power the pump puts into the system. The useful, or actual, power $\dot{W}_f$ is assumed to be constant and allows H_P to be represented in the manner

$$H_P(Q) = \frac{\dot{W}_f}{\gamma Q} \qquad (11.4.15)$$

This equation is particularly useful when the specific operating characteristics of a pump are unknown.

11.4.2 Linearization of System Energy Equations

Equation 11.4.10 is a general relation that can be applied to any path or closed loop in a network. If it is applied to a closed loop, ΔH is set equal to zero, and if no pump exists in the path or loop, H_P is equal to zero. Note that Eq. 11.4.9 can be considered to be a subset of Eq. 11.4.10. In

---

[2]In Eq. 11.4.13, the exponent x is used, since the Hazen–Williams formula is commonly employed in pipe network analysis. The equivalent length for use in the Hazen–Williams formulation is given by

$$L_e = 0.8106\, \frac{\Sigma\, K}{K_1}\, D^{0.87} C^{1.85} Q^{0.15}$$

Note that it is necessary to estimate the discharge when using this equation. The Darcy–Weisbach equation with $x = 2$ is recommended. An alternative to using the equivalent-length concept is to account for pipe friction and minor losses separately in each line, making use of Eq. 11.4.12.

the following development, Eq. 11.4.10 will be used to represent any loop or path in the network.

Define the function $\phi(Q)$ to contain the nonlinear terms $W(Q)$ and $H_P(Q)$ in the form

$$
\begin{aligned}
\phi(Q) &= W(Q) - H_P(Q) \\
&= \bar{R}Q^x - H_P(Q)
\end{aligned}
\tag{11.4.16}
$$

Equation 11.4.16 can be expanded in a Taylor series as

$$
\phi(Q) = \phi(Q_0) + \frac{d\phi}{dQ}\bigg|_{Q_0}(Q - Q_0) + \frac{d^2\phi}{dQ^2}\bigg|_{Q_0}\frac{(Q - Q_0)^2}{2} + \cdots \tag{11.4.17}
$$

in which Q_0 is an estimate of Q. To approximate $\phi(Q)$ accurately, Q_0 should be chosen so that the difference $(Q - Q_0)$ is numerically small. Retaining the first two terms on the right-hand side of Eq. 11.4.17, and using Eq. 11.4.16, yields

$$
\phi(Q) \simeq \bar{R}Q_0^x - H_P(Q_0) + \left[x\bar{R}Q_0^{x-1} - \frac{dH_P}{dQ}\bigg|_{Q_0}\right](Q - Q_0) \tag{11.4.18}
$$

Note that the approximation to $\phi(Q)$ is now linear with respect to Q.

The parameter G is introduced as

$$
G = x\bar{R}Q_0^{x-1} - \frac{dH_P}{dQ}\bigg|_{Q_0} \tag{11.4.19}
$$

Using Eq. 11.4.14 to represent the pump head, Eq. 11.4.19 becomes

$$
G = x\bar{R}Q_0^{x-1} - (a_1 + 2a_2Q_0) \tag{11.4.20}
$$

Alternatively, with Eq. 11.4.15 substituted into Eq. 11.4.19, one has

$$
G = x\bar{R}Q_0^{x-1} + \frac{\dot{W}_f}{\gamma Q_0^2} \tag{11.4.21}
$$

Substituting Eq. 11.4.19 into Eq. 11.4.18 gives

$$
\begin{aligned}
\phi(Q) &= \bar{R}Q_0^x - H_P(Q_0) + (Q - Q_0)G \\
&= W_0 - H_{P0} + (Q - Q_0)G
\end{aligned}
\tag{11.4.22}
$$

in which $W_0 = W(Q_0)$ and $H_{P0} = H_P(Q_0)$. Finally, Eq. 11.4.22 is substituted into Eq. 11.4.10 to produce the linearized loop or path energy equation

$$\Sigma(\pm)_i[(W_0)_i - (H_{P0})_i] + \Sigma[Q_i - (Q_0)_i]G_i + \Delta H = 0 \quad (11.4.23)$$

The second term does not contain the plus or minus sign since G_i is a monotonically increasing function of the flow correction. Since Q in the relations above can assume positive or negative values, Q_0^x and Q_0^{x-1} are often replaced by $Q_0|Q_0|^{x-1}$ and $|Q_0|^{x-1}$, respectively, in solution algorithms. This is done, for example, in Eq. 11.4.23, which forms the basis for the Hardy Cross and linear methods of solution outlined below.

11.4.3 Hardy Cross Method

The Hardy Cross method of analysis is a simplified version of the iterative linear analysis. In Eq. 11.4.23, let $(Q_0)_i$ be the estimates of discharge from the previous iteration, and let Q_i be the new estimates of the discharge. Define a flow adjustment ΔQ for each loop to be

$$\Delta Q = Q_i - (Q_0)_i \qquad (11.4.24)$$

The adjustment is applied independently to all pipes in a given loop. Hence Eq. 11.4.23 can be written as

$$\Sigma(\pm)_i[(W_0)_i - (H_{P0})_i] + \Delta Q \Sigma G_i + \Delta H = 0 \qquad (11.4.25)$$

Solving for ΔQ, one has

$$\Delta Q = \frac{-\Sigma(\pm)[(W_0)_i - (H_{P0})_i] - \Delta H}{\Sigma G_i} \qquad (11.4.26)$$

It is necessary for the algebraic sign of Q to be positive in the direction of normal pump operation; otherwise, the pump curve will not be represented properly and Eq. 11.4.26 will be invalid. Furthermore, it is important that the discharge Q through the pump remain within the limits of the data used to generate the curve.

For a closed loop in which no pumps or fixed-grade nodes are present, Eq. 11.4.26 reduces simply to the form

$$\Delta Q = \frac{-\Sigma(\pm)_i(W_0)_i}{\Sigma G_i} \qquad (11.4.27)$$

The Hardy Cross iterative solution is outlined in the following steps:

1. Assume an initial estimate of the flow distribution in the network that satisfies continuity, Eq. 11.4.8. The closer the initial estimates are to the correct values, the fewer will be the iterations

necessary for convergence. One guideline to use is that in a pipe element, as $\bar{R}$ increases, Q decreases.

2. For each loop or path, evaluate ΔQ with Eq. 11.4.26 or 11.4.27. The numerators should approach zero as the loops or paths become balanced.

3. Update the flows in each pipe in all loops and paths, that is, from Eq. 11.4.24,

$$Q_i = (Q_0)_i + \Sigma \Delta Q \qquad (11.4.28)$$

The term $\Sigma \Delta Q$ is used, since a given pipe may belong to more than one loop; hence the correction will be the sum of corrections from all loops to which the pipe is common.

4. Repeat steps 2 and 3 until a desired accuracy is attained. One possible criterion to use is

$$\frac{\Sigma |Q_i - (Q_0)_i|}{\Sigma |Q_i|} \leq \varepsilon \qquad (11.4.29)$$

in which ε is an arbitrarily small number. Typically, $0.001 < \varepsilon < 0.005$.

The Hardy Cross method of analysis is a simplified version of the method of successive approximations applied to a set of linearized equations. It does not require the inversion of a matrix; hence it can be used to solve relatively small networks using either a calculator or spread sheet algorithm on a microcomputer. The continuity relations, Eq. 11.4.8, are in a sense "decoupled" from the solution of the energy relations, Eq. 11.4.26; continuity is satisfied initially with assumed flows and remains satisfied throughout the solution process. Essentially, one computes a correction ΔQ to the flows Q_i in each loop separately, and then applies ΔQ to the entire network to bring flow through the loops into closer balance. In effect this is a superposition type of solution. Since the flow corrections ΔQ are applied to each loop independently, convergence may not be rapid.

Computer codes are available for solving large networks using the Hardy Cross method. In Section 11.5 one will find a program written in BASIC along with documentation that can readily be used to analyze network flows and hydraulic grade lines. Examples 11.7 and 11.8 illustrate the use of the program.

EXAMPLE 11.6

For the piping system shown, determine the flow distribution and piezometric heads at the junctions using the Hardy Cross method of solution.

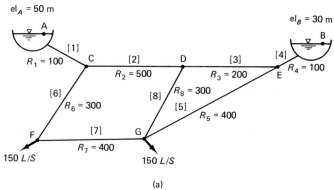

(a)

Solution

There are five junctions ($J = 5$), eight pipes ($P = 8$), and two fixed grade nodes ($F = 2$). Hence the number of closed loops is $L = 8 - 5 - 2 + 1 = 2$, plus one pseudoloop. The three loops and assumed flow directions are shown below.

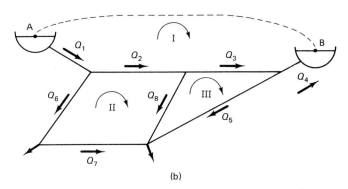

(b)

Equation 11.4.26 is applied to loop I:

$$\Delta Q_1 = \frac{-(\pm W_4 \pm W_3 \pm W_2 \pm W_1) - (z_A - z_B)}{G_4 + G_3 + G_2 + G_1}$$

Apply Eq. 11.4.27 to loops II and III:

$$\Delta Q_{II} = \frac{-(\pm W_2 \pm W_8 \pm W_7 \pm W_6)}{G_2 + G_8 + G_7 + G_6}$$

$$\Delta Q_{III} = \frac{-(\pm W_3 \pm W_5 \pm W_8)}{G_3 + G_5 + G_8}$$

The problem is solved using a spreadsheet algorithm, and is shown in the following two tables. In the table labeled "spreadsheet layout" the correct sign is attributed to each W automatically by use of the relation

$$\pm W = \bar{R} Q |Q|^{x-1}$$

in column F. In addition, the initial values of each Q take on a positive or negative sign depending on the assumed flow direction relative to the positive clockwise direction for each loop. The initial flow estimates, shown under the heading "iteration 1," are chosen to satisfy continuity. Note that flow adjustments to each pipe element are made after all of the ΔQ's have been computed; for example, relative to loop I, $Q_2 = (Q_0)_2 + \Delta Q_I - \Delta Q_{II}$. After four iterations the absolute magnitude of the ΔQ's are all less than 0.001, and the left-hand side of Eq. 11.4.29 is 0.0051.

The final values of Q_i are shown in the following figure, along with the final flow directions. Flow rates are in liters per second.

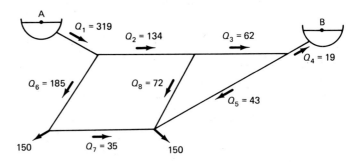

The piezometric heads are evaluated by computing the energy drop along designated paths, beginning at a known fixed-grade node, in this case, node A:

$$H_C = H_A - R_1 Q_1^2 = 50 - 100(0.319)^2 = 39.8 \text{ m}$$

$$H_D = H_C - R_2 Q_2^2 = 39.8 - 500(0.134)^2 = 30.8 \text{ m}$$

$$H_E = H_D - R_3 Q_3^2 = 30.8 - 200(0.062)^2 = 30.0 \text{ m}$$

$$H_F = H_C - R_6 Q_6^2 = 39.8 - 300(0.185)^2 = 29.5 \text{ m}$$

$$H_G = H_D - R_8 Q_8^2 = 30.8 - 300(0.072)^2 = 29.2 \text{ m}$$

Note that there is negligible loss in element four.

Spreadsheet layout

	A	B	C	D	E	F	G	H
1	Example 11.6							
2	Hardy Cross Method for Solution of a Pipe Network							
3								
4								
5						Iteration		
6							D5+1	
7		Component	R	2R	Q	RQ|Q|	2R|Q|	Q
8		delta H				20		
9								
10	Loop I	Pipe 4	100	2*C10	-.02	100*E10*ABS(E10)	200*ABS(E10)	E10+G17
11		Pipe 3	200	2*C11	-.06	200*E11*ABS(E11)	400*ABS(E11)	E11+G17-G36
12		Pipe 2	500	2*C12	-.13	500*E12*ABS(E12)	1000*ABS(E12)	E12+G17-G27
13		Pipe 1	100	2*C13	-.32	100*E13*ABS(E13)	200*ABS(E13)	E13+G17
14						--------	--------	
15						SUM(F9:F13)	SUM(G9:G13)	
16								
17						delta Q =	-F15/G15	
18								
19								
20	Loop II	Pipe 2	500	2*C20	.13	500*E20*ABS(E20)	1000*ABS(E20)	E20+G27-G17
21		Pipe 8	300	2*C21	.07	300*E21*ABS(E21)	600*ABS(E21)	E21+G27-G36
22		Pipe 7	400	2*C22	-.04	400*E22*ABS(E22)	800*ABS(E22)	E22+G27
23		Pipe 6	300	2*C23	-.19	300*E23*ABS(E23)	600*ABS(E23)	E23+G27
24						--------	--------	
25						SUM(F20:F23)	SUM(G20:G23)	
26								
27						delta Q =	-F25/G25	
28								
29								
30	Loop III	Pipe 3	200	2*C30	.06	200*E30*ABS(E30)	400*ABS(E30)	E30+G36-G17
31		Pipe 5	400	2*C31	.04	400*E31*ABS(E31)	800*ABS(E31)	E31+G36
32		Pipe 8	300	2*C32	-.07	300*E32*ABS(E32)	600*ABS(E32)	E32+G36-G27
33						--------	--------	
34						SUM(F30:F32)	SUM(G30:G32)	
35								
36						delta Q =	-F34/G34	

533

Spreadsheet solution

	A	B	C	D	E	F	G	H
3								
4								
5						Iteration	1	
6								
7		Component	R	2R	Q	RQ¦Q¦	2R¦Q¦	Q
8								
9		delta H				20		
10		Pipe 4	100	200	-.02	-.04	4	-.022477
11	Loop I	Pipe 3	200	400	-.06	-.72	24	-.063600
12		Pipe 2	500	1000	-.13	-8.45	130	-.137352
13		Pipe 1	100	200	-.32	-10.24	64	-.322477
14						--------	--------	
15						.55	222	
16								
17						delta Q = -.002477		
18								
19								
20		Pipe 2	500	1000	.13	8.45	130	.1373517
21	Loop II	Pipe 8	300	600	.07	1.47	42	.0737518
22		Pipe 7	400	800	-.04	-.64	32	-.035126
23		Pipe 6	300	600	-.19	-10.83	114	-.185126
24						--------	--------	
25						-1.55	318	
26								
27						delta Q = .0048742		
28								
29								
30		Pipe 3	200	400	.06	.72	24	.0635999
31	Loop III	Pipe 5	400	800	.04	.64	32	.0411224
32		Pipe 8	300	600	-.07	-1.47	42	-.073752
33						--------	--------	
34						-.11	98	
35								
36						delta Q = .0011224		

	I		J		K		L		M		N		O		P		Q	

Iteration	2		Iteration	3		Iteration	4	
RQ|Q|	2R|Q|	Q	RQ|Q|	2R|Q|	Q	RQ|Q|	2R|Q|	Q
	20			20			20	
-.050524	4.495495	-.019494	-.038003	3.898876	-.019839	-.039358	3.967758	-.018935
-.808990	25.43997	-.062044	-.769885	24.81749	-.062745	-.787382	25.09792	-.062371
-9.43274	137.3517	-.133466	-8.90658	133.4660	-.135276	-9.14981	135.2761	-.134274
-10.3992	64.49550	-.319494	-10.2077	63.89888	-.319839	-10.2297	63.96776	-.318935
--------	--------		--------	--------		--------	--------	
-.691430	231.7827		.0778641	226.0812		-.206230	228.3095	

delta Q = .0029831   delta Q = -.000344   delta Q = .0009033

9.432744	137.3517	.1334660	8.906582	133.4660	.1352761	9.149805	135.2761	.1342744
1.631797	44.25106	.0714222	1.530341	42.85334	.0725312	1.578235	43.51875	.0719037
-.493528	28.10063	-.036028	-.519219	28.82273	-.034563	-.477833	27.65019	-.034661
-10.2815	111.0755	-.186028	-10.3820	111.6171	-.184563	-10.2190	110.7376	-.184661
--------	--------		--------	--------		--------	--------	
.2895450	320.7789		-.464268	316.7591		.0311857	317.1826	

delta Q = -.000903   delta Q = .0014657   delta Q = -.000098

.8089901	25.43997	.0620437	.7698850	24.81749	.0627448	.7873822	25.09792	.0623708
.6764223	32.89796	.0425494	.7241789	34.03948	.0429060	.7363705	34.32481	.0434353
-1.63180	44.25106	-.071422	-1.53034	42.85334	-.072531	-1.57823	43.51875	-.071904
--------	--------		--------	--------		--------	--------	
-.146384	102.5890		-.036277	101.7103		-.054482	102.9415	

delta Q = .0014269   delta Q = .0003567   delta Q = .0005293

EXAMPLE 11.7

Determine the flow and piezometric head distributions for the tree system shown in Fig. 11.7a. Use the Hazen–Williams friction formula and assume that $C = 110$ for all pipe elements. The useful input power of the pump is 40.4 kW.

Solution

The computer program listed in Section 11.5 is used for the solution. The numbering scheme and assumed flow directions are shown below. Pipe element numbers are enclosed in brackets and junction numbers are encircled. The required data are listed in statements 2000 to 2290. Note that the assumed initial flows satisfy continuity.

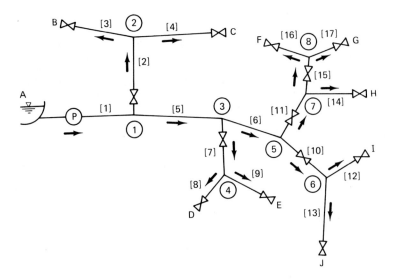

```
2000 REM ***** DATA SET EXAMPLE 11.7 (SI UNITS) *****
2010 DATA 17,8,1,SI,HW,1.E-6,1,1000,1000,.001
2020 DATA 1,195,100,130,0,40
2030 DATA 2,158,50,130,5,7
2040 DATA 3,115,50,130,23,3.5
2050 DATA 4,155,50,130,23,3.5
2060 DATA 5,188,100,130,0,33
2070 DATA 6,130,100,130,0,20
2080 DATA 7,117,100,130,10,13
2090 DATA 8,59,50,130,23,6.5
2100 DATA 9,82,50,130,23,6.5
2110 DATA 10,165,100,130,2,10
2120 DATA 11,102,100,130,0,10
2130 DATA 12,64,50,130,23,5
2140 DATA 13,127,50,130,23,5
2150 DATA 14,114,50,130,23,3
```

```
2160 DATA 15,78,100,130,0,7
2170 DATA 16,55,50,130,23,3.5
2180 DATA 17,55,50,130,23,3.5
2190 DATA 1,UP,40.4
2200 DATA 30,1,1,1,2,2,-2,5,1,7,2,-7,6,1,10,2,-10,11,1,15
2210 DATA -7,3,1,2,3
2220 DATA -3,3,1,2,4
2230 DATA 6,4,1,5,7,8
2240 DATA 4,4,1,5,7,9
2250 DATA 6,5,1,5,6,10,13
2260 DATA 4,5,1,5,6,10,12
2270 DATA -5,5,1,5,6,11,14
2280 DATA -10,6,1,5,6,11,15,16
2290 DATA -8,6,1,5,6,11,15,17
```

The computed results are shown below. Five iterations were required to converge to a relative flow change below 0.001. The flow rates are in liters per second, and the piezometric heads, listed in the column labeled HYD. GRADE LINE, are in meters.

TRIAL NO.	RELATIVE FLOW CHANGE	AVERAGE FLOW CHANGE
1	3.003E-02	3.121E-01
2	8.718E-03	9.054E-02
3	4.174E-03	4.333E-02
4	1.942E-03	2.015E-02
5	9.208E-04	9.556E-03

PIPE NO.	FLOW RATE
1	39.051
2	6.499
3	3.162
4	3.337
5	32.552
6	19.987
7	12.565
8	6.714
9	5.851
10	9.839
11	10.148
12	5.373
13	4.466
14	3.419
15	6.729
16	3.130
17	3.599

NODE NO.	HYD. GRADE LINE
1	89.622
2	47.574
3	58.047
4	53.372
5	49.200
6	46.018
7	47.222
8	46.515

EXAMPLE 11.8

For the distribution system shown in Fig. 11.7b, find the flows in the pipe elements and the piezometric heads at the junctions. Base the friction losses on the Darcy–Weisbach formulation, assuming an absolute roughness of 0.15 mm for all pipes and $\nu = 10^{-6}$ m^2/s.

Solution

The computer program listed in Section 11.5 is used for the solution. The numbering scheme and assumed flow directions are shown below. Head-discharge data for the pump are listed in line 2190 of the data set. Discharges for the pump and pipe elements are entered in liters per second, and pipe diameters and roughness elements are entered in millimeters.

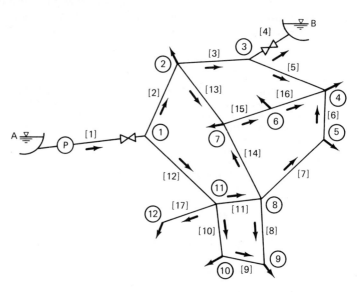

```
2000 REM ***** DATA SET EXAMPLE 11.8 (SI UNITS) *****
2010 DATA 17,12,1,SI,DW,1.E-6,1,1000,1000,.001
2020 DATA 1,3000,600,.15,10,1100
2030 DATA 2,1520,450,.15,0,675
2040 DATA 3,1520,400,.15,0,400
2050 DATA 4,305,150,.15,5,370
2060 DATA 5,1680,350,.15,0,30
2070 DATA 6,1070,300,.15,0,50
2080 DATA 7,1680,350,.15,0,150
2090 DATA 8,1380,300,.15,0,25
2100 DATA 9,760,150,.15,0,30
2110 DATA 10,1100,300,.15,0,85
2120 DATA 11,670,380,.15,0,285
2130 DATA 12,2000,450,.15,0,425
2140 DATA 13,1520,350,.15,0,135
2150 DATA 14,1680,300,.15,0,110
2160 DATA 15,900,350,.15,0,105
```

```
2170 DATA 16,1200,300,.15,0,20
2180 DATA 17,1200,400,.15,0,55
2190 DATA 1,HQ,180,177,850,171,1700
2200 DATA 15,1,1,1,2,1,3,1,5,1,-6,2,6,-16
2210 DATA 1,-15,1,-14,1,8,1,-9,1,-10,1,17
2220 DATA 0,5,2,13,-14,-11,-12
2230 DATA 0,5,3,5,-16,-15,-13
2240 DATA 0,5,15,16,-6,-7,14
2250 DATA 0,4,11,8,-9,-10
2260 DATA -46,4,1,2,3,4
```

Computed output is shown below. Nine iterations were required to achieve a relative flow change below 0.001.

TRIAL NO.	RELATIVE FLOW CHANGE	AVERAGE FLOW CHANGE
1	2.967E-01	6.022E+01
2	1.434E-01	2.548E+01
3	1.017E-01	1.705E+01
4	5.935E-02	9.805E+00
5	2.619E-02	4.303E+00
6	1.013E-02	1.660E+00
7	3.917E-03	6.416E-01
8	1.502E-03	2.460E-01
9	5.743E-04	9.403E-02

PIPE NO.	FLOW RATE
1	802.864
2	446.577
3	164.805
4	72.864
5	91.941
6	18.835
7	118.835
8	39.652
9	15.348
10	70.348
11	230.940
12	356.287
13	141.771
14	72.452
15	74.224
16	10.776
17	55.000

NODE NO.	HYD. GRADE LINE
1	148.307
2	119.809
3	112.383
4	107.148
5	107.506
6	107.004
7	108.862
8	116.102
9	114.229
10	119.841
11	124.319
12	123.604

11.4.4 Other Linearized Methods

With the advent of microcomputer technology, it is possible to analyze networks with as many as 500 pipes with relative ease. The Hardy Cross method has been utilized for such systems, but its convergence characteristics can be slow. Several other methods (Epp and Fowler, 1970; Gessler, 1981; Jeppson, 1982; Wood, 1981) all based on linear approximations, are discussed in this section.

Solutions Based on Discharges as Unknowns

In Eq. 11.4.23, let the individual discharges Q_i be considered as unknowns. Then the equation can be written as

$$[\Sigma(\pm)_i[(W_0)_i - (H_{P0})_i] - \Sigma\, G_i(Q_0)_i + \Delta H] + \Sigma\, G_iQ_i = 0 \quad (11.4.30)$$

For a piping network, Eq. 11.4.30 is written for L interior loops and $(F - 1)$ paths connecting fixed-grade nodes. In addition, the continuity relation, Eq. 11.4.8, applied at each node provides J additional equations. Equations 11.4.8 and 11.4.30 consist of a set of $J + L + F - 1 = P$ linear equations to be solved simultaneously for the P unknowns Q_i. In effect, one is making a flow adjustment ΔQ_i for each pipe element and solving for them simultaneously. This method has been termed the **linear method.** A Newton–Raphson approximation applied to the system of equations will produce identical results.

Advantages of the linear method over the Hardy Cross technique are that initial flows can be arbitrarily chosen, continuity is automatically satisfied, and convergence is more rapid. It is necessary to manipulate a large sparse matrix; however, a number of commercial algorithms are available for microcomputer use. In addition to the handling of pumps, pressure-reducing valves can be incorporated into the analysis.

Another linearized method—which can be classified as being in between the Hardy Cross and linear method—is one in which the flow adjustments in all defined loops and paths are solved simultaneously. Such a formulation requires the simultaneous solution of $L + F - 1$ equations. As opposed to the linear method, the size of the matrix is reduced, but convergence may not be as rapid.

Solutions Based on Piezometric Head as Unknowns

A solution treating the piezometric head at each junction as an unknown can be attributed to Cross (1936). The method is termed the **single node method** and is based on a Taylor series expansion of Q_i about H_i using the relation

$$\begin{aligned} W_i &= H_i - H_j \\ &= \bar{R}_iQ_i^x \end{aligned} \quad (11.4.31)$$

The result is

$$Q_i = (Q_0)_i + \frac{H_i - (H_0)_i}{G_i} \qquad (11.4.32)$$

in which $G_i = x\bar{R}_i \, [(Q_0)_i]^{x-1}$, and no pumps are present. In the two relations above, H_i is the piezometric head at the given junction node i, and H_j is the head at all the adjacent nodes. The variable Q_i refers to the discharge in the pipe element between nodes i and j. Substitution of Eq. 11.4.32 into Eq. 11.4.8, and solving for $\Delta H_i = H_i - (H_0)_i$ results in

$$\Delta H_i = \frac{-\Sigma(\pm)_i (Q_0)_i + Q_e}{\Sigma \, G_i^{-1}} \qquad (11.4.33)$$

The piezometric heads H_j at all nodes adjacent to node i are considered to be fixed. In the solution, a single trial requires that Eq. 11.4.33 be solved for each junction in the network. If necessary, pumps and pressure reducing valves can be incorporated. Even though convergence is slow for this method, it allows a reduced number of equations to be solved in comparison with the solutions based on discharges as unknowns.

A second method, which involves treating piezometric heads as unknowns rather than the discharges, is designated as the **simultaneous node method.** In Eq. 11.4.22, assuming no pumps in the path, the discharge Q can be expressed in terms of ϕ and Q_0 as

$$Q = \left(1 - \frac{1}{x}\right) Q_0 + \frac{\phi(Q)}{G} \qquad (11.4.34)$$

Equation 11.4.34 is substituted into the continuity relation, Eq. 11.4.8, yielding

$$\Sigma(\pm)_i \left[\left(1 - \frac{1}{x}\right)(Q_0)_i + \frac{\phi(Q_i)}{G_i}\right] - Q_e = 0 \qquad (11.4.35)$$

The subscript i refers to the pipe elements common to a given junction. Define $\phi(Q_i) = H_i - H_j$, where H_i is the piezometric head at the junction node and H_j are the piezometric heads at the adjacent nodes. Then Eq. 11.4.35 can be written in the form

$$\Sigma(\pm)_i \frac{H_i}{G_i} + \Sigma(\pm)_i \left[\left(1 - \frac{1}{x}\right)(Q_0)_i - \frac{H_j}{G_i}\right] - Q_e = 0 \qquad (11.4.36)$$

The relation is written for all junction nodes in the network. The result will be a set of simultaneous linear equations with piezometric heads H_i as unknowns. The number of unknowns is equal to J, the number of interior junctions present in the network. Pumps, pressure reducing valves, and orifices can readily be incorporated.

Solutions Based on Diameters as Unknowns

Up to this point, all of the analytical procedures related to pipe network analysis have treated the discharges or piezometric heads at junctions as the unknowns. Those formulations require that the piping geometry be specified; in particular, it is necessary that the piping diameters be known before the calculations begin. From a design viewpoint, it would be more desirable to treat the diameters as unknowns. The network relations can be formulated in such a manner, thereby producing a category 3 solution (see Section 7.6.4) for a piping system.

The discharge in any pipe can be expressed as a truncated Taylor series expanded about the pipe diameter:

$$Q = V \frac{\pi}{4} D^2$$
$$\simeq V \frac{\pi}{4} D_0^2 + V \frac{\pi}{2} D_0(D - D_0)$$

(11.4.37)

in which D_0 is an estimate of the unknown diameter D, and V is the velocity. In the formulation, the velocity is assumed to be known; one convenient way to accomplish this is to specify a global, or constant, velocity to be satisfied throughout the entire network. Substitution of Eq. 11.4.37 into Eqs. 11.4.8 to 11.4.10 results in a system of equations that are linear with respect to the unknown diameters. Given the specified flow demands, global velocity, and initial estimates of the pipe diameters, an iterative solution will produce pipe diameters accurate to within a desired tolerance. Ormsbee and Wood (1986) provide details of the solution, and in addition to considering a known global velocity term, they show that the velocity can be treated as an unknown along with the pipe diameters.

11.5 COMPUTER PROGRAM FOR SOLUTION OF FLOW IN PIPE NETWORKS USING THE HARDY CROSS METHOD OF FLOW ADJUSTMENT

In this section we present a BASIC program that solves the network flow equations using the Hardy Cross linear analysis method. The program is adapted from the publication "Algorithms for Pipe Network Analysis and Their Reliability," with permission of the author, Professor Don Wood, of the University of Kentucky. The program allows for the input of minor losses, use of either the Darcy–Weisbach or Hazen–Williams pipe frictional loss, and the provision of pumps either as useful power or a three-point, head-discharge data set. Either SI or English units are permissible.

In the program the W(N, M) array stores the required geometric data for the network. It is necessary to know the flow direction in each pipe element at all times because the required head summations depend on that flow direction. This process is handled by initially designating an assumed flow direction in each pipe element. As the calculations proceed, the direction of the flow rate is positive if it is in the initial assumed direction and negative if it reverses direction. If the flow through a pump reverses direction, the problem must be reformulated. Input geometric data require signs based on the initial assumed flow directions. The algorithm requires input data specifying assigned pipe numbers for the paths of pipe sections making up primary loops and pipe sections between fixed-grade nodes. The pipe numbers are input with a positive sign if the assumed flow direction is in the path direction and with a negative sign if it is opposite. This is illustrated using the loop shown in Fig. 11.8 with a path direction (clockwise) as specified and the assumed flow directions noted. The data input would be 10, 7, −6, 8, −3.

The program listing is given in Fig. 11.9. Pertinent notation is given below.

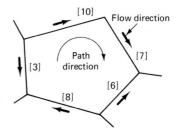

Figure 11.8 Pipe loop illustrating sign convention.

P2 = number of paths

O(N) = $\Sigma\, K/2gA^2$ for pipe N

C(N) = R for pipe N (Eq. 11.2.4 or Eq. 11.2.7)

G2 = hydraulic grade line at a reference fixed-grade node

E(N) = energy difference for path N

G1 = gradient for pipe element based on Q (Eq. 11.4.19)

H1 = piezometric head change for pipe element based on Q (Eq. 11.4.16)

Q1 = absolute value of flow rate

I = absolute value of pipe element number

T(N) = piezometric head change for pipe element N

W(M,N) = array for storing geometric data

Y(N) = number of pipe elements in path between junction N and junction N − 1 (line 470), or number of pipes in path N (line 530)

B1 = numerator for flow correction factor (Eq. 11.4.26)

E1 = denominator for flow correction factor (Eq. 11.4.26)

Q2 = ΔQ (Eq. 11.4.26)

```
10 CLS
20 REM ***** HARDY CROSS PIPE NETWORK ANALYSIS *****
30 DIM D(20),C(20),O(20),Z(20),V(20),R(20)
40 DIM E(20),Q(20),H(20),T(20),Y(20)
50 DIM W(20,20),CC(20),CK(20),A0(20),A1(20),A2(20)
60 OPEN "DATA.OUT" FOR OUTPUT AS #1
70 READ P,J,NP,UN$,FR$,NU,SG,C1,C2,EPS
80 IF UN$="EN" GOTO 120
90      G=9.806
100     K1=10.59
110     GOTO 140
120     G=32.2
130     K1=4.72
140 P2=P-J
150 IF FR$="HW" THEN X=1.852 ELSE X=2
160 FOR N=1 TO P
170 READ N1,L1,D1,R(N1),M1,Q1
180 Q(N1)=Q1/C1
190 D(N1)=D1/C2
200 O(N1)=.8106*M1/(G*D(N1)^4)
210 CK(N1)=0
220 Z(N1)=0
230 V(N1)=Q(N1)
240 IF FR$="HW" GOTO 280
250 R(N1)=R(N1)/C2
260 CC(N1)=0.8106*L1/(G*D(N1)^5)
270 GOTO 290
280 C(N1)=K1*L1/(R(N1)^X*D(N1)^4.87)
290 NEXT N
300 IF NP=0 GOTO 450
310 FOR N=1 TO NP
320 READ NPU, TYP$
330 IF TYP$="HQ" GOTO 370
340 READ P1
350 IF UN$="EN" THEN Z(NPU)=P1*550/(SG*62.4) ELSE Z(NPU)=P1/(SG*9.806)
360 GOTO 440
370 READ X0,X1,Y1,X2,Y2
380 Y1=Y1/C1
390 Y2=Y2/C1
400 CK(NPU)=1
410 A0(NPU)=X0
420 A2(NPU)=(X0+(X1/X0-1)*Y2-X2)/(Y1*Y1*Y2/X0-Y2*Y2)
430 A1(NPU)=(X1-X0-A2(NPU)*Y1*Y1)/X0
440 NEXT N
450 READ G2
460 FOR N=1 TO J
470 READ Y(N)
480 FOR M=1 TO Y(N)
490 READ W(N,M)
500 NEXT M
```

Figure 11.9 Computer program for pipe networks. (Adapted from Wood, 1981.)

Note in line 760 the initial head changes H_i are added algebraically and the correct sign is obtained by multiplying the head change by the flow rate and pipe number (with sign) and dividing by the absolute value of those quantities. An example data set is appended to the program following line 1190.

```
510 NEXT N
520 FOR N=J+1 TO P
530 READ E(N),Y(N)
540 FOR M=1 TO Y(N)
550 READ W(N,M)
560 NEXT M
570 NEXT N
580 PRINT#1, "TRIAL NO.    RELATIVE FLOW CHANGE        AVERAGE FLOW CHANGE"
590 FOR T5=1 TO 100
600 Q8=0
610 Q9=0
620 FOR N=J+1 TO P
630 B1=E(N)
640 E1=0
650 FOR M=1 TO Y(N)
660 I=ABS(W(N,M))
670 Q1=ABS(Q(I))
680 IF FR$="HW" GOTO 700
690 C(I)=CC(I)*1.325*(LOG(.27*R(I)/D(I)+5.74/(Q1/(.7854*NU*D(I)))^.9))^(-2)
700 G1=X*C(I)*Q1^(X-1)+2*O(I)*Q1+Z(I)/Q1^2
710 H1=C(I)*Q1^X+O(I)*Q1^2-Z(I)/Q1
720 IF CK(I)=0 GOTO 750
730      G1=G1-A1(I)-A2(I)*Q1
740      H1=H1-A0(I)-A1(I)*Q1-A2(I)*Q1*Q1
750 E1=E1+G1
760 B1=B1-H1*W(N,M)*Q(I)/(Q1*I)
770 NEXT M
780 Q2=B1/E1
790 FOR M=1 TO Y(N)
800 I=ABS(W(N,M))
810 Q(I)=Q(I)+Q2*W(N,M)/I
820 NEXT M
830 NEXT N
840 FOR N=1 TO P
850 Q8=Q8+ABS(ABS(Q(N))-ABS(V(N)))
860 Q9=Q9+ABS(Q(N))
870 V(N)=Q(N)
880 NEXT N
890 Q5=Q8/Q9
900 Q8=Q8*C1/P
910 PRINT#1, T5,:PRINT#1, USING "##.###^^^^                      ";Q5,Q8
920 IF Q5<EPS GOTO 940
930 NEXT T5
940 FOR N=1 TO P
950 Q1=ABS(Q(N))
960 IF FR$="HW" GOTO 980
970 C(N)=CC(N)*1.325*(LOG(.27*R(N)/D(N)+5.74/(Q1/(.7854*NU*D(N)))^.9))^(-2)
980 T(N)=Q(N)*(C(N)*Q1^(X-1)+O(N)*Q1)-Z(N)/Q(N)
990 IF CK(N)=1 THEN T(N)=T(N)-A0(N)-Q(N)*(A1(N)+A2(N)*Q1)
1000 NEXT N
```

Figure 11.9 *continued*

The data requirements are summarized below. An initial flow rate that satisfies continuity is required and the flow direction for these assumed flow rates are indicated. Pipe element numbers are input for paths of pipes for the energy equation and paths of pipes between junction nodes; these are entered with a positive sign if the initial flow direction is

```
1010 FOR N=1 TO J
1020 FOR M=1 TO Y(N)
1030 I=ABS(W(N,M))
1040 G2=G2-T(I)*I/W(N,M)
1050 NEXT M
1060 H(N)=G2
1070 NEXT N
1080 PRINT#1,
1090 PRINT#1,
1100 PRINT#1, "PIPE NO.        FLOW RATE "
1110 FOR N=1 TO P
1120 Q(N)=C1*ABS(Q(N))
1130 PRINT#1, N,:PRINT#1, USING "#####.###";Q(N)
1140 NEXT N
1150 PRINT#1,
1160 PRINT#1, "NODE NO.    HYD. GRADE LINE"
1170 FOR N=1 TO J
1180 PRINT#1, N,:PRINT#1, USING "#####.###";H(N)
1190 NEXT N
2000 REM ***** DATA SET EXAMPLE 11.9 (EN UNITS) *****
2010 DATA 7,2,0,EN,HW,1.E-6,1,449,12,.001
2020 DATA 1,200,4,110,0,500
2030 DATA 6,300,2,110,4,100
2040 DATA 3,200,4,110,0,300
2050 DATA 2,150,2,110,2,100
2060 DATA 4,100,2,110,11,100
2070 DATA 5,200,2,110,6,100
2080 DATA 7,80,4,110,3,100
2090 DATA 188.46,1,1,1,3
2100 DATA 102.3,2,1,2
2110 DATA 123.84,3,1,3,4
2120 DATA 99.23,3,1,3,7
2130 DATA 155.38,3,1,3,6
2140 DATA 88.46,2,1,5
```

Figure 11.9 *continued*

in the path direction and negative if it is opposite. Required data:

1. First line of data: number of pipes (P); number of junction nodes (J); number of pumps (NP); "EN" or "SI" (English or metric units); "DW" or "HW" (Darcy–Weisbach or Hazen–Williams friction formula); kinematic viscosity; specific gravity; flow conversion factor; conversion factor for diameter and roughness; convergence criterion.[3]

2. Next P lines (one for each pipe element): pipe element number; length; diameter; absolute roughness (if DW) or roughness coefficient (if HW); minor loss coefficient ΣK; initial flow rate estimate (must be greater than 0).

3. Next NP lines (skip if no pumps are present): pipe element number:
   (a) if useful power is used, enter "UP" and useful pump power (horsepower or kilowatts); or

[3]The convergence criterion is given by Eq. 11.4.29.

(b) if head-discharge data are used, enter "HQ" and shutoff head (head at zero discharge), intermediate head, intermediate discharge, final head, final discharge.

4. Next line: reference hydraulic grade line for grade calculations; number of pipe elements from reference grade line to junction node 1; pipe element numbers (with sign) in path from reference grade line to junction node 1 (repeat for each pipe); number of pipes between junction nodes 1 and 2; pipe element numbers (with sign) in path from junction nodes 1 to 2 (repeat for each pipe); repeat last two items for each junction node in pipe system.

5. Next P-J lines (one for every loop and pseudoloop): hydraulic grade line difference between starting and ending node (zero for loop); number of pipe elements in path, pipe element numbers (with sign) in path from starting to ending node.

In addition to the example problem that follows, two networks are analyzed in Examples 11.7 and 11.8. Example 11.7 employs a pump input with useful power; Example 11.8 makes use of a pump input with a head-discharge curve.

EXAMPLE 11.9

Determine the flow distribution in the branching or "tree-like" network (Wood, 1981). Water is the flowing liquid: $\gamma = 62.4$ lb/ft^3. The source of flow is a large pipeline (location A) maintained at a constant pressure of 60 psi. Each branch is at a location where the hydraulic grade line is known.

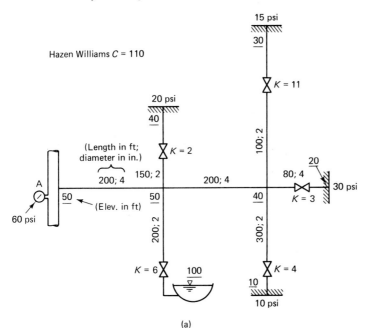

(a)

Solution

The numbering scheme is shown in the following figure. The terminal hydraulic grade lines are designated as H_A to H_F; an example calculation at location B is given as

$$H_B = \left(\frac{p}{\gamma} + z\right)_B$$

$$= \frac{20 \times 144}{62.4} + 40 = 86.15 \text{ ft}$$

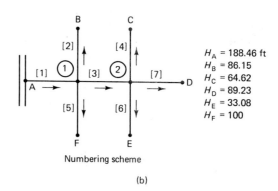

$H_A = 188.46$ ft
$H_B = 86.15$
$H_C = 64.62$
$H_D = 89.23$
$H_E = 33.08$
$H_F = 100$

Numbering scheme

(b)

The input data are listed in statements 2010 to 2140 in Fig. 11.9. The computed solution is tabulated below. The units for flow rate are in gallons per minute and the hydraulic grade line is given in feet.

TRIAL NO.	RELATIVE FLOW CHANGE	AVERAGE FLOW CHANGE
1	2.707E-01	6.372E+01
2	6.055E-02	1.445E+01
3	1.751E-02	4.181E+00
4	5.526E-03	1.318E+00
5	1.468E-03	3.500E-01
6	3.176E-04	7.569E-02

PIPE NO.	FLOW RATE
1	609.417
2	95.726
3	449.307
4	82.282
5	64.385
6	80.668
7	286.357

NODE NO.	HYD. GRADE LINE
1	130.438
2	97.443

(c)

PROBLEMS

11.1. Using the results from Example 11.2, tabulate and plot to correct vertical scale the hydraulic and energy gradelines. Assume a 10 m length between the two elbows.

11.2. Consider the simple pumping system shown below. Assume the following parameters to be known:

Reservoir elevations, z_1, z_2

Pipe length, L

Pipe roughness, e

Sum of minor loss coefficients, ΣK

Kinematic viscosity, ν

Discharge, Q

Develop a numerical solution to find the required diameter making use of a trial-and-error procedure. Use an equivalent length to represent the minor losses. Determine the diameter for the following data:

(a) $z_2 - z_1 = 120$ ft, $L = 1500$ ft, $e = 0.003$ ft, $\Sigma K = 2.5$, water at 50°F, $Q = 15,000$ gal/min, pump characteristic curve:

H_P (ft)	490	486	473	453	423	406
Q (gal/min)	0	4000	8000	12,000	16,000	18,000

(b) $z_2 - z_1 = 40$ m, $L = 500$ m, $e = 1$ mm, $\Sigma K = 2.5$, water at 20°C, $Q = 1.1$ m³/s, pump characteristic curve:

H_P (m)	160	158	154	148	138
Q (L/s)	0	400	800	1200	1600

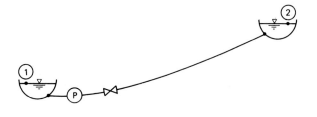

11.3. Gasoline is being pumped at 400 L/s in a pipeline from location A to B. The pipe follows the topography as shown, with the highest elevation shown at location C. The only contributions to minor losses are the two valves located at the ends of the pipe. If $S = 0.81$, $\nu = 4.26 \times 10^{-7}$ m²/s, $p_v = 55.2$ kPa absolute, and $p_{atm} = 100$ kPa,

(a) Determine the necessary power to be delivered to the system to meet the flow requirement.

(b) What is the maximum elevation possible at location C without causing vapor pressure conditions to exist?

(c) Sketch the hydraulic grade line.

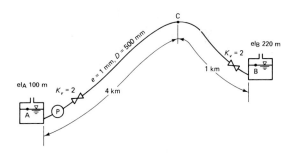

11.4. For the three pipes in series, minor losses are proportional to the discharge squared, and the Hazen–Williams formula is used to account for friction losses. With the data given, use Newton's method to determine the discharge. Note that minor losses can be neglected in order to initially estimate Q. The piezometric heads at A and B are $(p/\gamma + z)_A = 250$ m and $(p/\gamma + z)_B = 107$ m.

Pipe	L (m)	D (mm)	ΣK	C (Hazen–Williams)
1	200	200	2	100
2	150	250	3	120
3	300	300	0	90

11.4E. Work Problem 11.4 with the following data. $(p/\gamma + z)_A = 820$ ft and $(p/\gamma + z)_B = 351$ ft.

Pipe	L (ft)	D (in.)	ΣK	C (Hazen–Williams)
1	600	8	2	100
2	300	10	3	120
3	900	12	0	90

11.5. A long oil pipeline consists of the three segments shown below. Each segment has a booster pump that is used primarily to overcome pipe friction. The two reservoirs are at the same elevation.

(a) Derive a single equation to determine the discharge in the system if the resistance factors $\bar{R}_i$ for each pipe and the useful power $\dot{W}_{fi}$ for each pump are known.

(b) Determine the discharge for the data given in the figure. The unit weight of oil is $\gamma = 8830$ N/m³.

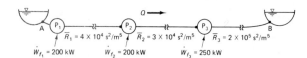

11.6. A liquid ($\rho = 680$ kg/m³) is delivered from a storage tank to a discharge station through a pipe 450 m long and 300 mm in diameter. The pump provides 10 kW of useful power. Determine the discharge in the piping system. Assume a constant friction factor of $f = 0.015$ and include minor losses.

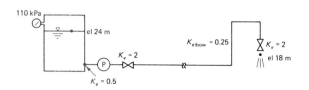

11.7. Water at 20°C is being pumped through the three pipes in series as shown. The power delivered to the pump is 1920 kW, and the pump efficiency is 0.82. Compute the discharge.

Pipe	L (m)	D (mm)	e (mm)	ΣK
1	200	1500	1	2
2	300	1000	1	0
3	120	1200	1	10

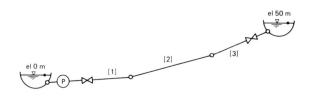

11.8. Find the water flow distribution in the parallel system and the required pumping power if the discharge through the pump is $Q_1 = 3$ m³/s. The pump efficiency is 0.75. Assume constant friction factors.

Pipe	L (m)	D (mm)	f	ΣK
1	100	1200	0.015	2
2	1000	1000	0.020	3
3	1500	500	0.018	2
4	800	750	0.021	4

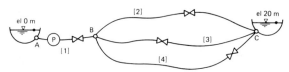

11.9. Determine the water flow distribution and the piezometric head at the junction using an ad hoc approach. Assume constant friction factors. The pump characteristic curve is $H_P = a - bQ^2$.

(a) $a = 20$ m, $b = 30$ s^2/m^5, $z_1 = 10$ m, $z_2 = 20$ m, $z_3 = 18$ m.

Pipe	L (m)	D (cm)	f	K
1	30	24	0.020	2
2	60	20	0.015	0
3	90	16	0.025	0

(b) $a = 55$ ft, $b = 0.1$ sec^2/ft^5, $z_1 = 20$ ft, $z_2 = 50$ ft, $z_3 = 45$ ft.

Pipe	L (ft)	D (in.)	f	K
1	100	10	0.020	2
2	200	8	0.015	0
3	300	6	0.025	0

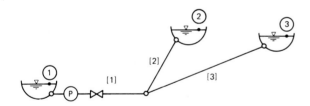

11.10. Work Problem 11.9 using a numerical approach:
(a) Newton's method.
(b) False position method.

11.11. Find the flow distribution in the parallel network. Assume constant friction factors. The change in hydraulic grade line between A and B is $(p/\gamma + z)_A - (p/\gamma + z)_B = 50$ m.

Pipe	L (m)	D (mm)	e (mm)	Σ K
1	600	1000	0.1	2
2	1000	1200	0.15	0
3	550	850	0.2	4
4	800	1000	0.1	1

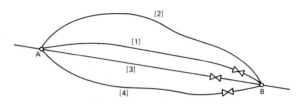

11.12. The water sprinkling system shown below is applied from a large diameter pipe with constant internal pressure $p_0 = 300$ kPa. The system is positioned in a horizontal plane. Determine the flow distribution Q_1, Q_2, Q_3, Q_4 for the given data. Valve losses are included in the $\bar{R}$ values. (*Hint:* If you are clever, no trial-and-error solution is necessary!)

Pipe	$\bar{R}$ (s^2/m^5)
1	1.6×10^4
2	5.3×10^5
3	1.0×10^6
4	1.8×10^6

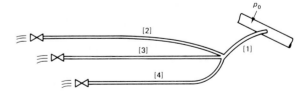

11.13. A proposed water irrigation system consists of one main pipe with a pump and three pipe branches. Each branch is terminated by an orifice, and each orifice has the same elevation. It is apparent that the flow distribution can be solved by treating the piping arrangement as a branched system. However, the piping can also be treated as a parallel system in order to determine the flows.

(a) Identify the equations and unknowns to satisfy the solution for a parallel system. Why is it possible to treat the irrigation system as a parallel piping problem?

(b) Why would the solution to a parallel system be preferred to that of a branching system?

(c) Determine the flow distribution and sketch the hydraulic grade line.

(d) What part of the piping would one change to approximately double the discharge, assuming that the individual lengths and pump curve could not be altered?

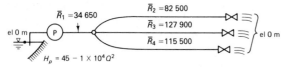

$\bar{R}_1 = 34\ 650$

$\bar{R}_2 = 82\ 500$

$\bar{R}_3 = 127\ 900$

$\bar{R}_4 = 115\ 500$

el 0 m

P

el 0 m

$H_p = 45 - 1 \times 10^4 Q^2$

(H_p in m; Q in m³/s; $\bar{R}$ in s²/m⁵, orifice loss included in $\bar{R}$)

11.14. Determine the required head (H_P) and discharge (Q_1) of water to be handled by the pump for the system shown. The discharge in pipe 2 is $Q_2 = 35$ L/s in the direction shown.

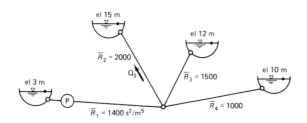

el 15 m

el 12 m

$\bar{R}_2 = 2000$

Q_2

$\bar{R}_3 = 1500$

el 10 m

el 3 m

P

$\bar{R}_1 = 1400$ s²/m⁵

$\bar{R}_4 = 1000$

11.15. Determine the flow distribution of water in the parallel piping system.

Pipe	L (m)	D (mm)	f	Σ K
1	30	50	0.020	3
2	40	75	0.025	5
3	60	60	0.022	1

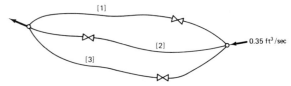

[1]

[2]

[3]

600 L/min

11.15E. Determine the flow distribution of water in the parallel piping system.

Pipe	L (ft)	D (in.)	f	Σ K
1	90	2	0.020	3
2	120	3	0.025	5
3	180	2.5	0.022	1

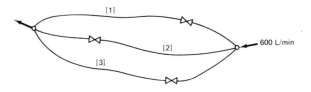

[1]

[2]

[3]

0.35 ft³/sec

11.16. Water is being pumped in the piping system. The pump curve is approximated by the relation $H_P = 150 - 5Q_1^2$, with H_P in meters and Q_1 in m³/s. The pump efficiency is $\eta = 0.75$. Compute the flow distribution and find the required pump power.

Pipe	$\bar{R}$ (s²/m⁵)
1	400
2	1000
3	1500

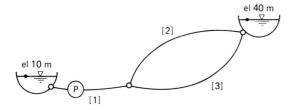

11.17. A pipeline consists of two pipe segments in series. The specific gravity of the fluid is 0.81. If pump A has a constant power input of 1 MW, find the discharge, the pressure head in pumps A and B, and the required power for pump B. The minimum allowable pressure on the suction side of pump B is 150 kPa, and both pumps have an efficiency of 0.76.

Pipe	L (m)	D (mm)	Σ K	f
1	5000	750	2	0.023
2	7500	750	10	0.023

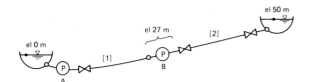

11.18. Determine the flow distribution of water in the system shown. The equivalent roughness for all elements is 0.1 mm.

Pipe	L (m)	D (mm)	Σ K
1	1000	200	3
2	200	25	0
3	250	25	0
4	340	30	2
5	420	40	0
6	500	175	5

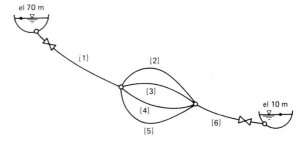

11.19. Determine the flow distribution of water in the branching pipe system:
(a) Without a pump in line 1.
(b) Include a pump located in line 1 adjacent to the lower reservoir. The pump characteristic curve is given by $H_P = 250 - 0.4Q - 0.1Q^2$, with H_P in meters and Q in m³/s.
The Hazen–Williams coefficient $C = 130$ for all pipes.

Pipe	L (m)	D (mm)
1	200	500
2	600	300
3	1500	300
4	1500	400

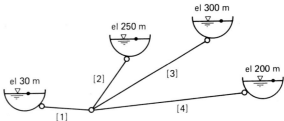

11.20. Determine the flow distribution of water in the system. Assume constant friction factors, with $f = 0.02$. The head-discharge relation for the pump is $H_P = 60 - 10Q^2$, where H_P is in meters and the discharge is in cubic meters per second.

Pipe	L (m)	D (mm)	ΣK
1	100	350	2
2	750	200	0
3	850	200	0
4	500	200	2
5	350	250	2

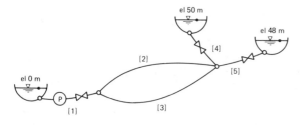

11.21. For the network shown, perform the following:
(a) Identify all of the unknown parameters and write the required system equations. Subsequently, reduce them to one equation in one unknown. Assume $x = 2$.
(b) Prepare the network for a Hardy Cross analysis by making a sketch, identifying necessary loops and paths, and writing the appropriate equations.

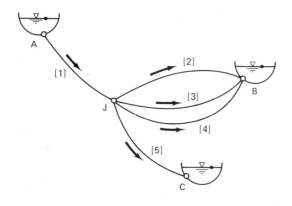

11.22. The branching system is to be solved by using the Hardy Cross method.
(a) Write the equations for ΔQ_I, ΔQ_II, W, and G.
(b) Perform two iterations by filling in the blanks in the accompanying table. After two iterations, sketch the hydraulic grade line.

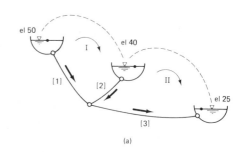

(a)

Loop	Pipe	$\overline{R}$	$2\overline{R}$	Q	W	G	Q	W	G	Q
I	1	2	4	-3.5						
	2	2	4	0						

$$\Delta Q_\mathrm{I} = \qquad\qquad \Delta Q_\mathrm{I} =$$

II	2	2	4	0						
	3	2	4	-3.5						

$$\Delta Q_\mathrm{II} = \qquad\qquad \Delta Q_\mathrm{II} =$$

(b)

11.23. The solution of flow of water in a pipe network is shown. Compute:
(a) The hydraulic grade line throughout the system.
(b) The pressure at each node.

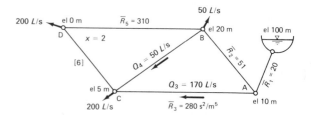

11.24E. For the configuration shown in Problem 11.21, determine the water flow distribution and the hydraulic grade line at J. The system data are tabulated below.

Reservoir	Elevation (ft)	Line	L (ft)	D (in.)	f	ΣK
A	650	1	800	8	0.015	0
B	575	2	600	3	0.020	2
C	180	3	650	3	0.020	2
		4	425	3	0.025	3
		5	1000	4	0.015	3

11.25. Water is flowing in the piping system. Determine the flow distribution using the Hardy Cross method.

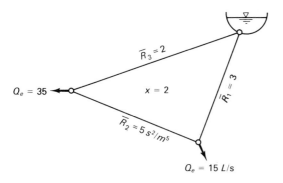

11.26. Determine the water flow distribution in the piping system and the pressures at the internal nodes employing the Hardy Cross method of solution.

Pipe	L (m)	D (mm)	e (mm)	ΣK
1	500	300	0.15	0
2	600	250	0.15	0
3	50	150	0.15	10
4	200	250	0.15	2
5	200	300	0.15	2

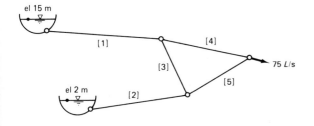

11.27. Discuss briefly the difference between the Hardy Cross, the linear, the single node, and the simultaneous node methods for the solution of flow in pipe networks.

11.28. Determine the discharge in the piping system **(a)** with an exact solution; **(b)** using the Hardy Cross method. Water is the flowing liquid and the losses are proportional to the velocity squared ($x = 2$). The pump characteristic curve is $H_P = 100 - 826Q^2$ (H_P in meters and Q in m^3/s).

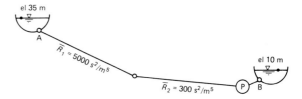

11.29. A piping system delivers water to two side-branch sprinklers (C and E) and to a lower reservoir (F). The water is supplied by an upper reservoir (A). The sprinklers are represented hydraulically as orifices with relatively high K values. Determine the flow distribution using the Hardy Cross method.

Line	L (m)	D (mm)	e (mm)	ΣK
1	200	100	0.1	2
2	150	50	0.1	30
3	500	100	0.1	0
4	35	50	0.1	35
5	120	100	0.1	2

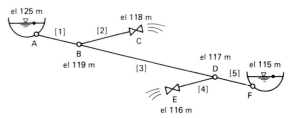

11.30. In Problem 11.29, if a pump supplying 10 kW of useful power was inserted near A, what changes in flow and pressure would occur?

11.31E. The 12 pipe system represents a low-pressure region that is connected to a high-pressure system by pressure regulators set at 50 psi. Determine the pressure and water flow distribution for the demands shown. Assume Hazen-Williams $C = 120$ for all pipes. (Courtesy of D. Wood.)

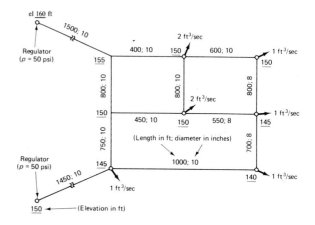

11.32. Determine the flow distribution for the 14 pipe water supply system. The characteristic curve for the pump is represented by the following data (courtesy of D. Wood):

H_P (m)	166	132	18
Q (L/s)	0	600	1000

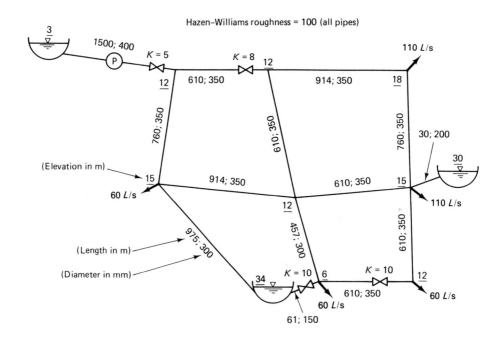

11.33E. The layout of the required water demands for a proposed industrial complex are shown. Design an appropriate network and determine the appropriate useful power for a pump to meet the demand needs. The design criteria are as follows:

(a) The lower reservoir supplies water to the system. The upper reservoir supplies water for emergency use only, such as fire, line breaks, and so on; under normal operation, there is no flow into or out of it.

(b) Under normal operation, pressures in the network can range between 80 and 120 psi.

(c) Two gate valves are to be placed in each line (one at each end) to isolate it in case of a break or for needed maintenance.

(d) All demands must be met in case of a single line break with minimum allowable pressures of 20 psi.

(e) Pipes are cast iron, with the following standard diameters available: 8, 12, 16, 20, 24, 30, 36, 42, and 48 in. Pipes can be placed anywhere within the region.

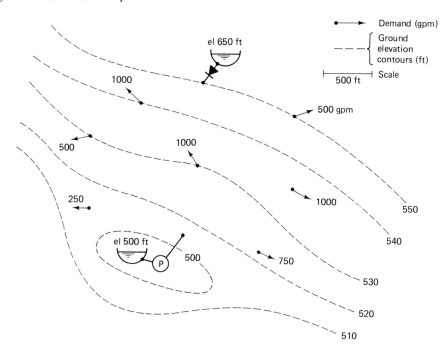

11.34E. Design an irrigation system for part or all of the 18-hole golf course shown. The requirements are as follows:

(a) Place the sprinkler couplings at approximately 100-ft intervals along the fairway centerline. The flow rate from each sprinkler is to be 40 gal/min. A maximum of two sprinklers on each fairway are to be in operation at any time.

(b) Place four sprinklers around the periphery of each green. The flow rate from each sprinkler is to be 40

gal/min. A maximum of three sprinklers on each green are to be in operation at any time.

(c) Water is to be pumped from the well shown. The maximum possible flow from the well is 2000 gal/min. Work with useful power in order to satisfy the system demand.

(d) System pressure is to be 80 ± 10 psi at all locations, and the design velocity in each pipe is not to exceed 5 ft/sec. Use smooth plastic piping and fittings throughout.

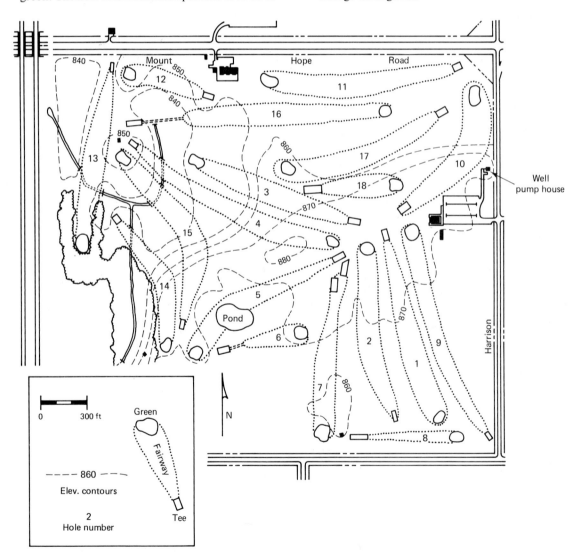

TWELVE

Turbomachinery

12.1 INTRODUCTION

Turbomachines are the commonly employed devices that either supply or extract energy from a flowing fluid by means of rotating propellers or vanes. A turbopump, more commonly called a pump, adds energy to a system, with the result that the pressure is increased; it also causes flow to occur or it increases the rate of flow. A turbine extracts energy from a system and converts it to some other useful form, typically, to electric power. Pumps are essential components of piping systems which are designed to convey liquids. Similarly, turbomachines are called blowers, fans, or compressors when performing work on air or other gases in ducts. A hydroturbine, or simply turbine, is a machine that generates power from high-pressure water; relatively large conduits or tunnels deliver fluid to closed turbines in order to generate power. Steam and air turbines are also of substantial engineering importance, but such devices are considered in other courses, such as thermodynamics.

The examples above are all those of turbomachines designed to facilitate or utilize internal flows. A wind turbine, on the other hand, makes use of the surrounding external flow to convert the energy contained in the natural movement of atmospheric air into useful electrical power. A propeller performs work on the surrounding fluid to provide thrust and propel an object along a desired path, while a stationary fan performs work to circulate air. All turbomachines are characterized as being capable of either adding or subtracting energy from the fluid by means of rotating propellers or vanes. There is no contained volume of fluid being transported, as in a positive-displacement piston pump.

The bulk of this chapter is devoted to pumps, in particular, those that convey liquids such as water or gasoline. The centrifugal, or radial-flow pump is presented in detail, and to a lesser extent, mixed-flow and axial-flow pumps are studied. Dimensional analysis, an important tool for the selection and design of turbomachines, is presented next, followed by a discussion of the proper selection and implementation of pumps in piping systems. Finally, turbines are introduced by considering their fundamental characteristics.

12.2 TURBOPUMPS

A turbopump consists of two principal parts: an **impeller,** which imparts a rotary motion to the liquid, and the pump **housing,** or **casing,** which directs the liquid into the impeller region and transports it away under a higher pressure. Figure 12.1 shows a typical single-suction **radial-flow pump.** The impeller is mounted on a shaft and is often driven by an electric motor. The casing includes the suction and discharge nozzles and houses the impeller assembly. The portion of the casing surrounding the impeller is termed the **volute.** Liquid enters through the suction nozzle to the impeller eye and travels along the **shroud,** developing a rotary motion

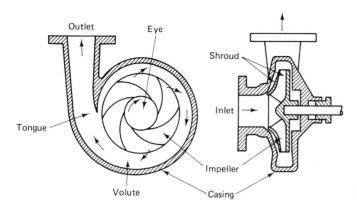

Figure 12.1   Single-suction pump.

due to the impeller vanes. It leaves the volute casing peripherally at a higher pressure through the discharge nozzle. Some single-suction impellers are open, with the front shroud removed. Double-suction impellers have liquid entering from both sides.

In a radial-flow pump, the impeller vanes are commonly curved backward and the impeller is relatively narrow. As the impeller becomes wider, the vanes have a double curvature, becoming twisted at the suction end. Such pumps convey liquids with a lower-pressure rise than radial-flow pumps and are called **mixed-flow pumps.** At the opposite extreme from the radial-flow pump is the **axial-flow pump;** it is characterized by the flow entering and leaving the impeller region axially, parallel to the shaft axis. Typically, an axial-flow pump delivers liquid with a relatively low pressure rise. For axial-flow pumps and some mixed-flow pumps, the impellers are open; that is, there is no shroud surrounding them. Various types of impellers are shown in Fig. 12.2.

12.2.1 Radial-Flow Pumps

A representative radial-flow pump is shown in Fig. 12.1; this type of pump is commonly called a **centrifugal pump,** and it is the most commonly used pump in existence today. An elementary analysis will be performed on a radial-flow pump, which will provide a theoretical relationship between the discharge and the developed head rise. In addition, a better understanding of the manner in which momentum exchange takes place in such

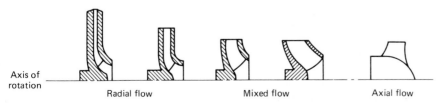

Figure 12.2   Various types of pump impellers.

a turbomachine will be provided. Actual pumps operate at efficiencies less than unity; that is, they do not perform at the theoretical, idealized conditions. It is necessary, therefore, to conduct experiments so that the true operating characteristics of a turbopump can be determined.

Elementary Theory

The actual flow patterns in a turbopump are highly three-dimensional with significant viscous effects and separation patterns taking place. To construct a simplified theory for the radial-flow pump, it is necessary to neglect viscosity and to assume idealized two-dimensional flow throughout the impeller region. Figure 12.3a defines a control volume that encompasses the impeller region. Flow enters through the inlet control surface and exits through the outlet surface. Note that a series of vanes exists within the control volume, and that they are rotating about the axis with an angular speed ω.

A portion of the control volume is shown at an instant in time in Fig. 12.3b. The idealized velocity vectors are diagrammed at the inlet, location 1, and the outlet, location 2. In the velocity diagrams, V is the absolute

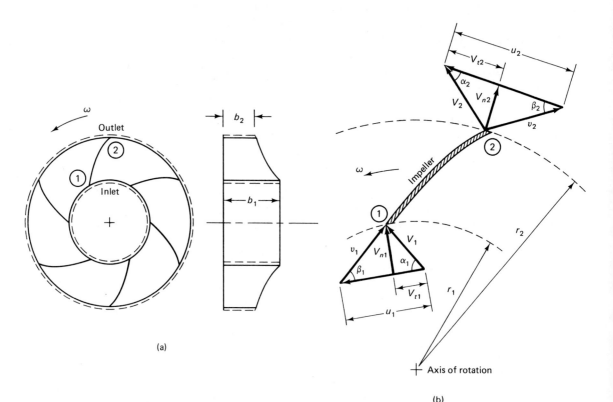

Figure 12.3 Idealized radial-flow impeller: (a) impeller control volume; (b) velocity diagrams at control surfaces.

fluid velocity, V_t is the tangential component of V, and V_n is the radial, or normal, component of V. The peripheral or circumferential speed of the blade is $u = \omega r$, where r is the radius of the control surface. The angle between V and u is α. The fluid velocity measured relative to the vane is v.

The relative velocity is assumed to be always tangent to the vane; that is, perfect guidance of the fluid throughout the control volume takes place. The angle between v and u is designated as β; since perfect guidance along the vane is assumed, β designates the blade angle as well.

The moment-of-momentum relation, Eq. 4.6.3, can be written for steady flow in the form

$$\Sigma \mathbf{M} = \int_{c.s.} \rho \mathbf{r} \times \mathbf{V}(\mathbf{V} \cdot \hat{n})\, dA \qquad (12.2.1)$$

Applied to the control volume of Fig. 12.3, this becomes

$$T = \rho Q(r_2 V_{t2} - r_1 V_{t1}) \qquad (12.2.2)$$

in which T is the torque acting on the fluid in the control volume, and the right-hand side represents the flux of angular momentum through the control volume.

The power delivered to the fluid is the product of ω and T:

$$\omega T = \rho Q(u_2 V_{t2} - u_1 V_{t1}) \qquad (12.2.3)$$

From the velocity vector diagrams in Fig. 12.3b, $V_t = V \cos \alpha$, so that Eq. 12.2.3 can be written as

$$\omega T = \rho Q(u_2 V_2 \cos \alpha_2 - u_1 V_1 \cos \alpha_1) \qquad (12.2.4)$$

For the idealized situation in which there are no losses, the delivered power must be equal to $\gamma Q H_t$, in which H_t is the theoretical pressure head rise across the pump. Then there results Euler's turbomachine relation,

$$\begin{aligned} H_t &= \frac{\omega T}{\gamma Q} \\ &= \frac{u_2 V_2 \cos \alpha_2 - u_1 V_1 \cos \alpha_1}{g} \end{aligned} \qquad (12.2.5)$$

Insight on the nature of flow through an impeller region can be obtained using Eq. 12.2.5. From the law of cosines we can write

$$v_1^2 = u_1^2 + V_1^2 - 2u_1 V_1 \cos \alpha_1$$
$$v_2^2 = u_2^2 + V_2^2 - 2u_2 V_2 \cos \alpha_2$$

These can be substituted into Eq. 12.2.5 to provide the relation

$$H_t = \frac{V_2^2 - V_1^2}{2g} + \frac{(u_2^2 - u_1^2) - (v_2^2 - v_1^2)}{2g} \qquad (12.2.6)$$

The first term on the right-hand side represents the gain in kinetic energy as the fluid passes through the impeller; the second term accounts for the increase in pressure across the impeller. This can be seen by applying the energy equation across the impeller and solving for H_t:

$$H_t = \frac{p_2 - p_1}{\gamma} + z_2 - z_1 + \frac{V_2^2 - V_1^2}{2g} \qquad (12.2.7)$$

Eliminating H_t between Eqs. 12.2.6 and 12.2.7 yields the expression

$$\frac{p_1}{\gamma} + z_1 + \frac{v_1^2 - u_1^2}{2g} = \frac{p_2}{\gamma} + z_2 + \frac{v_2^2 - u_2^2}{2g} = \text{const.} \qquad (12.2.8)$$

This relation has historically been called the Bernoulli equation in rotating coordinates. Since $z_2 - z_1$ is often much smaller than $(p_2 - p_1)/\gamma$, it can be eliminated, and thus the pressure difference is

$$p_2 - p_1 = \frac{\rho}{2}\left[(v_1^2 - u_1^2) - (v_2^2 - u_2^2)\right] \qquad (12.2.9)$$

Returning to Eq. 12.2.5, we see that a "best design" for a pump would be one in which the angular momentum entering the impeller is zero, so that maximum pressure rise can take place. Then in Fig. 12.3b, $\alpha_1 = 90°$, $V_{n1} = V_1$, and Eq. 12.2.5 becomes

$$H_t = \frac{u_2 V_2 \cos \alpha_2}{g} \qquad (12.2.10)$$

From the triangle geometry of Fig. 12.3, $V_2 \cos \alpha_2 = u_2 - V_{n2} \cot \beta_2$, so that Eq. 12.2.10 takes the form

$$H_t = \frac{u_2^2}{g} - \frac{u_2 V_{n2} \cot \beta_2}{g} \qquad (12.2.11)$$

Applying the continuity principle at the outlet region to the control volume provides the relation

$$V_{n2} = \frac{Q}{2\pi r_2 b_2} \qquad (12.2.12)$$

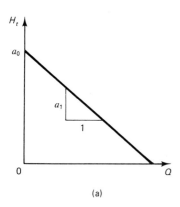

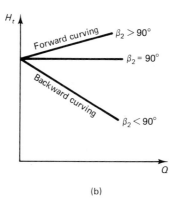

(a)   (b)

Figure 12.4 Ideal pump performance curves.

in which b_2 is the width of the impeller at location 2. Introducing Eq. 12.2.12 into Eq. 12.2.11, and recalling that $u_2 = \omega r_2$, one has the relation

$$H_t = \frac{\omega^2 r_2^2}{g} - \frac{\omega \cot \beta_2}{2\pi b_2 g} Q \qquad (12.2.13)$$

For a pump running at constant speed, Eq. 12.2.13 takes the form

$$H_t = a_0 - a_1 Q \qquad (12.2.14)$$

in which a_0 and a_1 are constants. Equation 12.2.13 is the **theoretical head curve** and is seen to be a straight line with a slope of $-a_1$, as shown in Fig. 12.4a. The effect of the blade angle β_2 is shown in Fig. 12.4b. A forward curving blade ($\beta_2 > 90°$) can be unstable and cause pump surge, where the pump oscillates in an attempt to establish an operating point. Backward-curving vanes ($\beta_2 < 90°$) are generally preferred.

EXAMPLE 12.1

A radial-flow pump has the following dimensions:

$$\beta_1 = 44° \qquad r_1 = 21 \text{ mm} \qquad b_1 = 11 \text{ mm}$$
$$\beta_2 = 30° \qquad r_2 = 66 \text{ mm} \qquad b_2 = 5 \text{ mm}$$

For a rotational speed of 2500 rev/min, assuming ideal conditions (frictionless flow, negligible vane thickness, perfect guidance), with $\alpha_1 = 90°$ (no prerotation), determine (a) the discharge, theoretical head, required power, and pressure rise across the impeller, and (b) the theoretical head-discharge curve. Use water as the fluid.

Solution

(a) Construct the velocity diagram at location 1 as shown.

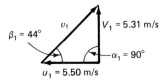

The rotational speed is converted to the appropriate units as

$$\omega = 2500 \, \frac{2\pi}{60} = 261.8 \text{ rad/s}$$

The impeller speed at r_1 is then

$$u_1 = \omega r_1$$
$$= 261.8 \times 0.021 = 5.50 \text{ m/s}$$

From the velocity diagram we see that

$$V_1 = u_1 \tan \beta_1$$
$$= 5.50 \tan 44° = 5.31 \text{ m/s}$$

and since $\alpha_1 = 90°$, $V_1 = V_{n1}$, or $V_{n1} = 5.31$ m/s. The discharge is computed to be

$$Q = 2\pi r_1 b_1 V_{n1}$$
$$= 2\pi \times 0.021 \times 0.011 \times 5.31 = 7.71 \times 10^{-3} \text{ m}^3/\text{s}$$

The normal component of velocity at location 2 is

$$V_{n2} = \frac{Q}{2\pi r_2 b_2}$$
$$= \frac{7.71 \times 10^{-3}}{2\pi \times 0.066 \times 0.005} = 3.72 \text{ m/s}$$

and the impeller speed at the outlet is

$$u_2 = \omega r_2$$
$$= 261.8 \times 0.066 = 17.28 \text{ m/s}$$

The velocity diagram at location 2 is now sketched as shown.

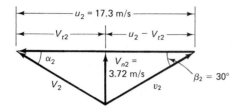

From the velocity diagram we see that

$$u_2 - V_{t2} = \frac{V_{n2}}{\tan \beta_2}$$

$$= \frac{3.72}{\tan 30°} = 6.44 \text{ m/s}$$

Therefore,

$$V_{t2} = u_2 - 6.44$$

$$= 17.28 - 6.44 = 10.84 \text{ m/s}$$

$$\alpha_2 = \tan^{-1} \frac{V_{n2}}{V_{t2}}$$

$$= \tan^{-1} \frac{3.72}{10.84} = 18.9°$$

$$V_2 = \frac{V_{t2}}{\cos \alpha_2}$$

$$= \frac{10.84}{\cos 18.9°} = 11.46 \text{ m/s}$$

The theoretical head is computed with Eq. 12.2.10:

$$H_t = \frac{u_2 V_2 \cos \alpha_2}{g}$$

$$= \frac{17.28 \times 11.46 \times \cos 18.9°}{9.81} = 19.1 \text{ m}$$

Hence the theoretical required power is

$$\dot{W}_P = \gamma Q H_t$$

$$= 9810 \times (7.71 \times 10^{-3}) \times 19.1 = 1440 \text{ W}$$

The pressure rise is determined from the energy equation as follows:

$$p_2 - p_1 = \left(H_t + \frac{V_1^2 - V_2^2}{2g} \right) \gamma$$

$$= \left[19.1 + \frac{(5.31)^2 - (11.46)^2}{2 \times 9.81} \right] \times 9810 = 1.36 \times 10^5 \text{ Pa}$$

(b) The theoretical head-discharge curve is Eq. 12.2.13. For the present example we have

$$H_t = \frac{(\omega r_2)^2}{g} - \frac{\omega \cot \beta_2}{2\pi b_2 g} Q$$

$$= \frac{(261.8 \times 0.066)^2}{9.81} - \frac{261.8 \cot 30°}{2\pi \times 0.005 \times 9.81} Q$$

$$= 30.4 - 1471Q$$

The curve is sketched as follows.

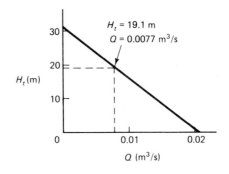

H_t (m)

$H_t = 19.1$ m
$Q = 0.0077$ m³/s

Q (m³/s)

Head-Discharge Relations: Performance Curves

For real fluid flow, the theoretical head curve cannot be achieved in practice, and it is necessary to resort to experimentation to determine the actual head-discharge curve. The energy equation written across a pump from the suction side (location 1, Fig. 12.3) to the discharge side (location 2) is

$$H_P = \left(\frac{p}{\gamma} + \frac{V^2}{2g} + z\right)_2 - \left(\frac{p}{\gamma} + \frac{V^2}{2g} + z\right)_1$$

$$= H_t - h_L \tag{12.2.15}$$

in which H_P is the actual head across the pump, and h_L represents the losses through the pump. The actual power delivered to the fluid, designated as $\dot{W}_f$, is

$$\dot{W}_f = \gamma Q H_P \tag{12.2.16}$$

while the power delivered to the impeller is $\dot{W}_P$, often termed the **brake power,** and is given by

$$\dot{W}_P = \omega T \tag{12.2.17}$$

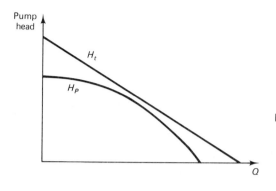

Figure 12.5 Comparison between theoretical and actual radial-flow pump performance curves.

If there were no losses, $\dot{W}_f$ would be equal to $\dot{W}_P$. Since in actuality $\dot{W}_f < \dot{W}_P$, the **pump efficiency**[1] η_P is defined as

$$\eta_P = \frac{\dot{W}_f}{\dot{W}_P} = \frac{\gamma Q H_P}{\omega T} \qquad (12.2.18)$$

One objective of pump design is to make the efficiency as high as possible. The theoretical head curve is compared with the actual head curve in Fig. 12.5. The difference between the two can be attributed to the following effects: (1) prerotation of the fluid before entering the impeller region, (2) separation due to imperfect guidance of fluid as it enters the impeller region, and (3) separation due to expansion in the flow passages. Leakage and high shear rates created by the differences in velocity between fluid particles in the volute and impeller contribute further to the losses. Finally, impeller blades are designed to be most efficient at the so-called design discharge; at any other discharge—that is, ''off-design''—the performance will deteriorate.

A representative pump and its performance curve are shown in Fig. 12.6; this is the curve supplied by a pump manufacturer. Typically, more than one impeller diameter will be available; in the diagram, four different impellers are available. In addition to the head-discharge performance

---

[1]More precisely, η_P defined by Eq. 12.2.18 is termed the **overall efficiency.** It is possible to define three additional efficiencies for pumps as follows:

Hydraulic efficiency: $\eta_H = \dfrac{H_P}{H_t}$

Volumetric efficiency: $\eta_V = \dfrac{Q}{Q + Q_L}$, Q_L = leakage flow rate

Mechanical efficiency: $\eta_M = \dfrac{\gamma H_t(Q + Q_L)}{T\omega}$

These are related to the overall efficiency by $\eta_P = \eta_H \eta_V \eta_M$.

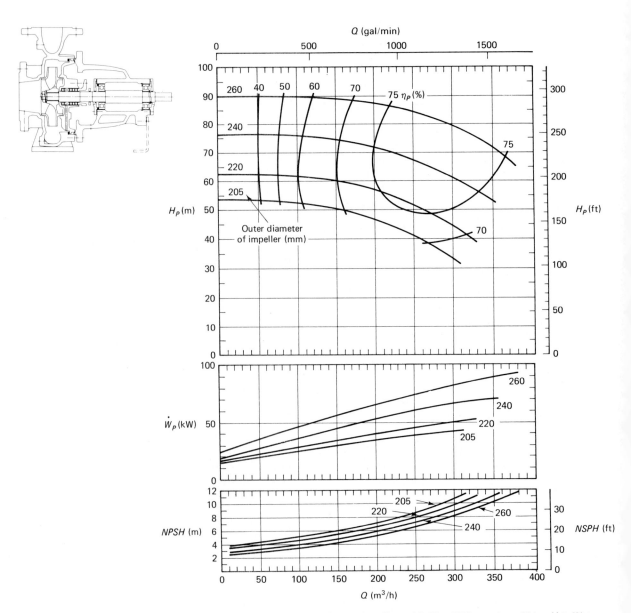

Figure 12.6 Radial-flow pump and performance curves for four different impellers with $N = 2900$ rpm ($\omega = 304$ rad/s). Water at 20°C is the pumped liquid. (Courtesy of Sulzer Bros. Ltd.)

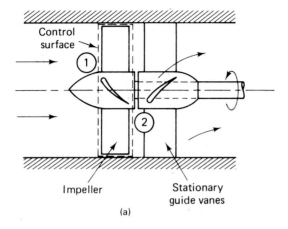

(a)

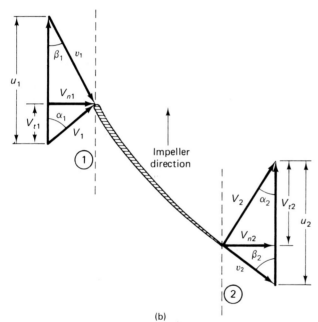

(b)

Figure 12.7 Idealized axial-flow impeller: (a) impeller control volume; (b) velocity diagrams at control surface.

curves, isoefficiency, power, and net positive suction head (*NPSH*) curves are usually provided. Net positive suction head is associated with concern for cavitation; it is discussed in Section 12.2.3.

12.2.2 Axial- and Mixed-Flow Pumps

The velocity diagram for an axial-flow pump is shown in Fig. 12.7. In the axial-flow pump, there is no radial flow and the liquid particles leave the

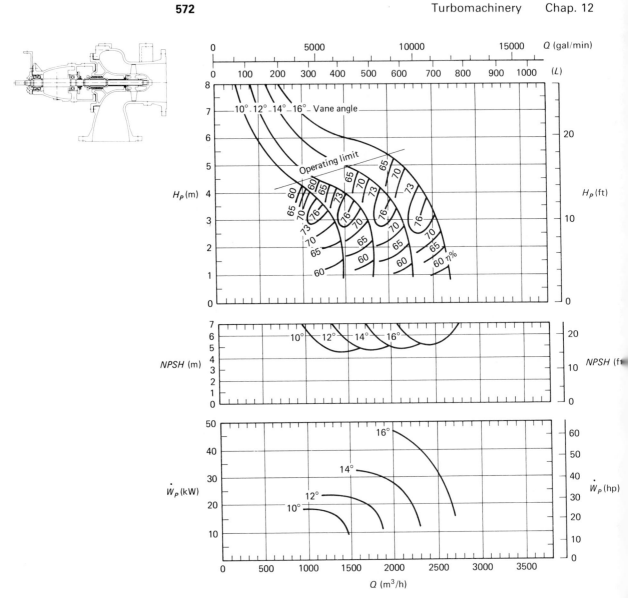

Figure 12.8 Axial-flow pump and performance curves for four different vane angles with N = 880 rpm (ω = 92.2 rad/s). Propeller diameter is 500 mm. Water at 20°C is the pumped liquid. (Courtesy of Sulzer Bros. Ltd.)

impeller at the same radius at which they enter, so that $u_1 = u_2 = u$. Furthermore, assuming a uniform flow, continuity considerations give $V_{n1} = V_{n2} = V_n$. Equation 12.2.5, which is valid for an axial-flow pump as well as a radial-flow pump, can be combined with the identities $V_2 \cos \alpha_2 = u - V_n \cot \beta_2$ and $V_1 \cos \alpha_1 = V_n \cot \alpha_1$ to produce

$$H_t = \frac{u^2}{g} - \frac{uV_n}{g} (\cot \alpha_1 + \cot \beta_2) \qquad (12.2.19)$$

This form of the turbomachine relation is useful when the ideal absolute velocity entrance angle α_1 is established by a fixed vane, or stator. If there is no prerotation, $\alpha_1 = 90°$ and the theoretical head relation, Eq. 12.2.19, becomes

$$H_t = \frac{u^2}{g} - \frac{uV_n \cot \beta_2}{g} \qquad (12.2.20)$$

This relation is identical to Eq. 12.2.11, which relates the theoretical head to the impeller outlet parameters for a radial flow pump. Note, however, that Eq. 12.2.11 is valid at the periphery of the impeller for the radial-flow pump, and that all fluid particles reach the maximum head at that location. By contrast, Eq. 12.2.20 pertains only at a specified radius for the axial-flow pump, since fluid particles enter and leave the control volume at their respective radii. In this case the head varies from a minimum at the axis to a maximum at the periphery, and the total pump head is an integrated average. Axial-flow impellers are designed to maintain a constant axial velocity throughout the impeller region; this requires that the vane angles increase gradually from the periphery to the axis, as well as from the inlet to the outlet region. Figure 12.8 shows a representative axial-flow pump, and its performance curves.

EXAMPLE 12.2

An axial-flow pump is designed with a fixed guide vane, or stator blade, located upstream of the impeller. The stator imparts an angle $\alpha_1 = 75°$ to the fluid as it enters the impeller region. The impeller has a rotational speed of 500 rpm with a blade exit angle of $\beta_2 = 70°$. The control volume has an outer diameter of $D_o = 300$ mm and an inner diameter of $D_i = 150$ mm. Determine the theoretical head rise and power required if 150 L/s of liquid ($S = 0.85$) is to be pumped.

Solution

First, the normal velocity component V_n is

$$V_n = \frac{Q}{A} = \frac{Q}{(\pi/4)(D_o^2 - D_i^2)}$$

$$= \frac{0.15}{(\pi/4)(0.3^2 - 0.15^2)} = 2.83 \text{ m/s}$$

The peripheral speed u of the impeller is based on an average radius:

$$u \simeq \omega \frac{D_o + D_i}{4}$$

$$= 500\left(\frac{2\pi}{60}\right)\frac{0.3 + 0.15}{4} = 5.89 \text{ m/s}$$

The theoretical head H_t is computed with Eq. 12.2.19 to be

$$H_t = \frac{u}{g}[u - V_n(\cot \alpha_1 + \cot \beta_2)]$$

$$= \frac{5.89}{9.81}[5.89 - 2.83(\cot 75° + \cot 70°)] = 2.46 \text{ m}$$

Finally, for the assumed ideal conditions, the required power is

$$\dot{W}_P = \gamma Q H_t$$

$$= (9810 \times 0.85) \times 0.15 \times 2.46 = 3080 \text{ W} \quad \text{or} \quad 3.1 \text{ kW}$$

Generally, flow in an impeller region consists of two components: the circular, vortex-like motion due to the action of the vanes, and the net flow through the impeller. The vortex motion is superposed on the radial outward flow in a radial-flow pump, and on the axial flow in an axial-flow pump. The action of the mixed-flow pump lies between these two types. The previous discussions relating to radial- and axial-flow pumps pertain to mixed-flow pumps as well. A mixed-flow pump and its characteristic curves are shown in Fig. 12.9.

As in the case for a radial-flow pump, the performance characteristics of axial- and mixed-flow pumps deviate from the idealized considerations. Generally, radial-flow pumps are designed to deliver relatively low discharges at high heads, and axial-flow pumps produce relatively large discharges at low heads. Mixed-flow pumps deliver heads and discharges

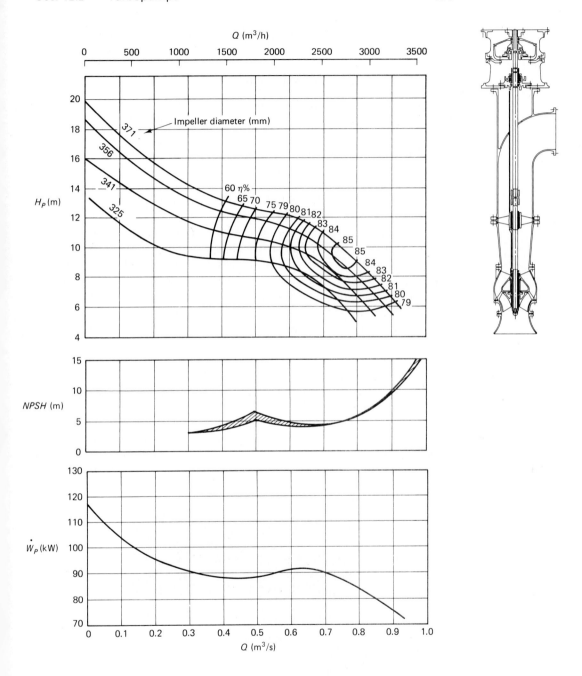

Figure 12.9   Mixed-flow pump and performance curves for four different impellers and vane angles with $N = 970$ rpm ($\omega = 102$ rad/s). Water at 20°C is the pumped liquid. (Courtesy of Sulzer Bros. Ltd.)

between these two extremes. A method of selecting a pump appropriate for required design considerations is presented in Section 12.4.

12.2.3 Cavitation in Turbomachines

Cavitation refers to conditions at certain locations within the turbomachine where the local pressure drops to the vapor pressure of the liquid, and as a result, vapor-filled cavities are formed. As the cavities are transported through the turbomachine into regions of greater pressure, they will collapse rapidly, generating extremely high localized pressures. Those bubbles that collapse close to solid boundaries can weaken the solid surface, and after repeated collapse cycles, pitting and further erosion and fatigue of the surface can occur. Signs of cavitation in turbopumps include noise and vibration and lowering of the head-discharge and efficiency curves. Regions most susceptible to damage in a turbomachine are those slightly beyond the low-pressure zones on the back side of impellers (see Fig. 12.10). In general, sudden changes in direction, sudden increases in area, and lack of streamlining are the culprits causing cavitation damage in turbopumps (Karassik, et al., 1986).

The proper design of turbopumps will have minimized the possibility for cavitation to occur. Under adverse operating conditions, however, the pressures may decrease and cavitation may occur. Two parameters are used to designate the potential for cavitation; these are the cavitation number and the net positive suction head. Their interrelationship and use are now discussed.

Consider a pump operating in the manner shown in Fig. 12.11. Location 1 is on the liquid surface on the suction side, and location 2 is the point of minimum pressure within the pump. Writing the energy equation from location 1 to location 2 and using an absolute pressure datum results

Figure 12.10 Cavitation bubble distribution in an impeller region. (Courtesy of Sulzer Bros. Ltd.)

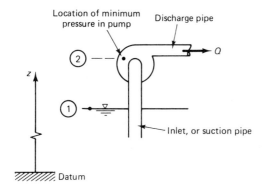

Location of minimum
pressure in pump

Discharge pipe

②

z

①

Inlet, or suction pipe

Datum

Figure 12.11 Cavitation setting for a pump.

in

$$\frac{V_2^2}{2g} = \frac{p_{\text{atm}} - p_2}{\gamma} - \Delta z - h_L \qquad (12.2.21)$$

in which h_L is the loss between location 1 and location 2, $\Delta z = z_2 - z_1$, and the kinetic energy at location 1 is assumed negligible. The minimum allowable pressure at location 2 is the vapor pressure, p_v. Substituting this into the expression, one can say that the left-hand side of Eq. 12.2.21 represents the maximum kinetic energy head possible at location 2 when cavitation is imminent. Thus the **net positive suction head** (*NPSH*) is defined as

$$NPSH = \frac{p_{\text{atm}} - p_v}{\gamma} - \Delta z - h_L \qquad (12.2.22)$$

The *NPSH* is also used for turbines; however, the sign of the h_L term in Eq. 12.2.22 changes, and location 1 refers to the liquid surface on the discharge side of the machine. The design requirement for a turbopump is thus established as follows:

$$NPSH \leq \frac{p_{\text{atm}} - p_v}{\gamma} - \Delta z - h_L \qquad (12.2.23)$$

The performance data supplied by turbomachinery manufacturers usually includes *NPSH* curves; these are developed by testing a given family in a laboratory environment. Figures 12.6, 12.8, and 12.9 show *NPSH* curves. The *NPSH* curve enables one to specify the required maximum value of Δz to be used for a given turbomachine; note that it is necessary to estimate h_L to obtain this.

Equation 12.2.22 can be divided by H_P, the total head across the pump to yield

$$\sigma = \frac{(p_{atm} - p_v)/\gamma - \Delta z - h_L}{H_P} \qquad (12.2.24)$$

in which σ is the **cavitation number.** This parameter is used as an alternative to the *NPSH* to establish design criteria for cavitation. A **critical cavitation number,** which signals that cavitation is imminent, is determined experimentally. Thus, for no cavitation to occur, σ must be greater than the critical cavitation number. The cavitation number is in dimensionless form, which is preferred to the dimensional form of the *NPSH*.

EXAMPLE 12.3

Determine the elevation that the 240-mm-diameter pump of Fig. 12.6 can be situated above the water surface of the suction reservoir without experiencing cavitation. Water at 15°C is being pumped at 250 m³/h. Neglect losses in the system. Use p_{atm} = 101 kPa.

Solution

From Fig. 12.6, at a discharge of 250 m³/h, the *NPSH* for the 240-mm-diameter impeller is approximately 7.4 m. For a water temperature of 15°C, p_v = 1666 Pa absolute, and γ = 9800 N/m³. Equation 12.2.22 with h_L = 0 is employed to compute Δz to be

$$\Delta z = \frac{p_{atm} - p_v}{\gamma} - NPSH - h_L$$

$$= \frac{101\ 000 - 1666}{9800} - 7.4 - 0 = 2.74 \text{ m}$$

Thus the pump can be placed approximately 2.7 m above the suction reservoir water surface.

12.3 DIMENSIONAL ANALYSIS FOR TURBOMACHINERY

The development and utilization of turbomachinery in engineering practice has benefited greatly from the application of dimensional analysis, probably to an extent more than any other area of applied fluid mechanics. It has enabled pump and turbine manufacturers to test and develop a relatively small number of turbomachines, and subsequently produce a series of commercial units that cover a broad range of head and flow demands. The material developed in Chapter 6, which covers similitude and dimensional analysis, is applied to turbomachines in this section.

12.3.1 Dimensionless Coefficients

The following parameters can be considered significant ones for a turbo-machine: power, rotational speed, outer diameter of the impeller, discharge, pressure change across the impeller, density of the fluid, and viscosity of the fluid. These are given in Table 12.1 along with the relevant dimensions. In functional form, the relation between these variables is given by

$$f(\dot{W}, \omega, D, Q, \Delta p, \rho, \mu) = 0 \qquad (12.3.1)$$

Using ω, ρ, and D as repeating variables, a set of convenient dimensionless groupings can be derived. They are given as follows:

$$C_{\dot{W}} = \frac{\dot{W}}{\rho \omega^3 D^5} \qquad \textbf{power coefficient} \qquad (12.3.2)$$

$$C_p = \frac{\Delta p}{\rho \omega^2 D^2} \qquad \textbf{pressure coefficient} \qquad (12.3.3)$$

$$C_Q = \frac{Q}{\omega D^3} \qquad \textbf{flow rate coefficient} \qquad (12.3.4)$$

$$\mathrm{Re} = \frac{\omega D^2 \rho}{\mu} \qquad \textbf{Reynolds number} \qquad (12.3.5)$$

It is customary to equate Δp to γH, where H represents either the pressure head rise in the case of a pump, or the pressure head drop for a turbine. Then Eq. 12.3.3 is replaced by

$$C_H = \frac{gH}{\omega^2 D^2} \qquad (12.3.6)$$

in which C_H is termed the **head coefficient.** Another dimensionless parameter for a pump can be found by grouping C_Q, $C_{\dot{W}}$, and C_H to form the

TABLE 12.1   TURBOMACHINERY PARAMETERS

Parameter	Symbol	Dimensions
Power	$\dot{W}$	ML^2/T^3
Rotational speed	ω	T^{-1}
Outer diameter of impeller	D	L
Discharge	Q	L^3/T
Pressure change	Δp	M/LT^2
Fluid density	ρ	M/L^3
Fluid viscosity	μ	M/LT

ratio

$$\frac{C_Q C_H}{C_{\dot{W}}} = \frac{\gamma QH}{\dot{W}} = \eta_P \quad \text{(pump)} \quad (12.3.7)$$

Hence the efficiency is also a similitude parameter. For a turbine the efficiency grouping is given by

$$\frac{C_{\dot{W}}}{C_Q C_H} = \frac{\dot{W}}{\gamma QH} = \eta_T \quad \text{(turbine)} \quad (12.3.8)$$

in which $\dot{W}$ is the output power.[2]

The interrelationships between the dimensionless parameters are expressed in the form of dimensionless performance curves, as shown in Figure 12.12; these are for the same turbomachine as shown in Fig. 12.6.[3] Dimensionless performance curves for the axial-flow pump of Fig. 12.8 and the mixed-flow pump of Fig. 12.9 are shown in Figs. 12.13 and 12.14, respectively. The vertical dashed lines designate the operating conditions at maximum or so-called "best" efficiency.

Note that the Reynolds number is not present as a parameter in Figs. 12.12 to 12.14, even though viscous effects are significant in turbomachinery flow. In addition to the Reynolds number, one could include a relative roughness ratio. It is known that viscous and roughness effects will alter pump behavior; however, because of the difficulties involved, it is usual practice not to make a correlation between the Reynolds number, roughness, and pump losses. The flow regime in pump passages is usually very turbulent, and the roughness varies widely for different pumps. Furthermore, the friction losses may not be as significant as losses due to eddy formation and separation caused by diffused flow occurring in the impeller and casing. As a result, the Re effect is usually not directly represented in similitude studies. Instead, the two following performance

---

[2]It is common industrial practice to use so-called **homologous units** in place of the dimensionless coefficients. The relationships between the two groupings are as follows:

Dimensionless coefficient	C_Q	C_H	$C_{\dot{W}}$
Homologous unit	$\dfrac{Q}{ND^3}$	$\dfrac{H}{N^2 D^2}$	$\dfrac{\dot{W}}{N^3 D^5}$

In the homologous units, density and gravitational acceleration have been eliminated, and the rotational speed is given in revolutions per minute, designated by N. Note that the homologous units are dimensional.

[3]In the caption, $(\omega_s)_P$ is the specific speed, which will be defined in Section 12.3.3.

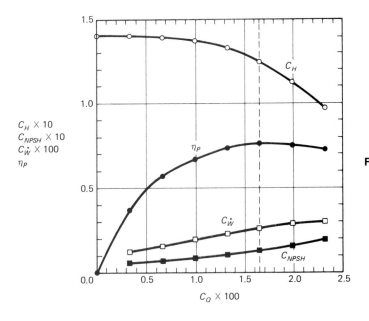

Figure 12.12 Dimensionless radial-flow pump performance curves for the pump presented in Fig. 12.6; $D = 240$ mm; $(\omega_s)_P = 0.61$.

characteristics are recognized: (1) larger pumps exhibit relatively smaller losses and higher efficiencies than smaller pumps of the same family since they have smaller roughness and clearance ratios; and (2) higher viscosity liquids will cause reductions in pump head and efficiency and an increase in power requirement for a given discharge.

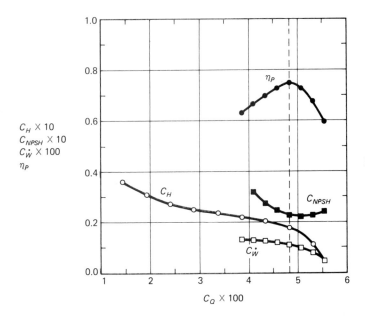

Figure 12.13 Dimensionless axial-flow pump performance curve for the pump presented in Fig. 12.8; $D = 500$ mm; Vane angle = 14°; $(\omega_s)_P = 4.5$.

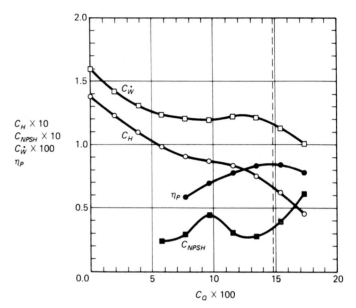

Figure 12.14 Dimensionless mixed-flow pump performance curve for the pump presented in Fig. 12.9; $D = 371$ mm; $(\omega_s)_P = 3.1$.

A dimensionless net positive suction head coefficient, C_{NPSH}, is also presented in Figs. 12.12 to 12.14. It is formed by replacing H in Eq. 12.3.6 with the *NPSH* defined by Eq. 12.2.22.

EXAMPLE 12.4

Determine the speed, size, and required power for the axial-flow pump of Fig. 12.13 to deliver 0.65 m³/s of water at a head of 2.5 m.

Solution

The pump data are obtained from Fig. 12.13. Reading from the figure we find that at the design, or maximum, efficiency of $\eta_P = 0.75$

$$C_Q = 0.048 \qquad C_H = 0.018 \qquad C_{\dot{W}} = 0.0012$$

Equations 12.3.4 and 12.3.6 are employed to determine the two unknowns ω and D. Rearrange Eq. 12.3.6 to solve for ωD:

$$\omega D = \sqrt{\frac{gH_p}{C_H}}$$

$$= \sqrt{\frac{9.81 \times 2.5}{0.018}} = 36.9 \text{ m/s}$$

Equation 12.3.4 is rearranged to include ωD as a known quantity, and solved for D:

$$D = \sqrt{\frac{Q}{C_Q \omega D}}$$

$$= \sqrt{\frac{0.65}{0.048 \times 36.9}} = 0.61 \text{ m}$$

Substituting $D = 0.61$ m into the relation $\omega D = 36.9$, one finds that

$$\omega = \frac{36.9}{D}$$

$$= \frac{36.9}{0.61} = 60.5 \text{ rad/s} \quad \text{or} \quad 578 \text{ rpm}$$

The density for water is $\rho = 1000$ kg/m³. Then the pump power is determined by using Eq. 12.3.2:

$$\dot{W}_P = C_{\dot{W}} \rho \omega^3 D^5$$

$$= 0.0012 \times 1000 \times 60.5^3 \times 0.61^5 = 2.2 \times 10^4 \text{W}$$

Thus the speed, size, and required power are approximately 580 rpm, 610 mm, and 22 kW, respectively.

___

12.3.2 Similarity Rules

Following the principles outlined in Chapter 6 for similitude, similarity relationships between any two pumps from the same geometric family can be developed. They are:

$$(C_{\dot{W}})_1 = (C_{\dot{W}})_2 \quad \text{or} \quad \frac{\dot{W}_2}{\dot{W}_1} = \frac{\rho_2}{\rho_1} \left(\frac{\omega_2}{\omega_1}\right)^3 \left(\frac{D_2}{D_1}\right)^5 \qquad (12.3.9)$$

$$(C_H)_1 = (C_H)_2 \quad \text{or} \quad \frac{H_2}{H_1} = \frac{g_1}{g_2} \left(\frac{\omega_2}{\omega_1}\right)^2 \left(\frac{D_2}{D_1}\right)^2 \qquad (12.3.10)$$

$$(C_Q)_1 = (C_Q)_2 \quad \text{or} \quad \frac{Q_2}{Q_1} = \frac{\omega_2}{\omega_1} \left(\frac{D_2}{D_1}\right)^3 \qquad (12.3.11)$$

These equations, called the turbomachinery similarity rules, are used to design or select a turbomachine from a family of geometrically similar units. Another use of them is to examine the effects of changing speed, fluid, or size on a given unit. One could even design a pump to deliver flow on the moon or on a space station!

In light of Eq. 12.3.7, one should also expect that $\eta_1 = \eta_2$, but as stated above, larger pumps are more efficient than smaller ones of the same geometric family. An acceptable empirical correlation (Stepanoff, 1957) relating efficiencies to size is

$$\frac{1 - (\eta_P)_2}{1 - (\eta_P)_1} = \left(\frac{D_1}{D_2}\right)^{1/4} \qquad (12.3.12)$$

This relation can also be used for turbines by replacing η_P by η_T.

EXAMPLE 12.5

Determine the performance curve for the mixed-flow pump whose characteristic curve is shown in Fig. 12.14. The required discharge is 2.5 m³/s with the pump operating at a speed of 600 rpm. At design efficiency, what are the power and the *NPSH* requirements? Water is being pumped.

Solution

From Fig. 12.14 the design efficiency is $\eta_P = 0.85$ and the corresponding design values for the coefficients are

$$C_Q = 0.15 \qquad C_H = 0.067 \qquad C_{\dot{W}} = 0.0115 \qquad C_{NPSH} = 0.035$$

The rotational speed in radians per second is

$$\omega = 600 \times \frac{\pi}{30} = 62.8 \text{ rad/s}$$

The pump diameter is computed with Eq. 12.3.4:

$$D = \left(\frac{Q}{\omega C_Q}\right)^{1/3}$$

$$= \left(\frac{2.5}{62.8 \times 0.15}\right)^{1/3} = 0.64 \text{ m}$$

With Eqs. 12.3.2 and 12.3.6, the required power and *NPSH* are computed:

$$\dot{W}_P = \rho \omega^3 D^5 C_{\dot{W}}$$

$$= 1000 \times 62.8^3 \times 0.64^5 \times 0.0115 = 3.1 \times 10^5 \text{ W}$$

$$NPSH = \frac{\omega^2 D^2}{g} C_{NPSH}$$

$$= \frac{62.8^2 \times 0.64^2}{9.81} \times 0.035 = 5.8 \text{ m}$$

Hence the required power is 310 kW and the required *NPSH* is 5.8 m. The performance curve is constructed using Eqs. 12.3.10 and 12.3.11:

$$H_2 = \frac{\omega_2^2 D_2^2}{g}(C_H)_1$$

$$= \frac{62.8^2 \times 0.64^2}{9.81}(C_H)_1 = 165(C_H)_1$$

$$Q_2 = \omega_2 D_2^3 (C_Q)_1$$

$$= 62.8 \times 0.64^3 (C_Q)_1 = 16.5(C_Q)_1$$

Values of $(C_H)_1$ and $(C_Q)_1$ are read from Fig. 12.14, along with η_P and placed in the table shown below. Columns 4 and 5 are Q_2 and H_2 computed from the similarity equations. At best efficiency $H_2 = 10.4$ m and $Q_2 = 2.54$ m³/s. The characteristic curve is plotted beneath the table.

$(C_Q)_1$	$(C_H)_1$	η_P	Q_2 (m³/s)	H_2 (m)
0	0.138		0	22.8
0.019	0.123		0.31	20.3
0.039	0.110		0.64	18.2
0.058	0.099		0.96	16.3
0.077	0.091	0.59	1.27	15.0
0.096	0.088	0.70	1.58	14.5
0.116	0.085	0.78	1.91	14.0
0.135	0.076	0.84	2.23	12.5
0.154	0.063	0.85	2.54	10.4
0.174	0.046	0.79	2.87	7.6

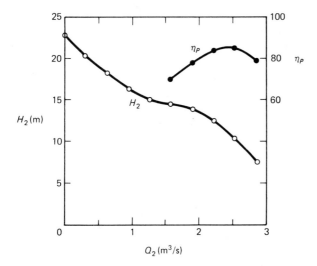

12.3.3 Specific Speed

It is possible to correlate a turbomachine of a given family to a dimensionless number that characterizes its operation at optimum conditions. Such a number is termed the **specific speed,** and it is determined in the following manner. The specific speed $(\omega_s)_P$ of a pump is a dimensionless parameter associated with its operation at maximum efficiency, with known ω, Q, and H_P. It is obtained by eliminating D between Eqs. 12.3.4 and 12.3.6 and expressing the rotational speed ω to the first power:

$$(\omega_s)_P = \frac{C_Q^{1/2}}{C_H^{3/4}} = \frac{\omega Q^{1/2}}{(gH_P)^{3/4}} \tag{12.3.13}$$

In Eq. 12.3.13, ω is usually based on motor requirements, and the values of Q and H_P are those at maximum efficiency.

A preliminary pump selection can be based on the specific speed. Figure 12.15 shows how the maximum efficiency and the discharge vary with $(\omega_s)_P$ for radial-flow pumps. The specific speed can be correlated

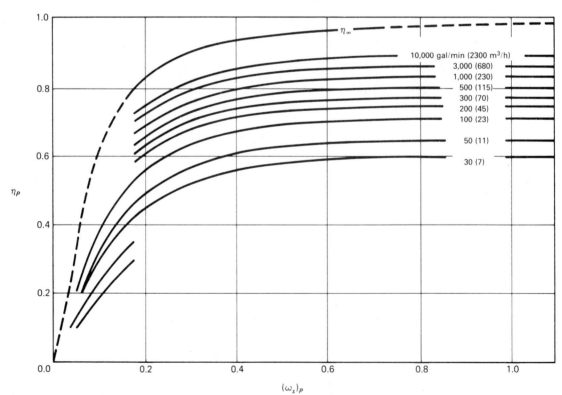

Figure 12.15 Maximum efficiency as a function of specific speed and discharge for radial-flow pumps. (Adapted with permission from Karassik et al., 1986.)

with the type of impeller shown in Fig. 12.2; it is given as follows:

$$(\omega_s)_P < 1 \qquad \text{radial-flow pump}$$

$$1 < (\omega_s)_P < 4 \qquad \text{mixed-flow pump} \qquad (12.3.14)$$

$$(\omega_s)_P > 4 \qquad \text{axial-flow pump}$$

For a turbine, the specific speed $(\omega_s)_T$ is a dimensionless parameter associated with a given family of turbines operating at maximum efficiency, with known ω, H_T, and $\dot{W}_T$. The diameter D is eliminated in Eqs. 12.3.2 and 12.3.6, and ω is expressed to the first power to obtain

$$(\omega_s)_T = \frac{C_{\dot{W}}^{1/2}}{C_H^{5/4}} = \frac{\omega(\dot{W}_T/\rho)^{1/2}}{(gH_T)^{5/4}} \qquad (12.3.15)$$

A comparison of $(\omega_s)_T$ for different turbines will be made in Section 12.5.[4]

[4]In place of Eqs. 12.3.13 and 12.3.15, it has been common industrial practice to use specific speeds in dimensional form. Hence, for a pump, the dimensional, or **homologous specific speed** is $(N_s)_P = NQ^{1/2}/H^{3/4}$, and for a turbine the corresponding homologous specific speed is $(N_s)_T = N\dot{W}^{1/2}/H^{5/4}$. For these expressions, Q is in gal/min, ft^3/sec, or m^3/s, H is in ft or m, $\dot{W}$ is in horsepower or kW, and N is in rpm.

EXAMPLE 12.6

Select a pump to deliver 1890 L/min of water with a pressure rise of 448 kPa. Assume a rotational speed not to exceed 3600 rpm.

Solution

To estimate the specific speed we need the following:

$$\omega = 3600 \times \frac{\pi}{30} = 377 \text{ rad/s}$$

$$H_P = \frac{\Delta p}{\rho g} = \frac{448 \times 10^3}{1000 \times 9.81} = 45.7 \text{ m}$$

$$Q = \frac{1.89}{60} = 0.0315 \text{ m}^3/\text{s}$$

Use Eq. 12.3.13 to find the specific speed to be

$$(\omega_s)_P = \frac{\omega\sqrt{Q}}{(gH_P)^{3/4}}$$

$$= \frac{377\sqrt{0.0315}}{(9.81 \times 45.7)^{3/4}} = 0.69$$

Turbomachinery   Chap. 12

From Eq. 12.3.14, this would indicate a radial-flow pump. The pump of Fig. 12.12 could be used, even though the specific speed (0.61) would result in a lower efficiency. With $(\omega_s)_P \approx 0.61$ (Fig. 12.12), the speed is estimated with Eq. 12.3.13 as

$$\omega = \frac{(\omega_s)_P(gH_P)^{3/4}}{\sqrt{Q}}$$

$$= \frac{0.61 \times (9.81 \times 45.7)^{3/4}}{\sqrt{0.0315}} = 335 \text{ rad/s}$$

Hence the required speed is $335 \times 30/\pi = 3200$ rpm, which does not exceed 3600 rpm. The required diameter is determined with use of Fig. 12.12, where at maximum efficiency, $C_Q = 0.0165$. This is substituted into Eq. 12.3.4 to determine the diameter:

$$D = \left(\frac{Q}{C_Q\omega}\right)^{1/3}$$

$$= \left(\frac{0.0315}{0.0165 \times 335}\right)^{1/3} = 0.18 \text{ m}$$

EXAMPLE 12.6 (English)

Select a pump to deliver 500 gal/min of water with a pressure rise of 65 psi. Assume a rotational speed not to exceed 3600 rpm.

Solution

To estimate the specific speed we need the following:

$$\omega = 3600 \times \frac{\pi}{30} = 377 \text{ rad/sec}$$

$$H_P = \frac{\Delta p}{\rho g} = \frac{65 \times 144}{1.94 \times 32.2} = 150 \text{ ft}$$

$$Q = \frac{500}{7.48 \times 60} = 1.11 \text{ ft}^3/\text{sec}$$

Use Eq. 12.3.13 to find the specific speed to be

$$(\omega_s)_P = \frac{\omega\sqrt{Q}}{(gH_P)^{3/4}}$$

$$= \frac{377\sqrt{1.11}}{(32.2 \times 150)^{3/4}} = 0.69$$

From Eq. 12.3.14, this would indicate a radial-flow pump. The pump of Fig. 12.12 could be used, even though the specific speed (0.61) would result in a lower efficiency. With $(\omega_s)_P \approx 0.61$ (Fig. 12.12), the speed is estimated with Eq. 12.3.13

as

$$\omega = \frac{(\omega_s)_P (gH_P)^{3/4}}{\sqrt{Q}}$$

$$= \frac{0.61 \times (32.2 \times 150)^{3/4}}{\sqrt{1.11}} = 335 \text{ rad/sec}$$

Hence the required speed is $335 \times 30/\pi = 3200$ rpm, which does not exceed 3600 rpm. The required diameter is determined with use of Fig. 12.12, where at maximum efficiency, $C_Q = 0.0165$. This is substituted into Eq. 12.3.4 to determine the diameter:

$$D = \left(\frac{Q}{C_Q \omega}\right)^{1/3}$$

$$= \left(\frac{1.11}{0.0165 \times 335}\right)^{1/3} = 0.59 \text{ ft}$$

12.4 USE OF TURBOPUMPS IN PIPING SYSTEMS

The appropriate selection of one or more pumps to meet the flow demands of a piping system requires, in addition to a fundamental understanding of turbopumps, a hydraulic analysis of the pumps integrated into the piping system. A single pipe and pump arrangement is analyzed in Section 7.6.7. In Example 7.16 a system demand curve is obtained, and a trial-and-error solution to finding the discharge and required pump head is shown. The technique is extended to simple piping systems in Section 11.3, wherein it is shown that one can employ either a pump performance curve or assume a constant power input to the pump. Section 11.4 deals with piping networks. In those systems, the pump performance curve can be represented by a polynomial relation (Eq. 11.4.14) and incorporated into the linearized energy equations, or alternatively, a constant pump power requirement can be employed (Eq. 11.4.15).

In this section we look more closely at the manner in which pumps are selected. The procedure involves not only satisfying the required flow and pressure requirements for a piping system, but also making sure that cavitation is avoided and that the most efficient pump is selected.

12.4.1 Matching Pumps to System Demand

Consider a single pipeline that contains a pump to deliver fluid between two reservoirs. The **system demand curve** is defined as

$$H_P = (z_2 - z_1) + \left(f\frac{L}{D} + \Sigma K\right)\frac{Q^2}{2gA^2} \tag{12.4.1}$$

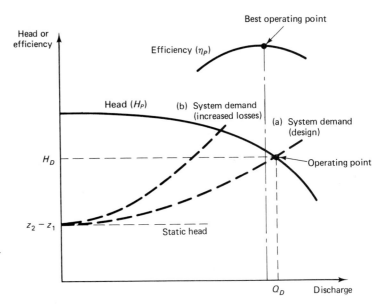

Figure 12.16 Pump characteristic curve and system demand curve.

where atmospheric pressure is assumed at each reservoir, and the upstream and downstream reservoirs are at elevations z_1 and z_2, respectively. Note that z_2 need not be greater than z_1, and that the friction factor can vary as the discharge (i.e., Reynolds number) varies. Equation 12.4.1 is represented as curve (a) in Fig. 12.16. The first term on the right-hand side is the static head and the second term is the head loss due to pipe friction and minor losses. The steepness of the demand curve is dependent on the sum of the loss coefficients in the system; as the loss coefficients increase, signified by curve (b) in Fig. 12.16, the pumping head required for a given discharge is increased. Piping systems may experience short-term changes in the demand curve such as throttling of valves, and over the long term, aging of pipes may cause the loss coefficients to increase permanently. In either case the system demand curve could change from (a) to (b) as shown.

For a given design discharge, Eq. 12.3.14 allows the selection of a pump based on the specific speed. Once the type of pump is determined, an appropriate size is selected from a manufacturer's characteristic curve; a representative curve is shown in Fig. 12.16. The intersection of the characteristic curve with the desired system demand curve will provide the **design head** H_D and **design discharge** Q_D. It is desirable to have the intersection occur at or close to the point of maximum efficiency of the pump, designated as the **best operating point.**

12.4.2 Pumps in Parallel and in Series

In some instances, pumping installations may have a wide range of head or discharge requirements, so that a single pump may not meet the re-

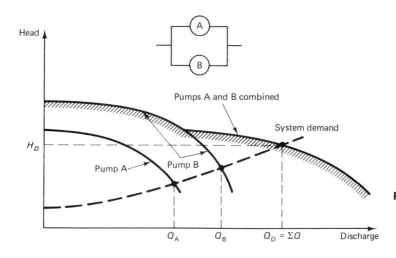

Figure 12.17 Characteristic curves for pumps operating in parallel.

quired range of demands. In these situations, pumps may be staged either in series or in parallel to provide operation in a more efficient manner. In this discussion, it is assumed that the pumps are placed at a single location with short lines connecting the separate units.

Where a large variation in flow demand is required, two or more pumps are placed in a parallel configuration (Fig. 12.17). Pumps are turned on individually to meet the required flow demand; in this way operation at higher efficiency can be attained. It is not necessary to have identical pumps, but individual pumps, when running in parallel, should not be operating in undesirable zones. For parallel pumping the combined characteristic curve is generated by recognizing that the head across each pump is identical, and the total discharge through the pumping system is ΣQ, the sum of the individual discharges through each pump. Note the existence of three operating points in Fig. 12.17, in which pump A or pump B is used separately, or in which pumps A and B are combined. Other design operating points could be obtained by throttling the flow or by changing the pump speeds. The overall efficiency of pumps in parallel is

$$\eta_P = \frac{\gamma H_D \, \Sigma Q}{\Sigma \dot{W}_P} \qquad (12.4.2)$$

in which $\Sigma \dot{W}_P$ is the sum of the individual power required by each pump.

For high head demands, pumps placed in series will produce a head rise greater than those of the individual pumps (Fig. 12.18). Since the discharge through each pump is identical, the characteristic curve is found by summing the head across each pump. Note that it is not necessary that the two pumps be identical. In Fig. 12.18 the system demand curve is such that pump A operating alone cannot deliver any liquid because its shutoff head is lower than the static system head. There are

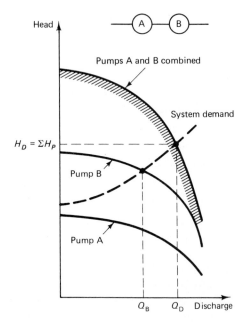

Figure 12.18 Characteristic curves for pumps operating in series.

two operating points, either with pump B alone or with pumps A and B combined. The overall efficiency is

$$\eta_P = \frac{\gamma(\Sigma H_P)Q_D}{\Sigma \dot{W}_P} \qquad (12.4.3)$$

in which ΣH_P is the sum of the individual heads across each pump.

EXAMPLE 12.7

Water is pumped between two reservoirs in a pipeline with the following characteristics: $D = 300$ mm, $L = 70$ m, $f = 0.025$, $\Sigma K = 2.5$. The radial-flow pump characteristic curve is approximated by the formula

$$H_P = 22.9 + 10.7Q - 111Q^2$$

where H_P is in meters and Q is in m³/s.

Determine the discharge Q_D and pump head H_D for the following situations:
(a) $z_2 - z_1 = 15$ m, one pump placed in operation; (b) $z_2 - z_1 = 15$ m, with two identical pumps operating in parallel; and (c) $z_2 - z_1 = 25$ m.

Solution

(a) The system demand curve (Eq. 12.4.1) is developed first:

$$H_P = (z_2 - z_1) + \left(f\frac{L}{D} + \Sigma K\right)\frac{Q^2}{2gA^2}$$

$$= 15 + \left(\frac{0.025 \times 70}{0.3} + 2.5\right)\frac{Q^2}{2 \times 9.81(\pi/4 \times 0.3^2)^2}$$

$$= 15 + 85Q^2$$

To find the operating point, equate the pump characteristic curve to the system demand curve,

$$15 + 85Q_D^2 = 22.9 + 10.7Q_D - 111Q_D^2$$

Reduce and solve for Q_D in the manner

$$195Q_D^2 - 10.7Q_D - 7.9 = 0$$

$$\therefore \quad Q_D = \frac{1}{2 \times 195}[10.7 + \sqrt{10.7^2 + 4 \times 195 \times 7.9}] = 0.23 \text{ m}^3/\text{s}$$

Using the system demand curve, H_D is computed as

$$H_D = 15 + 85 \times 0.23^2 = 19.5 \text{ m}$$

(b) For two pumps in parallel, the characteristic curve is

$$H_P = 22.9 + 10.7\left(\frac{Q}{2}\right) - 111\left(\frac{Q}{2}\right)^2$$

$$= 22.9 + 5.35Q - 27.75Q^2$$

Equate this to the system demand curve and solve for Q_D:

$$15 + 85Q_D^2 = 22.9 + 5.35Q_D - 27.75Q_D^2$$

$$112.8Q_D^2 - 5.35Q_D - 7.9 = 0$$

$$\therefore \quad Q_D = \frac{1}{2 \times 112.8}[5.35 + \sqrt{5.35^2 + 4 \times 112.8 \times 7.9}] = 0.29 \text{ m}^3/\text{s}$$

The design head is calculated to be

$$H_D = 15 + 85 \times 0.29^2 = 22.2 \text{ m}$$

(c) Since $z_2 - z_1$ is greater than the single pump shutoff head (i.e., $25 > 22.9$ m), it is necessary to operate with two pumps in series. The combined pump curve is

$$H_D = 2(22.9 + 10.7Q - 111Q^2)$$

$$= 45.8 + 21.4Q - 222Q^2$$

The system demand curve is changed since $z_2 - z_1 = 25$ m. It becomes

$$H_D = 25 + 85Q^2$$

Equating the two relations above and solving for Q_D and H_P results in

$$25 + 85Q_D^2 = 45.8 + 21.4Q_D - 222Q_D^2$$

or

$$307Q_D^2 - 21.4Q_D - 20.8 = 0$$

$$\therefore \quad Q_D = \frac{1}{2 \times 307}[21.4 + \sqrt{21.4^2 + 4 \times 307 \times 20.8}] = 0.30 \ \text{m}^3/\text{s}$$

and

$$H_D = 25 + 85 \times 0.30^2 = 32.7 \ \text{m}$$

12.4.3 Multistage Pumps

Instead of placing several pumps in series, multistage pumps are available (Fig. 12.19). Basically, the impellers are all housed in a single casing and the outlet from one impeller stage ejects into the eye of the next. Such pumps can provide extremely high heads. For the pump shown in Fig. 12.19, the pressure head range is 1500 to 3400 m, and the discharge can vary from 4500 m^3/h down to 260 m^3/h. Up to eight stages can be selected and the maximum speed is about 8000 rpm.

12.5 TURBINES

In many parts of the world, where sufficient head and large flow rates are possible, hydroturbines are used to produce electrical power. In contrast to pumps, turbines extract useful energy from the water flowing in a piping system. The moving component of a turbine is called a **runner,** which consists of vanes or buckets that are attached to a rotating shaft.

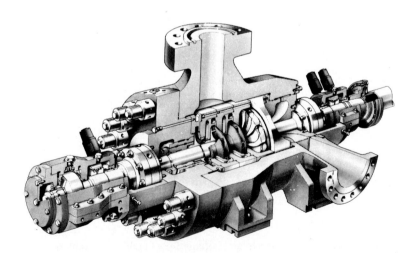

Figure 12.19   Four-stage centrifugal pump (Courtesy of Sulzer Brothers Ltd.).

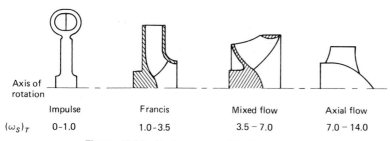

Axis of rotation

Impulse	Francis	Mixed flow	Axial flow
$(\omega_s)_T$ 0-1.0	1.0-3.5	3.5 – 7.0	7.0 – 14.0

Figure 12.20   Various types of turbine runners.

The energy available in the liquid is transferred to the shaft by means of the rotating runner, and the resulting torque transferred by the rotating shaft can drive an electric generator. Hydroturbines vary widely in size and capacity, ranging from microunits generating 5 kW to those in large hydroelectric installations that produce over 400 MW.

There are two types of turbines. The **reaction turbine** utilizes both flow energy and kinetic energy of the liquid; energy conversion takes place in an enclosed space at pressures above atmospheric conditions. Reaction turbines can be subdivided further, according to the available head, as either Francis or propeller type. The **impulse turbine** requires that the flow energy in the liquid be converted into kinetic energy by means of a nozzle before the liquid impacts on the runner; the energy is in the form of a high-velocity jet at or near atmospheric pressure. Turbines can be classified according to turbine specific speed, as shown in Figure 12.20; the Pelton wheel is a particular type of impulse turbine (see Section 12.5.2).

12.5.1 Reaction Turbines

In reaction turbines, the flow is contained in a **volute** that channels the liquid into the runner (Fig. 12.21). Adjustable **guide vanes** (also called **wicket gates**) are situated upstream of the runner; their function is to control the tangential component of velocity at the runner inlet. As a result, the fluid leaves the guide vane exit and enters the runner with an acquired angular momentum. As the fluid travels through the runner region, its angular momentum is reduced and it imparts a torque on the runner, which in turn drives the shaft to produce power. The flow exits from the runner into a diffuser, called the **draft tube,** which acts to convert the kinetic energy remaining in the liquid into flow energy.

Francis Turbine

In the Francis turbine, the incoming flow through the guide vanes is radial, with a significant tangential velocity component at the entrance to the runner vanes (Fig. 12.22). As the fluid traverses the runner the velocity takes on an axial component while the tangential component is reduced. As it leaves the runner, the fluid velocity is primarily axial with

(b)

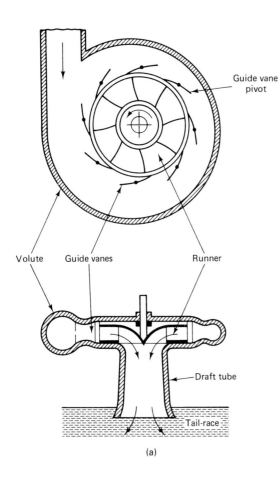

Guide vane
pivot

Volute   Guide vanes   Runner

Draft tube

Tail-race

(a)

Figure 12.21 Reaction turbine (Francis type): (a) schematic; (b) Francis spiral turbine for the Victoria Falls power station, Australia; and (c) runner for one of the turbines for the Itaipu power station, Brazil. (Courtesy of Voith Hydro, Inc.)

(c)

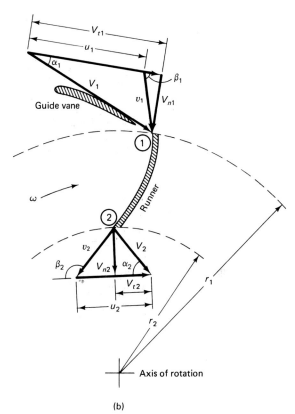

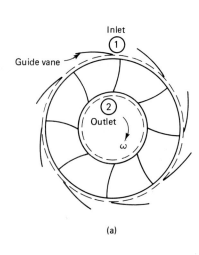

Figure 12.22 Idealized Francis turbine runner: (a) runner control volume; (b) velocity diagrams at control surfaces.

little or no tangential component. The pressure at the runner exit is below atmospheric.

The theoretical torque delivered to the runner is developed by applying Eq. 12.2.1 to the control volume shown in Fig. 12.22; the same assumptions leading to the pump torque relation, Eq. 12.2.2, apply. The resulting relation is

$$T = \rho Q(r_1 V_{t1} - r_2 V_{t2}) \tag{12.5.1}$$

Multiplying the torque by the angular speed ω gives the power delivered to the shaft:

$$\dot{W}_T = \omega T$$
$$= \rho Q(u_1 V_1 \cos \alpha_1 - u_2 V_2 \cos \alpha_2) \tag{12.5.2}$$

The fluid power input to the turbine is given by

$$\dot{W}_f = \gamma Q H_T \tag{12.5.3}$$

in which H_T is the actual head drop across the turbine. Thus the overall efficiency is given by

$$\eta_T = \frac{\dot{W}_T}{\dot{W}_f} = \frac{\omega T}{\gamma Q H_T} \tag{12.5.4}$$

The action of the guide vanes can be described by considering the velocity vector diagrams in Fig. 12.22b. Assume perfect guidance of the fluid along the guide vane; then the tangential velocity at the entrance to the runner is

$$V_{t1} = V_{n1} \cot \alpha_1 \tag{12.5.5}$$

From the velocity vector diagram the tangential velocity is also given by

$$V_{t1} = u_1 + V_{n1} \cot \beta_1 \tag{12.5.6}$$

The radial velocity component can be expressed in terms of the discharge Q and the width of the runner b_1:

$$V_{n1} = \frac{Q}{2\pi r_1 b_1} \tag{12.5.7}$$

Equations 12.5.5 to 12.5.7 are combined to eliminate V_{t1} and V_{n1}. Solving for α_1 gives

$$\alpha_1 = \cot^{-1}\left(\frac{2\pi r_1^2 b_1 \omega}{Q} + \cot \beta_1\right) \tag{12.5.8}$$

For a constant angular speed, in order to maintain the appropriate runner entry angle α_1, the guide vane angle is adjusted as Q changes.

Under normal operation, a turbine operates under a nearly constant head H_T so that the performance characteristics are viewed differently from those of a pump. For turbine operation under constant head, the important quantities are the variations of discharge, speed, and efficiency. The interrelationships among the three parameters are shown in the isoefficiency curve of Fig. 12.23. Note the drop in discharge as the speed increases at a given guide vane setting. The efficiency is reduced by the following mechanisms: (1) frictional head losses and draft tube head losses; (2) separation due to mismatch of flow entry angle with blade angle; (3) need to attain a certain turbine speed before useful power output is achieved; and (4) mechanical losses attributed to bearings, seals, and the like.

A representative dimensionless performance curve of a Francis turbine is shown in Fig. 12.24. The speed and head are kept constant, and the guide vanes are automatically adjusted as the discharge varies in order to attain peak efficiency.

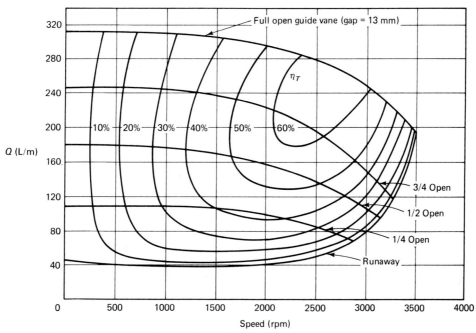

Figure 12.23   Isoefficiency curve for a model Francis turbine: $D = 82.5$ mm, $H_T = 15$ m. (Courtesy of Gilbert Gilkes and Gordon, Ltd.)

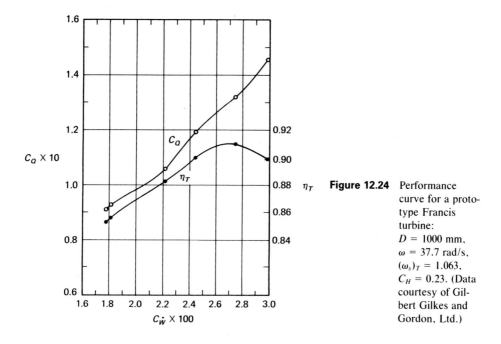

Figure 12.24   Performance curve for a prototype Francis turbine: $D = 1000$ mm, $\omega = 37.7$ rad/s, $(\omega_s)_T = 1.063$, $C_H = 0.23$. (Data courtesy of Gilbert Gilkes and Gordon, Ltd.)

EXAMPLE 12.8

A reaction turbine, whose runner radii are $r_1 = 300$ mm and $r_2 = 150$ mm, operates under the following conditions: $Q = 0.057$ m³/s, $\omega = 25$ rad/s, $\alpha_1 = 30°$, $V_1 = 6$ m/s, $\alpha_2 = 80°$, and $V_2 = 3$ m/s. Assuming ideal conditions, find the torque applied to the runner, the head on the turbine, and the fluid power. Use $\rho = 1000$ kg/m³.

Solution

The applied torque is computed using Eq. 12.5.1 to be

$$T = \rho Q(r_1 V_{t1} - r_2 V_{t2})$$

$$= \rho Q(r_1 V_1 \cos \alpha_1 - r_2 V_2 \cos \alpha_2)$$

$$= 1000 \times 0.057(0.3 \times 6 \times \cos 30° - 0.15 \times 3 \times \cos 80°)$$

$$= 84.4 \text{ N} \cdot \text{m}$$

Under ideal conditions, the power delivered to the shaft is the same as the fluid power input to the turbine (i.e., $\eta_T = 1$). Thus

$$\dot{W}_f = \dot{W}_T$$

$$= \omega T$$

$$= 25 \times 84.4 = 2110 \text{ W} \quad \text{or} \quad 2.11 \text{ kW}$$

The head on the turbine is found with the use of Eq. 12.5.3 to be

$$H_T = \frac{\dot{W}_f}{\gamma Q}$$

$$= \frac{2110}{9810 \times 0.057} = 3.77 \text{ m}$$

Axial-Flow Turbine

In an axial-flow turbine the flow is parallel to the axis of rotation (Fig. 12.25). Unlike the Francis turbine, the angular momentum of the liquid remains nearly constant and the tangential component of velocity is reduced across the blade. Both fixed-blade and pivoting-blade turbines are in use; the latter type, termed a Kaplan turbine, permits the blade angle to be adjusted to accommodate changes in head. Axial-flow turbines can be installed either vertically or horizontally. They are well suited for low-head installations.

Cavitation Considerations

The net positive suction head (*NPSH*) and the cavitation number, defined in Section 12.2.3, are applicable for turbines. In Eqs. 12.2.21 to 12.2.23 the sign of the loss term becomes positive. The cavitation number,

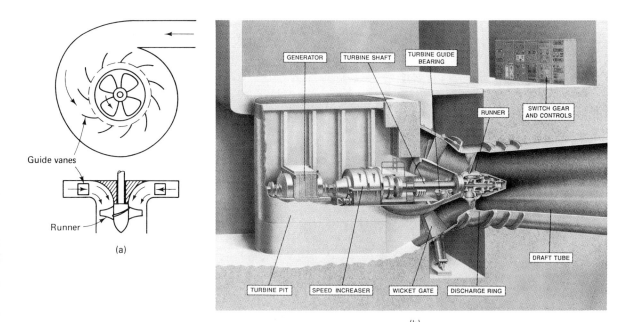

Guide vanes

Runner

(a)

(b)

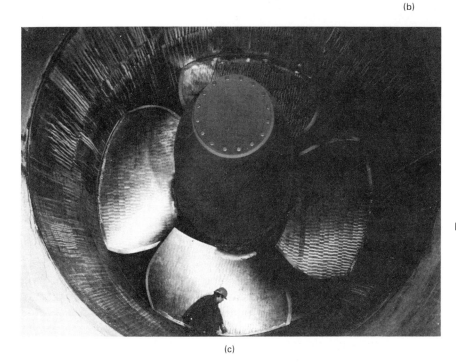

(c)

Figure 12.25 Axial-flow turbine
(Kaplan type):
(a) schematic;
(b) side view;
(c) turbine blades
in the Altenwörth
power station,
Austria. (Cour-
tesy of Voith
Hydro, Inc.)

defined by Eq. 12.2.24 for a pump, is given in the following form for a
turbine:

$$\sigma = \frac{(p_{\text{atm}} - p_v)/\gamma - \Delta z + h_L}{H_T} \qquad (12.5.9)$$

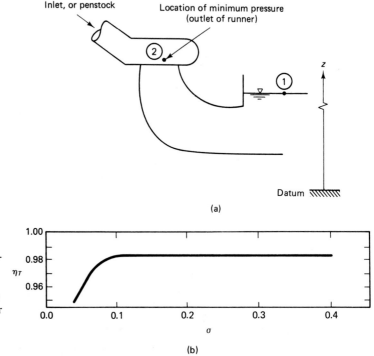

Figure 12.26 Cavitation considerations:
(a) schematic;
(b) representative
cavitation number
curve. (Courtesy
of Voith Hydro,
Inc.)

Typically, location 2 is defined at the outlet of the runner, and location 1
refers to the liquid surface at the draft tube outlet (Fig. 12.26a).

The cavitation number is commonly used to characterize the cavitation behavior of a turbine. Figure 12.26b shows a representative plot of
the cavitation number versus turbine efficiency, which is obtained experimentally by the turbine manufacturer. From such a plot operational procedures related to settings of the headwater and tailwater elevations can
be established, and in addition, admissible levels of the cavitation number
can be determined.

12.5.2 Impulse Turbines

The Pelton wheel (Fig. 12.27) is an impulse turbine that consists of three
basic components: one or more stationary inlet nozzles, a runner, and a
casing. The runner consists of multiple buckets mounted on a rotating
wheel. The pressure head upstream of the nozzle is converted into kinetic
energy contained in the water jet leaving the nozzle. As the jet impacts the
rotating buckets, the kinetic energy is converted into a rotating torque.
The buckets are shaped in a manner to divide the flow in half and turn its
relative velocity vector in the horizontal plane nearly 180°; the ideal angle
of 180° is not attainable since the exiting liquid must stay free of the
trailing buckets.

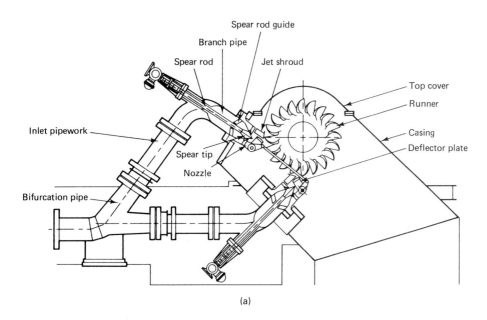

(a)

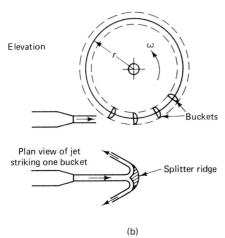

Elevation

Buckets

Plan view of jet
striking one bucket

Splitter ridge

(b)

Figure 12.27   Impulse turbine (Pelton type): (a) typical arrangement of twin-jet machine (Courtesy of Gilbert Gilkes and Gordon, Ltd.); (b) jet striking a bucket; (c) Pelton runner in the New Colgate power station, USA. (Courtesy Voith Hydro, Inc.)

(c)

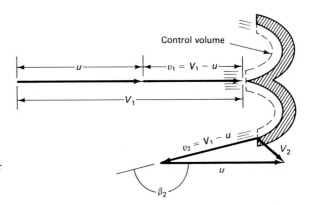

Figure 12.28 Velocity vector diagram for Pelton bucket.

The momentum equation, illustrated in the preceding sections, can be applied to the control volume shown in Fig. 12.28. Neglecting friction, the torque delivered to the wheel by the liquid jet is

$$T = \rho Q r (V_1 - u)(1 - \cos \beta_2) \tag{12.5.10}$$

in which Q is the discharge from all jets, and $u = r\omega$, r being the wheel radius as shown in Fig. 12.27b. The power delivered by the fluid to the turbine runner is

$$\dot{W}_T = \rho Q u (V_1 - u)(1 - \cos \beta_2) \tag{12.5.11}$$

Usually, β_2 varies between 160 and 168°. Differentiation of Eq. 12.5.11 with respect to u and setting it equal to zero shows that maximum power occurs when $u = V_1/2$. The jet velocity can be given in terms of the available head H_T:

$$V_1 = C_v \sqrt{2gH_T} \tag{12.5.12}$$

The **velocity coefficient** C_v accounts for the nozzle losses; typically, $0.92 \leq C_v \leq 0.98$. The efficiency is

$$\eta_T = \frac{\dot{W}_T}{\gamma Q H_T} \tag{12.5.13}$$

Substituting Eqs. 12.5.11 and 12.5.12 into this relation and rearranging produces

$$\eta_T = 2\phi(C_v - \phi)(1 - \cos \beta_2) \tag{12.5.14}$$

in which the **speed factor** ϕ is defined as

$$\phi = \frac{r\omega}{\sqrt{2gH_T}} \tag{12.5.15}$$

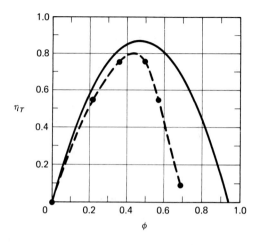

Figure 12.29 Speed factor versus efficiency for a laboratory-scale Pelton turbine: solid line, Eq. 12.5.14 ($C_v = 0.94$, $\beta_2 = 168°$); dashed line, experimental data.

Maximum efficiency occurs at $\phi = C_v/2$. Figure 12.29 shows Eq. 12.5.14 plotted for $C_v = 0.94$ and $\beta_2 = 168°$. In addition, data are plotted for a laboratory-scale Pelton wheel. The theoretical maximum efficiency is 0.874, while the experimental maximum efficiency is approximately 0.80. The efficiency is reduced because (1) the liquid jet does not strike the bucket with a uniform velocity, (2) losses occur when the jet strikes the bucket and splitter, (3) frictional loss is present due to the rotation of the bucket and splitter, and (4) there is frictional loss at the nozzle.

The interrelationships among Q, ω, and η_T are shown in the isoefficiency curves of Fig. 12.30. For a constant head and nozzle setting, the

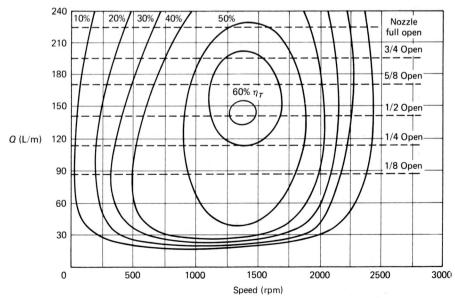

Figure 12.30 Isoefficiency curve for a laboratory-scale Pelton turbine: $D = 101.6$ mm, $H_T = 20$ m. (Courtesy of Gilbert Gilkes and Gordon, Ltd.)

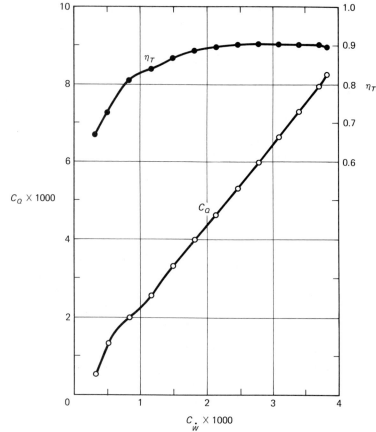

discharge remains constant, since the nozzle outlet is at atmospheric pressure. A representative dimensionless performance curve for a prototype Pelton turbine is shown in Fig. 12.31.

EXAMPLE 12.9

A Pelton turbine rotates at an angular speed of 400 rpm, developing 67.5 kW under a head of 60 m of water. The inlet pipe diameter at the base of the single nozzle is 200 mm. The operating conditions are $C_v = 0.97$, $\phi = 0.46$, and $\eta_T = 0.83$. Determine (a) the volumetric flow rate, (b) the diameter of the jet, (c) the wheel diameter, and (d) the pressure in the inlet pipe at the nozzle base.

Solution

(a) The discharge is computed from Eq. 12.5.13 to be

$$Q = \frac{\dot{W}_T}{\gamma H_T \eta_T}$$

$$= \frac{67\,500}{9810 \times 60 \times 0.83} = 0.138 \text{ m}^3/\text{s}$$

(b) From Eq. 12.5.12, the velocity of the jet is

$$V_1 = C_v \sqrt{2gH_T}$$
$$= 0.97\sqrt{2 \times 9.81 \times 60} = 33.3 \text{ m/s}$$

The area of the jet is the discharge divided by V_1, or

$$A_1 = \frac{Q}{V_1} = \frac{0.138}{33.3} = 4.14 \times 10^{-3} \text{ m}^2$$

Hence the jet diameter D_1 is

$$D_1 = \sqrt{\frac{4}{\pi} A_1}$$
$$= \sqrt{\frac{4}{\pi} \times 4.14 \times 10^{-3}} = 0.0726 \text{ m} \text{ or } 73 \text{ mm}$$

(c) Use Eq. 12.5.15 to compute the wheel diameter D to be

$$D = 2r$$
$$= \frac{2\phi}{\omega} \sqrt{2gH_T}$$
$$= \frac{2 \times 0.46}{400 \times \pi/30} \sqrt{2 \times 9.81 \times 60} = 0.754 \text{ m} \text{ or } 754 \text{ mm}$$

(d) The area of the inlet pipe is

$$A = \frac{\pi}{4} \times 0.20^2 = 0.0314 \text{ m}^2$$

The piezometric head just upstream of the nozzle is equal to H_T, so that the pressure at that location is

$$p = \gamma\left(H_T - \frac{Q^2}{2gA^2}\right)$$
$$= 9810\left(60 - \frac{0.138^2}{2 \times 9.81 \times 0.0314^2}\right)$$
$$= 5.79 \times 10^5 \text{ Pa} \text{ or } \text{approximately 580 kPa}$$

12.6 SELECTION AND OPERATION OF TURBINES

A preliminary selection of the appropriate type of turbine for a given installation is based on the specific speed. Figure 12.20 shows how the turbine impeller varies with $(\omega_s)_T$. Impulse turbines normally operate most economically at heads above 300 m, but small units can be used for heads as low as 60 m. Heads up to 300 m are possible for Francis units, and propeller turbines are normally used for heads lower than 30 m. Figure 12.32 illustrates the ranges of application for a variety of hydraulic turbines.

So-called minihydro and microhydro installations have recently been subject to renewed interest, not only in developing countries, but in industrially developed countries such as the United States, where the cost of energy has increased and incentive programs have been made available to promote their use. **Microhydro** has been classified as units that possess a capacity of less than 100 kW, and **minihydro** refers to system capacity from 100 to 1000 kW (Warnick, 1984). It is becoming common practice for manufacturers to develop a standardized turbine unit. In addition to specialized small Pelton, Francis, and propeller units, the **cross-flow turbine** is employed; basically, it is an impulse turbine that possesses a higher rotational speed than that of other impulse turbines. Another alternative is to use a commercially available pump and operate it in the reverse direction; in the turbine mode, the best efficiency operating point requires a larger head and flow than when the pump operates at best efficiency in the pumping mode.

Reversible pump/turbine units are used in pumped/storage hydro-power installations. Water is pumped from a lower reservoir to a higher storage reservoir during periods of low demand for conventional power plants. Subsequently, water is released from the upper reservoir, and the pumps are rotated in reverse to generate power during periods of high

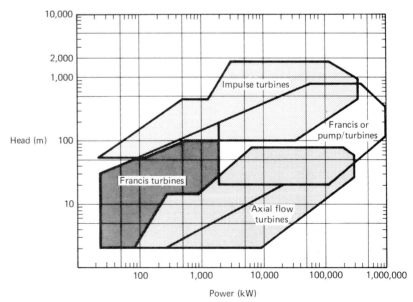

Figure 12.32 Application ranges for hydraulic turbines (Courtesy of Voith Hydro, Inc.)

Figure 12.33 Runner, distributor, and top cover of the pump turbine in the Kühtai power station, Austria. (Courtesy of Voith Hydro, Inc.)

electric power demand. Figure 12.33 shows a representative installation consisting of a Francis reversible pump/turbine unit combined with a generator/motor. Such units require special design considerations to operate efficiently in either mode. The problem of converting the flow from the pump mode to the turbine mode, or vice versa, is a very complicated transient problem that is beyond the scope of this book. Water with enormous mass moving through the entire system must be reversed in direction; a special technique must be developed to accomplish this.

EXAMPLE 12.10

A discharge of 2100 m³/s and a head of 113 m are available for a proposed pumped-storage hydroelectric scheme. Reversible Francis pump/turbines are to be installed; in the turbine mode of operation, $(\omega_s)_T = 2.19$, the rotational speed is 240 rpm, and the efficiency is 80%. Determine the power produced by each unit and the number of units required.

Solution

The power produced by each unit is found using the definition of specific speed, Eq. 12.3.15. Solving for the power, we have

$$\dot{W}_T = \rho \left[\frac{(\omega_s)_T}{\omega} (gH_T)^{5/4} \right]^2$$

$$= 1000 \left[\frac{2.19}{240 \times \pi/30} (9.81 \times 113)^{5/4} \right]^2 = 3.11 \times 10^8 \text{ W}$$

From Eq. 12.5.13, the discharge in each unit is

$$Q = \frac{\dot{W}_T}{\gamma H_T \eta_T}$$

$$= \frac{3.11 \times 10^8}{9800 \times 113 \times 0.8} = 351 \text{ m}^3/\text{s}$$

The required number of units is equal to the available discharge divided by the discharge in each unit, or $2100/351 = 5.98$. Hence, six units are required.

PROBLEMS

Elementary Theory

12.1. Determine the torque, power, and head input or output for each turbomachine shown. Is a pump or turbine implied? The following data are in common: outer radius 300 mm, inner radius 150 mm, $Q = 0.057$ m³/s, $\rho = 1000$ kg/m³, and $\omega = 25$ rad/s.

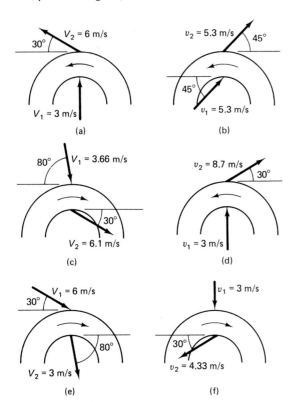

(a)

(b)

(c)

(d)

(e)

(f)

12.2. A centrifugal water pump rotates at 800 rpm. The impeller has uniform blade widths $b_1 = 50$ mm, $b_2 = 25$ mm, and radii $r_1 = 40$ mm, $r_2 = 125$ mm. The blade angles are $\beta_1 = 45°$, $\beta_2 = 30°$. Assuming no angular momentum of fluid at the blade entrance, determine the ideal flow rate, pressure head rise across the impeller, and the theoretical torque requirements.

12.2E. A centrifugal water pump rotates at 800 rpm. The impeller has uniform blade widths $b_1 = 2$ in., $b_2 = 1$ in., and radii $r_1 = 1.5$ in., $r_2 = 5$ in. The blade angles are $\beta_1 = 45°$, $\beta_2 = 30°$. Assuming no angular momentum of fluid at the blade entrance, determine the ideal flow rate, pressure head rise across the impeller, and the theoretical torque and power requirements.

12.3. The suction and discharge pipes of a pump are connected to a differential mercury manometer. The diameter of the suction pipe is 200 mm and the diameter of the discharge pipe is 150 mm. The specific gravity of the pumped liquid is $S = 0.81$, and the manometer reading is $H = 600$ mm. Determine the pressure head rise across the pump if the discharge is $Q = 115$ L/s. The centerline of the suction and discharge pipes are at the same elevation.

12.4. A centrifugal pump with the dimensions $r_2 = 60$ mm, $b_2 = 10$ mm, $\beta_2 = 60°$, is pumping kerosene ($S = 0.80$) at a rate of 7.5 L/s and rotating at 2000 rpm. For "best design" entry conditions, determine the theoretical pressure head rise and fluid power.

12.4E. A centrifugal pump with the dimensions $r_2 = 2.5$ in., $b_2 = 0.40$ in., $\beta_2 = 60°$, is pumping kerosene ($S = 0.80$) at a rate of 2 gal/sec and rotating at 2000 rpm. For "best design" entry conditions, determine the theoretical pressure head rise and fluid power.

12.5. An axial-flow pump has a stator blade positioned upstream of the impeller, and it provides an angle $\alpha_1 = 60°$ to the flow. The radius of the impeller tip is 285 mm and the hub radius is 135 mm. Determine the required average blade angle at the exit of the impeller if the pump is to deliver 0.57 m³/s of water with a theoretical head rise of 2.85 m. The rotational speed is 1500 rpm.

12.6. Water flows through the pump at a rate of 50 L/s. The allowable *NPSH* provided by the manufacturer at that flow is 3 m. Determine the maximum height Δz above the water surface that the pump can be located to operate without cavitating. Include all losses in the suction pipe.

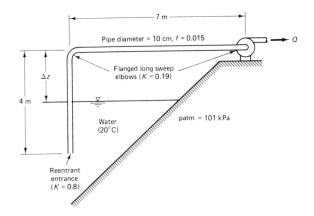

12.7. During a test on a water pump, no cavitation is detected when the gage pressure in the pipe at the pump inlet is −68.5 kPa and the water temperature is 25°C. The inlet pipe diameter is 100 mm, the head across the pump is 35 m, and the discharge is 50 L/s.
(a) Compute the *NPSH* and the cavitation number if the atmospheric pressure is 101 kPa.
(b) If the pump is to produce the same head and discharge at a location where the atmospheric pressure is 83 kPa, what is the necessary change in elevation of the pump relative to the inlet reservoir to avoid cavitation?

12.7E. During a test on a water pump, no cavitation is detected when the gage pressure in the pipe at the pump inlet is −9.9 psi, and the water temperature is 80°F. The inlet pipe diameter is 4 in., the head across the pump is 115 ft, and the discharge is 13 gal/sec.
(a) Compute the *NPSH* and the cavitation number if the atmospheric pressure is 14.7 psi.
(b) If the pump is to produce the same head and discharge at a location where the atmospheric pressure is 12 psi, what is the necessary change in elevation of the pump relative to the inlet reservoir to avoid cavitation?

Dimensional Analysis

12.8. Consider the 371-mm-diameter pump curve of Fig. 12.9. If the pump is run at 1200 rpm to deliver 0.8 m³/s of water, find the resulting head rise and fluid power.

12.9. In Problem 12.8, how high above the suction reservoir can the pump running at 1200 rpm be located if the water temperature is 50°C and the atmospheric pressure is 101 kPa?

12.10. A pump is needed to transport 150 L/s of oil ($S = 0.86$) with an increase in head across the pump of 22 m when running at 1800 rpm. What type of pump is best suited for this operation?

12.11. A pump is required to deliver 0.17 m³/s of water with a head increase of 104 m when operating at maximum efficiency at 2000 rpm. Determine the type of pump best suited for this duty.

12.12. Refer to Fig. 12.13, the dimensionless performance curves for an axial-flow pump. It is desired to deliver water at a rate of 1.25 m³/s and a speed of 750 rpm. Determine:
(a) The available head, diameter, power, and *NPSH* requirements.
(b) The actual head-discharge and power-discharge curves.

12.12E. Refer to Fig. 12.13, the dimensionless performance curves for an axial-flow pump. It is desired to deliver water at a rate of 45 ft³/sec and a speed of 750 rpm. Determine:
(a) The available head, diameter, power, and *NPSH* requirements.
(b) The actual head-discharge and power-discharge curves.

12.13. A pump running at 400 rpm discharges water at a rate of 85 L/s. The total head rise across the pump is 7.6 m and the efficiency is 0.7. A second pump, whose linear dimensions are two-thirds of the former, runs at 400 rpm under dynamically similar conditions; it is to be placed on a space satellite where simulated gravity produces a gravitational field which is 50% that of Earth's. What are the discharge, head rise, and power requirements for the second pump?

12.14. A pump is designed to operate under optimum conditions at 600 rpm when delivering water at 22.7 m³/min against a head of 19.5 m. What type of pump is recommended?

12.15. A pump operating under the conditions stated in Problem 12.14 has a maximum efficiency of 0.70. If the same pump is now required to deliver water at a head of 30.5 m at maximum efficiency, determine the rotational speed, the discharge, and the required power.

12.16. A manufacturer desires to double both the discharge and head on a geometrically similar pump. Determine the change in rotational speed and diameter under the new conditions.

12.17. A pump is required to move water at 0.182 m³/s against a head of 60.6 m. The approximate rotational speed is 1400 rpm. Using the appropriate dimensionless performance curve in the text, determine the actual speed and size of the pump to operate at peak efficiency.

12.17E. A pump is required to move water at 2900 gal/min against a head of 200 ft. The approximate rotational speed is 1400 rpm. Using the appropriate dimensionless performance curve in the text, determine the actual speed and size of the pump to operate at peak efficiency.

12.18. Select the type, size, and speed of pump to deliver a liquid ($\gamma = 8830$ N/m³) at 0.66 m³/s with fluid power requirements of 200 kW. (*Hint:* Assume that the pump speed can vary between 1000 and 2000 rpm.)

Use of Turbopumps

12.19. Recommend the appropriate machine to pump 1 m³/s of water if the impeller speed is 600 rpm.

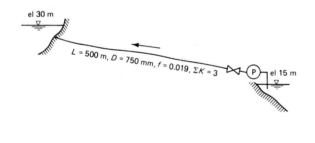

el 30 m

$L = 500$ m, $D = 750$ mm, $f = 0.019$, $\Sigma K = 3$ P   el 15 m

12.20. An oil ($S = 0.85$) is to be pumped through a pipe as shown. Using the characteristic curves of Fig. 12.12, determine the required impeller diameter, speed, and input power to deliver the oil at a rate of 14.2 L/s. Would you expect the efficiency of the pump to be the value shown on the curve?

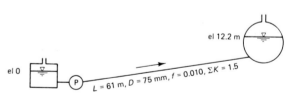

el 12.2 m

el 0   P $L = 61$ m, $D = 75$ mm, $f = 0.010$, $\Sigma K = 1.5$

12.21. The following performance curves were obtained from a test on a 216-mm double-entry centrifugal pump moving water at a constant speed of 1350 rev/min:

Q (m³/min)	0	0.454	0.905	1.36	1.81	2.27	2.72	3.80
H (m)	12.2	12.8	13.1	13.4	13.4	13.1	12.2	11.0
η_P	0	0.26	0.46	0.59	0.70	0.78	0.78	0.74

Plot H versus Q, η_P versus Q, and $\dot{W}_P$ versus Q. If the pump operates in a system whose demand curve is given by $5 + Q^2$, find the discharge and power required. In the demand curve, Q is given in cubic meters per minute.

12.22. With reference to the pump data in Problem 12.21, if the pump is run at 1200 rpm, find the discharge, head, and required power. Is the operating efficiency under these conditions the same as the pump running at 1350 rpm?

12.23. For the pump data presented in Problem 12.21, plot the dimensionless pump curves C_H versus C_Q, $C_{\dot W}$ versus C_Q, and η_P versus C_Q. What is the specific speed of the pump? (*Note:* For double entry pumps, use one-half the discharge when computing the specific speed.)

12.24. A closed-loop flow facility is to be used to study water flow in a hydraulics laboratory. It is constructed of 300-mm-diameter hydraulically smooth pipe. The loss coefficient at each of the four bends is 0.1, and the total length of the loop is 14 m. The design velocity is $V = 3$ m/s. It is proposed that an axial flow pump running at about 300 rpm be used. Verify that an axial-flow pump is the correct type to employ.

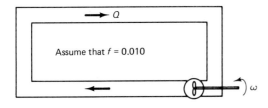

12.25. In Problem 12.24, assume that the only axial pump available is one from the family shown in Fig. 12.13. If the diameter of the impeller were restricted to the pipe diameter, what would be the rotational speed and resulting head rise across the pump?

12.26. The 240-mm-diameter pump represented in Fig. 12.6 is used to move water in a piping system whose demand curve is $62 + 270Q^2$, where Q is in cubic meters per second. Find the discharge, required input power, and the *NPSH* requirement for:

(a) One pump.

(b) Two pumps in parallel.

12.27. It is desired to pump 600 m³/h of liquid ammonia ($\rho = 607$ kg/m³) at a speed of 2500 rpm and a pressure rise of 1000 kPa.

(a) Determine the type of pump best suited for this duty.

(b) Using the appropriate dimensionless pump characteristic curve, find the diameter and number of stages required to meet the pumping requirements.

12.28. Crude oil ($S = 0.86$) is to be pumped through 5 km of 450-mm-diameter cast iron pipe. The elevation rise from the upstream to the downstream end is 195 m. If an available pump is the 240-mm-diameter radial-flow machine of Fig. 12.6, how many pumps in series are required to provide the most efficient operation? Find the required power.

12.28E. Crude oil ($S = 0.86$) is to be pumped through 3 miles of 18-in.-diameter cast iron pipe. The elevation rise from the upstream to the downstream end is 640 ft. If an available pump is the 240-mm-diameter radial-flow machine of Fig. 12.6, how many pumps in series are required to provide the most efficient operation? Find the required power.

Turbines

12.29. A Francis turbine has the following dimensions: $r_1 = 4.5$ m, $r_2 = 2.5$ m, $b_1 = b_2 = 0.85$ m, $\beta_1 = 75°$, $\beta_2 = 100°$. The angular speed is 120 rpm, and the discharge is 150 m³/s. For no separation at the entrance to the runner, determine the guide vane angle, theoretical torque, head, and power.

12.30. A model Francis turbine at 1/5 full scale develops 3 kW at 360 rev/min under a head of 1.8 m. Find the speed and power of the full-size turbine when operating under a head of 5.8 m. Assume that both units are operating at maximum efficiency.

12.31. A model is to be built for studying the performance of a turbine having a runner diameter of 1 m, maximum output of 2200 kW under a head of 50 m and

a speed of 240 rpm. Determine the diameter of the model runner and its speed if the corresponding model power is 9 kW and the head is 7.6 m.

12.31E. A model is to be built for studying the performance of a turbine having a runner diameter of 3 ft, maximum output of 2200 kW under a head of 150 ft and a speed of 240 rpm. Determine the diameter of the model runner and its speed if the corresponding model power is 9 kW and the head is 25 ft.

12.32. A hydroelectric site has available $H_T = 80$ m and $Q = 3$ m³/s. It is proposed to use a Francis turbine unit with the operating characteristics at optimum efficiency shown on Fig. 12.24. Determine the required speed and diameter of the machine.

12.33. Determine the power output, the type of turbine, and the approximate speed for the installation shown. Neglect all minor losses except those existing at the valve.

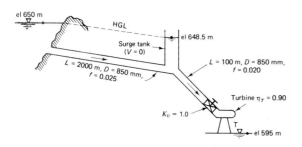

12.34. A water turbine is required to operate at 420 rpm under a net head of 3 m, a discharge of 0.312 m³/s, and with an efficiency of 0.9. Tests on a $\frac{1}{6}$-scale model running at 2000 rpm are to be carried out using water. For the model turbine, determine the head drop, flow rate, output power, and expected efficiency.

12.35. A Pelton wheel develops 4.5 MW under a head of 120 m at a speed of 200 rpm. The wheel diameter is eight times the jet diameter. Use the experimental data of Fig. 12.29 at maximum efficiency to determine the required flow, wheel diameter, diameter of each jet, the number of jets required, and the specific speed.

12.36. A reaction turbine installation typically includes a draft tube (Fig. 12.26a), whose function is to convert kinetic energy of the fluid exiting the runner into flow energy. Consider a turbine operating at $Q = 85$ m³/s and $H_T = 31.8$ m. The radius of the draft tube at the outlet of the runner is 2.5 m and the diameter at the end of the tube is 5.0 m. Losses in the draft tube can be neglected.

(a) For water at 20°C, what is the pressure at the runner outlet if in Fig. 12.26a, $z_2 - z_1 = 2.5$ m?

(b) What is the permissible $z_2 - z_1$ if the allowable cavitation number is $\sigma = 0.14$?

12.37E. The Ludington pumped-storage system on the eastern shore of Lake Michigan consists of six penstocks delivering a combined discharge of 73,530 ft³/sec in the power-producing mode of operation. Each turbine generates 427,300 hp at an efficiency of 0.85. What diameter penstock is required for the system to operate under design conditions? Classify the type of turbine employed.

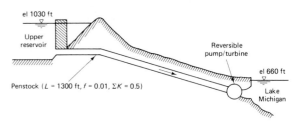

12.38. A Francis turbine of the same family as that represented in Fig. 12.24 is running at a speed of 480 rpm under a head of 9.5 m. Determine the runner diameter, discharge, and developed power if the turbine is operating at maximum efficiency.

12.39. The available head from reservoir level to the four nozzles of a twin-runner Pelton wheel is 305 m. The length of the supply pipe, or penstock, is 3 km with a friction factor $f = 0.02$ and $\Sigma K = 2.0$. The turbine develops 10.4 MW with an efficiency of 0.85.

(a) If the head across the turbine is 95% of the available head, what is the diameter of the penstock?

(b) If $C_v = 0.98$ for each nozzle, calculate the diameter of each jet.

12.40. In a projected low-head hydroelectric scheme, 282 m³/s of water is available under a head of 3.7 m. It is proposed to use Francis turbines with specific speed 2.42, for which the rotational speed is 50 rpm. Determine the number of units required and the power to be developed by each machine. Assume an efficiency of 0.9.

12.41. Repeat Problem 12.40 if propeller turbines are substituted for the Francis units. Assume a specific speed of 4.15 and an efficiency of 0.9.

THIRTEEN

Measurements in Fluid Mechanics

13.1 INTRODUCTION

In the engineering laboratory and in many industrial situations, the need to measure fluid properties and various flow parameters, such as pressure, velocity, and discharge, are exceedingly important. In many instances these needs are obvious. Examples include the flow rate in a pipe or irrigation channel; the contaminant or sediment load in a river; the peak pressures on the surface of a high-rise building or the flow patterns around that building; the drag on an automobile or truck traveling at high speeds; the velocity field about a commercial aircraft; the wind velocity profile above a hilly terrain; and the size and distribution of ocean or lake waves.

Fluid mechanics measurements are not exclusive to the engineer. Medical diagnosticians are interested in monitoring the functions of the cardiovascular and pulmonary systems in the human body. The movement of fluids in the agricultural, petroleum, gas, chemical, beverage, and water supply and wastewater industries require large investments annually; the uncertainties present in the measurement of these flows can have a significant impact on material and monetary considerations.

Many devices have been developed to measure flow parameters; obviously, each was designed to serve a specific purpose. Before undertaking the task of measurement, it is important to define clearly the need for measuring a particular parameter. Knowledge of the underlying fluid mechanics and the physical principles involved are essential to selecting an appropriate measuring instrument and conducting a successful measurement. In addition, we must recognize that there are differences in various types of measurements. Consider, for example, discharge readings resulting from a measurement of a velocity profile integrated over a cross section; the fluctuating velocity can be time-averaged at a specific location to produce a mean value. On the other hand, instantaneous measurements of velocity in a very small sampling volume may be required to observe the transient nature of an unsteady flow. The degree of sophistication in recording a parameter can range from a simple visual reading of a manometer or dial to high-speed digital sampling of voltages representing a number of variables.

The purpose of this chapter is to provide the reader with a basic introduction to the concepts and techniques applied by engineers who measure flow parameters either in the laboratory or in an industrial environment. Included is a general overview of experimental methods commonly used in fluid mechanics teaching and research laboratories; the treatment is not exhaustive, but references are provided for additional information. In the following two sections, methods and instrumentation employed to measure pressure, velocity, and discharge are presented. Next, a discussion of flow visualization is given, and the final section is devoted to data acquisition and methods of analyzing data.

13.2 MEASUREMENT OF LOCAL FLOW PARAMETERS

A local flow measurement means that a quantity is measured over a relatively small sample volume of the fluid. Usually, the volume is sufficiently small so that we can say that the measurement represents the magnitude of the quantity at a point in the flow field. Two significant local flow quantities are pressure and velocity; others are temperature, density, and viscosity. Only the first two are discussed in this section; the last three are not considered in this book.

13.2.1 Dynamic Response and Averaging

Flow measurements may be classified according to whether the flow is steady or unsteady. If the magnitude of a physical quantity remains constant with time, this value is referred to as the steady-state value. Conversely, if the quantity changes with time, the measurement is transient, or unsteady. Transient measurement demands a more highly specialized measuring apparatus. Most measuring instruments require a certain amount of time to respond to the physical quantity sensed. In transient measurements this response time should be much less than the time for a significant change in the physical quantity to occur. In steady-state measurements the only concern is to take a reading after the measuring system has time to respond to the flow conditions. In many situations, for example, turbulent flow, it is desired to obtain a time-average measurement of a quantity, even though the local flow conditions are unsteady; examples are Reynolds stress and turbulence intensity. In this situation, a time-averaging mechanism must be included in the measurement. In addition, an instrument must be sufficiently small to measure the spatial variation of the measured parameter, such as the wavelength of a pressure disturbance. Generally, the issue of spatial and temporal resolution must be resolved by selecting instrumentation compatible with the flow scale and with the scale of any flow disturbance; see Goldstein (1983) for further explanation.

13.2.2 Pressure

Fluid pressures are measured in many different ways. The type of instrument utilized depends on the levels of precision and detail required for the particular application. Virtually all pressure measurements are based either on the manometer principle or on the concept of pressure deforming a solid material such as a crystal, membrane, tube, or plate, and then converting that deformation to an electrical signal or mechanical readout. Pressure measurements are in either the static or dynamic mode. Time-dependent pressures result from unsteadiness in the flow and from pressure disturbances; the pressure disturbances are a result of either hydro-

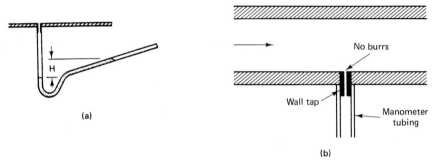

Figure 13.1 Manometer used to measure pressure: (a) inclined tube manometer; (b) piezometer opening

dynamic or acoustic perturbations. Often, static pressure measurements are based on time-averaged mean values, since unsteadiness is always present when the flow is turbulent. Several representative pressure gages and transducers are described below.

Manometer

The manometer was analyzed in Section 2.4.3. It is a simple, inexpensive instrument for measuring static pressure, and it has no moving mechanical parts. High accuracy can be attained by using an inclined tube manometer (Fig. 13.1a) or a micromanometer (Fig. 2.6). If the pressure tap is properly machined normal to the interior wall in a duct or pipe with no burrs, and the diameter is sufficiently small (usually about 1 mm diameter), the static pressure is measured accurately (Fig. 13.1b). Often the taps are placed circumferentially around the duct in the form of a piezometer ring to average out any type of flow irregularities.

Bourdon Gage

An inexpensive pressure gage for reading relatively large static pressures is the Bourdon tube (Fig. 13.2). It is a curved, flattened tube, which will expand outward when subjected to internal pressure. A mechanical

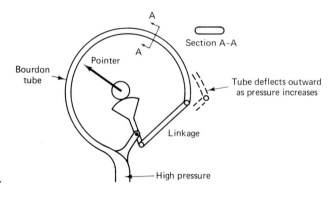

Figure 13.2 Bourdon tube pressure gage.

linkage attached to a dial gage will provide a direct reading of the pressure. When properly designed, these gages will provide good accuracy; however, they should not be used in situations where large pressure pulsations are present, since damage to the tube may result.

Pressure Transducer

Pressure transducers are electromechanical devices that are designed to measure either steady or unsteady hydrodynamic or acoustic pressures. They consist of flexible membranes or piezoelectric elements that respond to pressure changes (Fig. 13.3). All of them require accompanying electronic circuitry, which adds to their costs. Since their output is an electric signal, they can readily be interfaced with data acquisition systems for automatic storage and retrieval of data.

The condenser microphone (Fig. 13.3a) is commonly utilized in aeroacoustics. Basically, it is a capacitor whose capacitance varies as the membrane is deflected due to applied pressure. These transducers are quite delicate, since they measure very small pressures (100 μbar) and hence are not suitable for use in liquids. The piezoelectric transducer, commonly used with liquids, has a crystalline material (e.g., quartz) that produces an electric field when deformed (Fig. 13.3b). These rugged units

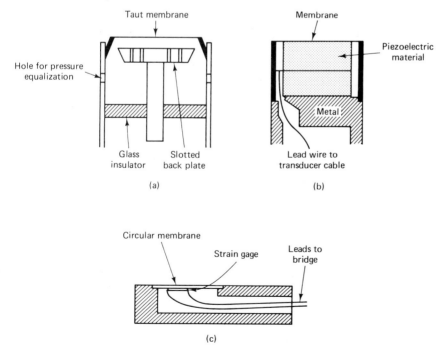

Figure 13.3   Pressure transducers: (a) condenser microphone; (b) piezoelectric pressure transducer; (c) strain-gage transducer. (Adapted from Goldstein, 1983.)

can be made very small (diameters down to 2.5 mm are possible) and can be mounted flush with the wall of a pipe. They respond very rapidly and can measure a wide range of dynamic pressures (full vacuum to 20 MPa). A strain gage transducer is another type of transducer employed with liquids; it makes use of a strain gage attached to the membrane that translates its deflection into an electrical signal (Fig. 13.3c). It is subject to DC drift and sensitivity changes; however, it is rugged, reliable, and insensitive to vibration.

13.2.3 Velocity

The fluid velocity is of primary interest in a fluid flow. By measuring the velocity we can calculate the flow rate, and perhaps even piece together an image of the streamline pattern, identifying regions of separated flow, regions of stagnated flow, and other flow characteristics. There are three categories of velocity measurements: local (at a point), spatially averaged, and velocity field measurements. The discussion in this section pertains to the first two; the third is covered in Section 13.4.

Local velocities are actually measured over a small region of flow. The required precision varies, depending on the desired application. For example, time-averaged velocities measured at various points in a flow cross section may be recorded so that the velocity distribution can be integrated to provide the discharge, or flow rate. On the other hand, it may be necessary to determine the history of the time-dependent turbulent velocity components at a particular location in a region of the flow.

Instruments for velocity measurements vary considerably in their complexity and cost, depending on the type of measurement required. The desire to measure unsteady turbulent velocity components on a relatively small local scale has popularized the use of the thermal anemometer as well as the laser-doppler velocimeter. Less complicated instruments that usually measure velocity over a large spatial region are, for example, the pitot-static probe and the propeller anemometer; they are more suitable for measuring velocities that either are steady or are slowly varying with time.

Particle Velocity Measurement

Neutrally-bouyant particles can be seeded in the flow to provide a visual image of the flow field; two of these, hydrogen bubbles in water and particles in air, are discussed in Section 13.4. Other tracer particles used in water are neutrally bouyant plastic beads, immiscible dyed droplets, and floats placed on a liquid surface or at a desired depth. It is important to ascertain if the particle is actually tracking the fluid motion. Very small particles that are natural impurities in the flow can be observed with appropriate optical equipment to measure velocities in many flows, especially microscopic flow fields.

Pitot-Static Probe

The pitot-static probe is analyzed in Section 3.4. It measures the local mean velocity by using the Bernoulli equation. Assuming frictionless, steady flow the velocity u is given by

$$u = \sqrt{\frac{2}{\rho}(p_T - p)} \qquad (13.2.1)$$

in which p_T is the total or stagnation pressure and p is the static pressure. Equation 13.2.1 is valid only if the probe does not greatly disturb the flow, and if it is aligned with the flow direction so that the velocity u is parallel to the pitot probe. A particular design, the so-called **Prandtl tube,** is shown in Fig. 13.4. It has the static pressure holes positioned along the horizontal tube so that the reduced pressure resulting from flow past the nose is balanced by the increased pressure due to the vertical stem. Typically, the Reynolds number based on the pitot-tube diameter should be in the neighborhood of 1000, so that viscous effects are not significant. If the flow is not parallel to the probe head, the measurement error is about 1% at a yaw angle of 5°, and if the yaw angle is greater than 5° the measurement error may be substantial (Goldstein, 1983). When the probe is placed close to a wall, the streamlines are deflected by the interaction of the probe with the wall; errors occur when the distance between the probe axis and the wall is less than two tube diameters. In a turbulent flow, the actual pressure is less than the sensed value; generally, the results are reliable if the fluctuation intensity is less than 10%.

Other probes of a similar design are the pitot or impact tube, Preston tube, static tube, and yaw meter. The pitot tube measures the total pressure, the static tube the static pressure, and the yaw meter, the direction of flow. The Preston tube is a special form of pitot tube that has been used for the measurement of wall shear stress in boundary layers over smooth

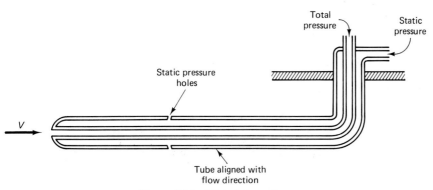

Figure 13.4   Pitot-static probe.

walls; basically, it is a hypodermic needle that senses the mean velocity adjacent to the wall. See Bean (1971) for details related to calibration and design.

Cup or Propeller Anemometer

Cup or propeller anemometers are employed to measure the velocity of either gases or liquids. They are relatively large, and as a result, the measurement is averaged spatially over a relatively large area. These anemometers are used for many applications where extreme precision is not required. One type of cup anemometer, the **current meter** (Fig. 13.5a), measures the velocity in water; similar devices can measure air velocity. Cup anemometers always rotate in one sense, and thus cannot provide the flow direction. On the other hand, the **propeller anemometer** will reverse rotation when the flow reverses (Fig. 13.5b); however, in contrast to cup anemometers, they must be aligned with the flow direction. A ducted propeller anemometer for measuring air velocities is shown in Fig. 13.5c. Unducted propellers are also employed; an example is the wind vane (Fig. 13.5d). Propellers are also placed inside pipes and used as flow meters. For all of these devices, counters or electromechanical readouts attached

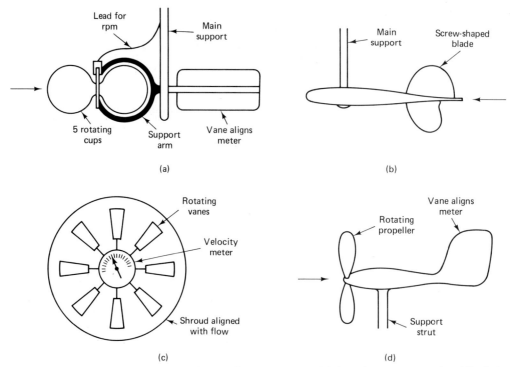

Figure 13.5   Anemometers: (a) current meter; (b) propeller anemometer; (c) ducted vane anemometer; (d) wind vane anemometer.

to sensors provide the rotational frequency, which is correlated to the velocity by means of a calibration test.

Thermal Anemometer

The **thermal** (hot wire) **anemometer** is employed to measure velocity in air. It consists of a small wire, typically made of tungsten, platinum, or platinum–iridium, which is insulated and mounted on supports (Fig. 13.6). The wire sensor is typically 2 mm in length and 5 μm in diameter, although smaller wires (0.2 mm long, 1 μm in diameter) have been used. When the wire is heated, the anemometer senses the changes in heat transfer as the flow speed varies; a representative wire temperature is 250°C.

The most common circuitry is one that operates in the constant-temperature mode (Fig. 13.6); earlier versions employed a constant-current arrangement. As the velocity past the probe is increased, the hot wire tends to cool, thereby attempting to lower the temperature and the resistance across the wire. The circuitry then increases the current to adjust the resistance and the probe temperature to their initial values, a process that appears to be instantaneous. Either the voltage of the amplifier or the sensor current is correlated with the velocity past the probe. The output is nonlinear with respect to velocity, and the sensitivity decreases as the velocity increases; often the signal is linearized electronically. The sensitivity as a percent of reading stays nearly constant so that a very wide range of velocities can be measured. When the fluctuation intensity is less than 20%, if the flow is incompressible, and if the fluid temperature and density remain constant, the hot-wire anemometer can be used to measure a velocity component. We can then calculate the mean velocity, fluctuation intensity, and turbulence spectrum. The single-wire anemometer senses only the normal component of velocity; if the wire is properly oriented, both the mean component and the fluctuations in the mean flow direction can be measured. Two-component measurements are made using a probe with wires crossed in an "×" fashion, to provide the correlation $\overline{u'v'}$ needed in turbulence measurements (u' and v' are the small

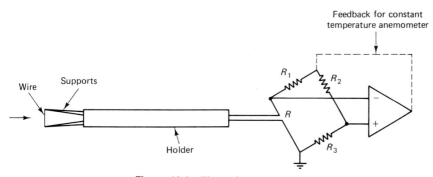

Figure 13.6 Thermal anemometer.

perturbations). Three components may be measured either by rotating the ×-probe or adding a third wire.

In liquids and gases that contain solid particles, the wire probe is replaced by a more rugged cylindrical film sensor. It operates on the same principle as the hot-wire probe, but it generates turbulence in its wake because of its larger size. This self-generated turbulence can limit the probe's capabilities for measuring actual fluctuation intensities in the flow field. In addition to measuring velocities, thermal anemometers have been adapted to monitor pressure and discharge.

Laser-Doppler Velocimeter

A device that can be used advantageously when it is desired not to have the probe immersed in the flow is the **laser-doppler velocimeter** (LDV). The most common LDV is the dual-beam type, shown in Fig. 13.7. The frequency of the scattered light is measured by the photodetector, which converts the receiving light to an electrical signal. A simplified way to understand the principle is as follows. At the intersection of the two beams a fringe pattern is formed with fringe spacing Δx. Light scattered from particles passing through the fringe pattern is modulated with a frequency f, which is directly proportional to the velocity component u normal to the fringes:

$$u = \frac{\Delta x}{\Delta t} = \frac{\lambda f}{2 \sin(\theta/2)} \qquad (13.2.2)$$

in which λ is the wavelength of the laser beam and θ is the beam angle. Detection of flow reversal is possible by using a moving fringe pattern. The ellipsoidal measurement volume is quite small, being on the order of 0.1 mm in diameter and 1.0 mm in length. A complete description of the

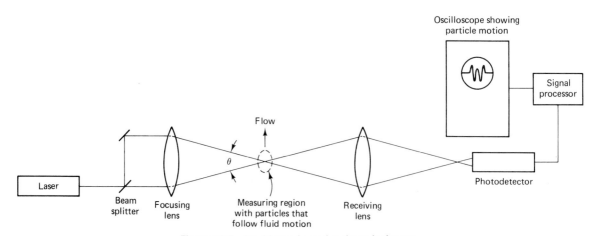

Figure 13.7 Dual-beam laser-doppler velocimeter.

phenomenon requires a detailed description of wave optics, which is beyond the scope of this book.

Velocities in both liquids and gases can be measured with an LDV. Liquids normally contain sufficient impurities which act as natural seeding agents, but gases usually are artificially seeded with extremely small liquid droplets or solid particles. Typically, an LDV can measure velocities ranging from 2 mm/s up to the supersonic range. The frequency response extends to 30 kHz, which enables turbulent velocity components to be measured. In contrast to the thermal anemometer, the LDV requires no calibration and measures flow reversal with relative ease. The initial cost of the LDV is high, and extensive electronic instrumentation is required to process the output signal: A counter-type signal processor is most commonly used. The LDV has been employed to measure supersonic flow, flow in pumps, natural free convection, steam and gas turbine flows, combusting flows, blood flows, and open-channel flows.

13.3 FLOW RATE MEASUREMENT

The measurement of flow rate—alternately termed discharge or volumetric flow—is one of the most common measurements made in flowing fluids. Numerous devices have been invented or adapted for the purpose of flow metering, ranging widely in sophistication, size, and accuracy. Basically, instruments for flow rate measurements can be divided into those that employ direct or "quantity" means of measurement, and those that are indirect, that is, the so-called "rate" meters. Quantity meters either weigh or measure a volume of fluid over a known time increment; examples are liquid weigh tanks, reciprocating pistons, and for gases, bellows and the liquid-sealed drum. These direct devices usually are relatively large and possess poor frequency response characteristics. However, they do provide high precision and accuracy, and as a result, are used most often as "primary" standards for the calibration of the indirect metering devices.

Indirect or rate meters consist of two components: the primary part, which is in contact with the fluid, and the secondary part, which converts the reaction of the primary part to a measurable quantity. They can be classified according to a characteristic operating principle: velocity–area measurement, pressure drop correlations, hydrodynamic drag, and the like. Rate meters are relatively low in cost, take up little space, and hence are commonly found in industrial and research laboratories. Several of the more common types are discussed below.

13.3.1 Velocity–Area Method

A number of the velocity measuring devices discussed in Section 13.2.3 may be inserted in a cross section of flow and used to measure the flow rate. Thus a thermal anemometer, a laser-doppler velocimeter, a pitot-static tube, or an anemometer can be inserted in a pipe or duct of known area and calibrated to measure the flow rate. In a manner that does not require calibration and for steady flow, a velocity distribution can be measured by systematically moving a velocity probe throughout the flow cross section, or by using a so-called "rake" where local velocities are measured simultaneously. Subsequently, a numerical integration of the velocity profile over the cross-sectional area will yield the discharge. This technique has been used in both closed conduit and open-channel flows.

13.3.2 Differential Pressure Meters

Differential pressure meters are widely used in industrial applications and laboratories because of their simplicity, reliability, ruggedness, and low cost. Three commonly used types are discussed in this section: the orifice meter, venturi meter, and flow nozzle. Their operation is based on the principle of an obstruction to flow being present in a duct or pipe, and consequently, a pressure differential will exist across the obstruction. This pressure drop can be correlated to the discharge by means of a calibration, and subsequently, the pressure-discharge curve can be used to determine the discharge by reading the differential pressure. In this section we deal only with the discharge of incompressible fluids in circular pipes.

   The fundamental discharge relation for the differential pressure meter can be described in the following manner. Figure 13.8 represents a thin-plate orifice meter, which can be considered as a representative differential pressure meter. Consider steady flow to occur in a circular duct, encounter the restrictive orifice with area A_0, and issue as a downstream jet. Downstream of the restriction, the streamlines converge to form a minimal flow area A_c, termed the vena contracta. Pressure taps are located at two positions: upstream of the restriction in the undisturbed flow region (location 1) and downstream at some location in the vicinity of the vena contracta (location 2). Assuming an ideal, frictionless, incompressible fluid, the Bernoulli equation applied along the center streamline from

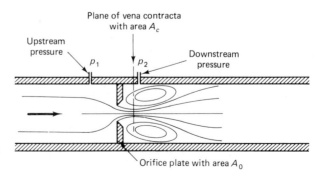

Figure 13.8   Flow through an orifice meter.

the upstream location to the vena contracta is

$$\frac{V_1^2}{2g} + \frac{p_1}{\gamma} + z_1 = \frac{V_c^2}{2g} + \frac{p_c}{\gamma} + z_c \tag{13.3.1}$$

Similarly, the continuity equation is

$$V_1 A_1 = V_c A_c \tag{13.3.2}$$

Combining Eqs. 13.3.1 and 13.3.2, and solving for V_c yields

$$V_c = \sqrt{\frac{2g(h_1 - h_c)}{1 - (A_c/A_1)^2}} \tag{13.3.3}$$

in which

$$h_1 = \frac{p_1}{\gamma} + z_1 \qquad h_c = \frac{p_c}{\gamma} + z_c \tag{13.3.4}$$

The ideal discharge Q_i is equal to the area multiplied by the average velocity at the vena contracta:

$$Q_i = A_c V_c$$
$$= \frac{A_c}{\sqrt{1 - (A_c/A_1)^2}} \sqrt{2g(h_1 - h_c)} \tag{13.3.5}$$

The actual discharge differs from the ideal for two primary reasons. Because of real fluid flow, friction causes the velocity at the centerline to be greater than the average velocity at each cross section. Second, the piezometric head h_c, evaluated at the vena contracta in the relation, is substituted with h_2, the known reading at the downstream pressure tap. Also, since the area of the vena contracta is unknown, it is convenient in Eq. 13.3.5 to replace A_c by $C_c A_0$, where C_c is the contraction coefficient. These anomalies are accounted for by introducing a discharge coefficient C_d, which is the product of the contraction coefficient and a velocity coefficient, so that the actual discharge Q is given by the relation

$$Q = \frac{C_d A_0}{\sqrt{1 - (C_c A_0/A_1)^2}} \sqrt{2g(h_1 - h_2)} \tag{13.3.6}$$

For a circular cross section, which is typical of most differential pressure meters, it is convenient to introduce the diameter ratio

$$\beta = \sqrt{\frac{A_0}{A_1}} = \frac{D_0}{D} \tag{13.3.7}$$

where D is the pipe diameter. A convenient way to express Eq. 13.3.6 is

$$Q = KA_0 \sqrt{2g(h_1 - h_2)} \tag{13.3.8}$$

in which K is the flow coefficient

$$K = \frac{C_d}{\sqrt{1 - C_c^2 \beta^4}} \tag{13.3.9}$$

A dimensional analysis would reveal that C_d and K are dependent on the Reynolds number. It is convenient to evaluate the Reynolds number either at the approach region or at the obstruction.

Orifice Meter

A thin plate orifice meter (Fig. 13.9) is typically manufactured in the range $0.2 \le \beta \le 0.8$. In the figure, two means of locating the pressure taps are shown: (1) flange taps, positioned 25 mm upstream and downstream of the orifice plate, and (2) taps placed one diameter upstream and one-half diameter downstream of the plate. The second arrangement is preferred, since it is capable of sensing a larger differential pressure, and it conforms to geometric similarity laws. A third arrangement, not shown in the figure, has the pressure taps located in the pipe wall immediately upstream and downstream of the orifice; taps placed at this location have been termed corner taps.

Figure 13.10 shows experimentally determined values of the flow coefficient K for orifices as a function of β and Re_0. These data were obtained using corner taps; however, data taken with flange taps or with $D:D/2$ taps would be indistinguishable when plotted on Fig. 13.10. If

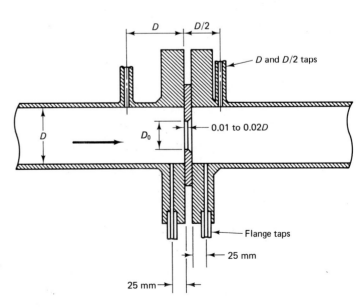

Figure 13.9 Details of a thin-plate orifice meter. (Adapted from Goldstein, 1983.)

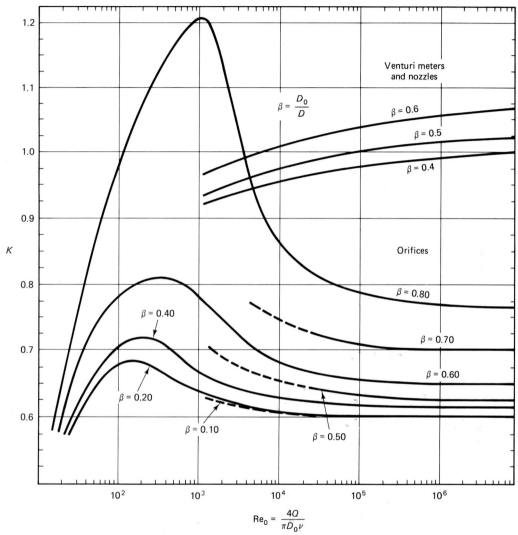

Figure 13.10 Flow coefficient K versus the Reynolds number for orifices, nozzles, and venturi meters. (Adapted with permission from Roberson and Crowe, 1990.)

greater precision is desired, numerical data for the different taps are provided in Bean (1971). Notice that for a given β, K becomes nearly constant at high Reynolds numbers, but as the Reynolds number becomes lower, K first increases to a maximum and then decreases. Maximum values of K occur at Reynolds numbers between 100 and 1000, depending on the value of β; here K is dominated by the reduced area of the vena contracta.

If the discharge is known, the Reynolds number at the orifice is known. Hence, with Fig. 13.10 one can read K directly, and subsequently determine $(h_1 - h_2)$ from Eq. 13.3.8. However, one would more likely use the figure in conjunction with Eq. 13.3.8 to determine the flow rate, given that $(h_1 - h_2)$ has been read from an attached manometer. In that situation, K is not known a priori, since it depends on Re_0. One can initially estimate K based on an assumed Re_0 (usually, assume Re_0 to be large), and subsequently by trial and error improve on that estimate by successive substitution into Eq. 13.3.8. However, if repeated readings are to be taken, it is more convenient to determine a calibration formula of the form

$$Q = C(h_1 - h_2)^m \tag{13.3.10}$$

in which C and m are constants determined by a "best fit" criterion. See Section 13.5.3 for details of one well-known criterion, the method of least squares.

The orifice has become the most widely used differential meter for measuring liquids. This is understandable, since it is relatively inexpensive and can easily be placed in an existing pipe. One disadvantage is that it produces a large head loss that cannot be recovered downstream of the orifice.

Venturi Meter

The venturi meter has a shape that attempts to mimic the flow patterns through a streamlined obstruction in a pipe. The classical, or Herschel, type of venturi meter is rarely used today, since its dimensions are rather large, making it cumbersome to install and expensive to fabricate. It is made up of a 21° conical inlet contraction, followed by a short cylindrical throat, leading to a 7° or 8° conical exit expansion. The discharge coefficient is nearly unity. By contrast, the contemporary venturi tube, shown in Fig. 13.11, consists of a standard flow nozzle inlet section (ISA 1932 standard (Bean, 1971)) and a conical exit expansion no greater than 30°. Its recommended range of Reynolds numbers is limited from 1.5×10^5 to 2×10^6.

Equation 13.3.8 is valid for the venturi as well as for the orifice; representative values of the flow coefficient K are shown in Fig. 13.10. Because of streamlining of the flow passage, the head loss in the venturi meter is much less than in the orifice. The vena contracta is not present, and as a result the discharge coefficient C_d remains close to unity.

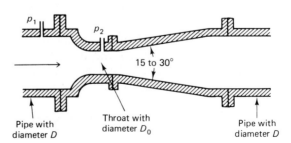

Figure 13.11   Venturi meter.

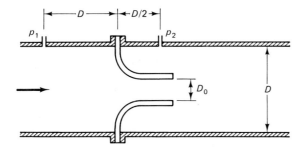

Figure 13.12   Flow nozzle.

Flow Nozzle

The flow nozzle is illustrated in Fig. 13.12. It consists of a standard-ized shape with pressure taps typically located one diameter upstream of the inlet and one-half diameter downstream. There are two standard shapes, either the long radius or the short radius. Due to the lack of an expansion section downstream of the nozzle, the total head loss is similar to that of an orifice, except that the vena contracta is nearly eliminated and the discharge coefficient is nearly unity. Similarly, when the pressure taps are located at $D:D/2$, the flow coefficient K varies with the Reynolds number in a manner nearly identical to that for the venturi meter, as shown in Fig.13.10. The flow nozzle has an advantage over the orifice plate in that it is less susceptible to erosion and wear, and relative to the venturi meter, it is less expensive and simpler to install.

Elbow Meter

A relatively simple meter can be fabricated by locating pressure taps at the outside and inside of a pipe elbow (Fig. 13.13). By applying the linear momentum equation to a control volume encompassing the fluid in the elbow (or Euler's equation normal to the streamlines), one can derive the flow equation

$$Q = KA\sqrt{\frac{R}{D}\frac{\Delta p}{\rho}} \qquad (13.3.11)$$

in which R is the radius of curvature of the elbow and Δp is the pressure difference generated by centrifugal force across the bend. The flow coeffi-

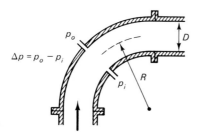

Figure 13.13   Elbow meter.

cient K can be most accurately determined by in-place calibration of the meter. However, a relation for K has been reported (Bean, 1971) for 90° elbows with pressure taps located in a radial plane 45° from the inlet:

$$K = 1 - \frac{6.5}{\sqrt{Re}} \qquad (13.3.12)$$

which is valid when $10^4 \leq Re = VD/\nu \leq 10^6$, and $R/D \geq 1.5$. The elbow meter costs much less than any of the differential pressure meters, and it usually does not add to the overall head loss of the piping system since elbows are often present in any case.

13.3.3 Other Types of Flow Meters

Turbine Meter

The turbine meter consists of a propeller mounted inside a duct that is spun by the flowing fluid. The propeller's angular speed is correlated to the discharge; the angular rotation is measured in a manner identical to the anemometer discussed in Section 13.2.3. Accuracies up to ±0.25% of the flow rate are attainable over a relatively large discharge range (Goldstein, 1983). Viscous effects become a limiting factor at the low end, whereas at the high end, accuracy is limited by the interaction between the blade tips and the electronic sensors. Each type of turbine meter must be calibrated individually. They are commonly employed to monitor flow rates in fuel supply lines.

Rotameter

The rotameter consists of a tapered tube in which the flow is directed vertically upward (Fig. 13.14). A float moves upward or downward

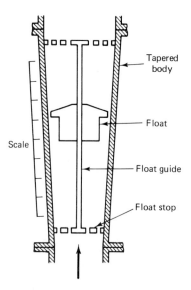

Figure 13.14 Rotameter.

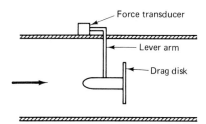

Figure 13.15   Target meter.

in response to the flow rate until a position is reached where the drag force on the float is in equilibrium with its submerged weight. Calibration consists of correlating the vertical elevation of the float with the discharge. With appropriate design, the float position can be made linearly proportional to the discharge, or if desired, another relation, such as position logarithmically proportional to discharge, can be formulated. The head loss depends on the friction loss of the tube plus the loss across the floating element. The rotameter does not provide accuracy as good as the differential pressure meters; typically it is in the range of 5% full scale.

Target Meter

Figure 13.15 shows a target meter, which consists of a disk suspended on a support strut immersed in the flow. The strut is connected to a lever arm, or alternatively, has a strain gage bonded to its surface. Fluid drag on the disk will cause the strut to flex slightly, and the recorded drag force can be related to the discharge. The assembly must be sufficiently stiff so that the drag can be measured without movement or rotation of the disk; otherwise, the drag characteristics would be altered. One advantage of this device is that it is possible to record flow reversal by measuring the sense of the drag force; hence it can be used as a bidirectional flow meter. Drag meters are fairly rugged, and can be used to measure discharges in sediment-laden fluids.

Electromagnetic Flowmeter

This is a nonintrusive device that consists of an arrangement of magnetic coils and electrodes encircling the pipe (Fig. 13.16). The coils are insulated from the fluid and the electrodes make contact with the fluid. Sufficient electrolytes are dissolved in the fluid so that it becomes capable of conducting an electric current. When it passes through the magnetic field generated by the coils, the liquid will create an induced voltage proportional to the flow. Electromagnetic flowmeters have been used for many applications, including blood flow and seawater measurements. Commercial models are available in a wide range of diameters; they are costly and can be used only with liquids.

Acoustic Flowmeter

The ultrasonic, or acoustic, flowmeter is based on one of two principles. The first one uses ultrasonic transmitters/receivers beamed across a

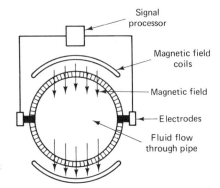

Figure 13.16 Electromagnetic flowmeter.

flow path (Fig. 13.17). The measured differences in the travel times (or inversely, the frequencies) are directly proportional to the average velocity, and consequently, the discharge. The second type is based on the Doppler effect, in that acoustic waves are transmitted into the flow field and subsequently scattered by seeded particles or contaminants. The Doppler shift recorded between the transmitter and the receiver is then related to the discharge. Like the electromagnetic flowmeter, acoustic flowmeters are nonintrusive, and have been used in both piping and free-surface flow systems.

Open-Channel Flowmeters

In Section 10.4.3, flow measurement in open-channel systems is presented. Discharge relations for the broad-crested weir and sharp-crested weir are developed, and additional methods of flow measurement are discussed. The reader is referred to that section for further information.

13.4 FLOW VISUALIZATION

Transparent flow fields have been visualized in many different ways. This is well illustrated by the hundreds of photographs presented in the publication *An Album of Fluid Motion* (Van Dyke, 1982). The book shows

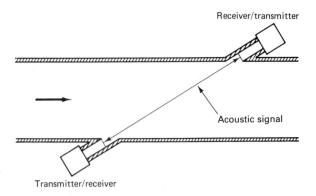

Figure 13.17 Acoustic flowmeter.

flows visualized with smoke, dye, bubbles, particles, shadowgraphs, schlieren images, interferometry, and other techniques. Usually, a method is selected that best shows the flow features of interest. Examples include particles used to visualize pathlines in liquid flows around submerged objects, dye released to study the mixing process in a stream, smoke released at the upwind end of a wind tunnel to study the development of a boundary layer, and airflow patterns above a solid surface visualized by coating that surface with a viscous liquid (streaky lines are formed which coincide with the streamlines near the surface). Several of these techniques are discussed in this section.

Pathlines, streaklines, and streamlines have been defined in Chapter 3. The purpose of defining them was to provide assistance in describing the motion of a flow field. A pathline is physically generated by following the motion of an individual particle such as a bubble or small neutrally bouyant sphere over a period of time; this could be achieved by time-lapse photography or video recording. On the other hand, if a trace from smoke, a train of extremely small bubbles, or dye continuously emanating from a stationary source is photographed or recorded, a streakline is observed. When the generation of a streakline is interrupted periodically, time streaklines are produced.

13.4.1 Tracers

Tracers are fluid additives that permit the observation of flow patterns. An effective tracer does not alter the flow pattern, but is transported with the flow and is readily observable. It is important that tracers are not affected by gravitational or centrifugal forces resulting from density differences. Furthermore, their size should be at least one order of magnitude smaller than the length scale of the flow field. Several of them are discussed below.

Hydrogen Bubbles

A very thin metallic wire can be placed in water to serve as the cathode of a direct-current circuit, with any suitable conductive material acting as an anode. When a voltage is supplied to the circuit, hydrogen bubbles will be released at the cathode and oxygen bubbles at the anode. The primary reaction is the electrolysis of water in the manner $2H_2O \rightarrow 2H_2 + O_2$. Usually, the hydrogen bubbles are used as the tracer since they are smaller than the oxygen bubbles, and more of them are formed. The bubbles are transported by the flow field away from the cathode wire as a continuous sheet. If the voltage is pulsed, discrete streaklines are formed; Fig. 13.18 shows such a pattern. Standard water supplies usually contain sufficient electrolytes to sustain a current, although addition of an electrolyte such as sodium sulfate greatly increases bubble generation. Use of extremely fine wire (0.025 to 0.05 mm in diameter) with adequate velocities (Reynolds numbers based on the wire diameter should be less than 20) may create bubbles sufficiently small to negate effects of bouyancy.

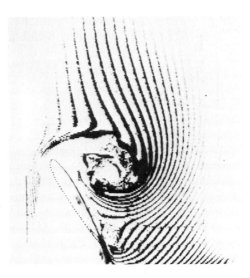

Figure 13.18 Hydrogen bubble streaklines showing separated flow around a rotating airfoil. (Courtesy of M. Koochesfahani.)

Chemical Indicators

Certain organic chemicals change color when the pH of water changes. For example, thymol blue solution is yellow at pH 8.0 and blue at pH 9.2; phenol red solution is yellow at pH 6.8 and red at pH 8.2. Thus a change in pH caused by the injection of a base solution changes the color of the liquid. The lifetime of the colored water depends on the molecular diffusion of the hydrogen ions and the turbulence. An advantage of this technique is that color residues can be eliminated by changing the pH. One application is in studies of solid–liquid phase transport phenomena; if the chemical indicator is in contact with a metal surface and the metal surface is given a negative charge, the solution in immediate contact with the plate will change color.

Particles in Air

Particles in air may be introduced as helium bubbles. By properly selecting the mixture of soap and water, one can produce helium-filled soap bubbles that are nonbouyant; it is possible to produce bubble diameters on the order of 4 mm. Solid particles or liquid droplets could also be utilized, but they must be extremely small to avoid gravitational effects. With these small sizes one must employ an extremely strong light source to visualize the flow.

Smoke has been used successfully to study the detailed structure of complex flow phenomena. It is the most popular agent used for flow visualization in wind tunnels. One injection technique is the so-called smoke-wire method, where the smoke is generated by vaporizing oil from a fine electrically heated wire. The method can be applied to flows where the Reynolds number based on the wire diameter is less than 20. An example of such a flow structure is given in Fig. 13.19. Smoke can also be

Figure 13.19   A turbulent wake behind a circular cylinder at Re = 1760. (Courtesy of R. Falco.)

released from a small-diameter tube or "rake" to create one or more streaklines.

Particle Image Velocimetry

Recent technological advances in simultaneous measurement of two- and three-dimensional flow domains have led to the definition of a generic pulsed light velocimeter (Fig. 13.20). It consists of a pulsed light source that illuminates small fluid-entrained particles over short exposure times, and a camera that is synchronized with the light to record the location of the particles. The velocity of each marker is then given by $\Delta x / \Delta t$, where Δx is the displacement of the marker and Δt is the time between exposures. The markers are usually particles that vary in size from 1 to 20 μm. In addition, progress has been made in the use of molecular

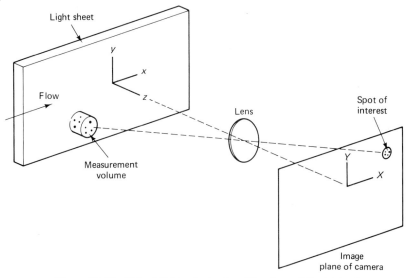

Figure 13.20   Pulsed light velocimeter. (Courtesy of R. Adrian.)

markers such as photochromic dyes in liquids, photochromic aerosols in gases, and molecular fluorescence in both liquids and gases. Particle image velocimetry is particularly useful in the study of unsteady flows, such as explosions, startup transients in channels and ducts, bubble growth and collapse, fluid–structure interaction, and in the structure of turbulent flow. A photograph and the resulting scaled velocity vectors are shown in Fig. 13.21.

Tufts and Oil Films

Flow visualization along a surface can be accomplished by attaching tufts to the surface, or if flow away from the surface is to be observed, they may be supported on wires. Surface tufts may be used to observe the transition from laminar to turbulent motion. They may also be used to study flow separation qualitatively, where the violent motion of the tufts or their tendency to point in the upstream direction identifies that separation is taking place. In air, surface flow patterns can also be visualized by distributing oil over the solid surface in such a manner that clear streaky patterns develop. Interpretation of the patterns is not always simple, particularly when flow reversals occur; the reason is that streak patterns do not indicate the flow direction.

Photography and Lighting

Many kinds of cameras are used for flow visualization, including still, stereo, and cinematic (Adrian, 1986). The video camera has become increasingly popular due to advances in resolution, and it is possible to interface certain models with a computer to obtain quantitative data. Stroboscopes, as well as laser, tungsten, and xenon lighting, are commonly employed to visualize tracers. Depending on the application, images can be produced using steady light to generate streak lines; alternatively, images can be produced using light pulses which are triggered by the camera shutter or a stroboscope to freeze the motion. If two-dimensional visualization is desired, the flow field can be illuminated by creating a narrow sheet of light directed through a window in the test section. Three-dimensional imaging can be accomplished by stereoscopic pictures or by use of holography.

13.4.2 Index of Refraction Methods

The three index of refraction techniques are schlieren, shadowgraph, and interferometry. They are used in attempting to visualize flows when the density differences are either naturally or artificially altered; both gas and liquid mediums can be employed. The methods depend on the variation of the index of refraction in a transparent fluid medium and the resulting deflection of a light beam directed through the medium. Most often the flows to be studied are two-dimensional, and the light beam is directed

(a)

(b)

Figure 13.21   Particle image velocimetry: (a) photograph of particle pathlines; (b) scaled velocity vectors. (Courtesy of R. Bouwmeester.)

normal to the flow field; the result is a measurement integrated over the test section. The index of refraction is a function of the thermodynamic state of the fluid and typically depends only on the density. Examples of flows that are investigated with these techniques are: high-speed flows with shock waves, convection flows, flame and combustion flow fields, mixing of fluids of different densities, and forced-convection flows with heat or mass transfer.

With a **schlieren** system, one measures the angle of the light beam after it has passed through the test section and emerges into the surrounding air (Fig. 13.22a); the angle is proportional to the first spatial derivative of the index of refraction. The light beam is turned in the direction of increasing index of refraction, or for most media, toward the region of higher density. A representative schlieren image is shown in Fig. 13.22b.

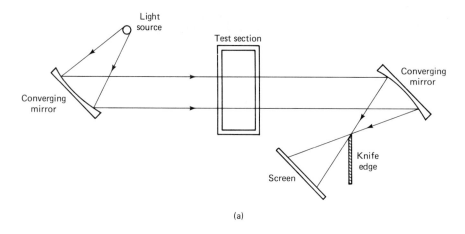

(a)

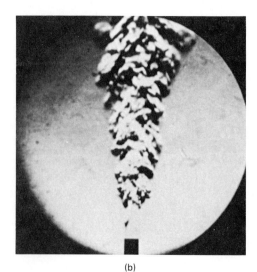

(b)

Figure 13.22 Schlieren system: (a) schematic; (b) helium jet entering atmospheric air. (Courtesy of R. Goldstein.)

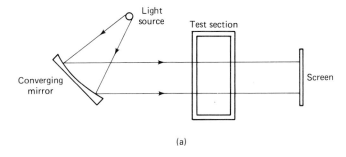

(a)

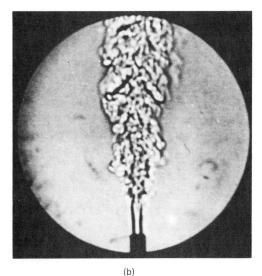

Figure 13.23 Shadowgraph system: (a) schematic; (b) helium jet
entering atmospheric air. (Courtesy of R. Goldstein.)

(b)

A **shadowgraph** system measures the linear displacement of perturbed
light (Fig. 13.23). In contrast to the schlieren imaging, the light pattern in a
shadowgraph is determined by the second derivative of the index of refraction.

   Shadowgraph and schlieren systems are usually employed for quali-

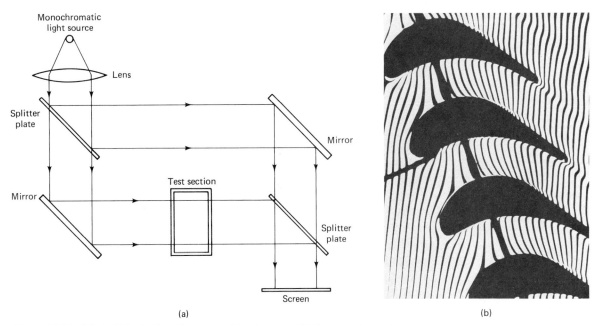

(a)   (b)

Figure 13.24 Mach–Zehnder interferometer: (a) schematic; (b) flow over heated gas-tur-
bine blades. (Courtesy of R. Goldstein.)

tative flow visualization, whereas interferometers are often used in quantitative studies. An interferometer responds to differences in optical path length; since a flow field with a variable density represents an optical disturbance, the phase of transmitted light waves is changed relative to undisturbed waves. A two-beam interferometer is designed to measure that phase difference (Fig. 13.24a). When the disturbed and undisturbed light beams are recombined, interference fringes appear on the screen, which are a measure of the density variation in the flow field. An example of an interferogram is shown in Fig. 13.24b.

13.5 DATA ACQUISITION AND ANALYSIS

This section begins with a brief introduction to automated acquisition of flow data using microcomputers. Next, methodology to deal with the uncertainty of data is presented, including the manner in which those uncertainties are propagated when one flow parameter is represented indirectly as the result of measuring others. Finally, since a number of flow relations involve exponential formulations, a means of curve fitting exponential data is presented.

13.5.1 Digital Recording of Data

Some of the measurements that have been described in previous sections are made simply by the direct reading of an instrument; those readings are probably steady-state, time-averaged, or perhaps root-mean-square values. Examples include reading a manometer attached to an orifice flowmeter, monitoring a digital rpm readout of a vane anemometer, and recording the pressure indicated on a Bourdon gage. The data acquired in this manner are probably sufficiently accurate for the desired goal; indeed, any further sophistication used to collect the data could be costly and unnecessary. On the other hand, there are situations that require more complex means of data acquisition. For instance, monitoring transient pressures in a pipeline using a strain-gage transducer, or observing turbulent velocity fluctuations in a free-stream flow, require that data be recorded continuously and accurately. Another illustration is the capture of the movement of a large number of moving particles in a two-dimensional plane in order to determine an unsteady flow field. Such measurements necessitate that data be recorded automatically, rapidly, and accurately.

The advent of the microcomputer has revolutionized automated acquisition of flow data (Malmstadt, et al, 1981). Many of the transducers in use provide a voltage output signal (an analog signal) that must be converted into a digital signal before the microcomputer can store and process the data. Readily available today are a number of plug-in analog input/output (I/O) boards specifically made for personal computers, which can be inserted into the central processing unit bus. These I/O boards consist of input multiplexers, an input amplifier, a sample-and-hold circuit, one or more output digital-to-analog converters, and interfacing and timing circuitry. In addition, some include built-in random-access memory, read-only memory, and real-time clocks. Compatible software is available so that programming is either unnecessary or requires minimal effort.

13.5.2 Uncertainty Analysis

When flow measurements are made, it is important to realize that no recorded value of a given parameter is perfectly accurate. Instruments do not measure the so-called "true value" of the parameter, but they will provide an estimate of that value. Along with each measurement, one must attempt to answer the important question: How accurate is that measurement? The inaccuracies associated with measurements can be considered as uncertainties rather than errors. An **error** is the fixed, unchangeable difference between the true value and the recorded value, whereas an **uncertainty** is the statistical value that an error may assume for any given measurement.

The purpose of this section is to describe a technique for propagating uncertainties of flow measurements into results. It has a variety of uses, among them helping us to decide whether calculations agree with collected data or lie beyond acceptable limits. Perhaps more important, it can help us to improve a given experiment or measuring procedure. It is not meant to be a substitute for accurate calibrations and sound experimental techniques; rather, it should be viewed as an additional aid for obtaining good, reliable results. Only a brief introduction is presented here; the reading of Coleman and Steele (1989) and Kline (1985) is recommended for a more thorough treatment of the subject.

Uncertainty can be divided into three categories: (1) uncertainty due to calibration of instruments, (2) uncertainty due to acquisition of data, and (3) uncertainty as a result of data reduction. The total uncertainty is the sum of bias and precision. **Bias** is the systematic uncertainty that is present during a test; it is considered to remain constant during repeated measurements of a certain set of parameters. There is no statistical formulation that can be applied to estimate bias, hence its value must be based on estimates. Calibrations and independent measurements will aid in its estimation. **Precision** is alternately termed **repeatability;** it is found by taking repeated measurements from the parameter population and using the standard deviation as a precision index. For the situation in which repeated measurements are not taken (which occurs quite often for flow measurements), a single value must be used; however, less precision will be obtained. Figure 13.25 illustrates how uncertainties are defined. A sampling of data (solid line) is shown that is assumed to be distributed in a Gaussian manner. The parameter x_i is one particular measured value, or measurand, $<x_i>$ is the average of all the measured values, and δx_i is the

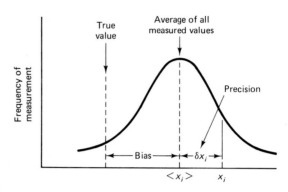

Figure 13.25 Uncertainties associated with a sampling of data.

precision associated with x_i. The uncertainty for the measurand can be given as

$$x_i = <x_i> \pm \delta x_i \qquad (13.5.1)$$

Furthermore, it is necessary to state the bias and the probability P of recording the measurand with the stated precision.

In addition to defining uncertainties in the data, it is necessary to propagate those uncertainties into the results. A two-step process is described as follows.

1. *Describe the uncertainty for the measurands.* Each measurand is assumed to be an independent observation, and it must come from a Gaussian population. Also, the precision for each measurand must be quoted at the same probability P. For example, consider an experiment to calibrate a weir, in which the measurands are the water level upstream of the weir, and the discharge. The water level is measured using a water-level gage, and the discharge is recorded by measuring the time it takes to fill a known volume. Thus the two measurands are independent. In addition, based on experience and perhaps some repeated measurements, one can reasonably state that (1) there is no bias, and (2) the recorded precision for each measurand has a probability of $P = 0.95$; that is, 20 out of every 21 readings of that measurand would yield the recorded value plus or minus the stated precision.

2. *Compute the overall uncertainty interval.* If we let the parameter R represent the result computed from the n measurands $x_1, \ldots, x_i, \ldots, x_n$, the overall uncertainty interval δR is defined using the root sum square,

$$\delta R = \left\{ \sum_{i=1}^{n} \left[\frac{\partial R}{\partial x_i} \delta x_i \right]^2 \right\}^{1/2}$$

$$\simeq \left\{ \sum_{i=1}^{n} [R(x_i + \delta x_i) - R(x_i)]^2 \right\}^{1/2} \qquad (13.5.2)$$

The overall uncertainty interval is an estimate of the standard deviation of the population of all possible experiments like the one being conducted.

EXAMPLE 13.1

Discharge and pressure data are collected for water flow through an orifice meter in a pipe. The orifice diameter is 35.4 mm and the diameter of the pipe is 50.8 mm. The original data are reduced to the form shown in the accompanying table. The second column is the discharge Q; the precision δQ (zero bias, $P = 0.95$) is given in the third column; and the pressure head data $\Delta h = h_1 - h_2$ are presented in the fourth column. In addition, the precision for each of the pressure measurements is determined to be the same, namely, $\delta(\Delta h) = 0.025$ m (zero bias, $P = 0.95$). The fifth and sixth columns show the Reynolds numbers and the flow coefficient K computed using Eq. 13.3.8. Determine the uncertainties in the result K as a result of the estimated uncertainties in the measurands Q and Δh.

Data no. i	Q (m^3/s)	δQ (m^3/s)	Δh (m)	Re	K	δK
1	0.00119	0.00010	0.139	42 800	0.732	0.085
2	0.00162	0.00012	0.265	58 300	0.722	0.062
3	0.00200	0.00015	0.403	71 900	0.723	0.058
4	0.00223	0.00015	0.517	80 200	0.711	0.051
5	0.00251	0.00015	0.655	90 300	0.711	0.045
6	0.00276	0.00017	0.781	99 300	0.716	0.046
7	0.00294	0.00018	0.907	105 700	0.708	0.045
8	0.00314	0.00019	1.008	112 900	0.717	0.045
9	0.00330	0.00016	1.134	118 700	0.711	0.035
10	0.00346	0.00017	1.247	124 400	0.711	0.036
11	0.00373	0.00017	1.386	134 200	0.727	0.034
12	0.00382	0.00017	1.512	137 400	0.713	0.032
13	0.00408	0.00018	1.663	146 700	0.726	0.033
14	0.00424	0.00019	1.852	152 500	0.715	0.032
15	0.00480	0.00021	2.293	172 600	0.727	0.032

Solution

Equation 13.3.8 is used in conjunction with Eq. 13.5.2 to propagate the uncertainties from Q and Δh into K. For example, consider the data for $i = 6$. Calculation of δk proceeds as follows. The orifice area is

$$A_0 = \frac{\pi}{4} 0.0354^2 = 9.84 \times 10^{-4} \text{ m}^2$$

and the diameter ratio is

$$\beta = \frac{35.4}{50.8} = 0.697 \simeq 0.7$$

With Eq. 13.3.8, values of K are determined for $(Q, \Delta h)$, $(Q + \delta Q, \Delta h)$, and $(Q, \Delta h + \delta(\Delta h))$:

$$K(Q, \Delta h) = \frac{0.00276}{9.84 \times 10^{-4}\sqrt{2 \times 9.81 \times 0.781}}$$

$$= 0.716$$

$$K(Q + \delta Q, \Delta h) = \frac{0.00276 + 0.00017}{9.84 \times 10^{-4}\sqrt{2 \times 9.81 \times 0.781}}$$

$$= 0.761$$

$$K(Q, \Delta h + \delta(\Delta h)) = \frac{0.00276}{9.84 \times 10^{-4}\sqrt{2 \times 9.81 \times (0.781 + 0.025)}}$$

$$= 0.705$$

These are then substituted into Eq. 13.5.2, with the variable K replacing the result R in that relation:

$$\delta K = \{[K(Q + \delta Q, \Delta h) - K(Q, \Delta h)]^2 + [K(Q, \Delta h + \delta(\Delta h)) - K(Q, \Delta h)]^2\}^{1/2}$$

$$= [(0.761 - 0.716)^2 + (0.705 - 0.716)^2]^{1/2}$$

$$= 0.046$$

The results for all measurands are similarly evaluated and shown in the last column in the table. In addition, they are plotted in the accompanying figure, where the vertical lines, or uncertainty bands, associated with each value of K have the magnitude of $2\delta K$. The bands can be interpreted as the estimated precision for K, based on zero bias and $P = 0.95$ for the two measurands. Note that the precision δK stabilizes at approximately 0.032 at the higher Reynolds numbers, and that the magnitude of δK is greater at lower Reynolds numbers. An alternative way to express this is that the uncertainty of K decreases with increasing Rey-

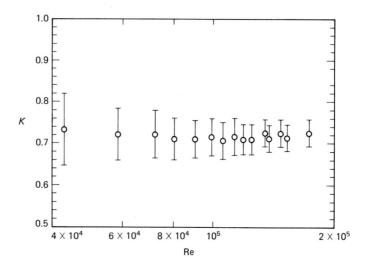

nolds number. The data in the figure should be compared with the orifice curve labeled $\beta = 0.7$ in Fig. 13.10.

___

13.5.3 Regression Analysis

A significant number of flow-measuring devices correlate the discharge with pressure drop or head according to the general formula

$$Y = CX^m \qquad (13.5.3)$$

in which Y is the derived result, X is the measurand, and C and m are constants. By taking a number of independent measurements of X and Y, estimates of C and m can be found. A systematic approach to determine C and m is to find the best estimates based on the method of least squares. If the logarithm of each side of Eq. 13.5.3 is taken, it becomes

$$\ln Y = \ln C + m \ln X \qquad (13.5.4)$$

which is of the form

$$y = b + mx \qquad (13.5.5)$$

By association we observe that Eq. 13.5.4 is linear provided that $y = \ln Y$, $b = \ln C$, and $x = \ln X$.

Consider the data set X_i, Y_i, $i = 1, \ldots, n$. The objective is to generate a straight line through the logarithms of the data (x_i, y_i) such that the errors of estimation are small. That error is defined as the difference between the observed value y_i and the corresponding value on the line; in Fig. 13.26, s_i is the error associated with the data point (x_i, y_i). The method of least squares requires that to minimize the error, the sum of the squares of the errors of all the data be as small as possible. From Eq. 13.5.5,

$$S = \sum_{i=1}^{n} s_i^2$$

$$\qquad (13.5.6)$$

$$= \sum_{i=1}^{n} [y_i - (b + mx_i)]^2$$

The parameter S is minimized by forming the partial derivatives of S with respect to b and m, setting the resulting algebraic equations to zero, and solving for b and m. It is left as an exercise for the reader to derive the results.

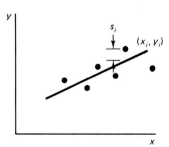

Figure 13.26 Method of least squares.

$$m = \frac{\Sigma\, x_i y_i - (\Sigma\, x_i\, \Sigma\, y_i)/n}{\Sigma\, x_i^2 - (\Sigma\, x_i)^2/n} \qquad (13.5.7)$$

$$b = \frac{\Sigma\, y_i - m\, \Sigma\, x_i}{n} \qquad (13.5.8)$$

$$C = e^b \qquad (13.5.9)$$

A simple computer algorithm can be prepared to evaluate m and C based on a set of input data $(X_i,\ Y_i)$.

EXAMPLE 13.2

With the data given in Example 13.1, perform a least squares regression to determine the coefficient C and exponent m in Eq. 13.5.3. Compare the result with Eq. 13.3.8.

Solution

Equations 13.5.7 to 13.5.9 are employed to evaluate C, b, and m. In Eqs. 13.5.7 and 13.5.8, x is replaced by $\ln \Delta h = \ln(h_1 - h_2)$, and y is replaced by $\ln Q$. The calculations are tabulated below.

i	$x_i = \ln \Delta h$	$y_i = \ln Q$	x_i^2	$x_i y_i$
1	−1.9733	−6.7338	3.8939	13.2878
2	−1.3280	−6.4253	1.7636	8.5328
3	0.9088	6.2146	0.8259	5.6478
4	−0.6597	−6.1058	0.4352	4.0280
5	−0.4231	−5.9875	0.1790	2.5333
6	−0.2472	−5.8925	0.06111	1.4566
7	−0.09761	−5.8293	0.009528	0.5690
8	0.007968	−5.7635	0.00006349	−0.0459
9	0.1258	−5.7138	0.01583	−0.7188
10	0.2207	−5.6665	0.04871	−1.2506
11	0.3264	−5.5913	0.1065	−1.8250
12	0.4134	−5.5675	0.1709	−2.3016
13	0.5086	−5.5017	0.2587	−2.7982
14	0.6163	−5.4632	0.3798	−3.3670
15	0.8299	−5.3391	0.6887	−4.4309
Σ	−2.5886	−87.7954	8.8374	19.3173

Substitute the summed values into Eqs. 13.5.7 to 13.5.9:

$$m = \frac{19.3173 - (-2.5886)(-87.7954)/15}{8.8374 - (-2.5886)^2/15} = 0.497$$

$$b = \frac{-87.7954 - 0.497(-2.5886)}{15} = 5.7673$$

$$C = \exp(-5.7673) = 0.00313$$

Hence the discharge is correlated to Δh by the regression curve

$$Q = 0.00313 \ \Delta h^{0.497}$$

Based on the analysis performed in Example 13.1, the average value of K is 0.718. The area of the orifice is $A_0 = 9.84 \times 10^{-4}$ m^2. Substituting these into Eq. 13.3.8, we find that

$$Q = 0.00313 \ \Delta h^{0.500}$$

Thus the two results agree quite closely. The data and regression line are plotted in the accompanying figure.

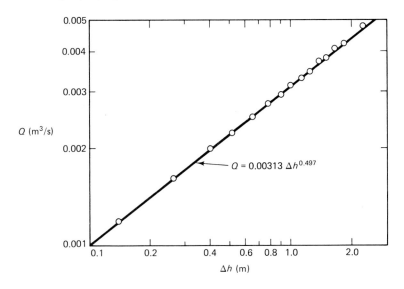

PROBLEMS

13.1. A 20° inclined manometer attached to a piezometer opening is used to measure the pressure at the wall of airflow. If the reading is measured to be 4 cm of mercury, calculate the pressure at the wall.

13.2. A pitot-static probe is to be used for repeated measurements of the velocity of an atmospheric airstream. It is desired to relate the velocity V to the manometer reading h in centimeters of mercury (i.e., $V = C\sqrt{h}$). Calculate the value of the constant C for a laboratory at 20°C situated at **(a)** sea level, and **(b)** 2000 m elevation.

13.3. A pitot-static probe attached to a manometer that has water as the manometer fluid is proposed as the measuring device for an airflow with a velocity of 8 m/s. What manometer reading is expected? Comment as to the use of the proposed device.

13.4. Water at 20°C flows through a 2-mm-diameter tube into a calibrated tank. If 2L are collected in 10 minutes, calculate the flow rate (m^3/s), the mass flux (kg/s), and the average velocity (m/s). Is the flow laminar, turbulent, or can't you tell?

13.4E. Water at 75°F flows through a 0.1-in-diameter tube into a calibrated tank. If 2 quarts are collected in 10 minutes, calculate the flow rate (ft^3/sec), the mass flux (slug/sec), and the average velocity (ft/sec). Is the flow laminar, turbulent, or can't you tell?

13.5. A velocity traverse in a rectangular channel that measures 10 cm by 100 cm is represented by the following data. Estimate the flow rate and the average velocity assuming a symmetric plane flow.

y (cm)	0	0.5	1	2	3	4	5
u (m/s)	0	5.2	8.1	9.2	9.8	10	10

13.6. A velocity traverse in a 10-cm-diameter pipe is represented by the following data. Estimate the flow rate and the average velocity assuming a symmetric flow.

r (cm)	0	1	2	3	4	4.5
u (m/s)	10	10	9.8	9.2	8.1	5.2

13.7. The flow rate of water in a 12-cm-diameter pipe is measured with a 6-cm-diameter venturi meter to be 0.09 m³/s. What is the expected deflection on a water-mercury manometer? Assume a water temperature of 20°C.

13.8. The flow rate of 20°C water flow in a 24-cm-diameter pipe is desired. If a water-mercury manometer reads 12 cm, calculate the discharge if the manometer is connected to:
(a) A 15-cm-diameter orifice.
(b) A 15-cm-diameter nozzle.

13.9. Calculate the flow rate of 40°C water in the pipes shown.

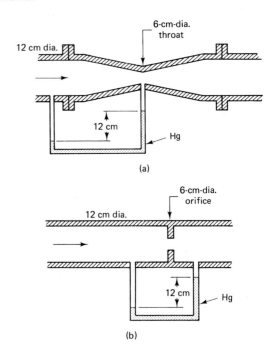

(a)

(b)

13.10. A pressure difference across an elbow in a water pipe is measured. Estimate the flow rate in the pipe. The pressure difference, pipe diameter, and radius of curvature are, respectively:
(a) 80 kPa, 10 cm, 20 cm.
(b) 80 kPa, 5 cm, 20 cm.
(c) 10 psi, 4 in., 8 in.
(d) 10 psi, 2 in., 8 in.
Use a water temperature of 20°C.

13.11. Experimental data used to calibrate discharge of water in a venturi meter are tabulated below. The diameter of the meter throat is 33.3 mm and the diameter at the upstream tap location of the differential manometer is 54.0 mm. The precision δQ of the discharge measurements has zero bias and $P = 0.95$, and the precision for all the manometer measurements is $\delta(\Delta h) = 2$ mm Hg, with zero bias and $P = 0.95$. Note that the manometer readings are given in mm Hg. With use of Eq. 13.3.8, determine the uncertainties in K as a result of the uncertainties in the two measurands. Plot the results in a fashion similar to those in Example 13.1, and compare the curve with Fig. 13.10.

13.12. With the numerical results obtained from Example 13.1, plot the relative uncertainty $\delta K/K$ versus Re on a linear scale. What conclusions can you draw?

13.13. Using the data from Problem 13.11, perform a least squares regression and determine the coefficient C and exponent m in Eq. 13.5.3. Compare the result with Eq. 13.3.8. Note that the manometer readings are given in mm Hg.

13.14. Establish a depth–discharge relation of the form $Q = CY^m$ for a 60° V-notch sharp-crested weir using the data in the following table. In the equation, Y is the vertical distance from the weir notch to the undisturbed upstream water surface, and Q is the discharge (see Fig. 10.10). Compare the result with Eq. 10.4.27.

Data no.	Q	δQ	Δh	
i	(m³/s)	(m³/s)	(mm Hg)	Re
1	0.00168	0.00011	13	64 200
2	0.00208	0.00014	22	79 500
3	0.00249	0.00016	32	95 100
4	0.00281	0.00019	40	107 400
5	0.00328	0.00022	54	125 300
6	0.00355	0.00024	62	135 600
7	0.00372	0.00025	70	142 100
8	0.00402	0.00023	82	153 600
9	0.00415	0.00024	91	158 600
10	0.00444	0.00025	100	169 700
11	0.00479	0.00027	113	183 000
12	0.00493	0.00028	123	188 400
13	0.00523	0.00030	134	199 800
14	0.00533	0.00035	140	203 700
15	0.00509	0.00029	140	194 500

Data no.	Q	Y
i	(ft³/sec)	(ft)
1	0.1390	0.387
2	0.1300	0.378
3	0.1280	0.374
4	0.1140	0.357
5	0.0963	0.336
6	0.0937	0.320
7	0.0809	0.312
8	0.0781	0.306
9	0.0751	0.302
10	0.0697	0.297
11	0.0518	0.263
12	0.0435	0.244
13	0.0418	0.238
14	0.0319	0.206

Appendix

A. UNITS AND CONVERSIONS

TABLE A.1 U.S. ENGINEERING UNITS, SI UNITS, AND THEIR CONVERSION FACTORS

Quantity	English units (U.S. system)	International system (SI)[a]	Conversion factor
Length	inch	millimeter	1 in. = 25.4 mm
	foot	meter	1 ft = 0.3048 m
	mile	kilometer	1 mile = 1.609 km
Area	square inch	square centimeter	$1\ \text{in}^2 = 6.452\ \text{cm}^2$
	square foot	square meter	$1\ \text{ft}^2 = 0.09290\ \text{m}^2$
Volume	cubic inch	cubic centimeter	$1\ \text{in}^3 = 16.39\ \text{cm}^3$
	cubic foot	cubic meter	$1\ \text{ft}^3 = 0.02832\ \text{m}^3$
	gallon		$1\ \text{gal} = 0.003789\ \text{m}^3$
Mass	pound-mass	kilogram	$1\ \text{lb}_m = 0.4536\ \text{kg}$
	slug		1 slug = 14.59 kg
Density	slug/cubic foot	kilogram/cubic meter	$1\ \text{slug/ft}^3 = 515.4\ \text{kg/m}^3$
Force	pound-force	newton	1 lb = 4.448 N
Work or torque	foot-pound	newton-meter	1 ft-lb = 1.356 N · m
Pressure	pound/square inch	newton/square meter (pascal)	1 psi = 6895 Pa
	pound/square foot		1 psf = 47.88 Pa
Temperature	degree Fahrenheit	degree Celsius	$\_°\text{F} = \frac{9}{5}\_°\text{C} + 32$
	degree Rankine	kelvin	$\_°\text{R} = \frac{9}{5}\_\text{K}$
Energy	British thermal unit	joule	1 Btu = 1055 J
	calorie		1 cal = 4.186 J
	foot-pound		1 ft-lb = 1.356 J
Power	horsepower	watt	1 hp = 745.7 W
	foot-pound/second		1 ft-lb/sec = 1.356 W
Velocity	foot/second	meter/second	1 ft/sec = 0.3048 m/s
Acceleration	foot/second squared	meter/second squared	$1\ \text{ft/sec}^2 = 0.3048\ \text{m/s}^2$
Frequency	cycle/second	hertz	1 cps = 1.000 Hz
Viscosity	pound-second/square foot	newton-second/square meter	$1\ \text{lb-sec/ft}^2 = 47.88\ \text{N-s/m}^2$

[a]The reversed initials in this abbreviation come from the French form of the name: Système International.

TABLE A.2 CONVERSIONS

Length	Force	Mass	Velocity
1 cm = 0.3937 in.	1 lb = 0.4536 kg	1 oz = 28.35 g	1 mph = 1.467 ft/sec
1 m = 3.281 ft	1 lb = 0.4448×10^6 dyn	1 lb = 0.4536 kg	1 mph = 0.8684 knot
1 km = 0.6214 mile	1 lb = 32.17 pdl	1 slug = 32.17 lb	1 ft/sec = 0.3048 m/s
1 in. = 2.54 cm	1 kg = 2.205 lb	1 slug = 14.59 kg	1 m/s = 3.281 ft/sec
1 ft = 0.3048 m	1 pdl = 0.03108 lb	1 kg = 2.205 lb	
1 mile = 1.609 km	1 dyn = 2.248×10^{-6} lb	1 kg = 0.06852 slug	
1 mile = 5280 ft	1 lb = 4.448 N		
1 mile = 1760 yd	1 N = 0.2248 lb		

Work and power	Pressure	Volume	Flow rate
1 Btu = 778.2 ft-lb	1 psi = 2.036 in. Hg	1 ft^3 = 28.32 L	1 ft^3/min = 4.719×10^{-4} m^3/s
1 J = 10^7 ergs	1 psi = 27.7 in. H_2O	1 ft^3 = 7.481 gal (U.S.)	1 ft^3/sec = 0.02832 m^3/s
1 cal = 3.088 ft-lb	14.7 psi = 22.92 in. Hg	1 gal (U.S.) = 231 in^3	1 m^3/s = 35.31 ft^3/sec
1 cal = 0.003968 Btu	14.7 psi = 33.93 ft H_2O	1 gal (Brit.) = 1.2 gal (U.S.)	1 gal/min = 0.002228 ft^3/sec
1 kWh = 3413 Btu	14.7 psi = 1.0332 kg/cm^2	1 m^3 = 1000 L	1 ft^3/sec = 448.9 gal/min
1 Btu = 1.055 kJ	14.7 psi = 1.0133 bar	1 ft^3 = 0.02832 m^3	
1 ft-lb = 1.356 J	1 kg/cm^2 = 14.22 psi	1 m^3 = 35.31 ft^3	
1 hp = 550 ft-lb/sec	1 in. Hg = 0.4912 psi		
1 hp = 0.7067 Btu/sec	1 ft H_2O = 0.4331 psi		
1 hp = 0.7455 kW	1 psi = 6895 Pa		
1 W = 1 J/s	1 psf = 47.88 Pa		
1 W = 1.0×10^7 dyn-cm/s	10^5 Pa = 1 bar		

B. FLUID PROPERTIES

TABLE B.1   PROPERTIES OF WATER

Temperature, °C	Density ρ, kg/m^3	Viscosity μ, (N · s/m^2)	Kinematic viscosity ν, m^2/s	Surface tension σ, N/m	Vapor pressure, kPa	Bulk modulus B, Pa
0	999.9	1.792×10^{-3}	1.792×10^{-6}	0.0762	0.588	204×10^7
5	1000.0	1.519	1.519	0.0754	0.882	206
10	999.7	1.308	1.308	0.0748	1.176	211
15	999.1	1.140	1.141	0.0741	1.666	214
20	998.2	1.005	1.007	0.0736	2.45	220
30	995.7	0.801	0.804	0.0718	4.30	223
40	992.2	0.656	0.661	0.0701	7.40	227
50	988.1	0.549	0.556	0.0682	12.22	230
60	983.2	0.469	0.477	0.0668	19.60	228
70	977.8	0.406	0.415	0.0650	30.70	225
80	971.8	0.357	0.367	0.0630	46.40	221
90	965.3	0.317	0.328	0.0612	68.20	216
100	958.4	0.284×10^{-3}	0.296×10^{-6}	0.0594	97.50	207×10^7

TABLE B.1—ENGLISH   PROPERTIES OF WATER

Temperature (°F)	Density (slug/ft^3)	Viscosity (lb-sec/ft^2)	Kinematic viscosity (ft^2/sec)	Surface tension (lb/ft)	Vapor pressure (psi)	Bulk modulus (psi)
32	1.94	3.75×10^{-5}	1.93×10^{-5}	0.518×10^{-2}	0.089	293,000
40	1.94	3.23	1.66	0.514	0.122	294,000
50	1.94	2.74	1.41	0.509	0.178	305,000
60	1.94	2.36	1.22	0.504	0.256	311,000
70	1.94	2.05	1.06	0.500	0.340	320,000
80	1.93	1.80	0.93	0.492	0.507	322,000
90	1.93	1.60	0.83	0.486	0.698	323,000
100	1.93	1.42	0.74	0.480	0.949	327,000
120	1.92	1.17	0.61	0.465	1.69	333,000
140	1.91	0.98	0.51	0.454	2.89	330,000
160	1.90	0.84	0.44	0.441	4.74	326,000
180	1.88	0.73	0.39	0.426	7.51	318,000
200	1.87	0.64	0.34	0.412	11.53	308,000
212	1.86	0.59×10^{-5}	0.32×10^{-5}	0.404×10^{-2}	14.7	300,000

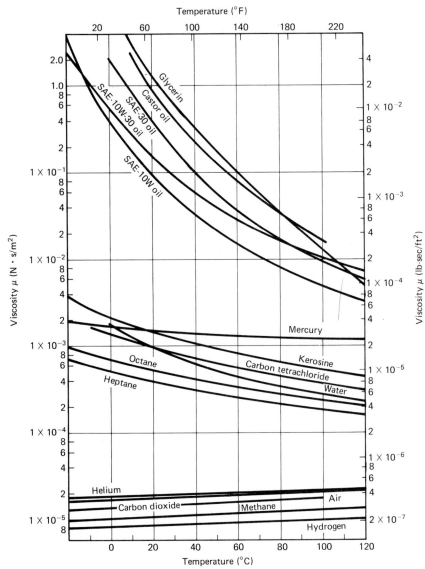

Figure B.1 Viscosity as a function of temperature. (From Fox, R. W., and McDonald, A. T., *Introduction to Fluid Mechanics,* 2nd Ed., John Wiley & Sons, 1978.)

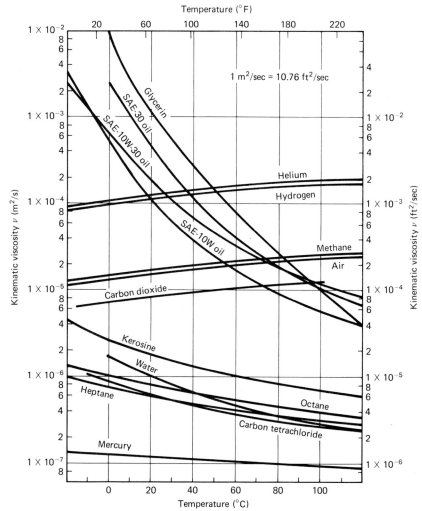

Figure B.2 Kinematic viscosity (at atmospheric pressure) as a function of temperature. (From Fox, R. W., and McDonald, A. T., *Introduction to Fluid Mechanics,* 2nd ed., John Wiley & Sons, 1978.)

TABLE B.2 PROPERTIES OF AIR
AT ATMOSPHERIC PRESSURE

Temperature T (°C)	Density ρ (kg/m³)	Viscosity μ (N · s/m²)	Kinematic viscosity ν (m²/s)	Velocity of sound c (m/s)
−50	1.582	1.46×10^{-5}	0.921×10^{-5}	299
−30	1.452	1.56×10^{-5}	1.08×10^{-5}	312
−20	1.394	1.61	1.16	319
−10	1.342	1.67	1.24	325
0	1.292	1.72	1.33	331
10	1.247	1.76	1.42	337
20	1.204	1.81	1.51	343
30	1.164	1.86	1.60	349
40	1.127	1.91	1.69	355
50	1.092	1.95	1.79	360
60	1.060	2.00	1.89	366
70	1.030	2.05	1.99	371
80	1.000	2.09	2.09	377
90	0.973	2.13	2.19	382
100	0.946	2.17×10^{-5}	2.30	387
200	0.746	2.57	3.45	436
300	0.616	2.93×10^{-5}	4.75×10^{-5}	480

TABLE B.2—ENGLISH PROPERTIES OF AIR
AT ATMOSPHERIC PRESSURE

Temperature (°F)	Density (slug/ft³)	Viscosity (lb-sec/ft²)	Kinematic viscosity (ft²/sec)	Velocity of sound (ft/sec)
−20	0.00280	3.34×10^{-7}	11.9×10^{-5}	1028
0	0.00268	3.38	12.6	1051
20	0.00257	3.50	13.6	1074
40	0.00247	3.62	14.6	1096
60	0.00237	3.74	15.8	1117
68	0.00233	3.81	16.0	1125
80	0.00228	3.85	16.9	1138
100	0.00220	3.96	18.0	1159
120	0.00213	4.07	18.9	1180
160	0.00199	4.23	21.3	1220
200	0.00187	4.50	24.1	1258
300	0.00162	4.98	30.7	1348
400	0.00144	5.26	36.7	1431
1000	0.000844	7.87×10^{-7}	93.2×10^{-5}	1839

TABLE B.3   PROPERTIES OF THE STANDARD ATMOSPHERE

Altitude (m)	Temperature (K)	Pressure (kPa)	Density (kg/m³)	Velocity of sound (m/s)
0	288.2	101.3	1.225	340
500	284.9	95.43	1.167	338
1,000	281.7	89.85	1.112	336
2,000	275.2	79.48	1.007	333
4,000	262.2	61.64	0.8194	325
6,000	249.2	47.21	0.6602	316
8,000	236.2	35.65	0.5258	308
10,000	223.3	26.49	0.4136	300
12,000	216.7	19.40	0.3119	295
14,000	216.7	14.17	0.2278	295
16,000	216.7	10.35	0.1665	295
18,000	216.7	7.563	0.1216	295
20,000	216.7	5.528	0.0889	295
30,000	226.5	1.196	0.184	302
40,000	250.4	0.287	4.00×10^{-3}	317
50,000	270.7	0.0798	1.03×10^{-3}	330
60,000	255.8	0.0225	3.06×10^{-4}	321
70,000	219.7	0.00551	8.75×10^{-5}	297
80,000	180.7	0.00103	2.00×10^{-5}	269

TABLE B.3—ENGLISH   PROPERTIES OF THE ATMOSPHERE

Altitude (ft)	Temperature (°F)	Pressure (lb/ft²)	Density (slugs/ft³)	Velocity of sound (ft/sec)
0	59.0	2116	0.00237	1117
1,000	55.4	2014	0.00231	1113
2,000	51.9	1968	0.00224	1109
5,000	41.2	1760	0.00205	1098
10,000	23.4	1455	0.00176	1078
15,000	5.54	1194	0.00150	1058
20,000	−12.3	973	0.00127	1037
25,000	−30.1	785	0.00107	1016
30,000	−48.0	628	0.000890	995
35,000	−65.8	498	0.000737	973
36,000	−67.6	475	0.000709	971
40,000	−67.6	392	0.000586	971
50,000	−67.6	242	0.000362	971
100,000	−51.4	23.2	3.31×10^{-5}	971

TABLE B.4 PROPERTIES OF IDEAL GASES AT 300 K

Gas	Chemical formula	Molar mass	$R \dfrac{\text{ft-lb}}{\text{slug-}°\text{R}}$	$R \dfrac{\text{kJ}}{\text{kg} \cdot \text{K}}$	$c_p \dfrac{\text{ft-lb}}{\text{slug-}°\text{R}}$	$c_p \dfrac{\text{kJ}}{\text{kg} \cdot \text{K}}$	k
Air	—	28.97	1716	0.287	6012	1.004	1.40
Argon	Ar	39.94	1244	0.2081	3139	0.5203	1.667
Carbon Dioxide	CO_2	44.01	1129	0.1889	5085	0.8418	1.287
Carbon Monoxide	CO	28.01	1775	0.2968	6238	1.041	1.40
Ethane	C_2H_6	30.07	1653	0.2765	10,700	1.766	1.184
Helium	He	4.003	12,420	2.077	31,310	5.193	1.667
Hydrogen	H_2	2.016	24,660	4.124	85,930	14.21	1.40
Methane	CH_4	16.04	3100	0.5184	13,330	2.254	1.30
Nitrogen	N_2	28.02	1774	0.2968	6213	1.042	1.40
Oxygen	O_2	32.00	1553	0.2598	5486	0.9216	1.394
Propane	C_3H_8	44.10	1127	0.1886	10,200	1.679	1.12
Steam	H_2O	18.02	2759	0.4615	11,150	1.872	1.33

$c_v = c_p - R, \quad k = c_p/c_v$

TABLE B.5 PROPERTIES OF COMMON LIQUIDS AT ATMOSPHERIC PRESSURE AND APPROXIMATELY 60 TO 70°F (16 TO 21°C)

Liquid	Specific weight, γ		Density, ρ		Surface tension,[a] σ		Vapor pressure, p_v	
	lb/ft³	N/m³	slugs/ft³	kg/m³	lb/ft	N/m	psia	kPa
Alcohol, ethyl	49.3	7,744	1.53	789	0.0015	0.022	—	—
Benzene	56.2	8,828	1.75	902	0.0020	0.029	1.50	10.3
Carbon tetrachloride	99.5	15,629	3.09	1,593	0.0018	0.026	12.50	86.2
Gasoline	42.4	6,660	1.32	680	—	—	—	—
Glycerin	78.6	12,346	2.44	1,258	0.0043	0.063	2×10^{-6}	1.4×10^{-5}
Kerosene	50.5	7,933	1.57	809	0.0017	0.025	—	—
Mercury	845.5	132,800	26.29	13,550	0.032	0.467	2.31×10^{-5}	1.59×10^{-4}
SAE 10 oil	57.4	9,016	1.78	917	0.0025	0.036	—	—
SAE 30 oil	57.4	9,016	1.78	917	0.0024	0.035	—	—
Turpentine	54.3	8,529	1.69	871	0.0018	0.026	7.7×10^{-3}	5.31×10^{-2}
Water	62.4	9,790	1.94	998	0.0050	0.073	0.34	2.34

[a]In contact with air

C. PROPERTIES OF AREAS AND VOLUMES

TABLE C.1   AREAS

	Sketch	Area	Centroid	Second moment
Rectangle		bh	$\bar{y} = h/2$	$\bar{I} = bh^3/12$
Triangle		$bh/2$	$\bar{y} = h/3$	$\bar{I} = bh^3/36$
Circle		$\pi D^2/4$	$\bar{y} = r$	$\bar{I} = \pi D^4/64$
Semicircle		$\pi D^2/8$	$\bar{y} = 4r/3\pi$	$I_x = \pi D^4/128$
Ellipse		πab	$\bar{y} = b$	$\bar{I} = \pi ab^3/4$
Semiellipse		$\pi ab/2$	$\bar{y} = 4b/3\pi$	$I_x = \pi ab^3/8$

TABLE C.2   VOLUMES

	Sketch	Surface area	Volume	Centroid
Cylinder		$\pi Dh + \pi D^2/2$	$\pi D^2 h/4$	$\bar{y} = h/2$
Sphere		πD^2	$\pi D^3/6$	$\bar{y} = r$
Cone		$\pi(r^2 + r\sqrt{r^2 + h^2})$	$\pi D^2 h/12$	$\bar{y} = h/4$
Hemisphere		$3\pi D^2/4$	$\pi D^3/12$	$\bar{y} = 3r/8$

D. COMPRESSIBLE-FLOW TABLES FOR AIR

TABLE D.1 ISENTROPIC FLOW

M	p/p_0	T/T_0	A/A^*	M	p/p_0	T/T_0	A/A^*
0	1.0000	1.0000	∞	.92	.5785	.8552	1.0056
.02	.9997	.9999	28.9421	.94	.5658	.8498	1.0031
.04	.9989	.9997	14.4815	.96	.5532	.8444	1.0014
.06	.9975	.9993	9.6659	.98	.5407	.8389	1.0003
.08	.9955	.9987	7.2616	1.00	.5283	.8333	1.000
.10	.9930	.9980	5.8218	1.02	.5160	.8278	1.000
.12	.9900	.9971	4.8643	1.04	.5039	.8222	1.001
.14	.9864	.9961	4.1824	1.06	.4919	.8165	1.003
.16	.9823	.9949	3.6727	1.08	.4800	.8108	1.005
.18	.9776	.9936	3.2779	1.10	.4684	.8052	1.008
.20	.9725	.9921	2.9635	1.12	.4568	.7994	1.011
.22	.9668	.9904	2.7076	1.14	.4455	.7937	1.015
.24	.9607	.9886	2.4956	1.16	.4343	.7879	1.020
.26	.9541	.9867	2.3173	1.18	.4232	.7822	1.025
.28	.9470	.9846	2.1656	1.20	.4124	.7764	1.030
.30	.9395	.9823	2.0351	1.22	.4017	.7706	1.037
.32	.9315	.9799	1.9219	1.24	.3912	.7648	1.043
.34	.9231	.9774	1.8229	1.26	.3809	.7590	1.050
.36	.9143	.9747	1.7358	1.28	.3708	.7532	1.058
.38	.9052	.9719	1.6587	1.30	.3609	.7474	1.066
.40	.8956	.9690	1.5901	1.32	.3512	.7416	1.075
.42	.8857	.9659	1.5289	1.34	.3417	.7358	1.084
.44	.8755	.9627	1.4740	1.36	.3323	.7300	1.094
.46	.8650	.9594	1.4246	1.38	.3232	.7242	1.104
.48	.8541	.9560	1.3801	1.40	.3142	.7184	1.115
.50	.8430	.9524	1.3398	1.42	.3055	.7126	1.126
.52	.8317	.9487	1.3034	1.44	.2969	.7069	1.138
.54	.8201	.9449	1.2703	1.46	.2886	.7011	1.150
.56	.8082	.9410	1.2403	1.48	.2804	.6954	1.163
.58	.7962	.9370	1.2130	1.50	.2724	.6897	1.176
.60	.7840	.9328	1.1882	1.52	.2646	.6840	1.190
.62	.7716	.9286	1.1657	1.54	.2570	.6783	1.204
.64	.7591	.9243	1.1452	1.56	.2496	.6726	1.219
.66	.7465	.9199	1.1265	1.58	.2423	.6670	1.234
.68	.7338	.9153	1.1097	1.60	.2353	.6614	1.250
.70	.7209	.9107	1.0944	1.62	.2284	.6558	1.267
.72	.7080	.9061	1.0806	1.64	.2217	.6502	1.284
.74	.6951	.9013	1.0681	1.66	.2151	.6447	1.301
.76	.6821	.8964	1.0570	1.68	.2088	.6392	1.319
.78	.6691	.8915	1.0471	1.70	.2026	.6337	1.338
.80	.6560	.8865	1.0382	1.72	.1966	.6283	1.357
.82	.6430	.8815	1.0305	1.74	.1907	.6229	1.376
.84	.6300	.8763	1.0237	1.76	.1850	.6175	1.397
.86	.6170	.8711	1.0179	1.78	.1794	.6121	1.418
.88	.6041	.8659	1.0129	1.80	.1740	.6068	1.439
.90	.5913	.8606	1.0089	1.82	.1688	.6015	1.461

TABLE D.1   ISENTROPIC FLOW   (*continued*)

M	p/p_0		T/T_0	A/A^*	M	p/p_0		T/T_0	A/A^*
1.84	.1637		.5963	1.484	2.82	.3574	-1	.3860	3.567
1.86	.1587		.5910	1.507	2.84	.3467	-1	.3827	3.636
1.88	.1539		.5859	1.531	2.86	.3363	-1	.3794	3.706
1.90	.1492		.5807	1.555	2.88	.3263	-1	.3761	3.777
1.92	.1447		.5756	1.580	2.90	.3165	-1	.3729	3.850
1.94	.1403		.5705	1.606	2.92	.3071	-1	.3696	3.924
1.96	.1360		.5655	1.633	2.94	.2980	-1	.3665	3.999
1.98	.1318		.5605	1.660	2.96	.2891	-1	.3633	4.076
2.00	.1278		.5556	1.688	2.98	.2805	-1	.3602	4.155
2.02	.1239		.5506	1.716	3.00	.2722	-1	.3571	4.235
2.04	.1201		.5458	1.745	3.02	.2642	-1	.3541	4.316
2.06	.1164		.5409	1.775	3.04	.2564	-1	.3511	4.399
2.08	.1128		.5361	1.806	3.06	.2489	-1	.3481	4.483
2.10	.1094		.5313	1.837	3.08	.2416	-1	.3452	4.570
2.12	.1060		.5266	1.869	3.10	.2345	-1	.3422	4.657
2.14	.1027		.5219	1.902	3.12	.2276	-1	.3393	4.747
2.16	.9956	-1	.5173	1.935	3.14	.2210	-1	.3365	4.838
2.18	.9649	-1	.5127	1.970	3.16	.2146	-1	.3337	4.930
2.20	.9352	-1	.5081	2.005	3.18	.2083	-1	.3309	5.025
2.22	.9064	-1	.5036	2.041	3.20	.2023	-1	.3281	5.121
2.24	.8785	-1	.4991	2.078	3.22	.1964	-1	.3253	5.219
2.26	.8514	-1	.4947	2.115	3.24	.1908	-1	.3226	5.319
2.28	.8251	-1	.4903	2.154	3.26	.1853	-1	.3199	5.420
2.30	.7997	-1	.4859	2.193	3.28	.1799	-1	.3173	5.523
2.32	.7751	-1	.4816	2.233	3.30	.1748	-1	.3147	5.629
2.34	.7512	-1	.4773	2.274	3.32	.1698	-1	.3121	5.736
2.36	.7281	-1	.4731	2.316	3.34	.1649	-1	.3095	5.845
2.38	.7057	-1	.4688	2.359	3.36	.1602	-1	.3069	5.956
2.40	.6840	-1	.4647	2.403	3.38	.1557	-1	.3044	6.069
2.42	.6630	-1	.4606	2.448	3.40	.1512	-1	.3019	6.184
2.44	.6426	-1	.4565	2.494	3.42	.1470	-1	.2995	6.301
2.46	.6229	-1	.4524	2.540	3.44	.1428	-1	.2970	6.420
2.48	.6038	-1	.4484	2.588	3.46	.1388	-1	.2946	6.541
2.50	.5853	-1	.4444	2.637	3.48	.1349	-1	.2922	6.664
2.52	.5674	-1	.4405	2.686	3.50	.1311	-1	.2899	6.790
2.54	.5500	-1	.4366	2.737	3.52	.1274	-1	.2875	6.917
2.56	.5332	-1	.4328	2.789	3.54	.1239	-1	.2852	7.047
2.58	.5169	-1	.4289	2.842	3.56	.1204	-1	.2829	7.179
2.60	.5012	-1	.4252	2.896	3.58	.1171	-1	.2806	7.313
2.62	.4859	-1	.4214	2.951	3.60	.1138	-1	.2784	7.450
2.64	.4711	-1	.4177	3.007	3.62	.1107	-1	.2762	7.589
2.66	.4568	-1	.4141	3.065	3.64	.1076	-1	.2740	7.730
2.68	.4429	-1	.4104	3.123	3.66	.1047	-1	.2718	7.874
2.70	.4295	-1	.4068	3.183	3.68	.1018	-1	.2697	8.020
2.72	.4165	-1	.4033	3.244	3.70	.9903	-2	.2675	8.169
2.74	.4039	-1	.3998	3.306	3.72	.9633	-2	.2654	8.320
2.76	.3917	-1	.3963	3.370	3.74	.9370	-2	.2633	8.474
2.78	.3799	-1	.3928	3.434	3.76	.9116	-2	.2613	8.630
2.80	.3685	-1	.3894	3.500	3.78	.8869	-2	.2592	8.789

TABLE D.1 ISENTROPIC FLOW (*continued*)

M	p/p_0	T/T_0	A/A^*	M	p/p_0	T/T_0	A/A^*
3.80	$.8629^{-2}$	.2572	8.951	4.48	$.3543^{-2}$	.1994	16.28
3.82	$.8396^{-2}$	.2552	9.115	4.50	$.3455^{-2}$	.1980	16.56
3.84	$.8171^{-2}$	.2532	9.282	4.52	$.3370^{-2}$	.1966	16.84
3.86	$.7951^{-2}$	.2513	9.451	4.54	$.3288^{-2}$	.1952	17.13
3.88	$.7739^{-2}$	.2493	9.624	4.56	$.3207^{-2}$	.1938	17.42
3.90	$.7532^{-2}$	.2474	9.799	4.58	$.3129^{-2}$	.1925	17.72
3.92	$.7332^{-2}$	.2455	9.977	4.60	$.3053^{-2}$	.1911	18.02
3.94	$.7137^{-2}$	.2436	10.16	4.62	$.2978^{-2}$	.1898	18.32
3.96	$.6948^{-2}$	.2418	10.34	4.64	$.2906^{-2}$	.1885	18.63
3.98	$.6764^{-2}$	.2399	10.53	4.66	$.2836^{-2}$	.1872	18.94
4.00	$.6586^{-2}$	.2381	10.72	4.68	$.2768^{-2}$	.1859	19.26
4.02	$.6413^{-2}$	.2363	10.91	4.70	$.2701^{-2}$	.1846	19.58
4.04	$.6245^{-2}$	.2345	11.11	4.72	$.2637^{-2}$	.1833	19.91
4.06	$.6082^{-2}$	.2327	11.31	4.74	$.2573^{-2}$	.1820	20.24
4.08	$.5923^{-2}$	.2310	11.51	4.76	$.2512^{-2}$	.1808	20.58
4.10	$.5769^{-2}$	.2293	11.71	4.78	$.2452^{-2}$	.1795	20.92
4.12	$.5619^{-2}$	.2275	11.92	4.80	$.2394^{-2}$	.1783	21.26
4.14	$.5474^{-2}$	.2258	12.14	4.82	$.2338^{-2}$	.1771	21.61
4.16	$.5333^{-2}$	.2242	12.35	4.84	$.2283^{-2}$	.1759	21.97
4.18	$.5195^{-2}$	.2225	12.57	4.86	$.2229^{-2}$	.1747	22.33
4.20	$.5062^{-2}$	.2208	12.79	4.88	$.2177^{-2}$	.1735	22.70
4.22	$.4932^{-2}$	.2192	13.02	4.90	$.2126^{-2}$	.1724	23.07
4.24	$.4806^{-2}$	.2176	13.25	4.92	$.2076^{-2}$	.1712	23.44
4.26	$.4684^{-2}$	.2160	13.48	4.94	$.2028^{-2}$	.1700	23.82
4.28	$.4565^{-2}$	.2144	13.72	4.96	$.1981^{-2}$	.1689	24.21
4.30	$.4449^{-2}$	.2129	13.95	4.98	$.1935^{-2}$	.1678	24.60
4.32	$.4337^{-2}$	.2113	14.20	5.00	$.1890^{-2}$	.1667	25.00
4.34	$.4228^{-2}$	.2098	14.45	6.00	$.0633^{-2}$	.1219	53.19
4.36	$.4121^{-2}$	.2083	14.70	7.00	$.0242^{-2}$	.0926	104.14
4.38	$.4018^{-2}$	.2067	14.95	8.00	$.0102^{-2}$	.0725	190.11
4.40	$.3918^{-2}$	.2053	15.21	9.00	$.0474^{-3}$	.0582	327.19
4.42	$.3820^{-2}$	.2038	15.47	10.00	$.0236^{-3}$	.0476	535.94
4.44	$.3725^{-2}$	.2023	15.74	∞	0	0	∞
4.46	$.3633^{-2}$	.2009	16.01				

TABLE D.2 NORMAL-SHOCK FLOW

M_1	M_2	p_2/p_1	T_2/T_1	p_{02}/p_{01}
1.00	1.000	1.000	1.000	1.000
1.02	.9805	1.047	1.013	1.000
1.04	.9620	1.095	1.026	.9999
1.06	.9444	1.144	1.039	.9997
1.08	.9277	1.194	1.052	.9994
1.10	.9118	1.245	1.065	.9989
1.12	.8966	1.297	1.078	.9982
1.14	.8820	1.350	1.090	.9973
1.16	.8682	1.403	1.103	.9961

TABLE D.2   NORMAL-SHOCK FLOW (*continued*)

M_1	M_2	p_2/p_1	T_2/T_1	p_{02}/p_{01}
1.18	.8549	1.458	1.115	.9946
1.20	.8422	1.513	1.128	.9928
1.22	.8300	1.570	1.141	.9907
1.24	.8183	1.627	1.153	.9884
1.26	.8071	1.686	1.166	.9857
1.28	.7963	1.745	1.178	.9827
1.30	.7860	1.805	1.191	.9794
1.32	.7760	1.866	1.204	.9758
1.34	.7664	1.928	1.216	.9718
1.36	.7572	1.991	1.229	.9676
1.38	.7483	2.055	1.242	.9630
1.40	.7397	2.120	1.255	.9582
1.42	.7314	2.186	1.268	.9531
1.44	.7235	2.253	1.281	.9476
1.46	.7157	2.320	1.294	.9420
1.48	.7083	2.389	1.307	.9360
1.50	.7011	2.458	1.320	.9298
1.52	.6941	2.529	1.334	.9233
1.54	.6874	2.600	1.347	.9166
1.56	.6809	2.673	1.361	.9097
1.58	.6746	2.746	1.374	.9026
1.60	.6684	2.820	1.388	.8952
1.62	.6625	2.895	1.402	.8877
1.64	.6568	2.971	1.416	.8799
1.66	.6512	3.048	1.430	.8720
1.68	.6458	3.126	1.444	.8640
1.70	.6405	3.205	1.458	.8557
1.72	.6355	3.285	1.473	.8474
1.74	.6305	3.366	1.487	.8389
1.76	.6257	3.447	1.502	.8302
1.78	.6210	3.530	1.517	.8215
1.80	.6165	3.613	1.532	.8127
1.82	.6121	3.698	1.547	.8038
1.84	.6078	3.783	1.562	.7948
1.86	.6036	3.870	1.577	.7857
1.88	.5996	3.957	1.592	.7765
1.90	.5956	4.045	1.608	.7674
1.92	.5918	4.134	1.624	.7581
1.94	.5880	4.224	1.639	.7488
1.96	.5844	4.315	1.655	.7395
1.98	.5808	4.407	1.671	.7302
2.00	.5774	4.500	1.688	.7209
2.02	.5740	4.594	1.704	.7115
2.04	.5707	4.689	1.720	.7022
2.06	.5675	4.784	1.737	.6928
2.08	.5643	4.881	1.754	.6835
2.10	.5613	4.978	1.770	.6742
2.12	.5583	5.077	1.787	.6649
2.14	.5554	5.176	1.805	.6557

TABLE D.2 NORMAL-SHOCK FLOW (*continued*)

M_1	M_2	p_2/p_1	T_2/T_1	p_{02}/p_{01}
2.16	.5525	5.277	1.822	.6464
2.18	.5498	5.378	1.839	.6373
2.20	.5471	5.480	1.857	.6281
2.22	.5444	5.583	1.875	.6191
2.24	.5418	5.687	1.892	.6100
2.26	.5393	5.792	1.910	.6011
2.28	.5368	5.898	1.929	.5921
2.30	.5344	6.005	1.947	.5833
2.32	.5321	6.113	1.965	.5745
2.34	.5297	6.222	1.984	.5658
2.36	.5275	6.331	2.002	.5572
2.38	.5253	6.442	2.021	.5486
2.40	.5231	6.553	2.040	.5401
2.42	.5210	6.666	2.059	.5317
2.44	.5189	6.779	2.079	.5234
2.46	.5169	6.894	2.098	.5152
2.48	.5149	7.009	2.118	.5071
2.50	.5130	7.125	2.138	.4990
2.52	.5111	7.242	2.157	.4911
2.54	.5092	7.360	2.177	.4832
2.56	.5074	7.479	2.198	.4754
2.58	.5056	7.599	2.218	.4677
2.60	.5039	7.720	2.238	.4601
2.62	.5022	7.842	2.259	.4526
2.64	.5005	7.965	2.280	.4452
2.66	.4988	8.088	2.301	.4379
2.68	.4972	8.213	2.322	.4307
2.70	.4956	8.338	2.343	.4236
2.72	.4941	8.465	2.364	.4166
2.74	.4926	8.592	2.386	.4097
2.76	.4911	8.721	2.407	.4028
2.78	.4896	8.850	2.429	.3961
2.80	.4882	8.980	2.451	.3895
2.82	.4868	9.111	2.473	.3829
2.84	.4854	9.243	2.496	.3765
2.86	.4840	9.376	2.518	.3701
2.88	.4827	9.510	2.540	.3639
2.90	.4814	9.645	2.563	.3577
2.92	.4801	9.781	2.586	.3517
2.94	.4788	9.918	2.609	.3457
2.96	.4776	10.06	2.632	.3398
2.98	.4764	10.19	2.656	.3340
3.00	.4752	10.33	2.679	.3283
3.02	.4740	10.47	2.703	.3227
3.04	.4729	10.62	2.726	.3172
3.06	.4717	10.76	2.750	.3118
3.08	.4706	10.90	2.774	.3065
3.10	.4695	11.05	2.799	.3012
3.12	.4685	11.19	2.823	.2960
3.14	.4674	11.34	2.848	.2910

TABLE D.2   NORMAL-SHOCK FLOW (*continued*)

M_1	M_2	p_2/p_1	T_2/T_1	p_{02}/p_{01}
3.16	.4664	11.48	2.872	.2860
3.18	.4654	11.63	2.897	.2811
3.20	.4643	11.78	2.922	.2762
3.22	.4634	11.93	2.947	.2715
3.24	.4624	12.08	2.972	.2668
3.26	.4614	12.23	2.998	.2622
3.28	.4605	12.38	3.023	.2577
3.30	.4596	12.54	3.049	.2533
3.32	.4587	12.69	3.075	.2489
3.34	.4578	12.85	3.101	.2446
3.36	.4569	13.00	3.127	.2404
3.38	.4560	13.16	3.154	.2363
3.40	.4552	13.32	3.180	.2322
3.42	.4544	13.48	3.207	.2282
3.44	.4535	13.64	3.234	.2243
3.46	.4527	13.80	3.261	.2205
3.48	.4519	13.96	3.288	.2167
3.50	.4512	14.13	3.315	.2129
3.52	.4504	14.29	3.343	.2093
3.54	.4496	14.45	3.370	.2057
3.56	.4489	14.62	3.398	.2022
3.58	.4481	14.79	3.426	.1987
3.60	.4474	14.95	3.454	.1953
3.62	.4467	15.12	3.482	.1920
3.64	.4460	15.29	3.510	.1887
3.66	.4453	15.46	3.539	.1855
3.68	.4446	15.63	3.568	.1823
3.70	.4439	15.81	3.596	.1792
3.72	.4433	15.98	3.625	.1761
3.74	.4426	16.15	3.654	.1731
3.76	.4420	16.33	3.684	.1702
3.78	.4414	16.50	3.713	.1673
3.80	.4407	16.68	3.743	.1645
3.82	.4401	16.86	3.772	.1617
3.84	.4395	17.04	3.802	.1589
3.86	.4389	17.22	3.832	.1563
3.88	.4383	17.40	3.863	.1536
3.90	.4377	17.58	3.893	.1510
3.92	.4372	17.76	3.923	.1485
3.94	.4366	17.94	3.954	.1460
3.96	.4360	18.13	3.985	.1435
3.98	.4355	18.31	4.016	.1411
4.00	.4350	18.50	4.047	.1388
4.02	.4344	18.69	4.078	.1364
4.04	.4339	18.88	4.110	.1342
4.06	.4334	19.06	4.141	.1319
4.08	.4329	19.25	4.173	.1297
4.10	.4324	19.45	4.205	.1276
4.12	.4319	19.64	4.237	.1254
4.14	.4314	19.83	4.269	.1234

TABLE D.2 NORMAL-SHOCK FLOW (*continued*)

M_1	M_2	p_2/p_1	T_2/T_1	p_{02}/p_{01}	
4.16	.4309	20.02	4.301	.1213	
4.18	.4304	20.22	4.334	.1193	
4.20	.4299	20.41	4.367	.1173	
4.22	.4295	20.61	4.399	.1154	
4.24	.4290	20.81	4.432	.1135	
4.26	.4286	21.01	4.466	.1116	
4.28	.4281	21.20	4.499	.1098	
4.30	.4277	21.41	4.532	.1080	
4.32	.4272	21.61	4.566	.1062	
4.34	.4268	21.81	4.600	.1045	
4.36	.4264	22.01	4.633	.1028	
4.38	.4260	22.22	4.668	.1011	
4.40	.4255	22.42	4.702	.9948	-1
4.42	.4251	22.63	4.736	.9787	-1
4.44	.4247	22.83	4.771	.9628	-1
4.46	.4243	23.04	4.805	.9473	-1
4.48	.4239	23.25	4.840	.9320	-1
4.50	.4236	23.46	4.875	.9170	-1
4.52	.4232	23.67	4.910	.9022	-1
4.54	.4228	23.88	4.946	.8878	-1
4.56	.4224	24.09	4.981	.8735	-1
4.58	.4220	24.31	5.017	.8596	-1
4.60	.4217	24.52	5.052	.8459	-1
4.62	.4213	24.74	5.088	.8324	-1
4.64	.4210	24.95	5.124	.8192	-1
4.66	.4206	25.17	5.160	.8062	-1
4.68	.4203	25.39	5.197	.7934	-1
4.70	.4199	25.61	5.233	.7809	-1
4.72	.4196	25.82	5.270	.7685	-1
4.74	.4192	26.05	5.307	.7564	-1
4.76	.4189	26.27	5.344	.7445	-1
4.78	.4186	26.49	5.381	.7329	-1
4.80	.4183	26.71	5.418	.7214	-1
4.82	.4179	26.94	5.456	.7101	-1
4.84	.4176	27.16	5.494	.6991	-1
4.86	.4173	27.39	5.531	.6882	-1
4.88	.4170	27.62	5.569	.6775	-1
4.90	.4167	27.85	5.607	.6670	-1
4.92	.4164	28.07	5.646	.6567	-1
4.94	.4161	28.30	5.684	.6465	-1
4.96	.4158	28.54	5.723	.6366	-1
4.98	.4155	28.77	5.761	.6268	-1
5.00	.4152	29.00	5.800	.6172	-1
6.00	.4042	41.83	7.941	.2965	-1
7.00	.3974	57.00	10.469	.1535	-1
8.00	.3929	74.50	13.387	.0849	-1
9.00	.3898	94.33	16.693	.0496	-1
10.00	.3875	116.50	20.388	.0304	-1
∞	.3780	∞	∞	0	

TABLE D.3 PRANDTL–MEYER FUNCTION

M	θ	μ	M	θ	μ
1.00	0	90.00	2.02	26.929	29.67
1.02	.1257	78.64	2.04	27.476	29.35
1.04	.3510	74.06	2.06	28.020	29.04
1.06	.6367	70.63	2.08	28.560	28.74
1.08	.9680	67.81	2.10	29.097	28.44
1.10	1.336	65.38	2.12	29.631	28.14
1.12	1.735	63.23	2.14	30.161	27.86
1.14	2.160	61.31	2.16	30.689	27.58
1.16	2.607	59.55	2.18	31.212	27.30
1.18	3.074	57.94	2.20	31.732	27.04
1.20	3.558	56.44	2.22	32.250	26.77
1.22	4.057	55.05	2.24	32.763	26.51
1.24	4.569	53.75	2.26	33.273	26.26
1.26	5.093	52.53	2.28	33.780	26.01
1.28	5.627	51.38	2.30	34.283	25.77
1.30	6.170	50.28	2.32	34.783	25.53
1.32	6.721	49.25	2.34	35.279	25.30
1.34	7.280	48.27	2.36	35.771	25.07
1.36	7.844	47.33	2.38	36.261	24.85
1.38	8.413	46.44	2.40	36.746	24.62
1.40	8.987	45.58	2.42	37.229	24.41
1.42	9.565	44.77	2.44	37.708	24.19
1.44	10.146	43.98	2.46	38.183	23.99
1.46	10.731	43.23	2.48	38.655	23.78
1.48	11.317	42.51	2.50	39.124	23.58
1.50	11.905	41.81	2.52	39.589	23.38
1.52	12.495	41.14	2.54	40.050	23.18
1.54	13.086	40.49	2.56	40.509	22.99
1.56	13.677	39.87	2.58	40.963	22.81
1.58	14.269	39.27	2.60	41.415	22.62
1.60	14.861	38.68	2.62	41.863	22.44
1.62	15.452	38.12	2.64	42.307	22.26
1.64	16.043	37.57	2.66	42.749	22.08
1.66	16.633	37.04	2.68	43.187	21.91
1.68	17.222	36.53	2.70	43.621	21.74
1.70	17.810	36.03	2.72	44.053	21.57
1.72	18.397	35.55	2.74	44.481	21.41
1.74	18.981	35.08	2.76	44.906	21.24
1.76	19.565	34.62	2.78	45.327	21.08
1.78	20.146	34.18	2.80	45.746	20.92
1.80	20.725	33.75	2.82	46.161	20.77
1.82	21.302	33.33	2.84	46.573	20.62
1.84	21.877	32.92	2.86	46.982	20.47
1.86	22.449	32.52	2.88	47.388	20.32
1.88	23.019	32.13	2.90	47.790	20.17
1.90	23.586	31.76	2.92	48.190	20.03
1.92	24.151	31.39	2.94	48.586	19.89
1.94	24.712	31.03	2.96	48.980	19.75
1.96	25.271	30.68	2.98	49.370	19.61
1.98	25.827	30.33	3.00	49.757	19.47
2.00	26.380	30.00	3.02	50.142	19.34

TABLE D.3 PRANDTL–MEYER FUNCTION (*continued*)

M	θ	μ	M	θ	μ
3.04	50.523	19.20	4.04	66.309	14.33
3.06	50.902	19.07	4.06	66.569	14.26
3.08	51.277	18.95	4.08	66.826	14.19
3.10	51.650	18.82	4.10	67.082	14.12
3.12	52.020	18.69	4.12	67.336	14.05
3.14	52.386	18.57	4.14	67.588	13.98
3.16	52.751	18.45	4.16	67.838	13.91
3.18	53.112	18.33	4.18	68.087	13.84
3.20	53.470	18.21	4.20	68.333	13.77
3.22	53.826	18.09	4.22	68.578	13.71
3.24	54.179	17.98	4.24	68.821	13.64
3.26	54.529	17.86	4.26	69.053	13.58
3.28	54.877	17.75	4.28	69.302	13.51
3.30	55.222	17.64	4.30	69.541	13.45
3.32	55.564	17.53	4.32	69.777	13.38
3.34	55.904	17.42	4.34	70.012	13.32
3.36	56.241	17.31	4.36	70.245	13.26
3.38	56.576	17.21	4.38	70.476	13.20
3.40	56.907	17.10	4.40	70.706	13.14
3.42	57.237	17.00	4.42	70.934	13.08
3.44	57.564	16.90	4.44	71.161	13.02
3.46	57.888	16.80	4.46	71.386	12.96
3.48	58.210	16.70	4.48	71.610	12.90
3.50	58.530	16.60	4.50	71.832	12.84
3.52	58.847	16.51	4.52	72.052	12.78
3.54	59.162	16.41	4.54	72.271	12.73
3.56	59.474	16.31	4.56	72.489	12.67
3.58	59.784	16.22	4.58	72.705	12.61
3.60	60.091	16.13	4.60	72.919	12.56
3.62	60.397	16.04	4.62	73.132	12.50
3.64	60.700	15.95	4.64	73.344	12.45
3.66	61.000	15.86	4.66	73.554	12.39
3.68	61.299	15.77	4.68	73.763	12.34
3.70	61.595	15.68	4.70	73.970	12.28
3.72	61.889	15.59	4.72	74.176	12.23
3.74	62.181	15.51	4.74	74.381	12.18
3.76	62.471	15.42	4.76	74.584	12.13
3.78	62.758	15.34	4.78	74.786	12.08
3.80	63.044	15.26	4.80	74.986	12.03
3.82	63.327	15.18	4.82	75.186	11.97
3.84	63.608	15.10	4.84	75.383	11.92
3.86	63.887	15.02	4.86	75.580	11.87
3.88	64.164	14.94	4.88	75.775	11.83
3.90	64.440	14.86	4.90	75.969	11.78
3.92	64.713	14.78	4.92	76.162	11.73
3.94	64.983	14.70	4.94	76.353	11.68
3.96	65.253	14.63	4.96	76.544	11.63
3.98	65.520	14.55	4.98	76.732	11.58
4.00	65.785	14.48	5.00	76.920	11.54
4.02	66.048	14.40			

E. FILMS OR VIDEOCASSETTES*

Flow Visualization (S. J. Kline)
Eulerian and Lagrangian Descriptions in Fluid Mechanics (J. L. Lumley)
Surface Tension in Fluid Mechanics (L. M. Trefethen)
Deformation of a Continuous Media (J. L. Lumley)
Vorticity (A. H. Shapiro)
Low-Reynolds-Number Flows (G. I. Taylor)
Flow Instabilities (E. L. Mollo-Christensen)
Turbulence (R. W. Stewart)
Turbulent Flow (S. Corrsin)
Pressure Fields and Fluid Acceleration (A. H. Shapiro)
Fundamentals of Boundary Layers (F. H. Abernathy)
Boundary Layer Control (D. C. Hazen)
The Fluid Dynamics of Drag (A. H. Shapiro)
Channel Flow of a Compressible Fluid (D. E. Coles)
Cavitation (P. Eisenberg)
Rotating Flows (D. Fultz)
Secondary Flow (E. S. Taylor)
Waves in Fluids (A. E. Bryson)

Additional films are available and can be identified by consulting The ASME Film Catalog of The *J. of Fluids Engineering,* June 1976.

*Available from:
 Encyclopaedia Britannica Education Corp.
 425 North Michigan Avenue
 Chicago, IL 60611

Bibliography

REFERENCES

Adrian, R. J., Multi-point Optical Measurements of Simultaneous Vectors in Unsteady Flow—A Review, *Int. J. Heat Fluid Flow*, Vol. 7, No. 2, June 1986, pp. 127–145.

Bakhmeteff, B. A., *Hydraulics of Open Channels*, McGraw-Hill Book Company, New York, 1932.

Bean, H. S., ed., *Fluid Meters: Their Theory and Application*, 6th ed., American Society of Mechanical Engineers, New York, 1971.

Benedict, R. P., *Fundamentals of Pipe Flow*, John Wiley & Sons, Inc., New York, 1980.

Chapra, S. C., and Canale, R. P., *Numerical Methods for Engineers*, 2nd ed., McGraw-Hill Book Company, New York, 1988.

Chow, V. T., *Open Channel Hydraulics*, McGraw-Hill Book Company, New York, 1959.

Coleman, H. W., and Steele, W. G., *Experimentation and Uncertainty Analysis for Engineers*, Wiley-Interscience, New York, 1989.

Cross, Hardy, Analysis of Flow in Networks of Conduits or Conductors, *University of Illinois Bulletin 286*, November 1936.

Epp, R., and Fowler, A. G., Efficient Code for Steady-State Flows in Networks, *J. Hydraulics Div., ASCE*, Vol. 96, No. HY1, January 1970, pp. 43–56.

Gessler, J., Analysis of Pipe Networks, Chapter 4 in *Closed-Conduit Flow*, M. H. Chaudhry and V. Yevjevich, eds., Water Resources Publications, Fort Collins, Colo., 1981.

Goldstein, R. J., ed., *Fluid Mechanics Measurements*, Hemisphere Publishing Corp., New York, 1983.

HEC-2 Water Surface Profiles, Hydrologic Engineering Center, U.S. Army Corps of Engineers, Davis, Calif., September 1990.

Henderson, F. M., *Open Channel Flow*, Macmillan Publishing Company, New York, 1966.

Jeppson, R. W., *Analysis of Flow in Pipe Networks*, Ann Arbor Science, Ann Arbor, Mich., 1982.

Karassik, I. J., Krutzsch, W. C., Fraser, W. H., and Messina, J. P., eds., *Pump Handbook*, 2nd ed., McGraw-Hill Book Company, New York, 1986.

King, H. W., and Brater, E. F., *Handbook of Hydraulics*, McGraw-Hill Book Company, New York, 1963.

Kline, S. J., ed., Proceedings of the Symposium on Uncertainty Analysis, *Journal of Fluids Eng.*, Vol. 107, No. 2, June 1985, pp. 153–178.

McBean, E. A., and Perkins, F. E., Convergence Schemes in Water Profile Computation, *J. Hydraulics Div., ASCE*, Vol. 101, No. HY10, October 1975, pp. 1380–1384.

Malmstadt, H. V., Christie, G. E., and Crouch, S. R., *Electronics and Instrumentation for Scientists*, Addison-Wesley Publishing Company, Reading, Mass., 1981.

Metcalf & Eddy, Inc., *Wastewater Engineering: Collection and Pumping of Wastewater*, McGraw-Hill Book Company, New York, 1981.

Ormsbee, L. E., and Wood, D. J., Hydraulic Design Algorithms for Pipe Networks, *J. Hydraulic Eng., ASCE,* Vol. 112, No. 12, December 1986, pp. 1195–1207.

Roberson, J. A., Cassidy, J. J., and Chaudhry, M. H., *Hydraulic Engineering,* Houghton Mifflin Co., Boston, Mass., 1988.

Roberson, J. A., and Crowe, C. T., *Engineering Fluid Mechanics,* 4th ed., Houghton Mifflin Company, Boston, Mass., 1990.

Stepanoff, A. J., *Centrifugal and Axial Flow Pumps,* 2nd ed., John Wiley & Sons, Inc., New York, 1957.

Swamee, P. K., and Jain, A. K., Explicit Equations for Pipe-Flow Problems, *J. Hydraulics Div., ASCE,* Vol. 102, No. HY5, May 1976, pp. 657–664.

U.S. Department of Interior, Bureau of Reclamation, *Design of Small Dams,* U.S. Government Printing Office, Washington D.C., 1974.

Van Dyke, M., *An Album of Fluid Motion,* Parabolic Press, Stanford, Calif., 1982.

Warnick, C. C., *Hydropower Engineering,* Prentice-Hall, Inc., Englewood Cliffs, N.J., 1984.

Wood, D. J., Algorithms for Pipe Network Analysis and Their Reliability, *Research Report No. 127,* Water Resources Research Institute, University of Kentucky, 1981.

GENERAL INTEREST

Daily, J. W., and Harleman, D. R. F., *Fluid Dynamics,* Addison-Wesley, Reading, Mass., 1968

Eskinazi, S., *Principles of Fluid Mechanics,* 2nd ed., Allyn and Bacon, Boston, Mass., 1968

Fox, R. W., and McDonald, A. T., *Introduction to Fluid Mechanics,* 3rd ed., McGraw-Hill, New York, 1985

Fung, Y. C., *Continuum Mechanics,* Prentice-Hall, Englewood Cliffs, N.J., 1969

Hanson, A. G., *Fluid Mechanics,* Wiley, New York, 1967

John, E. A. J., *Gas Dynamics,* Allyn and Bacon, Boston, Mass., 1969

Owczarek, J. A., *Introduction to Fluid Mechanics,* International, Scranton, Penn., 1968

Potter, M. C., and Foss, J. F., *Fluid Mechanics,* Great Lakes Press, Okemos, Mich., 1979

Roberson, J. A., and Crowe, C. T., *Engineering Fluid Mechanics,* 4th ed., Houghton-Mifflin, Boston, 1990

Sabersky, R. H., Acosta, A. J., and Hauptman, E. G., *Fluid Flow,* 3rd ed., Macmillan, New York, 1989

Schlichting, H., *Boundary Layer Theory,* 6th ed., McGraw-Hill, New York, 1968

Shames, I., *Mechanics of Fluids,* 2nd ed., McGraw-Hill, New York, 1982

Shapiro, A. H., *The Dynamics and Thermodynamics of Compressible Fluid Flow,* Ronald Press, New York, 1953

Streeter, V. L., and Wylie, E. B., *Fluid Mechanics*, 8th ed., McGraw-Hill, New York, 1985

Van Wylen, G. J., and Sonntag, R. E., *Fundamentals of Classical Thermodynamics*, 2nd ed., Wiley, New York, 1976

White, F. M., *Fluid Mechanics*, McGraw-Hill, New York, 2nd ed., New York, 1986

Yih, C. S., *Fluid Mechanics*, McGraw-Hill, New York, 1969

Yuan, S. W., *Foundations of Fluid Mechanics*, Prentice-Hall, Englewood Cliffs, N.J., 1967

Answers
to
Selected Problems

CHAPTER 1

1.2b) M/LT^2 **d)** ML^2/T^2 **f)** L^3/T
1.3b) F/L^2 **d)** FL **f)** L^3/T
1.4b) $kg \cdot m^2/s^2$ **d)** $kg/m \cdot s$ **f)** $m^2/K \cdot s^2$
1.5b) 572 GPa **d)** 17.6 cm^3 **f)** 76 mm^3
1.6b) 3.21×10^{-5} s **d)** 5.6×10^{-12} m^3
    f) 7.8×10^9 m^3
1.7b) 498.1 N
1.9b) 142.2 kPa **d)** 78.8 kPa
1.10b) 527 mm Hg **d)** 23.6 ft H$_2$O
1.12) 1000 kg/m^3, 9810 N/m^3
1.14) 1.01 kg/m^3, 1.19 kg/m^3, yes
1.16) 0.414 N $\cdot$ s/m^2
1.18) 0.21 Hp

1.20) 0.008 Pa, 0.016 Pa
1.22) -9550 cm^3
1.24b) 1526 m/s
1.26) 29.6 kPa, 59.3 kPa
1.28) -4.82 mm
1.30) $2\sigma\pi D$
1.34E) 0.1339 slug/ft^3, 2.01 slug
1.36) 7.67 m/s, 10.85 m/s
1.38) 55.8°C
1.40) 19.1°C
1.42) 804 kPa gage
1.44) 2830 m

CHAPTER 2

2.1b) 78.5 kPa **d)** 156 kPa
2.2b) 31.9 m **d)** 16.0 m
2.4b) 10.2 m
2.5b) 0.916 kPa
2.7b) 30.8 kPa
2.8b) 47.0 kPa **d)** 22.5 kPa
2.10) 3.92 kPa
2.12) 25.4 kPa
2.14) 14.0 kPa
2.15b) 18.84 kPa **d)** 3.06 psi
2.16) 17.89 kPa
2.18) 5.87 kPa
2.20) 6934 N
2.22) 523 N
2.24b) (1.001, 1.333) m, 588.6 N

2.26b) 0.6667 m
2.27b) 1.732 m **d)** 5.196 ft
2.28b) 369.8 kN, (0.849, -0.167) m
    d) 330 kN, (2.4, 0.343) m
2.29b) it will topple
2.30b) it will topple
2.32b) 111,000 lb
2.34) 70.07 kN
2.36b) 1996 lb, 0.955
2.37b) -23.4 kN
2.38b) 2912 lb
2.39b) 1.336 m
2.40) 7645 cm^3, 13 080 N/m^3
2.42) 0.535 m
2.44b) 5.33 m

2.44E) a) 12.82 ft   **c)** 16.25 ft
2.45b) 0.959
2.46a) 1.089   **b)** 0.01886 kg
2.47b) It will not float with ends horizontal
2.48) Unstable if $0.2113 < S < 0.7887$
2.50b) 11.3°
2.52) stable
2.53b) 59 620 Pa   **d)** 7.45 psi
2.54b) 76.26 kPa
2.55b) 4.8 m/s²
2.56b) 57.46 kPa   **d)** 932 psf
2.57b) 1163 kN

2.58b) 74.72 kN
2.59b) −9000 Pa, −3114 Pa, 5886 Pa
   d) 0, −112 psf, −112 psf
   f) 182 psf, 433 psf, 251 psf
2.60b) 4935 Pa, 8859 Pa, 3924 Pa
   d) 93.5 psf, 171.5 psf, 78 psf
2.61b) −18 kPa, −14.08 kPa, 3.92 kPa
   d) −341 psf, −263 psf, 78 psf
2.62b) 9.34 rad/s   **d)** 9.57 rad/s
2.63b) 10.78 kPa   **d)** 57.9 kPa
2.64b) 35.78 kPa   **d)** 82.9 kPa
2.65b) 7210 N   **d)** 26.4 kN

CHAPTER 3

3.4) $udy - vdx$
3.6b) $8\hat{i} - 4\hat{j}$ **d)** $2\hat{i} - 108\hat{j} + 15\hat{k}$
3.7b) 0   **d)** $-2\hat{i} + 3\hat{k}$
3.8b) 0   **d)** $-4\hat{i} + 6\hat{k}$
3.9b) 0 everywhere
3.12) -9.11×10^{-4} kg/m³·s
3.14) -0.04 kg/m³·s
3.16) $-51.4 \times 10^{-5}\hat{j} + 0.0224\hat{i}$ m/s²
3.17b) 2D, $\mathbf{v}(r,\theta)$ **d)** 3D   **f)** 2D and unsteady
3.18b) inviscid   **d)** viscous   **f)** viscous

3.19b) separated   **d)** separated   **f)** not separated
3.20) turbulent
3.22) turbulent
3.24) laminar
3.25b) incompressible
3.26b) 104 m/s
3.28) 57.0 m/s
3.29b) 28.3 m/s   **d)** 90.1 ft/sec
3.30) 51.0 m
3.32) −30 kPa

CHAPTER 4

4.5b) Conservation of mass   **d)** energy equation
4.10) 1.736 m/s, 19.63 kg/s, 0.01963 m³/s
4.12) 107.1 m/s, 1158 m/s
4.14a) 3.167m   **c)** 7.162 m
4.15b) 5 m/s, 25.13 kg/s, 0.02513 m³/s
4.16a) 5 m/s, 320 kg/s, 0.32 m³/s
   c) 7.5 m/s, 480 kg/s, 0.48 m³/s
4.17b) $3(1 - 10000y^2)$ m/s
4.18) $(1 - 40000r^2)$ m/s
4.20) 8.32 cm
4.22) 0.05774 m/s
4.24) 0.0141 kg/m³·s
4.26) 2.57 kg/s
4.28) 0.61 m, 2.71 m
4.30) 9.86 m/s

4.32E) 60.44 psi
4.34) 32.1×10^6 Pa
4.35b) 534.7 kPa, 54.5 m
4.36b) 42.5 cfs, 39.3 cfs
4.38b) 0.485 m³/s
4.39b) 0.0784 m³/s
4.41b) 0.01258 m³/s   **d)** 2.796 cfs
4.42b) 2.21 ft
4.44) 60.4 m
4.46a) 0.131 m
4.48) 2960 W
4.50) 1.304 MW
4.52a) 299°C
4.54) 712 kW
4.56) 118.2 kW

4.58) 263 kW

4.60) 1.85 m or 2.22 m

4.62) 0.815 m

4.63b) 184.1 lb

4.64) 10.18 kN

4.65b) 1479 N   **d)** 2900 N

4.66) 6280 N

4.67b) 2280 N

4.68) 353 kN

4.70) 4420 N

4.72b) 6250 N

4.73b) 27.84 m/s

4.74b) 202 kg/s

4.75b) 393 N, 680 N

4.76) 147.3 kW

4.77b) 679 N, 392 N

4.78) 1790 kW

4.79b) 297 kW

4.80b) 47°, 29.5°, 3.6 MW

4.81b) 41.4°, 48.2°, 188.5 kW

4.82b) 9050 N, 273 kg/s, 47 kg/s

4.84) 13.33 m/s

4.86) 88 Hp

4.88) 10.3 m/s, 2.88 m/s

4.90) 2890 Pa, 818 Hp

4.92) 2000 N, 26.8 Hp

4.94) 618 kN

4.95b) 3.23 m   **d)** 11.54 ft

4.96b) 25.4 ft/sec

4.98) 1.16 m, 5.17 m/s

4.100) 141 N

4.102) 200 s^{-1}

4.104) 3780 N

4.105b) 2.1 N

4.107a) 54.9 kW/m   **c)** 3.74 kW/m

4.108) 0.562

4.110) 10.3 kW

4.112) 1.16 m^3/s

CHAPTER 5

5.3) $\rho \dfrac{du}{dx} + u \dfrac{d\rho}{dx}$

5.5) $u \dfrac{\partial \rho}{\partial x} + w \dfrac{\partial \rho}{\partial z} = 0, \quad \dfrac{\partial u}{\partial x} + \dfrac{\partial w}{\partial z} = 0$

5.7) -1380 kg/m$^3 \cdot$s

5.9) $-Ay$

5.10) $5y/(x^2 + y^2)$

5.12) $-20(1 - r^{-2}) \cos \theta$

5.14) -2.19 kg/m^4

5.15b) 2772 m/s^2

5.22) 3×10^{-8}

5.30) $-(2\mu + 3\lambda) \, \nabla \cdot \mathbf{V}$

5.32) $\rho c \, \partial T / \partial t = K \nabla^2 T$

5.34) 72 N/m$^2 \cdot$s, 18 N/m$^2 \cdot$s

CHAPTER 6

6.4) $\dfrac{\sigma}{\rho V^2 d} = \text{Const}$

6.6) $\dfrac{V}{\sqrt{gH}} = f\left(\dfrac{m}{\rho H^3}, \dfrac{\mu}{\rho \sqrt{gH^3}}\right)$

6.8) $\dfrac{F_D}{\mu V d} = f\left(\dfrac{l}{d}, \dfrac{\rho V d}{\mu}\right)$

6.10) $F_c = Cm\omega^2 R$

6.12) $V = C \dfrac{(dp/dx)d^2}{\mu}$

6.14) $\dfrac{V}{\sqrt{gH}} = f\left(\dfrac{\mu}{\rho \sqrt{gH^3}}, \dfrac{d}{H}\right)$

6.16) $\dfrac{Q}{\sqrt{gR^5}} = f\left(\dfrac{A}{R^2}, \dfrac{e}{R}, S\right)$

6.18) $\dfrac{F_D}{\rho V^2 d^2} = f\left(\dfrac{\mu}{V\rho d}, \dfrac{e}{d}, I\right)$

6.20) $\dfrac{F_D}{\rho V^2 d^2} = f\left(\dfrac{\mu}{V\rho d}, \dfrac{e}{d}, \dfrac{r}{d}, Cd^2\right)$

6.22) $\dfrac{F_L}{\rho V^2 l_c^2} = f\left(\dfrac{c}{V}, \dfrac{t}{l_c}, \alpha\right)$

6.24) $\dot{W} = \rho \omega^3 D^5 f\left(\dfrac{h}{D}, \dfrac{d_1}{D}, \dfrac{d_0}{D}\right)$

6.26) $\dfrac{Q}{\sqrt{gH^5}} = f\left(\dfrac{w}{H}, \dfrac{\mu}{\rho\sqrt{gH^3}}, \dfrac{\sigma}{\rho g H^2}\right)$

6.28) $\dfrac{T}{\rho\omega^2 h^5} = f\left(\dfrac{H}{h}, \dfrac{R}{h}, \dfrac{t}{h}, \dfrac{\mu}{\rho\omega h^2}\right)$

6.30) $\dfrac{T}{\rho\omega^2 R^5} = f\left(\dfrac{e}{R}, \dfrac{r}{R}, \dfrac{l}{R}, \dfrac{\mu}{\rho\omega R^2}\right)$

6.32) $\dfrac{\omega d}{V} = f\left(\dfrac{l}{d}, \dfrac{\mu}{\rho V d}\right)$

6.34a) 0.214 m³/s, 1400 kW

6.35a) 160 kg/s, 15000 kPa

6.36a) 45 N

6.40) 0.245 cm, 1.09

6.42) 6.1×10^{-9} m²/s (impossible)

6.44a) 0.00632 m³/s

6.46a) 1897 rpm   **b)** 120 kN·m

6.48) 261 kW

6.50) 4.16 N, 71.1 kW

6.52b) 188 m/s, 2120 N

6.54a) 6320 rpm

6.56) select 5 m/s, 5 motions/s

6.58) fl/U

6.60) $K/(\rho c_p U l) = (1/\text{Pr})(1/\text{Re})$

CHAPTER 7

7.2b) 0.0015 m/s

7.4b) 16.4 m   **d)** 45.1 m

7.6) 3.7 m, 0.72 m

7.10) 0.123°, 7.79×10^{-6} m³/s

7.12a) 0.0254 m³/s (not laminar)
   c) 3.24×10^{-5} m³/s (laminar)

7.14) 1.5×10^{-5} m³/s, 1.2 m

7.16) 2210, 1.1 Pa

7.18) 0.0102 m, 0.1 Pa, 52 m

7.24a) 0.0287 N·s/m²

7.25a) 5.47 m/s

7.26) 12.1 m³/s, 60.4 m/s, 20.1 Pa

7.28) 0.028 m³/s, 2030

7.30) 1.74×10^{-3} m³/s

7.31b) 13.6 Pa/m   **d)** 9.05 Pa/m

7.32b) -18.5 m/s   **d)** -12.3 m/s

7.34b) 152 N·m

7.36) 7.9 N·m

7.38) $\omega R^2/r$, 5.26×10^{-4} N·m

7.40) 0.0134 N·s/m², 1070

7.46) 358 Pa

7.48) 0.125 m²/s², -0.0125 m²/s, 1.0

7.50) 9, 8.3 m/s

7.52) -40.8 kPa/m

7.54a) 50 Pa   **b)** 6.6 m/s   **c)** 5.52 m/s
   d) 8.35×10^5 **e)** 0.0423 m³/s

7.56a) 60.1 kPa   **d)** 52.6 kPa

7.57b) 18.7 m   **d)** 12.6 m

7.58b) 37.5 kPa

7.59a) 1.36×10^6 Pa   **c)** 200 kPa   **e)** 141 kPa

7.60) 652 kPa

7.62) 147 kPa

7.63b) 0.013 m³/s   **d)** 0.039 m³/s

7.64b) 0.0033 m³/s

7.65b) 0.069 kg/s

7.66) 12.1 m³/s

7.67a) 0.032 m   **c)** 31 mm

7.68) 0.96 m

7.70) 0.000143 m³/s

7.72b) 50.7 kPa

7.73a) 52.6 kPa

7.74a) 0.36   **c)** 180

7.76a) 0.435

7.78a) 0.007 m³/s   **c)** 0.014 m³/s

7.79b) 0.011 m³/s

7.80a) 0.00089 m³/s   **b)** 7300 Pa

7.81b) 11.9 m³/s

7.82b) 28.2 kW

7.84a) 265 kW   **c)** 1.6 MW

7.86) 0.182 m³/s, 194 kW

7.88a) 195 kW   **b)** -81 kPa   **c)** 625 kPa

7.90) 1.05 MW

7.92) 0.921 Pa

7.93b) 1.63 m³/s

7.94b) 6.14 m³/s   **d)** 0.783 m³/s

7.94E b) 171 cfs   **d)** 21.8 cfs

7.95a) 0.794 m^3/s   **7.98a)** 1.19 m
7.96a) 1.71 m   **7.100)** 0.389 m

CHAPTER 8

8.2) 3.78×10^{-5} m, 0.0755 m
8.3b) 0.082 mm/s
8.4a) 7.9×10^6 (separated)   **c)** 7950 (separated)
8.5b) 4.58×10^{-5} N
8.8a) 1.94 N
8.10a) 30.4 m/s
8.12) 117 kN, 3.5 MN·m
8.13b) 1020 N, 19600 N·m
8.14) 40 m/s, 37 mm
8.16b) 4.72 m/s   **d)** 0.67 m/s
8.17b) 15.9 m/s
8.17E b) 52 fps
8.18) 140 fps
8.19b) 14.3 Hp
8.20) 9.5 m/s
8.22) 0.3 Hp
8.24) 8 Hz, 3500 Hz
8.26) 0.095 m/s
8.28) 3.1 Hp, 0.11 Hp
8.30) 20.6 m/s
8.32) 52 kN
8.34) 16.7%
8.36a) 34.5 m/s   **b)** 50 m/s   **c)** 18.4 Hp
8.41) $100y - 50x$, $100x + 50y$

8.44) $(3y^2 - 10y^3)/6$, 1.867×10^{-4} m^2/s
8.46) 2.098 m, 0.286 m, -280 Pa
8.48) $0.4\hat{i} + 1.2\hat{j}$ m/s
8.49b) $(-2.414, 0)$, $(0.414, 0)$
8.50b) -6.96 m/s, -9.28 m/s   **c)** 2.667
8.52) $(2, 232.7°)$, $(2, 307.3°)$; 61 Pa; -158 Pa
8.54) $(1.21, 270°)$, -177 Pa, -11.7 Pa
8.58a) 8.82 cm   **c)** 21.2 cm
8.60) $16/(2.67 - x)$, $256/(2.67 - x)^3$
8.62) $\rho \, \dfrac{d}{dx} (\theta U^2) + \rho \, \dfrac{dU}{dx} \, U\delta_d$
8.64) 30.8% low, 13% low
8.66b) $1.83\sqrt{\nu x/U_\infty}$, 6.4%; $0.731\sqrt{\nu x/U_\infty}$, 13.5%
8.68a) 26 mm   **b)** 5.63 mPa   **c)** 0.338 N
   d) 4.1 mm/s
8.69b) 45.3 mm, 2.15 N
8.70a) 0.6 Pa
8.72a) 1.58 N   **c)** 1.30 N
8.73b) 4.44 N
8.74a) 235 m, 0.0618 Pa
8.76a) 0.00211 **b)** 12.9 Pa   **c)** 0.0231 mm
   d) 66.6 mm
8.78) 163 kN, 0.89 m
8.82a) 12.4 mPa   **b)** 12.2 mm   **c)** 5.27 mm/s
   d) 0.04 m^3/s/m

CHAPTER 9

9.2) $-c\Delta V$
9.4b) 0.624 **d)** 0.677
9.6) 3776 m, 3.776 s
9.8a) 77.3 m/s
9.9a) 81.3 m/s
9.10E) 0.1025 slug/sec
9.12) 494.2 kPa abs, 4.29 kPa abs
9.14a) 0.1473 kg/s
9.15a) 0.1472 kg/s
9.16) 211.3 kPa abs, 7.29 kg/s

9.18) 333 m^3
9.20) 0.064 m, 0.0836 m, 520 m/s
9.22) 8.16 cm
9.24) 0.00221 m^2, $772°$C, 3670 kPa
9.26) 349 m/s
9.28b) 2.97, 0.477, 809.6 kPa, $475°$C, 3.771 kg/m^3
9.30) 454 kPa, 425 m/s
9.32) 391 kPa abs, $448°$C
9.34) 102.5 kPa abs, 0.471 kg/s, 0.302 kg/s
9.36) 574 m/s, 240 kPa abs, 99 m/s

9.38) 12.4 cm

9.40a) 120.4 kPa abs, 1.49, 620 m/s
  b) 0.631, 230 kPa abs, 303 m/s

9.42) 780 m/s

9.44) 1.8 kPa abs, $-146°C$, 1080 m/s, 32.4°

9.46a) 14.24 kPa, 26.4 kPa **b)** 2.56, 2.58
  c) 0.139 **d)** 0.0122

CHAPTER 10

10.2) 2.86 m

10.5) $\sqrt{3} \, y^2$

10.6b) 1.825 rad, 1.31 m, 2.28 m², 3.83 m
  d) 1.71 m, 5.08 m², 5.93 m

10.7) 86 cm, 180 cm

10.10a) 2.13 m, -0.17 m **c)** 0.28 m

10.11a) 2.07 m

10.12) 3.46 m

10.14a) 27 m³/s **c)** 18.4 m³/s

10.15b) 3.89 m

10.16a) 3.45 m²/s **b)** 3.55 m, 2.50 m

10.17b) 1.5 ft **d)** 3.48 ft

10.18a) 1.75 m

10.19) 1.21 m

10.23) 1.03 m, 28.8 m

10.24) 0.00107, 1.82 m, 8.7 m³/s, 1.07 m

10.27a) 3.87 m **b)** 363 N

10.28a) 2.52 m, 2.98 m/s **b)** 26.6 kN

10.30) 3.8 m/s, 1.96 m

10.32a) 1.47 m, 0.898 m **b)** 0.29 m
  d) $Fr_1 = 2.4$ (weak jump)

10.34) 2.55 m, 338 kW

10.36a) 41.5 m³/s **b)** 1.21 m

10.38) 0.21 m, 0.316 m, 0.088 m, 0.113 m

10.40a) 1.85 m, M_1 curve

10.42) $y_{01} = 1.82$ m, $y_{02} = 0.89$ m, $y_c = 1.19$ m

10.44) $y_c = 0.7$ m, $y_0 = 0.99$ m

10.46) 15.5 m³/s, S_3 curve

10.48) 7.72 m³/s, $y_c = 0.72$ m, $y_0 = 1.17$ m

CHAPTER 11

11.2a) 1.29 ft **b)** 42 cm

11.4) 0.265 m³/s

11.6) 0.316 m³/s

11.8) 1.78 m³/s, 0.285 m³/s, 0.935 m³/s, 1.07 MW

11.10) 0.198 m³/s, 0.138 m³/s, 0.060 m³/s

11.12) 0.01605 m³/s, 0.00707 m³/s, 0.00514 m³/s,
  0.00383 m³/s

11.14) 0.182 m³/s, 60.8 m

11.16) 0.412 m³/s, 0.227 m³/s, 0.185 m³/s, 803 kW

11.18) 5.91 L/s, 1.07 L/s, 0.95 L/s, 1.32 L/s,
  2.56 L/s, 5.91 L/s

11.20) 90 L/s, 46 L/s, 44 L/s, 20 L/s, 70 L/s

11.22) 2.3 in, 0.43 in, 2.73 out

11.24) 3.32 cfs, 0.39 cfs, 0.38 cfs, 0.41 cfs, 2.13 cfs

11.26) 126 L/s, 50.6 L/s, 38.8 L/s, 86.8 L/s,
  11.8 L/s

11.28) 0.111 m³/s

11.30) 21.1 L/s, 4.45 L/s, 16.6 L/s, 2.52 L/s,
  14.0 L/s

CHAPTER 12

12.1a) 88.9 N·m, 2200 W **c)** 34.2 N·m, 855 W
  e) 84.5 N·m, 2110 W

12.2) 0.0421 m³/s, 35.6 N·m, 2980 W, 7.22 m

12.4) 14.6 m, 859 W

12.6) 1.23 m

12.8) 17.6 m, 172 kW

12.10) 1.3, a mixed flow pump

12.12a) 0.688 m, 5.35 m, 6.84 m, 82 kW

12.14) 0.751, a radial pump

12.16) 1.19 ω_1, 1.19 D_1

12.18) 25.1 cm, 282 rad/s, 235 kW; 87.8 cm,
  59.1 rad/s, 262 kW

12.20) 15.3 cm, 240 rad/s, 2695 W

12.22) 2.44 m³/min, 9.96 m, 5.06 kW

12.24) 0.4 m, 0.212 m³/s, 31.4 rad/s, 5.19

12.26a) 64 m, 280 m³/hr, 64 kW, 8.3 m

12.28) 0.072 m³/s, 197 m, 3 pumps in series, 160 kW

12.30) 434 kW, 129 rpm

12.32) 63.1 cm, 92.6 rad/s

12.34) 1.89 m, 0.0688 m³/s, 115 W, 0.84

12.36a) 68 kPa **b)** 5.62 m

12.38) 40 cm, 0.418 m³/s, 35.5 kW

12.40) six units

CHAPTER 13

13.2a) $46.8\sqrt{h}$

13.4) 3.33×10^{-6} m³/s, 0.00333 kg/s, 1.06 m/s, laminar or turb

13.4E) 1.114×10^{-4} ft³/sec, 2.16×10^{-4} slug/sec, 2.04 fps, laminar

13.6) 0.0592 m³/s, 7.54 m/s

13.8a) 0.064 m³/s

13.9a) 0.0154 m³/s

13.10b) 0.035 m³/s **d)** 1.18 cfs

Index

683

PROPERTIES OF WATER

Temperature, °C	Density ρ, kg/m³	Viscosity μ, $(N \cdot s/m^2) \times 10^{-3}$	Kinematic viscosity ν, $m^2/s \times 10^{-6}$	Bulk modulus B, $Pa \times 10^7$	Surface tension σ, $N/m \times 10^{-2}$	Vapor pressure, kPa
0	999.9	1.792	1.792	204	7.62	0.588
5	1000.0	1.519	1.519	206	7.54	0.882
10	999.7	1.308	1.308	211	7.48	1.176
15	999.1	1.140	1.141	214	7.41	1.666
20	998.2	1.005	1.007	220	7.36	2.45
30	995.7	0.801	0.804	223	7.18	4.30
40	992.2	0.656	0.661	227	7.01	7.40
50	988.1	0.549	0.556	230	6.82	12.22
60	983.2	0.469	0.477	228	6.68	19.60
70	977.8	0.406	0.415	225	6.50	30.70
80	971.8	0.357	0.367	221	6.30	46.40
90	965.3	0.317	0.328	216	6.12	68.20
100	958.4	0.284	0.296	207	5.94	97.50

PROPERTIES OF AIR AT ATMOSPHERIC PRESSURE

Temperature T (°C)	Density ρ (kg/m³)	Viscosity μ (N · s/m²)	Kinematic viscosity ν (m²/s)
−30	1.452	1.56×10^{-5}	1.08×10^{-5}
−20	1.394	1.61×10^{-5}	1.16×10^{-5}
−10	1.342	1.67×10^{-5}	1.24×10^{-5}
0	1.292	1.72×10^{-5}	1.33×10^{-5}
10	1.247	1.76×10^{-5}	1.42×10^{-5}
20	1.204	1.81×10^{-5}	1.51×10^{-5}
30	1.164	1.86×10^{-5}	1.60×10^{-5}
40	1.127	1.91×10^{-5}	1.69×10^{-5}
50	1.092	1.95×10^{-5}	1.79×10^{-5}
60	1.060	2.00×10^{-5}	1.89×10^{-5}
70	1.030	2.05×10^{-5}	1.99×10^{-5}
80	1.000	2.09×10^{-5}	2.09×10^{-5}
90	0.973	2.13×10^{-5}	2.19×10^{-5}
100	0.946	2.17×10^{-5}	2.30×10^{-5}
200	0.746	2.57×10^{-5}	3.45×10^{-5}
300	0.616	2.93×10^{-5}	4.75×10^{-5}